J A

The literature references are given in square brackets in the text and collected in numerical order at the end of each chapter. The two series of *Zeitschrift für Physikalische Chemie*, referred to in the German edition as *Z. Physik. Chem.* and *Z. Physik. Chem. Neue Folge*, are given as *Z. Physik. Chem.* (Leipzig) and *Z. Physik. Chem.* (Frankfurt) respectively, in accordance with the recommendations of Chemical Abstracts.

The translation has been revised by the author.

Aachen, Germany   R. Haase
June 1968

PUBLISHER'S NOTE (1990)

For the Dover edition the author has provided an Appendix (beginning on page 497) containing 47 notes (keyed to the main text by superscript numbers) consisting of corrections of, comments on, and additions to the 1969 edition.

PREFACE

The thermodynamics of irreversible processes describes a relatively recent branch of macroscopic physics which has achieved great significance for its applications, particularly in physical chemistry. This monograph is therefore intended primarily for physical chemists and physicists. Chemists and engineers with fundamental training in physical chemistry will also be able to follow the text. The reader should master differential and integral calculus, the elements of vector analysis and linear algebra as well as the fundamentals of mechanics and electrodynamics.

Two things were particularly important to me in the completion of this book: the critical presentation of the fundamentals and the detailed description of their experimental applications.

We will first show the connection between the conceptual structure of classical thermodynamics and that of the thermodynamic–phenomenological theory of irreversible processes. This requires a recapitulation of the most important formal principles of classical thermodynamics. In doing so, we will strive for a more general and more exact formulation of these basic laws. In this way, we will demonstrate how classical thermodynamics and the thermodynamics of irreversible processes merge at their roots. We will also show the limits of validity of the thermodynamic–phenomenological theory as clearly as possible through fundamental considerations and simple examples.

The application of the theory extends from problems involving chemical reactions and relaxation phenomena in homogeneous systems to processes and systems within electromagnetic fields and to anisotropic materials. Those regions for which precise experimental data are available will be dealt with in detail. In other cases, it is only possible to work out representative examples or to sketch results of the theory concisely.

In all major sections, and particularly in those which deal with a special subject (such as electrokinetic effects, electrical conduction, thermoelectric effects, thermal diffusion, viscosity, etc.), symbols will be identified separately, so that each major topic can be understood independently. Thus there is no need for a table of notation and the annoying cross-referencing.

iii

THERMODYNAMICS OF IRREVERSIBLE PROCESSES

by Rolf Haase

DOVER PUBLICATIONS, INC.
NEW YORK

Published in Canada by General Publishing Company, Ltd., 30 Lesmill Road, Don Mills, Toronto, Ontario.

Published in the United Kingdom by Constable and Company, Ltd., 3 The Lanchesters, 162–164 Fulham Palace Road, London W6 9ER.

This Dover edition, first published in 1990, is a corrected, slightly enlarged republication of the English translation, revised by the author and published by Addison-Wesley Publishing Company, Reading, Mass., 1969 (in the "Addison-Wesley Series in Chemical Engineering"). The original German edition was published by Dr. Dietrich Steinkopff Verlag, Darmstadt, Germany, 1963, under the title *Thermodynamik der irreversiblen Prozesse*.

Manufactured in the United States of America
Dover Publications, Inc., 31 East 2nd Street, Mineola, N.Y. 11501

Library of Congress Cataloging-in-Publication Data

Haase, R. (Rolf)
    Thermodynamics of irreversible processes / by Rolf Haase.
        p.   cm.
    "Corrected, slightly enlarged republication of the English translation . . . published by Addison-Wesley Publishing Company, Reading, Mass., 1969 (in the 'Addison-Wesley series in chemical engineering')"—T.p. verso.
    ISBN 0-486-66356-6 (pbk.)
    1. Thermodynamics.   2. Irreversible processes.   I. Title.
QD504.H26   1990
541.3'69—dc20   90-3687
                                                                              CIP

CONTENTS

Chapter 3   PROCESSES IN HETEROGENEOUS (DISCONTINUOUS) SYSTEMS

Chapter 4   PROCESSES IN CONTINUOUS SYSTEMS

A   Fundamental Principles

viiiCONTENTS

UNIVERSAL CONSTANTS

Speed of light in a vacuum	$c_0 = 2.99792 \times 10^8$ m s^{-1}.
Elementary charge (proton charge)	$e = 1.60218 \times 10^{-19}$ C.
Avogadro constant	$L = 6.02214 \times 10^{23}$ mol^{-1}.
Faraday constant	$F = 9.64853 \times 10^4$ C mol^{-1}.
Boltzmann constant	$k = 1.38066 \times 10^{-23}$ J K^{-1}.
Gas constant	$R = 8.31451$ J K^{-1} mol^{-1}.

CONVERSION OF UNITS (see also page 425)

1 P (poise) $= 1$ g cm^{-1} s$^{-1} = 10^{-1}$ N m^{-2} s.

1 atm (standard atmosphere) $= 760$ Torr $= 1.01325$ bar
$= 1.01325 \times 10^6$ dyn cm^{-2}
$= 1.01325 \times 10^5$ N m^{-2}.

1 cal (thermochemical calorie) $= 4.184$ J.

CHAPTER 1

FUNDAMENTAL PRINCIPLES

1-1 INTRODUCTION

Classical thermodynamics is a part of classical macroscopic physics. In this it is similar to classical mechanics and electrodynamics whose presentation logically precedes that of thermodynamics. The "thermodynamics of irreversible processes," with which we are concerned here, should be described more exactly as "the thermodynamic–phenomenological theory of irreversible processes," for it consists of a thermodynamic and a phenomenological part. The thermodynamic part of the theory proceeds via the same method as classical thermodynamics; even the terms "entropy flow" and "entropy production," so characteristic of the new theory (§1–24), develop from a consistent elaboration and definition of the concepts of classical thermodynamics. The phenomenological part of the theory introduces new postulates from the point of view of macroscopic physics, that is, the "phenomenological equations" (§1–25) and the "Onsager reciprocity relations" (§1–26), which are justified, just as are the first and second laws of thermodynamics, by experience. However, from the point of view of kinetics or the statistical mechanics of irreversible processes, these postulates are, just as are both laws of thermodynamics, derived, in principle, from general molecular statistical considerations.

We will first (§1–2 to §1–22) develop the fundamentals of thermodynamics and then proceed (§1–23 to §1–27) to the concepts of "thermodynamics of irreversible processes." In doing this we will make the fundamental principles so general that there should be no discontinuity in their presentation during this transition. It should thus be emphasized that classical thermodynamics and thermodynamics of irreversible processes may be joined to yield a general theory for those properties of macroscopic systems which are not yet encompassed by classical mechanics and electrodynamics.

Those systems to be considered consist either of a single phase, that is, of a single macroscopic homogeneous body, or of a finite number of phases, so that we have the terms "heterogeneous" or "discontinuous"; alternatively, they represent "continuous" systems in which quantities such as pressure, density, concentration, etc., vary continuously in space from point to point. Accordingly, Chapter 2 deals with processes in homogeneous systems, Chapter 3 with processes in heterogeneous (discontinuous) systems, and

1

Chapter 4 with processes in continuous systems. Chapter 5 will deal with stationary states which are of significance in all kinds of systems.

In general, we will consider a *region* of a system to be either a single phase (a homogeneous system, or one phase of a heterogeneous system), or a volume element[1]* of a continuous system, that is, for example, a volume element defined arbitrarily in a gas contained in a gravitational field or in a liquid contained in a centrifuge. When dealing with anisotropic bodies or systems in external force fields (for example, in gravitational, centrifugal, or electromagnetic fields), we must consider that we are dealing in general with continuous systems.

A system which is so cut off from the outside world that no matter can pass through the boundary of the system is called *closed*. A system that can exchange matter with its surroundings will be called *open*. Accordingly, there are the expressions "closed region" and "open region." The term "isolated system" should not be confused with this for it describes a system which is isolated from *all* exchanges with the surroundings.

Those variables which describe the state of a system macroscopically are called *state variables*. Two classes of these can be distinguished, namely, *external coordinates* (the macroscopic speed of the entire system or of connected system parts and position coordinates in external force fields) and *internal variables* (pressure, volume, amounts of different substances, etc.). Definite values of the internal state variables characterize a definite *internal state*. A change in these variables means an *internal state change*.

The internal state variables are again divided into "intensive" and "extensive" quantities. An *intensive variable* is independent of the quantity of matter considered and has a definite value at each point of the materially filled space. In particular, for a homogeneous system, it has a constant value. An *extensive quantity* depends upon the amount of material in the region considered and is increased by n times when the amount of all substances present is increased by n times at a fixed value of the intensive state variables. Moreover, each extensive state variable of a system is equal to the sum of its values for the macroscopic parts into which the given system can be divided. Examples of intensive quantities are: pressure, density, electric field strength, etc., and for extensive quantities: total mass, volume, total electric moment, etc.

Those processes which actually take place in macroscopic systems in nature are called *actual processes* or *natural processes*. The limiting case, that is, a process in which there is "equilibrium" at each moment inside the system (cf. §1–17), neglecting such "dissipative effects" as friction, current transport, etc. (cf. §1–3), is called a *quasistatic process*.

A process will be called *reversible* when the state change which has taken place in the system can be reversed and thus return the system to its initial state without changes remaining in the total surroundings of the system, that

---

*The superscript numbers refer to the notes in the 1990 Appendix beginning on page 497.

is, in nature. A process will be called *irreversible* if the corresponding state change cannot be reversed and the system returned to its initial state without changes in nature. From the definition of the quasistatic process, it is obvious that all quasistatic processes must be reversible. However, we must first use the second law of thermodynamics (§1–8) to show that all actual (natural) processes are irreversible. We anticipate, nevertheless, the conceptual identities: "quasistatic equals reversible" and "actual equals irreversible" in order to follow the usual notation from the beginning.

In the following exposition, the basic concepts of mechanics and electrodynamics are considered as known. Typical quantities of importance in thermodynamics, such as "temperature," "heat," "internal energy," "entropy," will be concisely discussed and clearly characterized by their main properties. More detailed presentations of the basic principles of thermodynamics must be sought elsewhere[1]. We would like to point out, however, that the foundations of thermodynamics dealt with in this chapter are more general than in previous presentations.

References to monographs, articles, etc., dealing with the thermodynamics of irreversible processes are given at the end of §1–24.

1-2 EMPIRICAL TEMPERATURE (ZEROTH LAW)

Any boundary which surrounds a system and which is impermeable to matter and chemically unchangeable will, for the sake of conciseness, be designated as a "wall." We distinguish between two kinds of walls. These will be defined by the behavior of the system enclosed by them provided that there is an internal equilibrium.† Thus we shall be able to give a definition of the "empirical temperature" and of "energy" and "heat" (§1–4).

A *thermally conducting wall* is distinguished by the following property: a system in internal equilibrium and separated from a second system by such a wall will show changes of state even when the wall is absolutely rigid. During such a contact both states approach a condition which will be designated as a *thermal equilibrium*, which obviously has nothing to do with phenomena such as pressure and concentration equalization or with chemical reaction. (The second kind of wall, the *thermally insulating wall* will be discussed in §1–4.)

The most important empirical law on thermal equilibrium is the following: If two systems A and B are in thermal equilibrium with a third system then there is also thermal equilibrium between A and B.

---

† The exact definition of the term "equilibrium" is found in §1–17. Here it is sufficient to note that a "system in internal equilibrium" exhibits no changes of state of the system, as long as it does not come into contact with another system. The expression "*internal* equilibrium" should emphasize that there is no need for there to be equilibrium between the system and the outside world.

This general law will be called the *Zeroth Law of Thermodynamics*, following the designation of Fowler and Guggenheim. This law allows an intensive quantity which characterizes thermal equilibrium to be qualitatively fixed, and this we call the *empirical temperature*. Systems which are in thermal equilibrium with each other have the same empirical temperature, while those which are not in thermal equilibrium with each other are considered to have different empirical temperatures. Through a suitable scale, based upon the behavior of gases at sufficiently low pressures, and a fixed zero point we can finally define quantitatively the empirical temperature. We call the quantity so fixed the "empirical temperature on the Kelvin scale."[†]

Through the second law of thermodynamics (§1–8) a positive quantity, the "absolute temperature T" (or "thermodynamic temperature") dependent only on the empirical temperature of the region and independent of the particular material properties, will be introduced for each region of the system.[2] It can then be shown that the empirical temperature on the Kelvin scale is proportional to the absolute temperature. When the absolute temperature is also fixed numerically, through a suitable zero point, then those temperatures are identical. For the sake of simplicity, we will assume this identity and, even before considering the second law, will write the symbol T for the empirical temperature and speak merely about "temperature."

1-3 WORK

Consider a closed system[‡] on which an arbitrary macroscopic force acts. The force can have mechanical, electrical, or magnetic origin and can act on the entire system or on part of the system. Due to the force we may have a deformation, a displacement of an electric charge, or some similar consequence. Designating the force with the vector **K** and the corresponding infinitesimal displacement of the respective system part with the vector **ds**, then the work W done on the system is defined as follows:[||]

$$W \equiv \int \mathbf{K} \, \mathbf{ds}. \tag{1-3.1}$$

This definition is also valid in thermodynamics.

The infinitesimal quantity dW, the reversible work for an infinitesimal state change, is in general a nonexact differential because W is not a state

---

† Its unit is the degree Kelvin (°K) or simply the kelvin (K).

‡ Here, as well as in §1–4 and §1–5, only closed systems or regions are considered. Open systems will first be treated in §1–6.

|| Since in international literature the symbol A, used previously in German literature for work, is now used as the symbol for the affinity (§1–14), we designate work with W.

variable. The concept of "reversible work" arises when the process associated with the doing of the work occurs quasistatically, for then the process (a deformation, for example) will proceed under a condition of force balance and without friction, that is, infinitely slowly.[†]

Let us assume that a homogeneous, isotropic body of variable volume V is reversibly compressed or expanded under a variable pressure P. Then from Eq. (1) for the infinitesimal compression or expansion work (reversible volume work)

$$dW = -P \, dV. \tag{1-3.2}$$

This expression is valid also for a volume element of an isotropic, continuous system.

When a volume element of an anisotropic medium of volume V_0 (the reference volume for the definition of strain) is reversibly deformed then, from Eq. (1) for the infinitesimal deformation work[2], it can be shown that

$$dW = V_0 \sum_{i=1}^{3} \sum_{k=1}^{3} \tau_{ik} \, de_{ik} = V_0 \sum_{i=1}^{6} \tau_i \, de_i. \tag{1-3.3}$$

Here $\tau_{11} = \tau_1$, $\tau_{22} = \tau_2$, $\tau_{33} = \tau_3$, $\tau_{12} = \tau_{21} = \tau_4$, $\tau_{13} = \tau_{31} = \tau_5$, $\tau_{23} = \tau_{32} = \tau_6$ are the six stress components and $e_{11} = e_1$, $e_{22} = e_2$, $e_{33} = e_3$, $2e_{12} = 2e_{21} = e_4$, $2e_{13} = 2e_{31} = e_5$, $2e_{23} = 2e_{32} = e_6$ the six strain components.

If the surface area ω of an autonomous interfacial layer[‡] of volume V be infinitesimally changed, then for the reversible deformation work

$$dW = -P \, dV + \sigma \, d\omega. \tag{1-3.4}$$

Here P is the pressure[||] and σ the interfacial tension[4a].

If a phase or a volume element of an isotropic medium is electrified or magnetized reversibly, the electric or magnetic field strength being $\mathfrak{E}$ or $\mathfrak{H}$, the values of the pressure P and of the volume V being variable, then for the infinitesimal work done on the region[4b, 5, 6]

$$dW = -P \, dV + \mathfrak{E} \, d(\mathfrak{P}V) + \mathfrak{H} \, d(\mathfrak{J}V). \tag{1-3.5}$$

Here $\mathfrak{P}$ is the electric polarization vector and $\mathfrak{J}$ the magnetic polarization vector. Thus $\mathfrak{P}V$ or a quantity such as $\mathfrak{J}V$ defines the total electric or magnetic moment of the region (further information is found in §4–34).

If we are dealing with an anisotropic medium with electrification and magnetization then we must introduce the six strain and stress components

---

[†] Processes inside a system such as mass transport or chemical reactions could be irreversible in this case.

[‡] An interfacial layer is called "autonomous" if its state is determined by the variables of the layer itself, not by those of the adjoining "bulk phases." In the case of equilibrium, the interfacial layer can be considered autonomous otherwise only under special conditions[3].

[||] For the problematic concept "pressure" for surface phases, see p. 63.

which appeared in Eq. (3), the three components of the electric or magnetic field strength $\mathfrak{E}_1$, $\mathfrak{E}_2$, $\mathfrak{E}_3$ or $\mathfrak{H}_1$, $\mathfrak{H}_2$, $\mathfrak{H}_3$ and the three components of the electric polarization or magnetic polarization $\mathfrak{P}_1$, $\mathfrak{P}_2$, $\mathfrak{P}_3$ or $\mathfrak{I}_1$, $\mathfrak{I}_2$, $\mathfrak{I}_3$. For a volume element with reversible work done during infinitesimal state changes, we find[7]

$$dW = V_0 \sum_{i=1}^{6} \tau_i \, de_i + \sum_{j=1}^{3} \mathfrak{E}_j \, d(\mathfrak{P}_j V) + \sum_{j=1}^{3} \mathfrak{H}_j \, d(\mathfrak{I}_j V). \qquad (1\text{--}3.6)$$

In this relation, Eqs. (2), (3), and (5) are contained as special cases.

Expressions (2) to (6) are all of the same type: they refer to reversible deformation work or reversible electrification or magnetization work on a single region α and can be written in the general form

$$dW_i^\alpha = \sum_i L_i^\alpha \, dl_i^\alpha. \qquad (1\text{--}3.7)$$

Here l_i^α is a *work-coordinate* with an extensive character, and L_i^α the conjugate *work-coefficient* with an intensive character. (Also, the surface area of an interfacial layer is an extensive quantity in the sense of §1–1 because a change in mass at constant intensive variables, i.e. keeping constant the qualitative character of the phase, is only conceivable at given thickness of the layer.) By our definition, external forces such as gravitational forces, or centrifugal forces, or forces accelerating the system and quantities such as friction forces which depend only on changes of the external coordinates of the system or on irreversible processes *cannot* be interpreted as work-coefficients.

The *internal* state of a region is empirically determined by the variables L_i^α or l_i^α if the temperature and the composition of the region are given. It is evident that we need only introduce either the work-coefficients or the work-coordinates as independent internal state variables since there is an *equation of state*.† This requires a functional relation between a work-coordinate and the conjugate coefficient (with the other work-coefficients, as well as temperature and composition, constant).

---

† Apart from the known form of the equation of state for isotropic regions (T equals temperature, n_k equals amount of substance k)

$$V = V(P, T, n_k),$$

which for ideal gases yields, with R equal to the gas constant and $n \equiv \sum_k n_k$,

$$V = \frac{nRT}{P},$$

we also have another relation of this kind, the stress–strain relation (the generalized Hooke's law) for anisotropic bodies (cf. p. 463), or the statement (4–34.20) or (4–38.6) for the electric polarization vector as a function of the electric field strength, or the relation (4–34.21) for the dependence of the magnetic polarization on the magnetic field strength.

Table 1 presents a survey of the values of work-coordinates and conjugate work-coefficients in a few important special cases. Here we designate a region as "simple" if it exhibits neither electrification nor magnetization. The two first cases (isotropic media–one dimension) are special cases of the third case (anisotropic media) as can be derived from Eq. (3).

Table 1

Meaning of Work-coordinates l_i and Work-coefficients L_i in
Special Cases

Kind of region	l_i	L_i
Simple isotropic region	V (volume)	$-P$ (negative pressure)
Simple bar (unidirectional stress)	l (length)	K (force)
Simple anisotropic region	$V_0 e_i$ (product of reference volume and strain component)	τ_i (stress component)
Simple autonomous interfacial layer	V (volume) ω (surface area)	$-P$ (negative pressure) σ (interfacial tension)
Electrified isotropic region	V (volume) $\mathfrak{P}V$ (product of electric polarization and volume)	$-P$ (negative pressure) $\mathfrak{E}$ (electric field strength)
Magnetized isotropic region	V (volume) $\mathfrak{J}V$ (product of magnetic polarization and volume)	$-P$ (negative pressure) $\mathfrak{H}$ (magnetic field strength)

We will no longer consider the idealized limiting case of a reversible deformation, electrification or magnetization, but rather an actual change of state of a system. Here not only should the internal state variables be permitted to change but also the external coordinates of the region must be considered. If we, for example, compress a real gas from the volume V_1 to the volume V_2, we will have to employ a greater pressure than the corresponding gas pressure P, for the piston in the compression vessel chamber would otherwise not move with a speed different from zero. Friction is unavoidable. Thus we must perform more work than that which comes from Eq. (2) for which, in the limiting reversible case, the following expression is valid:

$$W_l = W_{\text{rev}} = -\int_{V_1}^{V_2} P \, dV.$$

The difference between the compression work actually performed on the gas and the above expression will be called the "dissipated work." If the gas is raised in a gravitational field at the same time then we must also do some "external work."

We will then quite generally set for the work dW^α performed on a region α during an infinitesimal state change:

$$dW^\alpha = dW_i^\alpha + dW_{diss}^\alpha + dW_a^\alpha, \qquad (1\text{–}3.8)$$

where dW_i^α is given in Eq. (7), dW_{diss}^α is the "dissipated work" and dW_a^α the "external work." Individually, dW_{diss}^α is connected to "dissipative effects" such as friction, turbulence or plastic flow during the deformation, hysteresis phenomena during the actual electrification or magnetization, electric conduction due to external sources of current, etc., while dW_a^α is related to the work of forces which accelerate the region or react against existing external force fields. "External work" can only bring about changes in the external coordinates (space-coordinates, speed) not, as through the first two terms of the right-hand side of Eq. (8), changes in the internal state variables of the regions.

The work which is performed on a system consisting of several regions is not necessarily equal to the sum of the amounts of work performed on each single region. This is shown by two examples.

The first example concerns a heterogeneous system which is surrounded by a rigid wall. Here no deformation work can be performed even if processes occur inside the system which lead to volume changes of certain phases. Therefore, although the actual volumes V^α are variable, it is not permissible to use an expression for the reversible compression or expansion work performed for the entire system obtained by summing Eq. (2) over all phases to yield

$$-\sum_\alpha P^\alpha \, dV^\alpha.$$

This expression only represents the volume work performed on the system if each phase α is separated from the rest of the system by rigid walls but from the outside world by a flexible wall on which a pressure P^α acts. However, the only real case where the pressures P^α have different values in the individual phases is that described above. Otherwise mechanical equilibrium, which leads to a constant pressure P, would adjust itself very swiftly in relation to other processes with which we are concerned and in this case the sum is reduced to the expression

$$-P \, dV \quad (V = \text{total volume}),$$

which again describes the reversible volume work. From this it follows that the above summed expression does not in principle but does in practice give the reversible compression or expansion work.[3]

The second example is based on a heterogeneous system which is arranged in a galvanic cell. Here, even in the reversible limiting case (i.e. at zero current strength), a work term appears which contains the e.m.f. of the cell. This work term is not found in Eq. (8) since it is impossible to make a galvanic cell out of a single phase. With all reversible cells the potential differences build up only at the phase boundaries.

For an arbitrary infinitesimal change of state of a system consisting of several regions, we write for the work done

$$dW = dW_a + dW_l + dW^*. \qquad (1\text{–}3.9)$$

From this, for a finite change of state of the system, we find

$$W = W_a + W_l + W^*. \qquad (1\text{–}3.10)$$

Here W_a signifies the acceleration work done on the entire system and work against external force fields, W_l the reversible deformation work as well as reversible electrification and magnetization work done on the entire system, and W^* the rest of the work done (friction work, electrical work with external current sources, etc.). This last term can also contain reversible portions (i.e. reversible electrical work in galvanic cells). From the above we have

$$dW_l = \sum_{\alpha} \sum_{i} L_i^{\alpha} \, dl_i^{\alpha}, \qquad (1\text{–}3.11)$$

where dl_i^{α} strictly describes only those changes of the work-coordinates of the individual regions which contribute to work on the entire system.

As is obvious from the above formula, the work-coefficients L_i^{α} and work-coordinates l_i^{α} are fixed, in that the *reversible* work done on a closed region with an infinitesimal change of the *internal* state will be equal to the expression $\sum_i L_i^{\alpha} \, dl_i^{\alpha}$.

Thus work can still be done on a system even with all work-coordinates being constant, e.g. the work of lifting in the gravitational field, the work of acceleration, friction work, electrical work, etc. This apparently paradoxical situation is only a consequence of our conventions. Our definitions are, however, useful throughout, for the work-coefficients and work-coordinates appear later in many thermodynamic functions.

1–4 ENERGY AND HEAT (FIRST LAW)

A *thermally insulating wall* is defined in such a way that a system enclosed by it on all sides, and in internal equilibrium, can only suffer a change of state if work is done on or by the system. A system surrounded by such a wall is called *thermally insulated* even if it is not in internal equilibrium. Any change of state taking place in a thermally insulated system is called *adiabatic*.

It is found from experience that the work done on a thermally insulated system W_{ad} depends only on the initial state (I) and the final state (II) of the system, but not on the path of the change from state (I) to (II). This is a partial statement of the *First Law of Thermodynamics*. This permits a definition of a state variable E:

$$W_{\mathrm{ad}} \equiv E_{\mathrm{II}} - E_{\mathrm{I}} \equiv \Delta E. \tag{1–4.1}$$

In the above, E designates the *energy* of the system. Since, according to experience, all state changes—at least in one direction—can be represented by a combination of adiabatic paths, the energy change ΔE can be defined for *any* change of state of the system. This clear definition of energy is due to Carathéodory (1909).

We consider now an arbitrary, nonadiabatic change of state in a closed system. If W is the work done on the system, then experience shows that

$$\Delta E \neq W. \tag{1–4.2}$$

Here we may have $\Delta E \neq 0$ even though $W = 0$ and the system is originally in internal equilibrium. This fact presents a further statement of the first law. There are therefore influences from the surroundings which lead to a change of state of the system and thus to a change of the energy but have nothing to do with the work done on the system. Since these influences occur particularly if there is a temperature difference between the system and the surroundings, the difference

$$Q \equiv \Delta E - W \tag{1–4.3}$$

is called the *heat* supplied to the system from the surroundings. This clear definition of heat is due to Born (1921).

We can now say: a "thermally conducting wall" (§1–2) allows an exchange of heat between the system enclosed by the wall and the surroundings, while a "thermally insulating wall" prevents such an exchange of heat.

We see from the above that in the general expression

$$\Delta E = E_{\mathrm{II}} - E_{\mathrm{I}} = Q + W \tag{1–4.4}$$

the energy E of the system is a state variable, while the heat supplied to the system Q and the work done on the system W are not state variables. Accordingly, in the expression for an infinitesimal state change

$$dE = dQ + dW, \tag{1–4.5}$$

dQ and dW are inexact differentials, while dE does represent an exact differential. The quantities W or dW in Eqs. (4) or (5) respectively are given individually in Eqs. (1–3.10) or (1–3.9) respectively.

Finally, the energy E of a system can be split up in the following way

$$E = E_{\mathrm{kin}} + E_{\mathrm{pot}} + U. \tag{1–4.6}$$

Here E_{kin} is the macroscopic *kinetic energy* of the system, E_{pot} the *potential energy* of the system in external force fields, and U the *internal energy* of the system. The quantities E_{kin} and E_{pot} are defined as in mechanics. The kinetic energy of a system, or of a region of mass m in a system whose macroscopic speed amounts to v is therefore $(m/2)v^2$ and depends on no other state variable apart from the speed. The potential energy of a system or system region is defined in that its increase is equal to the work which is done on the system or the system region which is transferred from one place, assuming for the time being a stationary external force field, to another place in this field, when only "conservative" force fields are considered.† Accordingly, the potential energy depends on no other state variable than the space-coordinates (with regard to external force fields). The internal energy of a system or system region is, finally, only a function of internal state variables (temperature, work-coefficients or work-coordinates, amounts of individual parts). It represents, moreover, an extensive quantity in the sense of §1–1. Therefore the internal energy of a system consisting of several regions is

$$U = \sum_\alpha U^\alpha, \qquad (1\text{–}4.7)$$

where U^α represents the internal energy of the region α and the sum is taken over all regions.

From Eqs. (1–3.10), (4), and (6) it follows that for an internal state change in a closed system ($W_a = 0$, $\Delta E = \Delta U$)

$$\Delta U = Q + W = Q + W_l + W^*. \qquad (1\text{–}4.8)$$

With all work-coordinates constant, the term W_l disappears according to Eq. (1–3.11). Thus we have

$$\Delta U = Q + W^* \quad \text{(constant work-coordinates)}. \qquad (1\text{–}4.9)$$

A further particular case arises for a thermally insulated system ($Q = 0$)

$$\Delta U = W^* \quad \text{(adiabatic state change with constant work-coordinates)}. \qquad (1\text{–}4.10)$$

For an isotropic system without electric or magnetic effects, the volumes of the individual regions represent the particular work-coordinates (cf. Table 1, p. 7). An internal state change at constant volume of the entire system will now be called an *isochoric process*, though the volumes of the individual regions may be variable. By our remarks on p. 8 the work term W_l disappears in Eq. (8), if the total volume of a system remains unchanged and if differences in pressure between the individual regions of the systems

---

† The field of a force **K** is called "conservative" if the condition curl **K** = 0 is valid (rotation-free force field or potential field). Examples are gravitational fields, centrifugal fields, electrostatic fields.

are ruled out.† Thus with the last assumption the relations (9) and (10) hold for isochoric state changes in isotropic systems without electric or magnetic polarization.[4]

Consider a system of this kind enclosed in a rigid wall. Then it is possible to do friction work, e.g. with the help of an agitator, on the system, and thereby to produce a value of W^* in Eqs. (9) or (10) different from zero. This is frequently referred to as "friction heat." This expression can be misunderstood and in certain cases it is quite nonsensical. If we consider the temperature as the single changeable internal state variable, ΔU will be determined by the temperature change alone. We distinguish now two extreme cases:

1. The process takes place isothermally, i.e. at constant temperature. Then $\Delta U = 0$; thus from Eq. (9): $Q = -W^*$. In this case, the friction work done on the system will be equal to the heat given off from the system to the surroundings.

2. The process occurs adiabatically. Then because of Eq. (10): $\Delta U = W^*$. The friction work is therefore equal to the increase in the internal energy, which results in a temperature change. In this case, it is nonsense to speak of "friction heat," since the system exchanges no heat with the surroundings.

Analogous considerations are valid for the term "Joule heat" (cf. p. 86).

For an *isolated system* (§1–1), the quantities Q and W disappear and Eqs. (4) and (6) become:

$$\Delta E = \Delta E_{kin} + \Delta E_{pot} + \Delta U = 0 \quad \text{(isolated system)}, \qquad (1\text{–}4.11)$$

where E now designates the energy of the entire system. In this form, the first law can be called the *law of conservation of energy*. Using such a concept, Eq. (4) is frequently interpreted in the following way: the energy E of an arbitrary system can only be increased through "energy transport" from the outside world (in form of work W and heat Q) but not through "energy production" inside the system. Here the difference between energy and entropy (cf. §1–8, §1–10, and §1–24) is emphasized; for the entropy such a conservation law does not hold.

"Potential energy" cannot generally be defined for systems in electromagnetic fields since these need not be conservative. Furthermore, it is difficult to reduce all influences of the surroundings to the concepts "work" and "heat" uniquely. The "energy" of such a system is fixed by the condition

---

† If an isochoric change of state of the system is realized by enclosing the system in rigid walls, the work term W_l always disappears. If a system of two homogeneous parts (phases ′ and ″) is allowed to have variable volumes (V' and V'') as well as different pressures ($P' \neq P''$), as does occur in osmotic experiments, then for a constant total volume, i.e. for isochoric processes in the most general sense, we have

$$V' + V'' = \text{const},$$

$$W_l = -\int (P'\, dV' + P''\, dV'') = \int (P'' - P')\, dV' \neq 0.$$

that the quantity in question—which naturally must first have the correct dimension—is described by a conservation law and reduces to the energy defined above when the electromagnetic field vanishes (§4–34).

1–5 ENTHALPY

We define for a region α (internal energy U^α, work-coefficients L_i^α, work-coordinates l_i^α) a new state variable

$$H^\alpha \equiv U^\alpha - \sum_i L_i^\alpha \, l_i^\alpha, \qquad (1\text{–}5.1)$$

which we will call the *enthalpy* of the region. The quantity

$$H \equiv \sum_\alpha H^\alpha \qquad (1\text{–}5.2)$$

is the enthalpy of the whole system. Obviously, the enthalpy is, as is the internal energy, an extensive property and a function of the internal state variables.

For isotropic regions without electric or magnetic polarization, the negative pressure $-P^\alpha$ and the volume V^α represent the only work-coefficients and the only work-coordinates respectively (cf. Table 1, p. 7). We then find from Eq. (1) the usual definition of enthalpy:

$$H^\alpha \equiv U^\alpha + P^\alpha V^\alpha. \qquad (1\text{–}5.3)$$

For an interfacial layer (cf. Table 1, p. 7), on the other hand (omitting the phase index α):

$$H \equiv U + PV - \sigma\omega, \qquad (1\text{–}5.3a)$$

where σ is the interfacial tension and ω the surface area. For an electrified isotropic region (cf. Table 1), we find

$$H = U + PV - \mathfrak{E}\mathfrak{P}V. \qquad (1\text{–}5.3b)$$

Here $\mathfrak{E}$ is the electric field strength and $\mathfrak{P}$ the electric polarization vector.†

The most important property of enthalpy follows from the application of Eqs. (1–3.11), (1–4.7), (1–4.8), as well as (1) and (2) to an internal state change in a closed system with all work-coefficients constant:

$$\Delta H = Q + W^* \quad \text{(constant work-coefficients).} \qquad (1\text{–}5.4)$$

Here Q is the heat supplied to the system, and W^* the work done on the system, excluding the work referring to reversible deformations, electrification

---

† In many works, the definition (3) of enthalpy is used for all cases. This is disadvantageous for two reasons:

(a) For anisotropic media, pressure and volume do not appear as state variables, in particular, the term "pressure" has no meaning whatsoever.

(b) Many simple regularities, such as Eq. (4), lose their general validity.

and magnetization (cf. §1–3). For a thermally insulated system ($Q = 0$), we find:

$\Delta H = W^*$ (adiabatic state change at constant work-coefficients).   (1–5.5)

The relations (4) and (5) for the enthalpy are completely analogous to Eqs. (1–4.9) and (1–4.10) for the internal energy.

For isotropic systems without electrification or magnetization, constancy of work-coefficients means invariable pressures in all regions of the system. If the pressures are not only constant in time but also in space, we speak of an *isobaric process*. For isobaric processes in closed isotropic systems without electric or magnetic polarization, the statements (4) and (5) are thus valid.

If we consider an internal state change in an *isolated system*—which again should be isotropic and show no electrification or magnetization— and assume uniform pressure P ($P^\alpha = P$ for all phases α) then we are dealing with an *isochoric process* (example: chemical reactions in a thermally insulated vessel enclosed by a rigid wall). We then have, from Eqs. (1–4.7), (1–4.11), (2), and (3),

$$\Delta H = V \Delta P,$$

where

$$V = \sum_\alpha V^\alpha$$

is the constant total volume and ΔP is the total pressure increase of the system. From this we recognize that the enthalpy H, in contrast to the internal energy U, can change during internal state changes within isolated systems.

The quantities work, heat, energy, and enthalpy have the same dimension. Thus they may be measured, for example, in erg, J, or cal.

1–6 PARTIAL MOLAR QUANTITIES

Let us consider an arbitrary extensive property Z (e.g. volume, internal energy, enthalpy). Then for a single region (where we have omitted the index α for the sake of brevity):

$$Z = Z(T, L_i, n_k).$$ (1–6.1)

Here we have chosen the temperature T, the work-coefficients L_i, and the amounts n_k of the substances contained in the region† as independent variables.

---

† The quantity of matter may be measured either by the mass m_k of substance k or by the amount n_k of substance k. The basic unit for m_k is g (gramme), while for n_k it is mol. Thus the "amount of substance" is treated as an independent basic quantity. The term "number of moles" should not be used for n_k, but for the numerical value of n_k expressed in the unit mol.

Changes in amounts can occur not only by chemical reactions inside the region but also through exchange of matter with the surroundings. We now include *open* regions in our discussion. The fact that quantities such as the work-coefficients L_i, the work-coordinates l_i, the internal energy U, and the enthalpy H were originally only defined for closed regions (§1–3 to §1–5) does not introduce any difficulty. We can, for example, state the pressure P or the volume V of an open region without the term $-P\,dV$ necessarily meaning the reversible volume work done on the region during an infinitesimal state change. Similarly, quantities such as U or H have a definite meaning for open regions, since the change of the state variable is independent of the path the change of state follows. Thus particularly the change of an extensive quantity Z is independent of whether the changes in amount of the individual substances occur through chemical reactions inside the region or through exchange of matter with the outside world. On the other hand, the terms "work" and "heat" are ambiguous in open regions, as can be seen by means of simple examples[1a, 8]. Thus, of the above formulas, only those that do *not* contain the work W and the heat Q are valid for open regions or open systems, as, for example, Eqs. (1–4.7), (1–5.1), and (1–5.2).

We now define the *partial molar quantity* of substance k for the considered region:

$$Z_k \equiv \left(\frac{\partial Z}{\partial n_k}\right)_{T,L_i,n_j}, \tag{1-6.2}$$

where L_i stands for all work-coefficients and n_j for all amounts except n_k. If the pressure $-P$ is the only work-coefficient, Eq. (2) reduces to the familiar definition of a partial molar quantity following Lewis and Randall (1921):

$$Z_k \equiv \left(\frac{\partial Z}{\partial n_k}\right)_{T,P,n_j}. \tag{1-6.3}$$

Equation (3) is true only for an isotropic region without electrification and magnetization.

The partial molar quantities introduced by Eq. (2) are intensive variables. Consequently, they depend only on the temperature T, on the work-coefficients L_i, and on the composition of the region, i.e. the concentrations of the substances. Examples are: the partial molar volume V_k, the partial molar internal energy U_k, and the partial molar enthalpy H_k.

As composition variables we temporarily introduce two quantities: the *mole fraction* of species k

$$x_k \equiv \frac{n_k}{\sum_k n_k} = \frac{n_k}{n} \tag{1-6.4}$$

and the *molarity* of species k

$$c_k \equiv \frac{n_k}{V}, \tag{1–6.5}$$

where n is the total amount of substance and V the volume of the region.[5]

We further define the *molar quantity*

$$\bar{Z} \equiv \frac{Z}{n} \tag{1–6.6}$$

and the *generalized density*

$$Z_V \equiv \frac{Z}{V}, \tag{1–6.7}$$

both of which again represent intensive variables. Examples are: the molar volume $\bar{V}$, the molar internal energy $\bar{U}$, and the molar enthalpy $\bar{H}$ as well as the (internal) energy density U_V and the enthalpy density H_V.

Because of the general definition of extensive quantities (§1–1) it follows from Eq. (2) and using Euler's theorem for homogeneous functions of first degree that

$$Z = \sum_k n_k Z_k, \tag{1–6.8}$$

where the sum is taken over all species. With the help of Eqs. (4) to (7) we find that

$$\bar{Z} = \sum_k x_k Z_k \tag{1–6.9}$$

and

$$Z_V = \sum_k c_k Z_k. \tag{1–6.10}$$

If we put the volume V for Z in Eqs. (7) and (10) then we have the identity

$$\sum_k c_k V_k = 1. \tag{1–6.11}$$

We can write for the total differential of Z from Eqs. (1) and (2)

$$dZ = \frac{\partial Z}{\partial T} dT + \sum_i \frac{\partial Z}{\partial L_i} dL_i + \sum_k Z_k \, dn_k, \tag{1–6.12}$$

where the first sum is over all work-coefficients and the second over all species.

If we construct the total differential of expression (8) and compare it with Eq. (12), we obtain the relation:

$$\frac{\partial Z}{\partial T} dT + \sum_i \frac{\partial Z}{\partial L_i} dL_i - \sum_k n_k \, dZ_k = 0. \tag{1–6.13}$$

This equation, from which the generalized Gibbs–Duhem equation (§1–13) follows, can also be written in the form

$$\frac{\partial \bar{Z}}{\partial T} dT + \sum_i \frac{\partial \bar{Z}}{\partial L_i} dL_i - \sum_k x_k\, dZ_k = 0 \qquad (1\text{-}6.14)$$

with the help of Eqs. (4) and (6).

Because of the mathematical identities

$$d\left(\frac{Z}{n}\right) = Z d\left(\frac{1}{n}\right) + \frac{1}{n} dZ,$$

$$d\left(\frac{n_k}{n}\right) = n_k d\left(\frac{1}{n}\right) + \frac{1}{n} dn_k,$$

the following relation can be derived from Eqs. (4), (6), (8), and (12):

$$d\bar{Z} = \frac{\partial \bar{Z}}{\partial T} dT + \sum_i \frac{\partial \bar{Z}}{\partial L_i} dL_i + \sum_k Z_k\, dx_k. \qquad (1\text{-}6.15)$$

By comparison of the total differential of Eq. (9) with Eqs. (15) we again obtain Eq. (14).

Expressions for the concentration variables c_k in place of the x_k analogous to relations (14) and (15), can be obtained from Eqs. (5), (7), (8), (12), and (13):

$$\frac{1}{V}\left(\frac{\partial Z}{\partial T} dT + \sum_i \frac{\partial Z}{\partial L_i} dL_i\right) - \sum_k c_k\, dZ_k = 0, \qquad (1\text{-}6.16)$$

$$dZ_V = \frac{1}{V}\left(\frac{\partial Z}{\partial T} dT + \sum_i \frac{\partial Z}{\partial L_i} dL_i\right) + \sum_k Z_k\, dc_k. \qquad (1\text{-}6.17)$$

If the total differential of expression (10) is constructed and compared with Eq. (17), Eq. (16) is recovered. That the analogy between the relations (16) and (17) and the expressions (14) and (15) is not complete is due to the fact that the total amount of substance n is independent of the temperature T and of the work-coefficients L_i, while the volume V is a function of these variables.

1-7 HEAT IN OPEN SYSTEMS

The term "work" is indeterminate for open regions (cf. §1–6). It is possible, however, to use the expression w for that work which *would* be done on the region in question if it *were* closed. From Eqs. (1–3.7) and (1–3.8) we have for an infinitesimal state change

$$dw = \sum_i L_i\, dl_i + dW_{\text{diss}} + dW_a. \qquad (1\text{-}7.1)$$

Here the L_i and l_i denote the work-coefficients and work-coordinates respectively, which have definite meanings for open regions also, while the summation term in Eq. (1) need not have, as it does in closed regions, the physical meaning of reversible deformation, electrification, and magnetization work (cf. §1–6). The term dW_{diss} or dW_a is the work done on the region during an infinitesimal state change related to dissipative effects (cf. p. 8) or to forces, respectively, which accelerate the region or act against external force fields. These two terms are defined for open regions also since they can always be calculated from their outside sources, e.g. electrical work, friction work, acceleration work, lifting work in a gravitational field, etc. In the simplest case, namely, for a static isotropic region without electrification or magnetization as well as without dissipative effects and external force fields, Eq. (1) can be reduced to the expression

$$dw = -P\,dV,$$

where P is the pressure and V the volume of the open region, without $-P\,dV$ having to be the work actually done on the region.

We again establish the term "heat" for open regions by definition, since this term will be necessary in the structure of thermodynamics of irreversible processes.

It is useful to define the "heat" dQ absorbed by an open phase from the surroundings during an infinitesimal state change as follows:[†]

$$dQ \equiv dE - dw - \sum_k H_k\,d_e n_k. \qquad (1\text{–}7.2)$$

Here E is the (total) energy of the phase, w the quantity defined above, H_k the partial molar enthalpy of species k, and $d_e n_k$ the infinitesimal increase of the amount n_k of substance k due to "external" causes, i.e. not by chemical reactions inside the phase but by exchange of matter with the outside world, that is, either with neighboring regions or with the surroundings of the entire system.

As is obvious by comparison with Eq. (1–4.5), for zero transfer of matter through the boundaries of the system ($dw = dW$, $d_e n_k = 0$), Eq. (2) is reduced to the classical expression for the heat in closed systems.

Other definitions of "heat" in open systems introduce other intensive state variables in place of the partial molar enthalpy in Eq. (2). The final results of thermodynamics of irreversible processes will not be affected by such varying conventions; the definition (2), however, proves to have the advantage of simplicity in calculations.

A further advantage of definition (2) arises in that the "heat" so defined is independent of the arbitrary zero point of the internal energy. In order

[†] The formulation for a volume element of a continuous system is more complicated (cf. §4–6).

to show this, we split the change of amount n_k of species k into two parts of which the first $(d_r n_k)$ refers to internal causes (chemical reactions inside the phase) and the second $(d_e n_k)$ to external causes (mass exchange with the surroundings):

$$dn_k = d_r n_k + d_e n_k,$$

$$d_r n_k = \sum_r \nu_{kr} \, d\xi_r.$$

Here ν_{kr} is the stoichiometric number of species k in the chemical reaction r (considered positive or negative if the species in question is produced or consumed respectively in the reaction) and ξ the extent of the reaction r (cf. §1–14). The summation is taken over all chemical reactions. Thus we obtain from Eq. (1–6.12) for the change of the internal energy U of the phase:

$$dU = \frac{\partial U}{\partial T} \, dT + \sum_i \frac{\partial U}{\partial L_i} \, dL_i + \sum_k \sum_r \nu_{kr} U_k \, d\xi_r + \sum_k U_k \, d_e n_k,$$

where T is the temperature and U_k the partial molar internal energy of species k. While the coefficients of dT, dL_i, and $d\xi_r$ are completely determined ($\sum_k \nu_{kr} U_k$ is, for example, the differential energy of reaction of the chemical reaction r), the last term contains arbitrary additive constants a_k in the partial molar internal energies U_k. There thus remains in dU an undetermined expression of the form

$$\sum_k a_k \, d_e n_k.$$

From Eqs. (1–5.1) and (1–6.2), however, the same expression appears in the last term of Eq. (2) with the opposite sign, while

$$dE_{\text{pot}} + dE_{\text{kin}} = dE - dU$$

(see Eq. 1–4.6) and dw (see Eq. 1) are not affected by the energy zero points. This same property of invariance would also be found, it is true, if in Eq. (2) the partial molar internal energy U_k were used in place of the partial molar enthalpy H_k.

For *internal* state changes ($dE = dU$, $dW_a = 0$), we obtain from Eqs. (1) and (2) an equation which we will use later, namely:

$$dQ = dU - \sum_i L_i \, dl_i - dW_{\text{diss}} - \sum_k H_k \, d_e n_k. \qquad (1\text{–}7.3)$$

We now consider an important special case of Eq. (3); the phase is allowed to be isotropic and no electrification, magnetization, or dissipative effects are present. Then there follows[9]:

$$dQ = dU + P \, dV - \sum_k H_k \, d_e n_k. \qquad (1\text{–}7.4)$$

On the basis of the last equation we can see that the definition of heat chosen here for open systems is physically meaningful. We consider a single open phase, e.g. a homogeneous fluid mixture of constant temperature and under constant pressure, to which is added (e.g. by withdrawing a partition or by dropping out from a burette) more liquid of the same temperature, the same pressure, and the same composition. Then, because of Eqs. (1–5.3) and (1–6.3) and in the absence of chemical reactions, we obtain

$$dU + P\,dV = d(U + PV) = dH = \sum_k H_k\,d_e n_k.$$

From this together with Eq. (4)

$$dQ = 0.$$

This result is felt to be "reasonable." It would not have been obtained if, in place of H_k in Eqs. (2), (3), and (4), other quantities had been used.

It should be noted that a definition analogous to (2) when applied to continuous systems leads to the establishment of a "heat flux," which is distinguished from other "heat fluxes" (defined elsewhere) by many simple properties, for example, by invariance both towards shifts of energy zero and also towards changes of the reference velocity (§4–7). This heat flow is known in the literature mainly under the name "reduced heat flux"[10–12].

In heterogeneous (discontinuous) systems, Eq. (2) is applied to each phase α of the system. The quantities $d_e n_k^\alpha$ and dQ^α will be split into two parts, the first of which ($d_a n_k^\alpha$ or $d_a Q^\alpha$, respectively) refers to the mass or heat exchange with the surroundings of the entire system while the second ($d_u n_k^\alpha$ or $d_i Q^\alpha$, respectively) designates the transfer of mass or heat from other phases of the system to the phase in question, that is, connected with processes inside the total system. We thus obtain

$$d_e n_k^\alpha = d_a n_k^\alpha + d_u n_k^\alpha, \tag{1–7.5}$$

$$dn_k^\alpha = d_r n_k^\alpha + d_e n_k^\alpha, \tag{1–7.6}$$

$$dm = \sum_\alpha \sum_k M_k\,d_a n_k^\alpha, \tag{1–7.7}$$

$$dQ^\alpha = d_a Q^\alpha + d_i Q^\alpha, \tag{1–7.8}$$

$$dQ = \sum_\alpha d_a Q^\alpha. \tag{1–7.9}$$

Here $d_r n_k^\alpha$ is the infinitesimal increase of n_k^α due to chemical reactions inside the phase (cf. above), m the total mass of the system, M_k the molar mass of species k, and dQ the heat absorbed by the whole system out of the surroundings during an infinitesimal state change. The last quantity, in the case of a closed system ($dm = 0$), is given by Eq. (1–4.5). The decomposition (8) is clearly determined[6] with Eq. (9).†

1–8 ENTROPY AND ABSOLUTE TEMPERATURE (SECOND LAW)

Through the *Second Law of Thermodynamics* two functions are simultaneously added to the list of internal state variables, namely, the *entropy* and the *absolute temperature*. We proceed from the following statement[1b] of the second law which, compared to all previous formulations, has the advantage of greater generality and clarity:

For each region α whose internal state is described by the internal energy U^α (or the empirical temperature), the work-coordinates l_i^α (or the work-coefficients L_i^α), and the amounts of substances, there is an extensive state variable S^α, the *entropy* of the region, which is dependent on all of the above variables and has the following properties:

(a) The differential of the entropy of the region at constant amounts of substances is given by

$$T^\alpha \, dS^\alpha = dU^\alpha - \sum_i L_i^\alpha \, dl_i^\alpha. \qquad (1\text{–}8.1)$$

Here T^α is an intensive positive quantity which represents a universal function of the empirical temperature of the region and will be called the *absolute temperature* of the region.

(b) For an arbitrary state change $I \to II$ of the system, writing the entropy of the whole system as

$$S \equiv \sum_\alpha S^\alpha, \qquad (1\text{–}8.2)$$

the following relations hold:

$$\Delta S = S_{II} - S_I = \Delta_a S + \Delta_i S, \qquad (1\text{–}8.3)$$

$$\Delta_i S = 0 \quad \text{(reversible change)}, \qquad (1\text{–}8.4)$$

$$\Delta_i S > 0 \quad \text{(irreversible change)}, \qquad (1\text{–}8.5)$$

$$\Delta_i S < 0 \quad \text{(impossible change)}. \qquad (1\text{–}8.6)$$

---

† The quantity $d_r n_k^\alpha$ has no analog in the corresponding expression for the heat. The amount of substance can be increased or decreased by processes inside a single phase, namely, by chemical reactions; the *heat*, however, according to its definition is not a state variable which could increase or decrease, but represents a quantity which is absorbed by the phase in question from the surroundings or given off by the phase to the surroundings, where the "surroundings" can mean both neighboring phases or the environment of the whole system. Let us denote expressions which relate to the inside of *whole* systems generally by an inferior i; so the quantity $d_i Q^\alpha$ can be compared to the expression

$$d_i n_k^\alpha = d_u n_k^\alpha + d_r n_k^\alpha.$$

Here the decomposition of the entropy change in Eq. (3) is established by the condition

$$\Delta_a S = 0 \text{ when the system is thermally insulated.} \qquad (1\text{-}8.7)$$

Statement (a) introduces, with the differential relation (1) for each system region, the entropy S^α and the absolute temperature T^α. Equation (1) still remains valid if S^α is multiplied by an arbitrary positive quantity and T^α is divided by the same quantity. The arbitrariness in the definition of S^α and T^α is removed by fixing the scale for T^α. Since it can be shown that T^α is proportional to the empirical temperature of the region (measured on the Kelvin scale, §1–2), the identity of both temperatures can be established by an appropriate choice of a fixed point[1c]. This identity will always be assumed.

The statement (a) can be formulated in a mathematically stricter form according to Carathéodory in the following way: there is, for each region of given mass and composition, and for the incomplete differential

$$dU^\alpha - \sum_i L_i^\alpha \, dl_i^\alpha$$

an integrating factor T^α, which depends only on the empirical temperature of the region and makes the above expression into an exact differential, namely, into the differential of the state variable S^α.

In systems with nonautonomous interfacial layers (cf. footnote to p. 5), the differential relation (1) cannot be written for each individual region of the system. On the other hand, the expressions for the total entropy S, namely, the relations (3) to (7), remain valid even here. We must now define S by requiring that the relation (1–10.10) is true in the case of a closed system (cf. §1–10).

Statement (b) refers to the entropy increase $\Delta S = S_{II} - S_I$ of the whole system for a state change $I \rightarrow II$. Thermal insulation of a system means nothing more than a prevention of mass and heat exchange with the surroundings, i.e. an adiabatic path for the state change. Thus $\Delta_a S$ represents the entropy increase of the system due to mass and heat exchange with the outside world and $\Delta_i S$ the entropy increase of the system due to processes taking place inside the system.† While $\Delta_a S$ and ΔS can be positive, negative, or zero for an arbitrary state change, $\Delta_i S$ is, according to (4) to (6), never negative, and for an actual (irreversible) path of the process inside the system is always positive.

---

† Many authors interpret $\Delta_a S$ as the part of the entropy change of the system due to "interaction with the surroundings." This statement is misleading, since the work done on or by the system also represents such an interaction, although, in Eq. (7), it contributes to $\Delta_i S$ and not to $\Delta_a S$. This is recognized quite clearly in the case of "dissipated work," which, for adiabatic state changes, calls for irreversible processes inside the system and appears explicitly in $\Delta_i S$, as is obvious from our remarks in §1–10.

If the processes which are interesting for a particular problem occur inside an individual region, the summation sign in Eq. (2) is redundant so that the expressions (3) to (7) are valid for the entropy of the region. For continuous systems, in which the "region" represents a volume element, it is possible in this way to have a "local" formulation of the second law (§4–8).

If only *adiabatic* processes are considered then relations (3) to (7) immediately result in the classic statements:

$$\Delta S = 0 \quad \text{(reversible adiabatic change)}, \qquad (1\text{–}8.8)$$

$$\Delta S > 0 \quad \text{(irreversible adiabatic change)}, \qquad (1\text{–}8.9)$$

$$\Delta S < 0 \quad \text{(impossible adiabatic change)}. \qquad (1\text{–}8.10)$$

As was shown in §1–1, the designation "irreversible" has here been introduced as a synonym for "actual." It remains to be proved that the concept "irreversibility" (§1–1) connected with the above designation is correct.

Let us consider an arbitrary actual (natural) process in a system leading to a change of its state. We then bring the system in question, by an arbitrary path, back to the initial state by a process which can be either a quasistatic (reversible) or an actual process. Then the entropy of the system is again at its original value. If we now extend the system to make it an "isolated system" (§1–1) by including all those parts of the surroundings which undergo a change, the total system so established represents a special case of a thermally insulated system. Relations (8) and (9) therefore hold. Thus the entropy of the total system has increased. Since the original system suffered no entropy change, the entropy increase is due to a state change in the surroundings of the original system. Thus, in returning the original system to its initial state, changes in nature have remained: the actual process is irreversible in the sense of the definition in §1–1.

The statement that all natural processes are irreversible, is connected very closely to the principle that a "perpetual motion of the second kind" is impossible. This principle, in the traditional representations of thermodynamics, stands at the head of the statements on the second law. Since, however, all statements which are necessary for the modern formulation of the second law cannot be derived with the help of this principle, we have reversed the order of presentation.

As is obvious, our statement of the second law does not only contain all traditional formulations of scientific thermodynamics (cf. above and §1–10), but also all statements of technical thermodynamics (efficiency of steam engines, motors, refrigerators, etc.) as special cases.

The concept of absolute temperature is due to W. Thomson (Lord Kelvin) (1848). Entropy was introduced by Clausius (1854, 1865). Several authors (Gibbs, Planck, Carathéodory, and de Donder, among others) subsequently extended and modified the formulations.

The "Nernst Heat Theorem" (Nernst, 1906) and the "Law of Unattainability of Absolute Zero" (Nernst, 1912) are frequently regarded as equivalent. Either of these theorems is then called the "Third Law of Thermodynamics" and is thus compared to the first and second laws of thermodynamics. It can, however, be shown[13] that the law of unattainability of absolute zero follows in certain cases from the first and second laws, while in other cases it must be considered as a consequence of the Nernst heat theorem. Moreover, the exact formulation[14] of the Nernst heat theorem is far more complicated and less fundamental than the original statement of the theorem by Nernst. Thus the term "third law of thermodynamics" is inappropriate in any case.[7]

1–9 CHEMICAL POTENTIALS AND THE GIBBS EQUATION

Equation (1–8.1) states the dependence of the entropy S^α of a region α on the internal energy U^α and the work-coordinates l_i^α at constant amounts of substances n_k^α. We can therefore write

$$T^\alpha \left(\frac{\partial S^\alpha}{\partial U^\alpha} \right)_{l_i^\alpha, n_k^\alpha} = 1, \qquad (1\text{–}9.1)$$

$$T^\alpha \left(\frac{\partial S^\alpha}{\partial l_i^\alpha} \right)_{U^\alpha, l_j^\alpha, n_k^\alpha} = -L_i^\alpha, \qquad (1\text{–}9.2)$$

where l_j^α in Eq. (2) denotes all work-coordinates apart from l_i^α, L_i^α signifies the work-coefficients belonging to the l_i^α, and T^α is the absolute temperature.

Since the entropy also depends on the amounts of substances n_k^α from §1–8, we introduce, by analogy with the work-coefficients, the following expression:

$$\mu_k^\alpha \equiv -T^\alpha \left(\frac{\partial S^\alpha}{\partial n_k^\alpha} \right)_{U^\alpha, l_i^\alpha, n_j^\alpha}, \qquad (1\text{–}9.3)$$

where n_j^α stands for all amounts of substances apart from n_k^α. The intensive state variable μ_k^α will be called the *chemical potential* of species k in the region α. This fundamental quantity has already been defined in a similar way by Gibbs (1875).†

---

† The differential quotient in Eq. (3) must not be confused with the *partial molar entropy* S_k^α of species k in region α. For this the following is valid from Eq. (1–6.2):

$$S_k^\alpha = \left(\frac{\partial S^\alpha}{\partial n_k^\alpha} \right)_{T^\alpha, L_i^\alpha, n_j^\alpha}.$$

The connection between the chemical potential and the partial molar entropy will be derived in §1–12.

With Eqs. (1) to (3) we can formulate the general expression for the differential of the entropy of a region of variable amount and composition, and thus generalize Eq. (1–8.1):

$$T^\alpha \, dS^\alpha = dU^\alpha - \sum_i L_i^\alpha \, dl_i^\alpha - \sum_k \mu_k^\alpha \, dn_k^\alpha. \tag{1–9.4}$$

Here the first sum is taken over all work-coefficients and work-coordinates respectively and the second sum over all species. This fundamental relation we call the *generalized Gibbs equation*. Since changes in amount and composition can occur both through chemical reactions inside the region and through mass exchange with the surroundings, the equation holds for both closed and open regions. It is also in no way restricted to reversible state changes because it contains only differentials of state variables (cf. §1–23).

As is obvious from Eqs. (1–7.1) and (1–7.2), the term

$$\sum_i L_i^\alpha \, dl_i^\alpha$$

and the expression

$$dU^\alpha - \sum_i L_i^\alpha \, dl_i^\alpha$$

have the meaning of work done on the region and heat supplied to the region respectively only if the region is *closed* and if an infinitesimal change of the *internal* state *without* dissipative effects is considered.

In the simplest case, i.e. if only the volume V^α appears as the work-coordinate and correspondingly the negative pressure $-P^\alpha$ as the appropriate work-coefficient, Eq. (4) is reduced to the classical *Gibbs equation*:

$$T^\alpha \, dS^\alpha = dU^\alpha + P^\alpha \, dV^\alpha - \sum_k \mu_k^\alpha \, dn_k^\alpha, \tag{1–9.5}$$

which is valid for any isotropic region without electrification and magnetization.

The quantities present in Eq. (4) as differentials are extensive; the other quantities, however, are intensive state variables. Thus it follows that

$$T^\alpha S^\alpha = U^\alpha - \sum_i L_i^\alpha l_i^\alpha - \sum_k \mu_k^\alpha n_k^\alpha. \tag{1–9.6}$$

The transition from Eq. (4) to Eq. (6) can either be regarded intuitively as an integration over the quantity of material present while keeping all intensive quantities, i.e. temperature, pressure, composition, etc., constant, or formally as an application of Euler's theorem to homogeneous functions of the first degree.

Designating any state variable of the region by y (leaving out the index α), we have

$$T d\left(\frac{S}{y}\right) = TS\, d\left(\frac{1}{y}\right) + \frac{T}{y}\, dS.$$

From this with the help of Eqs. (4) and (6) we obtain

$$T d\left(\frac{S}{y}\right) = \left(U - \sum_i L_i l_i - \sum_k \mu_k n_k\right) d\left(\frac{1}{y}\right)$$
$$+ \frac{1}{y}\left(dU - \sum_i L_i\, dl_i - \sum_k \mu_k\, dn_k\right).$$

Thus we have

$$T d\left(\frac{S}{y}\right) = d\left(\frac{U}{y}\right) - \sum_i L_i\, d\left(\frac{l_i}{y}\right) - \sum_k \mu_k\, d\left(\frac{n_k}{y}\right). \tag{1–9.7}$$

If, for example, we substitute the total amount of substance n or the volume V of the region for y, we then find by considering Eqs. (1–6.4) to (1–6.7) for the differential of the molar entropy $\bar{S}$ or the entropy density S_V:

$$T\, d\bar{S} = d\bar{U} - \sum_i L_i\, d\bar{l}_i - \sum_k \mu_k\, dx_k \tag{1–9.8}$$

or

$$T\, dS_V = dU_V - \sum_i L_i\, dl_{iv} - \sum_k \mu_k\, dc_k. \tag{1–9.9}$$

Here $\bar{U}$ represents the molar internal energy, $\bar{l}_i$ are molar work-coordinates (e.g. molar volume $\bar{V}$, molar electric moment $\mathfrak{P}\bar{V}$), x_k the mole fraction of species k, U_V the density of the internal energy, l_{iv} a work-coordinate referring to unit volume (e.g. electric polarization $\mathfrak{P}$), and c_k the molarity of species k.

In the case of an isotropic region without electric or magnetic polarization, the volume V or the negative pressure $-P$ again appear, as in Eq. (5), as the single work-coordinate or as the single work-coefficient respectively. We then obtain from Eqs. (7) to (9) the following forms of the Gibbs equation:

$$T d\left(\frac{S}{y}\right) = d\left(\frac{U}{y}\right) + P\, d\left(\frac{V}{y}\right) - \sum_k \mu_k\, d\left(\frac{n_k}{y}\right), \tag{1–9.10}$$

$$T\, d\bar{S} = d\bar{U} + P\, d\bar{V} - \sum_k \mu_k\, dx_k, \tag{1–9.11}$$

$$T\, dS_V = dU_V - \sum_k \mu_k\, dc_k. \tag{1–9.12}$$

For isotropic regions without electrification and magnetization which contain only a single species, it is found from Eqs. (5) and (11) that

$$T\, dS = dU + P\, dV - \mu\, dn$$

and

$$T\, d\bar{S} = d\bar{U} + P\, d\bar{V}.$$

The last equation is actually the form of the Gibbs equation used by Planck. (For mixtures Planck used another representation.) If we assume isotropic regions without electric and magnetic polarization whose mass and composition are constant then the following relation is derived from Eq. (5):

$$T \, dS = dU + P \, dV.$$

This simple formula was already known to Clausius.

Two further special forms of the generalized Gibbs equation should now be mentioned, since they are needed later.

For an isotropic region with electrification and magnetization, there follows from Eq. (9) and Table 1 (p. 7):

$$T \, dS_V = dU_V - \mathfrak{E} \, d\mathfrak{P} - \mathfrak{H} \, d\mathfrak{J} - \sum_k \mu_k \, dc_k. \tag{1–9.13}$$

Here $\mathfrak{E}$ is the electric field intensity, $\mathfrak{P}$ the electric polarization, $\mathfrak{H}$ the magnetic field intensity, and $\mathfrak{J}$ the magnetic polarization. Equation (13) will be used in §1–12 and in §4–34.

For anisotropic regions without electric and magnetic polarization, we start from Eq. (7) and put $y = m$ (mass of the reference volume element of volume V_0). With $m/V_0 = \rho$ (density of the region) we then get according to Table 1 (p. 7):

$$T \, d\tilde{S} = d\tilde{U} - \frac{1}{\rho} \sum_{i=1}^{6} \tau_i \, de_i - \sum_k \bar{\mu}_k \, d\chi_k. \tag{1–9.14}$$

Here $\tilde{S} = S/m$ is the specific entropy, $\tilde{U} = U/m$ is the specific internal energy, τ_i and e_i are the stress and strain components respectively, $\chi_k \ (= M_k n_k/m$, $M_k = $ molar mass of species k) the mass fraction of substance k, and

$$\bar{\mu}_k = \frac{\mu_k}{M_k} \tag{1–9.15}$$

the specific chemical potential of species k. Equation (14) will be necessary in §4–37.

The generalized Gibbs equation is a direct consequence of the general formulation of the second law in §1–8. When taken with the statements (1–8.2) to (1–8.7) it yields one of the most important results for the thermodynamics of irreversible processes. A comprehensive discussion of the applicability of this fundamental relation to nonequilibrium states is found in §1–23.

1-10 RELATION BETWEEN ENTROPY AND HEAT

We now derive the relation between the entropy change of an arbitrary heterogeneous (discontinuous) system and the heat supplied to the individual parts of the system from the surroundings. We consider the most general case, namely, that of an open and nonisothermal system where arbitrary heat and matter exchange is possible both between the individual system parts (phases) and also between the system and the surroundings[15].

From Eqs. (1–7.5) and (1–7.6) the infinitesimal increase in the amount n_k^α of substance k in the phase α can be split up in the following way:

$$dn_k^\alpha = d_a n_k^\alpha + d_u n_k^\alpha + d_r n_k^\alpha, \qquad (1\text{–}10.1)$$

where the first term of the right-hand side refers to material exchange with the surroundings of the entire system, the second term to transport of matter to or from neighboring phases of the system, and the third term to chemical reactions inside the phase α.

The heat dQ^α supplied to the phase α during an infinitesimal *internal* state change is, according to Eqs. (1–7.3), (1–7.5), and (1–7.8), given by the expression

$$dQ^\alpha = d_a Q^\alpha + d_i Q^\alpha = dU^\alpha - \sum_i L_i^\alpha \, dl_i^\alpha - dW_{\text{diss}}^\alpha$$

$$- \sum_k H_k^\alpha (d_a n_k^\alpha + d_u n_k^\alpha). \qquad (1\text{–}10.2)$$

Here $d_a Q^\alpha$ refers to the heat exchange with the surroundings of the entire system and $d_i Q^\alpha$ to the heat transferred to or from neighboring phases of the system. U^α signifies the internal energy of phase α and H_k^α the partial molar enthalpy of species k in the phase α. The L_i^α and l_i^α are the work-coefficients and work-coordinates respectively of phase α. The term dW_{diss}^α is, finally, the "dissipated work" (electrical work, friction work, etc.) done on the phase α during an infinitesimal change.

The generalized Gibbs equation (1–9.4) for the differential of the entropy S^α of the phase α is

$$T^\alpha \, dS^\alpha = dU^\alpha - \sum_i L_i^\alpha \, dl_i^\alpha - \sum_k \mu_k^\alpha \, dn_k^\alpha. \qquad (1\text{–}10.3)$$

Here T^α is the absolute temperature of the phase α and μ_k^α the chemical potential of species k in the phase α. Moreover, from the relation (1–12.6) derived in §1–12† we have

$$\mu_k^\alpha = H_k^\alpha - T^\alpha S_k^\alpha,$$

where S_k^α represents the partial molar entropy of substance k in the phase α.

---

† The derivation of Eq. (1–12.6) does not assume the results of this section, but can be obtained without any reference to the physical interpretation of the function G from Eqs. (1–6.2), (1–9.4), and (1–11.2) by mathematical means.

By combining Eqs. (1) and (2) with (3) and the last equation we find that

$$T^\alpha \, dS^\alpha = d_a Q^\alpha + T^\alpha \sum_k S_k^\alpha \, d_a n_k^\alpha + dW_{\text{diss}}^\alpha$$

$$+ d_i Q^\alpha + T^\alpha \sum_k S_k^\alpha \, d_u n_k^\alpha - \sum_k \mu_k^\alpha \, d_r n_k^\alpha. \qquad (1\text{--}10.4)$$

The first two terms on the right side are obviously connected to the heat and material exchange of the whole system with the outside world or surroundings of the whole system, while the following terms refer to the dissipative effects (current transport, friction, etc.), the heat transfer to or from neighboring phases of the system, the exchange of matter with neighboring phases, and the chemical reactions inside the considered phase; that is, they depend on processes inside the entire system.[8]

For the change in entropy S of the whole system during an arbitrary state change I to II, we obtain, with (1–8.2) to (1–8.5),

$$S = \sum_\alpha S^\alpha, \qquad (1\text{--}10.5)$$

$$\Delta S = S_{\text{II}} - S_{\text{I}} = \int_{\text{I}}^{\text{II}} dS = \sum_\alpha \int_{\text{I}}^{\text{II}} dS^\alpha = \Delta_a S + \Delta_i S, \qquad (1\text{--}10.6)$$

$$\Delta_i S \geqslant 0, \qquad (1\text{--}10.7)$$

where the inequality is valid for irreversible (actual) paths of the processes inside the system and the equality for the limiting reversible case.

From the condition (1–8.7) both of the first terms of the right side of Eq. (4) are part of $\Delta_a S$; the remaining terms are part of $\Delta_i S$. This follows also from the interpretation of the individual terms above. We derive from Eqs. (4) to (6)

$$\Delta_a S = \sum_\alpha \int_{\text{I}}^{\text{II}} \left(\frac{d_a Q^\alpha}{T^\alpha} + \sum_k S_k^\alpha \, d_a n_k^\alpha \right). \qquad (1\text{--}10.8)$$

From (6), (7), and (8) we find the inequality

$$\Delta S = S_{\text{II}} - S_{\text{I}} \geqslant \sum_\alpha \int_{\text{I}}^{\text{II}} \left(\frac{d_a Q^\alpha}{T^\alpha} + \sum_k S_k^\alpha \, d_a n_k^\alpha \right) \qquad (1\text{--}10.9)$$

and this is valid for any internal state change of a discontinuous system. The analogous relation to (9) for continuous systems will be derived in §4–11. From (9) all classical formulations of the second law follow. Moreover, (9) contains the desired general relation between the entropy change (ΔS) of a heterogeneous system and the infinitesimal amounts of heat $d_a Q^\alpha$ which are transferred to the individual phases from the surroundings of the system.

For a *closed system* ($d_a n_k^\alpha = 0$ for all species k and all phases α), we find from (9) that

$$\Delta S = S_{\text{II}} - S_{\text{I}} \geqslant \sum_\alpha \int_{\text{I}}^{\text{II}} \frac{d_a Q^\alpha}{T^\alpha}. \qquad (1\text{–}10.10)$$

Should a definition of heat different from that in Eq. (1–7.2) be employed, and the partial molar enthalpy H_k^α in Eq. (2) be replaced by some other intensive variable ζ_k^α (cf. §1–7) then, instead of $T_k^\alpha S_k^\alpha$ in Eqs. (4), (8), and (9), the quantities $\zeta_k^\alpha - \mu_k^\alpha$ appear. The inequality (10) remains, however, unchanged, and is therefore invariant to a change in the definition of the "heat" for open phases, although the individual phases could also be open in a closed system, i.e. arbitrary heat and matter exchange with neighboring phases can take place.

For a *thermally insulated system*, that is, for an *adiabatic* state change, there results from (10) with $d_a Q^\alpha = 0$ (for all phases α):

$$\Delta S \geqslant 0 \quad \text{(adiabatic change)}, \qquad (1\text{–}10.11)$$

which is in agreement with (1–8.8) and (1–8.9).

For an *isothermal* change of state in a closed system, it follows from (10) and Eq. (1–7.9) that

$$\Delta S \geqslant \frac{Q}{T} \quad \text{(isothermal change)}, \qquad (1\text{–}10.12)$$

where T is the absolute temperature of the system (constant in space and time) and Q is the heat transferred to the system from the surroundings.

The relation (10) will, in many traditional works on thermodynamics, be stated at the beginning of the exposition of the second law, and most often in the form

$$\Delta S = \int \frac{dQ_{\text{rev}}}{T}, \qquad (1\text{–}10.13)$$

$$\Delta S > \int \frac{dQ_{\text{irrev}}}{T}, \qquad (1\text{–}10.14)$$

where the first equation is used immediately as "the definition of entropy." Such a presentation leads to conceptual difficulties. Also, the formulation (14) has often given cause for misunderstanding of the meaning of the quantities dQ_{irrev} and T.

That the classical relations (13) and (14) are to be interpreted in the sense of (10) was already noted by Gibbs who wrote in the year 1875 in the introduction to his work *On the Equilibrium of Heterogeneous Substances*[16a], "the difference of entropy is the limit of all the possible values of the integral $\int dQ/T$ (dQ denoting the element of the heat received from external sources, and T the temperature of the part of the system receiving it)."

1–11 HELMHOLTZ FUNCTION AND GIBBS FUNCTION

We will define two further extensive state variables, the *Helmholtz function F* and the *Gibbs function G* (cf. Eqs. 1–5.1 and 1–5.2):

$$F^\alpha \equiv U^\alpha - T^\alpha S^\alpha, \tag{1–11.1}$$

$$G^\alpha \equiv H^\alpha - T^\alpha S^\alpha = U^\alpha - \sum_i L_i^\alpha l_i^\alpha - T^\alpha S^\alpha = F^\alpha - \sum_i L_i^\alpha l_i^\alpha, \tag{1–11.2}$$

$$F \equiv \sum_\alpha F^\alpha, \qquad G \equiv \sum_\alpha G^\alpha. \tag{1–11.3}$$

The function G^α is identical with the quantity usually called the "Gibbs function" (or the "Gibbs free energy")

$$G^\alpha \equiv U^\alpha + P^\alpha V^\alpha - T^\alpha S^\alpha = F^\alpha + P^\alpha V^\alpha \tag{1–11.4}$$

if we are dealing with an isotropic region without electric or magnetic polarization.

From Eq. (1–9.6) and Eq. (2) we have for a single region α:

$$G^\alpha = \sum_k \mu_k^\alpha n_k^\alpha. \tag{1–11.5}$$

From Eqs. (1–6.4) to (1–6.7) there follows for the molar Gibbs function of the region

$$\bar{G}^\alpha = \sum_k \mu_k^\alpha x_k^\alpha \tag{1–11.6}$$

and for the density of the Gibbs function of the region

$$G_V^\alpha = \sum_k \mu_k^\alpha c_k^\alpha. \tag{1–11.7}$$

From Eq. (6) it is obvious that for systems of one component the chemical potential is the same as the molar Gibbs function of the region in question.

If we consider now an *isothermal* change of the *internal* state of a *closed* system then we find from Eq. (1–10.12):

$$T \Delta S \geqslant Q \quad \text{(isothermal change)}, \tag{1–11.8}$$

where Q denotes the heat supplied to the system at temperature T which is constant in space and in time. The inequality is valid for any irreversible (actual) path of the process and the equality for the reversible limiting case.

From Eqs. (1) and (3) the increase of the Helmholtz function F during an isothermal processes is given by

$$\Delta F = \Delta U - T \Delta S. \tag{1–11.9}$$

From this, with the help of Eq. (1-4.8) and (8) for an isothermal change of the internal state of a closed system, we derive

$$\Delta F \leqslant W \quad \text{(isothermal change)} \tag{1-11.10}$$

with

$$W = W_l + W^*, \tag{1-11.11}$$

where W is the total work done on the system, W_l the reversible work of deformation, electrification, and magnetization, and W^* the remaining work. With all work-coordinates constant there follows from Eq. (1-3.11) and (10) and (11):

$$\Delta F \leqslant W^* \quad \text{(isothermal change with constant work-coordinates).} \tag{1-11.12}$$

From Eqs. (2) and (3) we find for the increase of the Gibbs function G for an isothermal process:

$$\Delta G = \Delta H - T\,\Delta S. \tag{1-11.13}$$

From this we obtain, with the help of Eqs. (1-5.4) and (8), for an isothermal internal state change of a closed system with all work-coefficients constant:

$$\Delta G \leqslant W^* \quad \text{(isothermal change with constant work-coefficients).} \tag{1-11.14}$$

From statements (10), (12), and (14) the quantities F and G introduced formally above receive a simple physical interpretation.

A special case which frequently arises is a closed system in which processes such as phase changes, mixing and separation processes, or chemical reactions occur, which take place isothermally and at constant work-coefficients. With the exception of chemical reactions in galvanic cells, the work done on the system consists in practice only of reversible deformation, electrification and magnetization work (in particular of reversible volume work), and this depends on the changes in work-coordinates (in particular on the volume changes) connected with the processes in question. We obtain in this case from Eq. (14):

$$\Delta G \leqslant 0. \tag{1-11.15}$$

For isotropic systems without electrification and magnetization, there results from Eqs. (1-5.3), (1-5.4), (4), (13), and (15):

$$\Delta H = \Delta U + P\,\Delta V = Q, \tag{1-11.16}$$

$$\Delta G = \Delta H - T\,\Delta S = \Delta F + P\,\Delta V = \Delta U - T\,\Delta S + P\,\Delta V \leqslant 0. \tag{1-11.17}$$

Here P is the pressure (constant in position and time) and ΔV the volume increase† appearing during the isothermal–isobaric process considered. For chemical reactions, these equations are related to the stoichiometric conversion and ΔU is then called the "energy of reaction," ΔH the "enthalpy of reaction," ΔS the "entropy of reaction," and ΔG the "Gibbs function of reaction."

1-12 CHARACTERISTIC FUNCTIONS AND FUNDAMENTAL EQUATIONS

If we consider the generalized Gibbs equation (1–9.4) in more detail then we recognize that all thermodynamic quantities (including H, F, and G) can be derived from the entropy $S(U, l_i, n_k)$ and its derivatives with respect to appropriate independent variables. The function of state so chosen is called, after Massieu, a *characteristic function* for the variables in question and the appropriate differential relation, after Gibbs, a *fundamental equation*. Thus the entropy S is the characteristic function for the variables U, l_i, n_k and, accordingly, the generalized Gibbs equation (1–9.4) represents a fundamental equation. If, for the sake of simplicity, we omit the index α in all following relations, which are valid for any single region, we obtain from Eq. (1–9.4), taking account of Eqs. (1–5.1), (1–11.1), and (1–11.2),

$$dU = T\,dS + \sum_i L_i\,dl_i + \sum_k \mu_k\,dn_k, \qquad (1\text{–}12.1)$$

$$dH = T\,dS - \sum_i l_i\,dL_i + \sum_k \mu_k\,dn_k, \qquad (1\text{–}12.2)$$

$$dF = -S\,dT + \sum_i L_i\,dl_i + \sum_k \mu_k\,dn_k, \qquad (1\text{–}12.3)$$

$$dG = -S\,dT - \sum_i l_i\,dL_i + \sum_k \mu_k\,dn_k. \qquad (1\text{–}12.4)$$

These differential relations are fundamental equations, so that the functions

$$U(S, l_i, n_k), \quad H(S, L_i, n_k), \quad F(T, l_i, n_k), \quad G(T, L_i, n_k)$$

---

† The relations (15) and (17) are not limited to the assumption $W^* = 0$, if the changes of the work-coordinates represent only attendant phenomena of the considered process. Then we can allow, theoretically, at least, the processes to occur so that only reversible deformation, electrification and magnetization work, will be done. Since the initial and final states of the system for the abstract experiment $W^* = 0$ coincide with the corresponding states for the actual experiment $W^* \neq 0$, ΔG always has the same value, so that the statements (15) and (17) remain valid.

represent characteristic functions. Generalizing a mnemotechnic scheme of
Guggenheim, we can symbolically write

$$
\begin{array}{ccc}
S & U & l \\
H & & F \\
L & G & T
\end{array}
$$

From Eqs. (1) to (4) we can next derive an expression for the chemical
potential which supplements the original definition (1–9.3):

$$
\mu_k = \left(\frac{\partial U}{\partial n_k}\right)_{S,l_i,n_j} = \left(\frac{\partial H}{\partial n_k}\right)_{S,L_i,n_j} = \left(\frac{\partial F}{\partial n_k}\right)_{T,l_i,n_j} = \left(\frac{\partial G}{\partial n_k}\right)_{T,L_i,n_j}. \tag{1–12.5}
$$

After comparison with Eq. (1–6.2) we find that the chemical potential can be
denoted as the "partial molar Gibbs function" and, as an *intensive* state
variable, depends on the temperature, the work-coefficients, and the composition, but not on the amount of material in the region considered.

From Eqs. (1–6.2), (1–11.2), (4), and (5) there follow some important
relations for the chemical potential μ_k:

$$
\mu_k = H_k - TS_k, \tag{1–12.6}
$$

$$
\left(\frac{\partial \mu_k}{\partial T}\right)_{L_i,n_k} = -S_k, \tag{1–12.7}
$$

$$
\left(\frac{\partial(\mu_k/T)}{\partial T}\right)_{L_i,n_k} = -\frac{H_k}{T^2}, \tag{1–12.8}
$$

$$
\left(\frac{\partial \mu_k}{\partial L_i}\right)_{T,L_j,n_k} = -\left(\frac{\partial l_i}{\partial n_k}\right)_{T,L_i,n_j}, \tag{1–12.9}
$$

where H_k is the partial molar enthalpy and S_k the partial molar entropy of
species k. The derivative on the right-hand side of Eq. (9) denotes a partial
molar work-coordinate.†

For isotropic regions without electric and magnetic polarization,

$$
\sum_i L_i l_i = -PV. \tag{1–12.10}
$$

If we introduce the partial molar volume of species k

$$
\left(\frac{\partial V}{\partial n_k}\right)_{T,P,n_j} \equiv V_k, \tag{1–12.11}
$$

we then find from Eq. (9)

$$
\left(\frac{\partial \mu_k}{\partial P}\right)_{T,n_k} = V_k. \tag{1–12.12}
$$

---

† Equation (1–12.9) produces among other things the *teinochemischen Bezie-
hungen* of Kuhn[17].

For the above case, the fundamental equations (1) to (4) take on a more simple form. That is, from Eq. (10) we find

$$\sum_i L_i \, dl_i = -P \, dV, \qquad \sum_i l_i \, dL_i = -V \, dP. \qquad (1\text{–}12.13)$$

From this we obtain, with Eqs. (1) to (4),

$$dU = T \, dS - P \, dV + \sum_k \mu_k \, dn_k, \qquad (1\text{–}12.14)$$

$$dH = T \, dS + V \, dP + \sum_k \mu_k \, dn_k, \qquad (1\text{–}12.15)$$

$$dF = -S \, dT - P \, dV + \sum_k \mu_k \, dn_k, \qquad (1\text{–}12.16)$$

$$dG = -S \, dT + V \, dP + \sum_k \mu_k \, dn_k. \qquad (1\text{–}12.17)$$

These are the classical Gibbs fundamental equations.

If we write the Gibbs equation in the form (1–9.11), and take account of the definitions (1–5.3), (1–6.6), (1–11.1), and (1–11.4) as well as the following identity from Eq. (1–6.4):

$$\sum_{k=1}^{N} dx_k = 0, \qquad (1\text{–}12.18)$$

where x_k is the mole fraction of species k and N the number of species, and if we choose the mole fractions $x_1, x_2, \ldots, x_{N-1}$ as the independent composition variables, then we obtain the following relations instead of Eqs. (14) to (17):

$$d\overline{U} = T \, d\overline{S} - P \, d\overline{V} + \sum_{k=1}^{N-1} (\mu_k - \mu_N) \, dx_k, \qquad (1\text{–}12.19)$$

$$d\overline{H} = T \, d\overline{S} + \overline{V} \, dP + \sum_{k=1}^{N-1} (\mu_k - \mu_N) \, dx_k, \qquad (1\text{–}12.20)$$

$$d\overline{F} = -\overline{S} \, dT - P \, d\overline{V} + \sum_{k=1}^{N-1} (\mu_k - \mu_N) \, dx_k, \qquad (1\text{–}12.21)$$

$$d\overline{G} = -\overline{S} \, dT + \overline{V} \, dP + \sum_{k=1}^{N-1} (\mu_k - \mu_N) \, dx_k. \qquad (1\text{–}12.22)$$

Here $\overline{U}$, $\overline{S}$, etc., are the molar state variables and μ_N the chemical potential of that component N whose mole fraction x_N appears as a dependent variable. From Eqs. (19) to (22) we find

$$\mu_k - \mu_N = \left(\frac{\partial \overline{U}}{\partial x_k}\right)_{S,V,x} = \left(\frac{\partial \overline{H}}{\partial x_k}\right)_{S,P,x}$$

$$= \left(\frac{\partial \overline{F}}{\partial x_k}\right)_{T,V,x} = \left(\frac{\partial \overline{G}}{\partial x_k}\right)_{T,P,x} \quad (k = 1, 2, \ldots, N - 1). \quad (1\text{–}12.23)$$

Here the inferior x denotes that all independent mole fractions apart from x_k are constant.

For isotropic regions with electrification and magnetization, we have (cf. Table 1, p. 7)

$$\sum_i L_i l_{iV} = -P + \mathfrak{E}\mathfrak{P} + \mathfrak{H}\mathfrak{I}. \quad (1\text{–}12.24)$$

The term, l_{iV} is a work-coordinate referred to unit volume, $\mathfrak{E}$ or $\mathfrak{H}$ the electric or magnetic field strength, and $\mathfrak{P}$ or $\mathfrak{I}$ the electric or magnetic polarization. We now continue from the generalized Gibbs equation in the form (1–9.13) and write for the differential of the entropy density S_V (U_V = density of the internal energy)

$$T \, dS_V = dU_V - \mathfrak{E} \, d\mathfrak{P} - \mathfrak{H} \, d\mathfrak{I} - \sum_k \mu_k \, dc_k. \quad (1\text{–}12.25)$$

Here c_k denotes the molarity of species k. With Eq. (24) there follows from Eqs. (1–5.1), (1–6.7), (1–11.1), and (1–11.2):

$$H_V = U_V + P - \mathfrak{E}\mathfrak{P} - \mathfrak{H}\mathfrak{I} \quad \text{(enthalpy density)}, \quad (1\text{–}12.26)$$

$$F_V = U_V - TS_V \quad \text{(density of the Helmholtz function)}, \quad (1\text{–}12.27)$$

$$G_V = H_V - TS_V \quad \text{(density of the Gibbs function)}. \quad (1\text{–}12.28)$$

From Eqs. (25) to (28) we obtain the fundamental equations

$$dU_V = T \, dS_V + \mathfrak{E} \, d\mathfrak{P} + \mathfrak{H} \, d\mathfrak{I} + \sum_k \mu_k \, dc_k, \quad (1\text{–}12.29)$$

$$dH_V = T \, dS_V + dP - \mathfrak{P} \, d\mathfrak{E} - \mathfrak{I} \, d\mathfrak{H} + \sum_k \mu_k \, dc_k, \quad (1\text{–}12.30)$$

$$dF_V = -S_V \, dT + \mathfrak{E} \, d\mathfrak{P} + \mathfrak{H} \, d\mathfrak{I} + \sum_k \mu_k \, dc_k, \quad (1\text{–}12.31)$$

$$dG_V = -S_V \, dT + dP - \mathfrak{P} \, d\mathfrak{E} - \mathfrak{I} \, d\mathfrak{H} + \sum_k \mu_k \, dc_k. \quad (1\text{–}12.32)$$

From Eq. (32) there results

$$\left(\frac{\partial \mu_k}{\partial P}\right)_{T,\mathfrak{E},\mathfrak{H},c_k} = 0, \tag{1–12.33}$$

$$d\mu_k = -\left(\frac{\partial \mathfrak{P}}{\partial c_k}\right)_{T,\mathfrak{E},\mathfrak{H},c_j} d\mathfrak{E}[T, \mathfrak{H}, c_k = \text{const}], \tag{1–12.34}$$

$$d\mu_k = -\left(\frac{\partial \mathfrak{J}}{\partial c_k}\right)_{T,\mathfrak{E},\mathfrak{H},c_j} d\mathfrak{H}[T, \mathfrak{E}, c_k = \text{const}]. \tag{1–12.35}$$

The chemical potential μ_k—as well as $\mathfrak{P}$ and $\mathfrak{J}$ (cf. Eq. 32)—depend according to this representation on the temperature T, the molarities $c_1, c_2, \ldots, c_N$ of the N species as well as on the electric field strength $\mathfrak{E}$ and the magnetic field strength $\mathfrak{H}$. (The pressure P disappears as the independent variable, otherwise we must eliminate one of the concentrations according to Eq. 1–6.11.) The relations (25), (26), (34), and (35) will be useful to us later (§4–34).

As a further example, we consider an elastic rod or wire of length l under an (internal) tensile force K (cf. p. 43), which can simultaneously be polarized in an electric field in the direction of its length. In this case, we can continue directly from the fundamental equations (1) to (4) and obtain (cf. Table 1, p. 7):

$$\sum_i L_i l_i = Kl + \mathfrak{E}\mathfrak{P}', \tag{1–12.36}$$

where $\mathfrak{E}$ and $\mathfrak{P}'$ denote the components of the electric field strength and of the total electric moment (product of electric polarization and volume) in the direction of its length. If for simplicity we assume the chemical content to be fixed, we have the following fundamental equations:

$$dU = T\,dS + K\,dl + \mathfrak{E}\,d\mathfrak{P}', \tag{1–12.37}$$

$$dH = T\,dS - l\,dK - \mathfrak{P}'\,d\mathfrak{E}, \tag{1–12.38}$$

$$dF = -S\,dT + K\,dl + \mathfrak{E}\,d\mathfrak{P}, \tag{1–12.39}$$

$$dG = -S\,dT - l\,dK - \mathfrak{P}'\,d\mathfrak{E}. \tag{1–12.40}$$

These relations contain the link between the thermoelastic, pyroelectric, and piezoelectric phenomena for the simplest (one-dimensional) case. Thus it follows from Eq. (40) that

$$\left(\frac{\partial l}{\partial \mathfrak{E}}\right)_{T,K} = \left(\frac{\partial \mathfrak{P}'}{\partial K}\right)_{T,\mathfrak{E}}. \tag{1–12.41}$$

The left-hand side of Eq. (41) refers to "electrostriction," the right side to "piezoelectricity."

1–13 GIBBS–DUHEM EQUATION

If we substitute the Gibbs function G in place of Z in the differential relation (1–6.13), we find with the help of Eq. (1–12.4) the *generalized Gibbs–Duhem equation*, valid for any arbitrary region:

$$S\, dT + \sum_i l_i\, dL_i + \sum_k n_k\, d\mu_k = 0. \tag{1–13.1}$$

With Eqs. (1–6.4) to (1–6.7) we can write Eq. (1) also in the form

$$\bar{S}\, dT + \sum_i l_i\, dL_i + \sum_k x_k\, d\mu_k = 0 \tag{1–13.2}$$

or

$$S_V\, dT + \sum_i l_{iv}\, dL_i + \sum_k c_k\, d\mu_k = 0. \tag{1–13.3}$$

For isotropic regions without electrification and magnetization, we obtain, with Eq. (1–12.13) and from Eqs. (1) to (3), the different forms of the classical *Gibbs–Duhem equation*:

$$S\, dT - V\, dP + \sum_k n_k\, d\mu_k = 0, \tag{1–13.4}$$

$$\bar{S}\, dT - \bar{V}\, dP + \sum_k x_k\, d\mu_k = 0, \tag{1–13.5}$$

$$S_V\, dT - dP + \sum_k c_k\, d\mu_k = 0. \tag{1–13.6}$$

Equation (4) is due to Gibbs (1875) and Duhem (1886). Equation (5) constitutes one of the starting points for the thermodynamics of equilibrium of several phases[1d].

For isotropic regions with electric and magnetic polarization, we obtain from Eq. (1–12.24) and Eq. (3):

$$S_V\, dT - dP + \mathfrak{P}\, d\mathfrak{E} + \mathfrak{J}\, d\mathfrak{H} + \sum_k c_k\, d\mu_k = 0. \tag{1–13.7}$$

This differential relation will be used later (§4–34). For $\mathfrak{P} = 0$, $\mathfrak{J} = 0$, it is reduced to Eq. (6).

A further example refers to an autonomous interfacial phase, of which the volume V and the surface area ω represent the work-coordinates, the negative pressure $-P$ and the interfacial tension σ being the appropriate work-coefficients (cf. Table 1, p. 7). For this case, we write Eq. (1), divide by ω, and set

$$\frac{S}{\omega} \equiv S_\omega, \quad \frac{V}{\omega} \equiv \tau, \quad \frac{n_k}{\omega} \equiv \Gamma_k, \tag{1–13.8}$$

where S_ω is the entropy per unit area of the interfacial phase, τ the thickness of the surface layer, and Γ_k the "surface concentration" of species k in the interfacial phase. Then we find

$$S_\omega \, dT - \tau \, dP + d\sigma + \sum_k \Gamma_k \, d\mu_k = 0. \tag{1–13.9}$$

This relation provides a starting point for the thermodynamics of interfacial equilibrium[4c].

1–14 AFFINITY

We consider again a single region. In this, an arbitrary number of chemical reactions can take place. Let ν_{kr} be the *stoichiometric number* of species k in the chemical reaction r and ξ_r the *extent of this reaction*. Then, for an infinitesimal increase in the amount n_k of species k in a *closed* region (cf. p. 19), we have

$$dn_k = \sum_r \nu_{kr} \, d\xi_r. \tag{1–14.1}$$

Here the summation extends over all chemical reactions.

As an example, consider the reaction

$$2\,H_2 + O_2 \longrightarrow 2\,H_2O$$

which takes place inside a specified region. We then obtain

$$\nu_{H_2} = -2, \quad \nu_{O_2} = -1, \quad \nu_{H_2O} = 2.$$

The extent of reaction ξ in a closed region, that is, in the absence of any exchange of matter with the surroundings, is given by

$$d\xi = \frac{1}{\nu_k} \, dn_k = -\tfrac{1}{2} \, dn_{H_2} = -dn_{O_2} = \tfrac{1}{2} \, dn_{H_2O}.$$

The quantity ξ is therefore a "reaction variable". ξ has the dimension of n_k. It may thus be measured in the unit "mol."

With de Donder[18] we define the *affinity* A_r of the chemical reaction r as a linear combination of the chemical potentials of the substances taking part in the reaction:

$$A_r \equiv -\sum_k \nu_{kr} \mu_k, \tag{1–14.2}$$

where μ_k denotes the chemical potential of the species k in the specified region (which can also be open) and the sum is over all species.

For the affinity A of the reaction considered above, we have

$$A = 2\mu_{H_2} + \mu_{O_2} - 2\mu_{H_2O}.$$

We will show in §1–19, §1–21, and §2–3 that such linear combinations of chemical potentials have simple properties. A has the dimension of μ_k. It may thus be measured in J mol^{-1} or cal mol^{-1}.

We find from the generalized Gibbs equation (1–9.4) with Eqs. (1) and (2), for a closed region:

$$T\,dS = dU - \sum_i L_i\,dl_i + \sum_r A_r\,d\xi_r. \tag{1–14.3}$$

It follows directly from this that

$$A_r = T\left(\frac{\partial S}{\partial \xi_r}\right)_{U,l_i,\xi_s}, \tag{1–14.4}$$

where ξ_s stands for the extents of all the reactions except ξ_r. This relation is analogous to the definition Eq. (1–9.3) for the chemical potential.

By inserting Eqs. (1) and (2) into the fundamental equations (1–12.1) to (1–12.4), we obtain for a closed region:

$$dU = T\,dS + \sum_i L_i\,dl_i - \sum_r A_r\,d\xi_r, \tag{1–14.5}$$

$$dH = T\,dS - \sum_i l_i\,dL_i - \sum_r A_r\,d\xi_r, \tag{1–14.6}$$

$$dF = -S\,dT + \sum_i L_i\,dl_i - \sum_r A_r\,d\xi_r, \tag{1–14.7}$$

$$dG = -S\,dT - \sum_i l_i\,dL_i - \sum_r A_r\,d\xi_r, \tag{1–14.8}$$

where the first differential relation is identical with Eq. (3). From this we immediately find

$$-A_r = \left(\frac{\partial U}{\partial \xi_r}\right)_{S,l_i,\xi_s} = \left(\frac{\partial H}{\partial \xi_r}\right)_{S,L_i,\xi_s}$$

$$= \left(\frac{\partial F}{\partial \xi_r}\right)_{T,l_i,\xi_s} = \left(\frac{\partial G}{\partial \xi_r}\right)_{T,L_i,\xi_s}, \tag{1–14.9}$$

analogous to Eq. (1–12.5).

If we are dealing with a closed isotropic region without electrification and magnetization then the fundamental equations (5) to (8) reduce to the following expressions:

$$dU = T \, dS - P \, dV - \sum_{r} A_r \, d\xi_r, \qquad (1\text{-}14.10)$$

$$dH = T \, dS + V \, dP - \sum_{r} A_r \, d\xi_r, \qquad (1\text{-}14.11)$$

$$dF = -S \, dT - P \, dV - \sum_{r} A_r \, d\xi_r, \qquad (1\text{-}14.12)$$

$$dG = -S \, dT + V \, dP - \sum_{r} A_r \, d\xi_r, \qquad (1\text{-}14.13)$$

which correspond to Eqs. (1-12.14) to (1-12.17).

If only one chemical reaction (extent of reaction ξ, affinity A) occurs in the considered region, it follows from Eq. (13) that

$$dG = -S \, dT + V \, dP - A \, d\xi. \qquad (1\text{-}14.14)$$

From this there results (cf. Eq. 9):†

$$A = -\left(\frac{\partial G}{\partial \xi}\right)_{T,P}. \qquad (1\text{-}14.15)$$

With the help of the definition Eq. (1-11.2) for the Gibbs function

$$G = H - TS \qquad (1\text{-}14.16)$$

we obtain from Eq. (15)

$$-A = h_{TP} - Ts_{TP}, \qquad (1\text{-}14.17)$$

where

$$h_{TP} \equiv \left(\frac{\partial H}{\partial \xi}\right)_{T,P} \qquad (1\text{-}14.18)$$

is "the enthalpy of reaction" and

$$s_{TP} \equiv \left(\frac{\partial S}{\partial \xi}\right)_{T,P} \qquad (1\text{-}14.19)$$

the "entropy of reaction."

---

† Equation (15) is a special case of a more general relation valid for any extensive function Z (cf. Eqs. 1, 2, and 1-6.12):

$$z_{TP} \equiv \left(\frac{\partial Z}{\partial \xi}\right)_{T,P} = \sum_{k} \nu_k Z_k \quad (Z_k = \text{partial molar quantity}).$$

We derive, further, by the application of Schwarz's law to Eqs. (14) and (19) that

$$\left(\frac{\partial A}{\partial T}\right)_{P,\xi} = \left(\frac{\partial S}{\partial \xi}\right)_{T,P} = s_{TP}, \tag{1–14.20}$$

$$\left(\frac{\partial A}{\partial P}\right)_{T,\xi} = -\left(\frac{\partial V}{\partial \xi}\right)_{T,P} \equiv v_{TP}. \tag{1–14.21}$$

Here v_{TP} is the volume change of reaction (for unit extent). From Eqs. (17), (18), and (20) we obtain finally:

$$\left(\frac{\partial (A/T)}{\partial T}\right)_{P,\xi} = \frac{1}{T^2}\left(\frac{\partial H}{\partial \xi}\right)_{T,P} = \frac{h_{TP}}{T^2}. \tag{1–14.22}$$

The relations (17), (20), (21), and (22) are analogous to the Eqs. (1–12.6), (1–12.7), (1–12.8), and (1–12.12).

A "differential" reaction quantity in the above sense (e.g. h_{TP}) is only identical (in absolute magnitude) with the "integral" reaction quantity (e.g. ΔH) defined on p. 33, and occurring in Eqs. (1–11.16) or (1–11.17), if the composition of the reacting system does not change during the chemical reaction or if that kind of change has no influence on the differential reaction quantity (e.g. h_{TP}) in question. Then there follows, e.g. from Eq. (18), for $\Delta \xi = \xi^\dagger (\equiv 1 \text{ mol})$:

$$(\Delta H)_{T,P} = \int h_{TP}\, d\xi = h_{TP}\, \Delta \xi = h_{TP}\xi^\dagger = \sum_k \nu_k H_k \xi^\dagger.$$

Here H_k denotes the partial molar enthalpy of species k. The quantity mentioned on p. 19 $(\sum_k \nu_{kr} U_k)$ as the "differential" energy of reaction of the chemical reaction r corresponds thus to h_{TP}. If we use the energy unit "cal" and the unit "mol" for amount of substance then we have to measure ΔH in cal but H_k and h_{TP} in cal mol^{-1}.

1–15 HEAT CAPACITY

We define a *heat capacity at constant work-coordinates* for any region of a system (temperature T, work-coordinates l_i, work-coefficients L_i, amounts of substances n_k, internal energy U, enthalpy H, entropy S) as

$$C_l \equiv \left(\frac{\partial U}{\partial T}\right)_{l_i, n_k} \tag{1–15.1}$$

and a *heat capacity at constant work-coefficients* as

$$C_L \equiv \left(\frac{\partial H}{\partial T}\right)_{L_i, n_k}. \tag{1–15.2}$$

With the help of the fundamental equations (1–12.1) and (1–12.2), we derive

$$C_l = T\left(\frac{\partial S}{\partial T}\right)_{l_i, n_k}, \qquad C_L = T\left(\frac{\partial S}{\partial T}\right)_{L_i, n_k}. \tag{1–15.3}$$

The usual unit for both C_l and C_L is J K^{-1} or cal K^{-1}.

As is obvious, the quantities C_l or C_L are reduced to the "heat capacity at constant volume C_V" and to the "heat capacity at constant pressure C_P" when the volume V is the only work-coordinate and the negative pressure $-P$ the only work-coefficient. We have therefore for an isotropic region without electrification and magnetization:

$$C_V = \left(\frac{\partial U}{\partial T}\right)_{V,n_k} = T\left(\frac{\partial S}{\partial T}\right)_{V,n_k}, \qquad (1\text{–}15.4)$$

$$C_P = \left(\frac{\partial H}{\partial T}\right)_{P,n_k} = T\left(\frac{\partial S}{\partial T}\right)_{P,n_k}. \qquad (1\text{–}15.5)$$

Another example deals with a rod, thread, or wire, which is extended in the direction of its length l by a tensile force K.† If we assume, for the sake of simplicity, that all other stress components disappear in comparison to the normal stress in the longitudinal direction belonging to K then l and K represent the only work-coordinate and the only work-coefficient respectively (cf. Table 1, p. 7). Thus we find from Eqs. (1) to (3), for the "heat capacity at constant strain":

$$C_l = \left(\frac{\partial U}{\partial T}\right)_{l,n_k} = T\left(\frac{\partial S}{\partial T}\right)_{l,n_k} \qquad (1\text{–}15.6)$$

and for the "heat capacity at constant tensile force":

$$C_k = \left(\frac{\partial H}{\partial T}\right)_{K,n_k} = T\left(\frac{\partial S}{\partial T}\right)_{K,n_k}. \qquad (1\text{–}15.7)$$

---

† The symbol K denotes the *internal* tensile force for which an equation of the form

$$K = K(T, l, n_k)$$

is valid as, in the analogous case of the pressure inside an isotropic medium, an equation of state in the usual form

$$P = P(T, V, n_k)$$

will apply.

The tensile force acting on the outside of the system is strictly equal to the internal tensile force only for an infinitely slow expansion. If we later, i.e. in Eqs. (15) and (16), postulate the absence of dissipative effects, then such an identity of the work-coefficients corresponding to the internal forces (internal tensile force, internal pressure, etc.) with the externally acting forces (external tensile force, external pressure, etc.) is generally assumed. Otherwise, "dissipated work" (cf. p. 8) must be introduced for an infinitesimal deformation, as in Eqs. (13) and (14). Then in the case of a one-dimensionally stressed rod which is expanded by the external force K' the following relation holds:

$$dW_{\text{diss}} = (K' - K)\, dl.$$

With the help of Schwarz's law there results from Eqs. (1–12.3) and (1–12. 4) respectively:

$$-\left(\frac{\partial S}{\partial l_i}\right)_{T,l_j,n_k} = \left(\frac{\partial L_i}{\partial T}\right)_{l_i,n_k}, \tag{1–15.8}$$

$$\left(\frac{\partial S}{\partial L_i}\right)_{T,L_j,n_k} = \left(\frac{\partial l_i}{\partial T}\right)_{L_i,n_k}, \tag{1–15.9}$$

where the inferiors l_j and L_j denote that all work-coordinates and work-coefficients except those quantities with respect to which we have differentiated are to be held constant.

With Eqs. (8) and (9) we obtain from Eq. (3) the general connection between the two heat capacities:

$$C_l - C_L = T \sum_i \left(\frac{\partial L_i}{\partial T}\right)_{l_i,n_k} \left(\frac{\partial l_i}{\partial T}\right)_{L_i,n_k}. \tag{1–15.10}$$

The best-known example again represents an isotropic region without electrification and magnetization, as has already been considered in Eqs. (4) and (5):

$$C_P - C_V = T \left(\frac{\partial P}{\partial T}\right)_{V,n_k} \left(\frac{\partial V}{\partial T}\right)_{P,n_k}. \tag{1–15.11}$$

For the one-dimensionally stressed rod, thread, or wire assumed in Eqs. (6) and (7), it follows from Eq. (10) that

$$C_l - C_K = T \left(\frac{\partial K}{\partial T}\right)_{l,n_k} \left(\frac{\partial l}{\partial T}\right)_{K,n_k}. \tag{1–15.12}$$

In comparing Eq. (11) with Eq. (12) it should be noticed that the quantity K corresponds to a negative pressure $-P$.

We now consider an infinitesimal internal change in a region of given amount and composition. Then it follows from Eqs. (1–3.7), (1–3.8), (1–4.8), and (1–5.1) that

$$dU = dH + \sum_i L_i \, dl_i + \sum_i l_i \, dL_i = dQ + dW_{\text{diss}} + \sum_i L_i \, dl_i,$$

where dQ is the heat supplied to the region and dW_{diss} the dissipated work done on the region. From this there results with Eqs. (1) and (2):

$$C_l \, dT + \sum_i \left[\left(\frac{\partial U}{\partial l_i}\right)_{T,l_j,n_k} - L_i\right] dl_i = dQ + dW_{\text{diss}}, \tag{1–15.13}$$

$$C_L \, dT + \sum_i \left[\left(\frac{\partial H}{\partial L_i}\right)_{T,L_j,n_k} + l_i\right] dL_i = dQ + dW_{\text{diss}}. \tag{1–15.14}$$

These equations describe three kinds of experiments:

1. Heat exchange with the surroundings without dissipative effects ($dW_{\text{diss}} = 0$) for fixed work-coordinates ($dl_i = 0$) or fixed work-coefficients ($dL_i = 0$). Then we have from Eq. (13) or (14):

$$C_l \, dT = dQ \quad \text{or} \quad C_L \, dT = dQ$$

respectively. This explains the name "heat capacity" for C_l or C_L respectively.

2. "Calorimetric" determination of C_l or C_L by measurement of W_{diss} (as electrical work from external current sources) and of the temperature change belonging to it, under the condition of thermal insulation and with constant work-coordinates or constant work-coefficients respectively. Then we find from Eqs. (13) and (14):

$$C_l \, dT = dW_{\text{diss}} \quad \text{or} \quad C_L \, dT = dW_{\text{diss}}$$

respectively.

3. Adiabatic change of work-coordinates or work-coefficients respectively, with no dissipative effects ($dQ = 0$, $dW_{\text{diss}} = 0$). Then we obtain from Eq. (13) or (14) respectively:

$$C_l \, dT + \sum_i \left[\left(\frac{\partial U}{\partial l_i} \right)_{T, l_j, n_k} - L_i \right] dl_i = 0, \tag{1–15.15}$$

$$C_L \, dT + \sum_i \left[\left(\frac{\partial H}{\partial L_i} \right)_{T, L_j, n_k} + l_i \right] dL_i = 0. \tag{1–15.16}$$

Contained in these formulas there are, among others, the equations for the "adiabatics" for isotropic bodies and the formulas for adiabatic magnetization and demagnetization as well as for the Joule–Gough effect (cf. below).

The Eqs. (15) and (16) are based only on the first law. We now use the second law in the form of the fundamental equations (1–12.1) and (1–12.2) and Eqs. (8) and (9). With this we derive for the derivatives appearing in Eqs. (15) and (16):

$$\left(\frac{\partial U}{\partial l_i} \right)_{T, l_j, n_k} = L_i - T \left(\frac{\partial L_i}{\partial T} \right)_{l_i, n_k}, \tag{1–15.17}$$

$$\left(\frac{\partial H}{\partial L_i} \right)_{T, L_j, n_k} = T \left(\frac{\partial l_i}{\partial T} \right)_{L_i, n_k} - l_i. \tag{1–15.18}$$

These formulas give the isothermal dependence of the internal energy and the enthalpy on the work-coordinates and work-coefficients. The best-known example is again an isotropic phase without electrification and magnetization for which

$$\left(\frac{\partial U}{\partial V} \right)_{T, n_k} = T \left(\frac{\partial P}{\partial T} \right)_{V, n_k} - P, \tag{1–15.19}$$

$$\left(\frac{\partial H}{\partial P} \right)_{T, n_k} = V - T \left(\frac{\partial V}{\partial T} \right)_{P, n_k}. \tag{1–15.20}$$

We apply Eqs. (16) and (18) to three practically important examples:

(a) Adiabatic reversible compression or expansion of isotropic media (cf. Eq. 20):

$$\left(\frac{dT}{dP}\right)_{\text{ad}} = \frac{T}{C_P}\left(\frac{\partial V}{\partial T}\right)_{P,n_k}. \tag{1–15.21}$$

This is the equation of the "adiabatics" or "isentropes" for the simplest case.

(b) Adiabatic reversible compression or expansion of one-dimensionally stressed rods, threads, or wires (Joule–Gough effect):

$$\left(\frac{dT}{dK}\right)_{\text{ad}} = -\frac{T}{C_K}\left(\frac{\partial l}{\partial T}\right)_{K,n_k}. \tag{1–15.22}$$

Since C_K is always positive, cooling or heating of a material occurs following an adiabatic increase of the tensile force if the linear expansion coefficient is positive (metal wire) or negative (rubber band).

(c) Adiabatic reversible magnetization or demagnetization of isotropic media at constant pressure (cf. Table 1, p. 7):

$$\left(\frac{dT}{d\mathfrak{H}}\right)_{\text{ad}} = -\frac{T}{C_L}\left(\frac{\partial(\mathfrak{I}V)}{\partial T}\right)_{P,\mathfrak{H},n_k}. \tag{1–15.23}$$

Here $\mathfrak{H}$ and $\mathfrak{I}$ are the magnetic field strength and the magnetic polarization respectively, and C_L the heat capacity at constant pressure and at constant field strength. The derivative on the right-hand side of Eq. (23) is negative for most paramagnetic salts. Thus, for $C_L > 0$, we find that a lowering of temperature results from an adiabatic demagnetization (method of Giauque for reaching the lowest temperatures).

We define the heat capacities C_l and C_L of a heterogeneous system by the relations

$$C_l \equiv \sum_\alpha C_l^\alpha; \qquad C_L \equiv \sum_\alpha C_L^\alpha, \tag{1–15.24}$$

where the index α denotes the phase α and the sums are over all phases of the system. With the help of Eqs. (1–4.7), (1–5.2), and (1–8.2) it follows from Eqs. (1) to (3) for a heterogeneous system of uniform temperature that

$$C_l = \left(\frac{\partial U}{\partial T}\right)_{l_i^\alpha,n_k^\alpha} = T\left(\frac{\partial S}{\partial T}\right)_{l_i^\alpha,n_k^\alpha}, \tag{1–15.25}$$

$$C_L = \left(\frac{\partial H}{\partial T}\right)_{L_i^\alpha,n_k^\alpha} = T\left(\frac{\partial S}{\partial T}\right)_{L_i^\alpha,n_k^\alpha}. \tag{1–15.26}$$

T is the absolute temperature common to all phases of the system, U, H, and S, the internal energy, enthalpy, and entropy respectively of the whole system. The inferior labeling in Eqs. (25) and (26) indicates that, in the differentiations, the amounts of substances and the work-coordinates or work-coefficients are to be held constant in each single phase.

1-16 COMPONENTS, CHEMICAL SPECIES, INTERNAL PARAMETERS, AND INTERNAL DEGREES OF FREEDOM

Thus far we have referred to the "amount of substance" and, accordingly, to the partial molar enthalpy, the chemical potential, etc., of a "substance" without elaborating these expressions. In the following, it is expedient to consider the meaning of a quantity such as n_k ("amount of substance k") more closely.

We denote those substances of a system the masses or amounts of which can be varied independently as *components* of the system. In comparison to this, we want to call all kinds of particles, in the sense of chemistry (atoms, molecules, radicals, ions, electrons), which appear at all in the system, *chemical species*. Thus, in a fluid mixture of chloroform $CHCl_3$ and nitrogen tetroxide N_2O_4 which dissociates according to

$$N_2O_4 \rightleftharpoons 2\,NO_2 \qquad (1\text{-}16.1)$$

there are three chemical species, but only two components, as soon as the dissociation equilibrium (1) has been attained.

The quantity n_k can thus denote either the amount of component k or the amount of species k. Here the expression "amount" is defined as the ratio of the mass of the substance in question to the molar mass calculated from the chemical formula.

Many conclusions of thermodynamics are not affected by the choice we have in considering either the components or the chemical species as the "substances" which make up the system, provided that possible chemical equilibria have been established[1c].

Apart from a few special cases, for which we operate with the concept of components, we will henceforth specify the meaning of "substance k" in the sense of chemical species k, for only thus can the chemical reactions and, in fact, all actual processes within the system be explicitly described. If, for example, in a solution of nitrogen tetroxide in chloroform we want to discuss the rate of the reaction

$$N_2O_4 \longrightarrow 2\,NO_2 \qquad (1\text{-}16.2)$$

or if we want to follow the diffusion of $CHCl_3$, N_2O_4, and NO_2 in a solution that exhibits concentration gradients—which is also of interest in the case of the equilibrium described by Eq. (1)—then we must consider the three species individually.

It is possible that, at fixed values of temperature, work-coefficients, and amounts of all species, there can still be internal changes in the state of a region of the system. Thus we observe, for example, volume changes within a "glass" or in one of the other "phases" under the above conditions. In these cases, the variables chosen for the determination of the "internal state" are obviously not sufficient. While such situations were previously not considered subject to thermodynamic descriptions at all, it has become evident in recent times that even "internally metastable states" can be understood with the help of thermodynamic concepts. The after-effects or relaxation phenomena appearing in such a case are formally described as unknown chemical reactions for which "extents of reactions" are allotted. These extents of reactions are called *internal parameters* and they complete the state variables of the region considered in such a way that the "internal state" is macroscopically defined although the relaxation mechanisms belonging to the internal parameters are in most cases not known. For further information, see §2–11.

Finally, we can proceed one step further and describe the adjustment of *internal degrees of freedom* (in the molecular physics sense) macroscopically, i.e. with the help of thermodynamics of irreversible processes. Such problems as the deformation of macromolecules during mechanical flow or the orientation of dipole molecules in electromagnetic fields are of this type. Then, even in the case of a single species, the amount of which is n, a quantity n_k should be introduced where the k characterizes a molecular state such as the length of a deformed molecule or the angle between the dipole direction of a polar molecule and the direction of an external electric field (see §4–40).

1–17 EQUILIBRIUM AND STATIONARY STATE

If all intensive quantities (temperature, pressure, concentrations, etc.) have time-independent values in a system, the system is then either in "equilibrium" or in a "stationary nonequilibrium state." In place of the last expression, we will use for brevity the simpler term "stationary state." The question now arises, how can we distinguish between both cases by a simple macroscopic criterion?

We consider as a first example a piece of metal in which there is a temperature difference between two points having a value which is constant in time. The temperature is therefore a function of position but does not depend on time. Here we have a "stationary state," for "equilibrium" requires the temperature to be uniform. If we isolate the metal piece from all influences of the surroundings and thus also from all heat sources, that is, we make it into an "isolated system," then the difference between the "stationary state" and "equilibrium" becomes obvious: in the first case, processes occur after the isolation (equalization of temperature); in the second case, they do not.

As a second example, let there be a simple gas in a gravitational field having a temperature and pressure which are constant in time but not in position. Here "equilibrium" requires a temperature which is constant in position and a pressure P which is variable in position and related to the height h above a fixed reference plane by the hydrostatic equilibrium condition

$$dP = -\rho g\, dh \quad (\rho = \text{density}, \ g = \text{acceleration due to gravity}). \quad (1\text{–}17.1)$$

If the temperature is not uniform or the pressure is not given by Eq (1) then we have a "stationary state" for values of temperature and pressure which are constant in time. If we isolate the gas from all influences of the outside world, excepting the gravitational field, then the difference between the "stationary state" and "equilibrium" becomes obvious: in the first case, processes still occur after the isolation (equalization of temperature and a development of the equilibrium pressure distribution); in the second case, they do not.

The number of these examples can be increased; we always come to the same conclusion: if a system with values of intensive state variables which are constant in time is isolated from all influences of the outside world, with the exception of time-invariant (stationary) external fields, then, in the case of a "stationary state," processes still occur after the isolation; in the case of an "equilibrium," they do not. We can regard this criterion as the definition of "stationary states" or "equilibria" of any complicated system.

Obviously, whether the intensive variables are time invariant is determined for "stationary states" by external influences and for "equilibrium" states by internal causes. From this it follows that the conditions for equilibrium in a system must be formulated with the help of the state variables of the system. These formulations will be developed subsequently. We will turn to the difficult problem of the thermodynamic characterization of stationary states later (Chapter 5).

For many systems, the concept of "local equilibrium" is of significance. Such a case appears either if the equilibrium conditions are only fulfilled for certain parts of the system (example: local heterogeneous equilibrium at the electrodes in irreversible galvanic cells, cf. p. 297) or if time-variable external fields are present such that the local equilibrium values of the state variables of the system adjust to these fields at every moment (example: local equilibrium for systems in time-variable electromagnetic fields, cf. p. 441).

1–18 GENERAL CRITERION FOR EQUILIBRIUM

Let there be a system in which arbitrary processes (heat, matter, and electricity transport, chemical reactions, etc.) can occur. We will consider a system surrounded by a rigid wall which is fixed in position and which is impermeable to matter, electric current, and heat but which does not preclude the possible influence of stationary (time-invariant) external force fields. Then according to §1–17 there is *equilibrium* inside the system if no state changes can take place in the system after this isolation. Since a system, cut off from the outside world in the way described, represents a special case of a thermally insulated system, the relations (1–8.8) to (1–8.10) must be valid. Thus equilibrium is characterized in that the entropy of the system cannot increase further under the stated subsidiary conditions. The equilibrium state is thus distinguished from all neighboring nonequilibrium states by the highest value of the entropy S. For a virtual displacement from equilibrium (into neighboring nonequilibrium states), we obtain the following *general criterion for equilibrium*[16b, 19]:

$$\delta S \leqslant 0, \tag{1–18.1}$$

where the symbol δ denotes a variation of the first order[1f]. If all virtual displacements are two-directional (possible in both directions) state changes then the equality sign holds in Eq. (1). If there are also, among the virtual displacements, unidirectional (only possible in one direction) state changes then the general form of Eq. (1) must be retained.

In the literature, the inequality in (1) is often omitted and two-directional virtual displacements are interpreted as "reversible changes" in the sense of Eq. (1–8.8). This is totally misleading, because a virtual displacement from equilibrium does not refer to a reversible but to an impossible change (cf. Eq. 1–8.10). The equality sign in the case of the two-directional changes is due to the fact that the highest value of the entropy causes a stationary point (in the simplest case, i.e. for an entropy curve in the plane, there is a maximum with a horizontal tangent).† Here each variation of the *first* order, which takes the system out of equilibrium into neighboring nonequilibrium states, disappears. Seen mathematically, two-directional virtual state changes are present if all subsidiary conditions are equalities, while for one-directional changes there are inequalities among the subsidiary conditions (cf. §1–22).

As a supplement to (1), there are also the subsidiary conditions to be taken into account, i.e. the isolation of the system described above and other conditions suggested by the particular problem. The mathematical formulation of the last conditions differs from case to case. The isolation of the system by a stationary rigid wall which is impermeable to matter, electric current, and heat, but permeable to time-independent external force fields has as one result that all influences from the outside are restricted to the influence of time-invariant fields which we will for simplicity assume to be potential (conservative) force fields (cf. §1–4). Then the influence of the external fields can be described globally by a variable potential energy of the system. It results, therefore, that for all virtual displacements to which Eq. (1) refers:

1. The macroscopic kinetic energy of the whole system is constant.
2. No work is done on the total system.

---

† It is assumed that the derivatives of S are continuous and bounded.

3. No heat flows out of the surroundings into the system.
4. The total volume of the system is constant.
5. No exchange of matter takes place with the outside world.

The first three assumptions mean that according to Eqs. (1–4.5) and (1–4.6)

$$\delta E^* \equiv \delta U + \delta E_{pot} = 0, \qquad (1\text{–}18.2)$$

where U denotes the total internal energy of the system and E_{pot} the potential energy of the total system (determined by stationary conservative force fields). Equation (2) then states that the internal energy U and the potential energy E_{pot} of the system can change by a redistribution of matter in the virtual state change in the external force fields considered, but that the sum $E^* = U + E_{pot}$ must remain constant for all virtual displacements. The fourth and fifth assumptions demand that the total volume and the total mass of the system be constant while the volumes of individual regions and the quantities of matter (masses or amounts) of the individual substances can vary. We will formulate these conditions from case to case.

If we ignore the third subsidiary condition (thermal isolation of the system) and instead prescribe that

$$\delta S = 0, \qquad (1\text{–}18.3)$$

we obtain the equilibrium criterion

$$\delta E^* \geqslant 0. \qquad (1\text{–}18.4)$$

Here conditions 1, 2, 4, and 5 remain valid as before. This equilibrium criterion, to which Gibbs on mathematical grounds gives preference over criterion (1), should be considered somewhat further, since the presentation of Gibbs may easily lead to misunderstandings.

Condition (3) indicates that consideration should be given to appropriate heat exchange with the surroundings, so that for all virtual displacements from equilibrium the system retains the same entropy S ("isentropic" state changes). Since the system is not thermally isolated any more, Eqs. (1–4.5) and (1–4.6) indicate that condition (2) cannot generally be fulfilled. It will be shown by an indirect proof that the relation (4) does actually hold.

We denote—following Fowler and Guggenheim—a system that lies under the conditions 1, 2, 4, and 5 as a "system of given configuration." Then we can formulate both of the equilibrium criteria (1) and (4) together with the subsidiary conditions in the following way:

$$\delta S \leqslant 0 \quad \text{for} \quad \delta E^* = 0 \quad \text{and given configuration,} \qquad (1\text{–}18.5)$$

$$\delta E^* \geqslant 0 \quad \text{for} \quad \delta S = 0 \quad \text{and given configuration.} \qquad (1\text{–}18.6)$$

Now, E^* and S can always be simultaneously raised or lowered for a given configuration by adding heat to the system (or system parts) or by removing heat from the system (cf. Eqs. 1–4.5 and 1–10.10). If there is a variation in state, for which, in contradiction to (5),

$$\delta S > 0, \qquad \delta E^* = 0 \quad \text{(given configuration)}$$

then it is possible, because of the last consideration, to withdraw heat from the system in its varied state so that the original value of the entropy S is again reached while E^* must decrease. There is therefore a virtual state change

$$\delta S = 0, \qquad \delta E^* < 0 \quad \text{(given configuration)},$$

which contradicts (6). If, on the other hand, (6) is not satisfied then there must be a variation, which is distinguished by the expression

$$\delta E^* < 0, \qquad \delta S = 0 \quad \text{(given configuration)}.$$

As a consequence of the above, a virtual state change exists in this case

$$\delta E^* = 0, \qquad \delta S > 0 \quad \text{(given configuration)},$$

which is in contradiction to (5). Thus criteria (5) and (6) are equivalent.

The original criteria for equilibrium stated by Gibbs are[16b]:

"I. For the equilibrium of any isolated system it is necessary and sufficient that in all possible variations of the state of the system which do not alter its energy, the variation of its entropy shall either vanish or be negative.

"II. For the equilibrium of any isolated system it is necessary and sufficient that in all possible variations in the state of the system which do not alter its entropy, the variation of its energy shall either vanish or be positive."

These theorems coincide with ours if "energy" denotes the quantity E^* and "isolated system" is interpreted as "system of given configuration."

In the derivations of particular equilibrium conditions (§1–19 to §1–21), we restrict ourselves, for the sake of brevity, to systems for which interfacial phenomena, anisotropic stresses as well as electric and magnetic polarization of matter do not have to be considered. We will, however, give a few results without proof which relate to the latter phenomena (cf. also §4–34).

Apart from the general criterion (1) for equilibrium and the appropriate subsidiary conditions (Eq. 2, etc.), the Gibbs equation (1–9.5) is to be used for systems which are composed of isotropic regions without electrification and magnetization. If we replace the differential symbol d by the variation symbol δ in this equation, we obtain a relation between the variations of the entropy S^α, the internal energy U^α, the volume V^α, and the amounts n_k^α inside each region α:

$$T^\alpha \, \delta S^\alpha = \delta U^\alpha + P^\alpha \, \delta V^\alpha - \sum_k \mu_k^\alpha \, \delta n_k^\alpha. \qquad (1\text{–}18.7)$$

Here T^α is the absolute temperature, P^α the pressure, and μ_k^α the chemical potential of species k. For a volume element of a continuous system, it is useful to write the Gibbs equation in the form (1–9.12)

$$T \, \delta S_V = \delta U_V - \sum_k \mu_k \, \delta c_k, \qquad (1\text{–}18.8)$$

where S_V and U_V denote the entropy density and the density of the internal energy respectively of the volume element and c_k the molarity of substance k.

The applicability of the Gibbs equation to virtual displacements from equilibrium will be established further in §1–23.

The entropy S of a heterogeneous (discontinuous) system is given by the relation (1–8.2)

$$S = \sum_{\alpha} S^{\alpha}, \tag{1–18.9}$$

where the sum is over all phases. For continuous systems, Eq. (9) is replaced by the relation

$$S = \int S_V \, d\tau. \tag{1–18.10}$$

Here $d\tau$ represents a volume element.[9] The integration is to be taken over the whole space occupied by the system.

1–19 EQUILIBRIUM IN HOMOGENEOUS SYSTEMS

We now consider a single isotropic phase without electrification and magnetization. The only processes in such a homogeneous system which come into consideration for a virtual displacement from equilibrium in the sense of §1–18 are virtual chemical reactions which are possible in both directions. We obtain therefore from Eq. (1–18.1) for a variation of the entropy S

$$\delta S = 0. \tag{1–19.1}$$

Since a change in the potential energy is not of interest here, the subsidiary conditions according to §1–18 (cf. Eqs. 1–14.1 and 1–18.2) are

$$\delta U = 0, \quad \delta V = 0, \quad \delta n_k = \sum_r \nu_{kr} \, \delta\xi_r. \tag{1–19.2}$$

Here U denotes the internal energy, V the volume, n_k the amount of substance k, ν_{kr} the stoichiometric number of species k in the reaction r (to be counted positive or negative if the species in question is produced or consumed respectively in the reaction), and ξ_r the extent of the reaction r. The sum is over all reactions.

The variations of the quantities in Eqs. (1) and (2) are connected by the Gibbs equation (1–18.7). Due to the last equation in (2) and the definition (1–14.2), Eq. (1–18.7) can be written in the following form (cf. Eq. 1–14.10):

$$T \, \delta S = \delta U + P \, \delta V + \sum_r A_r \, \delta\xi_r. \tag{1–19.3}$$

Here T is the absolute temperature, P the pressure, and A_r the affinity of the reaction r.

If we consider the first two equations in (2) then we find from (1) and (3) the equilibrium condition

$$T \, \delta S = \sum_r A_r \, \delta \xi_r = 0, \qquad (1\text{-}19.4)$$

in which all subsidiary conditions have already been incorporated.

If the reaction equations have been so formulated that all reactions are linearly independent then we can consider the variations of the different ξ_r as independent of each other. Otherwise we must represent the chemical changes in the form of independent reaction equations. Under these circumstances we can take any one reaction, e.g. the first reaction distinguished by the affinity A_1 and the extent of reaction ξ_1, and consider all other reactions $(2, 3, \ldots)$ as "frozen" $(\delta \xi_2 = \delta \xi_3 = \cdots = 0)$. We obtain then from (4) the condition

$$A_1 \, \delta \xi_1 = 0,$$

which must be satisfied for any arbitrary value of $\delta \xi_1$. We find, thus, that

$$A_1 = 0.$$

In an analogous manner,

$$A_2 = A_3 = \cdots = 0.$$

Thus the general equilibrium conditions for R independent chemical reactions in a homogeneous system ("homogeneous reactions") are:

$$A_r = 0 \quad (r = 1, 2, 3, \ldots, R). \qquad (1\text{-}19.5)$$

These R equations describe the *chemical equilibrium* in the formulation of de Donder[18].

If we consider the definition for the affinity (1-14.2)

$$A_r \equiv -\sum_k \nu_{kr} \mu_k, \qquad (1\text{-}19.6)$$

where μ_k denotes the chemical potential of species k and the sum is over all species, we obtain the Gibbs formulation of the conditions for chemical equilibrium:

$$\sum_k \nu_{kr} \mu_k = 0 \quad (r = 1, 2, 3, \ldots, R). \qquad (1\text{-}19.7)$$

Application of Eq. (7) to ideal gas mixtures or ideal dilute solutions leads to the classical "Law of Mass Action" (§2–5 and §2–8).

Another important special case of the equilibrium condition (7) results from consideration of the dissociation equilibrium of a dissolved electrolyte.

Let $A_{\nu_+} B_{\nu_-}$ be an electrolyte molecule which can dissociate into ν_+ cations A of charge number z_+ and ν_- anions B of charge number z_-:

$$A_{\nu_+} B_{\nu_-} \rightleftharpoons \nu_+ A + \nu_- B,$$

for example,

$$Fe_2(SO_4)_3 \rightleftharpoons 2\,Fe^{3+} + 3\,SO_4^{2-}$$

The charge number z_+ or z_- is a positive or negative integer. The quantities ν_+ and ν_- are the dissociation numbers. In the above example of ferric sulfate, we have

$$z_+ = 3, \quad z_- = -2, \quad \nu_+ = 2, \quad \nu_- = 3.$$

Electrical neutrality requires:

$$z_+\nu_+ + z_-\nu_- = 0. \tag{1–19.8}$$

We introduce the quantities

$$\nu \equiv \nu_+ + \nu_-, \tag{1–19.9}$$

$$\nu_\pm^\nu \equiv \nu_+^{\nu_+}\nu_-^{\nu_-}. \tag{1–19.10}$$

If n_+, n_-, and n_u are the amounts of substance of the cations, anions, and undissociated electrolyte molecules and if n_1 and n_2 denote the (stoichiometric) amounts of substance of the solvent and the electrolyte respectively then the *molalities* m_i of the species are defined as

$$m_i \equiv \frac{n_i}{M_1 n_1} \quad (i = +, -, u) \tag{1–19.11}$$

and the *molality* m of the electrolyte as

$$m \equiv \frac{n_2}{M_1 n_1}, \tag{1–19.12}$$

where M_1 represents the molar mass of the solvent. If α is the degree of dissociation of the electrolyte then we have

$$m_+ = \nu_+ \alpha m, \quad m_- = \nu_- \alpha m, \quad m_u = (1 - \alpha)m. \tag{1–19.13}$$

From Eqs. (9), (10), and (13) it follows that

$$m_+^{\nu_+}m_-^{\nu_-} = (\nu_\pm \alpha m)^\nu. \tag{1–19.14}$$

The quantity m serves as an independent composition variable. The usual unit for m is mol kg^{-1}.

The *practical activity coefficients* γ_i $(i = +, -, u)$ of the species are defined by the following relations:†

$$\mu_+ = \mu_+^\ominus + RT \ln (m_+ \gamma_+), \tag{1–19.15a}$$

$$\mu_- = \mu_-^\ominus + RT \ln (m_- \gamma_-), \tag{1–19.15b}$$

$$\mu_u = \mu_u^\ominus + RT \ln (m_u \gamma_u), \tag{1–19.15c}$$

---

† Other activity coefficients refer to different composition scales (e.g. molarity scale, mole fraction scale).

where $\mu_i^\ominus$ is a standard value of the chemical potential of species i $(i = +, -, u)$ dependent only on temperature and pressure, and R is the gas constant. (With $\gamma_i = \gamma^\dagger \equiv 1$ kg mol^{-1} we obtain the laws for ideal dilute solutions.) Here, because of the limiting law for infinite dilution,

$$\lim_{m \to 0} \left(\ln \frac{\gamma_i}{\gamma^\dagger} \right) = 0 \quad (i = +, -, u). \tag{1–19.16}$$

Further we define

$$\gamma_\pm^\nu \equiv \gamma_+^{\nu_+} \gamma_-^{\nu_-}, \tag{1–19.17}$$

$$\gamma \equiv \alpha \gamma_\pm. \tag{1–19.18}$$

The quantity $\gamma_\pm$ is known as the *mean practical activity coefficient of the electrolyte*. The symbol γ ("stoichiometric mean practical activity coefficient") denotes a measurable quantity (cf. below) which is found in all modern tables on electrolyte solutions and thus should be called the *conventional activity coefficient* of the electrolyte. γ_+, γ_-, γ_u, $\gamma_\pm$, and γ have the dimensions[10] of m^{-1}. Thus they may be measured in kg mol^{-1}.

It can be shown† that μ_u, i.e. the chemical potential of the undissociated portion of the electrolyte, is identical with μ_2, the chemical potential of component 2 (i.e. of the electrolyte as a whole):

$$\mu_u = \mu_2. \tag{1–19.19}$$

---

† Equation (19) can be derived from Eq. (1–11.5) which must always give the same expression for the Gibbs function G of the electrolyte solution regardless of whether the components (solvent = component 1, electrolyte = component 2) or the individual species (solvent molecule = species L, undissociated electrolyte molecule = species u, cations = species $+$, anions = species $-$) are chosen as the "substances" (cf. §1–16). We have, thus, from Eq. (1–11.5):

$$G = n_1\mu_1 + n_2\mu_2 = n_L\mu_L + n_u\mu_u + n_+\mu_+ + n_-\mu_-.$$

From this with Eqs. (13) and (20)

$$n_L = n_1, \quad n_u = (1 - \alpha)n_2, \quad n_+ = \nu_+\alpha n_2, \quad n_- = \nu_-\alpha n_2, \quad \mu_L = \mu_1,$$
$$\mu_u = \nu_+\mu_+ + \nu_-\mu_-,$$

we obtain the expression

$$G = n_1\mu_1 + n_2\mu_2 = n_1\mu_1 + n_2\mu_u.$$

We have therefore

$$\mu_2 = \mu_u.$$

For complete dissociation ($\alpha = 1$), we find from the above relations that[11]

$$\mu_2 = \nu_+\mu_+ + \nu_-\mu_-.$$

The general condition following from Eq. (7) for the dissociation equilibrium considered here

$$\mu_u = \nu_+\mu_+ + \nu_-\mu_- \qquad (1\text{-}19.20)$$

has meaning even if the electrolyte is completely dissociated ($\alpha = 1$) and the quantity μ_u is thus undefined provided we substitute μ_u by μ_2. For incomplete dissociation ($\alpha \neq 1$), Eq. (20) relating the chemical potentials and the concentrations of the individual species leads to a general law of mass action (see below).

With the abbreviation

$$\mu_u^\ominus - \nu_+\mu_+^\ominus - \nu_-\mu_-^\ominus \equiv RT \ln K_m, \qquad (1\text{-}19.21)$$

where K_m depends only on T and P, we find from Eqs. (15) and (20)

$$K_m = \frac{(m_+\gamma_+)^{\nu_+}(m_-\gamma_-)^{\nu_-}}{m_u\gamma_u} \qquad (1\text{-}19.22)$$

or with Eqs. (13), (14), (17), and (18)

$$K_m = \frac{\nu_\pm^\nu m^{\nu-1}\gamma^\nu}{(1-\alpha)\gamma_u}. \qquad (1\text{-}19.23)$$

Equation (22) or (23) is the generalized law of mass action for the dissociation equilibrium of an electrolyte which produces two kinds of ions. The dimensionless quantity K_m is called the *dissociation constant* (in the m-scale). Equation (22) is transformed to the classical law of mass action with $\gamma_i = \gamma^\dagger$ ($i = +, -, u$).

If we know γ (e.g. from e.m.f. measurements, cf. §4-19) and α (e.g. from optical data, cf. p. 264) as functions of the molality m then the following limiting law can be derived from Eqs. (16) and (23):

$$\ln K_m = \lim_{m \to 0}\left(\ln\frac{K_m\gamma_u}{\gamma^\dagger}\right) = \nu \ln \nu_\pm + \lim_{m \to 0}\left(\ln\frac{m^{\nu-1}\gamma^\nu}{(1-\alpha)\gamma^\dagger}\right). \quad (1\text{-}19.24)$$

An exact determination of K_m is possible with the help of this relation.

Finally, we find with

$$\nu_+\mu_+^\ominus + \nu_-\mu_-^\ominus \equiv \mu_2^\ominus \qquad (1\text{-}19.25)$$

from Eqs. (14), (15a), (15b), (17), (18), (19), and (20) the general expression for the chemical potential μ_2 of the electrolyte

$$\mu_2 = \mu_2^\ominus + \nu RT \ln (\nu_\pm m\gamma). \qquad (1\text{-}19.26)$$

The conventional activity coefficient γ becomes a measurable quantity on the basis of this formula, a result which also holds for complete dissociation (cf. above and §4–19). Since $(\mu_2 - \mu_2^{\ominus})$ can be experimentally determined, Eq. (26) may even be regarded as a definition for γ.†

For dissociation occurring in several steps, several dissociation constants are present. But even in this case the conventional activity coefficients γ are defined by Eq. (26). We then use for ν and $\nu_\pm$ those values which result for the dissociation in the *last* step. Thus, for aqueous sulfuric acid, in which we have the dissociation equilibria

$$H_2SO_4 \rightleftharpoons H^+ + HSO_4^-, \qquad HSO_4^- \rightleftharpoons H^+ + SO_4^{2-}$$

we only consider the gross equilibrium

$$H_2SO_4 \rightleftharpoons 2\,H^+ + SO_4^{2-}$$

for the above purpose and accordingly write from Eqs. (9) and (10) in Eq. (26):

$$\nu = 3, \qquad \nu_\pm = \sqrt[3]{4}.$$

1–20 EQUILIBRIUM IN HETEROGENEOUS (DISCONTINUOUS) SYSTEMS

We now consider a system consisting of two isotropic phases without electrification and magnetization. This is the simplest example of a heterogeneous or discontinuous system. External force fields such as gravitational or centrifugal fields are, as in homogeneous systems, to be excluded, so that both phases can be treated as macroscopically homogeneous bodies. Meanwhile, we now allow electric potential differences between the phases. These potential differences rest on the existence of charged particle species (ions or electrons) which can cross the phase boundary and have the character of "Galvani tensions."‡ The phases of our system can be separated from each

---

† More on the thermodynamic functions for electrolyte solutions of arbitrary concentrations can be found in Haase[20, 21].

‡ We designate "Galvani tension" after Lange[22] as the difference between the *internal* electric potentials for two neighboring phases, while the difference between the external electric potentials is called "Volta tension." The Volta tension can be measured, the Galvani tension cannot. Since, however, in the summation of the individual potential differences in a galvanic cell (also in a thermocouple) which leads to the measurable e.m.f. only the Galvani tensions are counted, only these appear in our formulas. The difference between Volta and Galvani tension depends on the (more or less fortuitous) surface charges at the phase boundaries. The Galvani tension at the metal/metal phase boundary is called "contact potential" and that at the metal/solution phase boundary "electrode potential." The measurable e.m.f. of a galvanic cell (or of a thermocouple) is always due to the addition of several types of potential differences. Thus the e.m.f. of a reversible galvanic cell without pressure gradients can be conceived as the sum of the contact and electrode potentials, while the e.m.f. of a thermocouple is composed of "thermoelectric potentials" and contact potentials.

other by a natural phase boundary or by a (rigid or flexible) membrane, which is permeable either to all species ("permeable membrane") or only to certain species ("semipermeable membrane"). Even natural phase boundaries can be "semipermeable," e.g. the interface metal/solution.

The virtual displacements from equilibrium to be considered here consist of a virtual transport of heat, matter, or electricity between the two phases. They only refer to two-directional changes.† We exclude chemical reactions; for homogeneous reactions have been dealt with in §1–19, and "heterogeneous reactions" are most simply regarded as superpositions of homogeneous reactions and passages of certain substances from one phase to the other.

We characterize the two phases by the simple prime (′) and the double prime (″) respectively, and we denote the internal energy by U, the potential energy which is determined by the electrostatic potential differences by E_{pot}, the volume by V, the pressure by P, the absolute temperature by T, the entropy by S, the amount and the chemical potential of substance k by n_k and μ_k respectively, and the total number of species by N. Then because of Eqs. (1–4.7), (1–18.1), (1–18.2), and (1–18.9) we have, for a virtual displacement from equilibrium (cf. §1–18),

$$\delta S = \delta S' + \delta S'' = 0 \qquad (1\text{–}20.1)$$

under the subsidiary conditions

$$\delta U + \delta E_{pot} = \delta U' + \delta U'' + \delta E_{pot} = 0, \qquad (1\text{–}20.2)$$

$$\delta V' + \delta V'' = 0, \qquad (1\text{–}20.3)$$

$$\delta n'_k + \delta n''_k = 0 \quad (k = 1, 2, \ldots, N). \qquad (1\text{–}20.4)$$

Moreover, the Gibbs equation (1–18.7) holds for each of the two phases:

$$T'\,\delta S' = \delta U' + P'\,\delta V' - \sum_{k=1}^{N} \mu'_k\,\delta n'_k, \qquad (1\text{–}20.5)$$

$$T''\,\delta S'' = \delta U'' + P''\,\delta V'' - \sum_{k=1}^{N} \mu''_k\,\delta n''_k. \qquad (1\text{–}20.6)$$

Although no "external force fields" are actually present, in Eq. (2) we divided the total internal energy of the system into the "true internal energy" U and the "potential electrostatic energy" E_{pot}, so that U' and U'' and thus also the chemical potentials μ'_k and μ''_k depend only on the temperature, on the pressure, and on the composition, but not on the electrical state of the

---

† This is in no way evident *a priori*, as Gibbs has already recognized. The inequality in (1–18.1) is to be excluded *a posteriori* on grounds which are connected with the limiting laws for infinite dilution[1g].

phase in question. This fictitious division permits the electric potentials φ' and φ'' to appear explicitly in the formulas (even if they can neither be measured individually nor as a difference) and the connection with the considerations in the next section will thus be clear. If the charge number z_i (positive or negative integer for cations or anions respectively) belongs to a charged species i for which the interface is permeable then we have, considering Eq. (4) and the definition of potential electrostatic energy ($F =$ Faraday constant),

$$\delta E_{pot} = \sum_i z_i F(\varphi' - \varphi'') \, \delta n_i' = \sum_i z_i F(\varphi'' - \varphi') \, \delta n_i''. \qquad (1\text{–}20.7)$$

Any chemical species which can cross the phase boundary is denoted by an inferior i; Eq. (7) is identically satisfied for uncharged species on account of $z_i = 0$. If an inferior j $(_j)$ refers to all species (charged and neutral) for which the phase boundary is impermeable then we have in place of Eq. (4) the following detailed equations of condition:

$$\delta n_i' + \delta n_i'' = 0, \qquad \delta n_j' = 0, \qquad \delta n_j'' = 0. \qquad (1\text{–}20.8)$$

If, finally, the two phases are divided from each other by a rigid membrane then the additional subsidiary conditions (cf. Eq. 3)

$$\delta V' = 0, \qquad \delta V'' = 0 \qquad (1\text{–}20.9)$$

must be considered for the variations in question.

For nonrigid phase boundaries, the conditions (2), (3), and (8) are necessary and sufficient. Combining these with the general criterion for equilibrium (1) and with Eqs. (5), (6), and (7), we obtain the following expression:

$$\delta S = \left(\frac{1}{T'} - \frac{1}{T''} \right) \delta U' + \left(\frac{P'}{T'} - \frac{P''}{T''} \right) \delta V'$$
$$- \sum_i \left(\frac{\mu_i'}{T'} + \frac{z_i F \varphi'}{T''} - \frac{\mu_i'' + z_i F \varphi''}{T''} \right) \delta n_i' = 0. \qquad (1\text{–}20.10)$$

The variations $\delta U'$ (virtual heat transfer between the two phases), $\delta V'$ (virtual increase or decrease of the volume of one phase), and $\delta n_i'$ (virtual mass and charge transfer between the two phases) are independent of each other, and so arbitrary values can be attributed to them. Thus we derive the following relations as the necessary and sufficient conditions for *heterogeneous equilibrium* between two phases separated from each other by a *nonrigid boundary*:

$$T' = T'' \quad \text{(thermal equilibrium)}, \qquad (1\text{–}20.11)$$

$$P' = P'' \quad \text{(mechanical equilibrium)}, \qquad (1\text{–}20.12)$$

$$\mu_i' + z_i F \varphi' = \mu_i'' + z_i F \varphi''. \qquad (1\text{–}20.13)$$

For uncharged species ($z_i = 0$), the last equation is reduced to the condition

$$\mu_i' = \mu_i'' \quad \text{(equilibrium for distribution of matter).} \quad (1\text{--}20.14)$$

For charged species (ions or electrons), we introduce the *electrochemical potential* of species i, following the notation of Guggenheim:

$$\eta_i \equiv \mu_i + z_i F\varphi. \quad (1\text{--}20.15)$$

We accordingly write:

$$\eta_i' = \eta_i'' \quad \text{(electrochemical equilibrium).} \quad (1\text{--}20.16)$$

Equations (11) to (16) contain the conditions of equilibrium for all heterogeneous systems with natural phase boundaries.

If the phase boundary is a rigid membrane then, according to Eq. (9), the variation $\delta V'$ in Eq. (10) does not appear, so that the condition (12) does not apply whereas the relations (11) and (13) still hold. Whether the pressures P' and P'' are different from each other depends on the kind of rigid membrane. If this is permeable to all species then both homogeneous subsystems are identical in equilibrium if they represent phases of the same kind (e.g. two gases or two liquids with complete miscibility). Then Eq. (12) is also valid. If, on the other hand, we are dealing with a semipermeable membrane then the two phases cannot be identical in equilibrium, but must show different compositions. Now, according to Eqs. (11) and (13), the temperatures and the electrochemical potentials of all species which can cross the membrane have the same value for both phases in equilibrium, and the chemical potentials depend only on temperature, pressure, and composition. Thus the pressures P' and P'' must generally be different, so that Eq. (12) ceases to hold. In this case, in which there is no mechanical equilibrium, we speak of "osmotic equilibrium" if the species of type i (for which the membrane is permeable) are all uncharged ($z_i = 0$ for all i), and of "Donnan equilibrium" if the above-mentioned species are, partly at least, ions. In the first case, no electric potential difference results; in the second case, there is, according to Eq. (13), a difference in the electrostatic potentials ($\varphi'' - \varphi'$) which is called the "Donnan potential."

Thus we find for *osmotic equilibrium*:

$$T' = T'', \quad (1\text{--}20.17)$$

$$P' \neq P'', \quad (1\text{--}20.18)$$

$$\mu_i' = \mu_i'', \quad (1\text{--}20.19)$$

while for *Donnan equilibrium* we have

$$T' = T'', \quad (1\text{--}20.20)$$

$$P' \neq P'', \quad (1\text{--}20.21)$$

$$\mu_i' + z_i F\varphi' = \mu_i'' + z_i F\varphi''. \quad (1\text{--}20.22)$$

The equilibrium pressure difference $P'' - P'$ can, in both cases, be indirectly ascertained from the other conditions. It is called *osmotic pressure* for the case when one of the phases consists of a pure substance.

Apart from the inequality (18) or (21)

$$P' \neq P''$$

we still have for both osmotic equilibrium and Donnan equilibrium the statement

$$\mu'_j \neq \mu''_j$$

for all species which cannot pass through the membrane.

A poorly conceived generalization of these considerations could lead to the conclusion that for a membrane impermeable to heat the statement

$$T' \neq T''$$

is valid, while the equilibrium conditions (13), (14), (19), or (22) still hold for all species which can pass through the boundary. That this is a wrong conclusion can easily be seen, for a membrane impermeable to heat is always simultaneously impermeable to matter, so that no distribution equilibrium can be achieved if the condition for thermal equilibrium cannot, in principle, be satisfied. If a temperature difference between two phases is maintained by interference from outside, i.e. by suitable heat exchange with the surroundings, then, with a phase boundary which is permeable to matter, a steady state in which there is no flow of matter through the membrane can be established. However, this situation does not refer to an equilibrium as is seen from the general criterion in §1–17 and will therefore be described not by the equilibrium conditions of classical thermodynamics but with the help of the methods of thermodynamics of irreversible processes (§3–9).

It can be recognized that there is a more fundamental significance to the conditions for thermal equilibrium than to those for mechanical equilibrium. In the discussion of the more general case of the continuous system (§1–21), we will indeed see that the conditions for mechanical equilibrium can be derived from the other equilibrium conditions when systems with natural phase boundaries (liquid/vapor, liquid/liquid for systems with phase separation, etc.) are excluded.

The equilibrium conditions (11) to (22) represent the precise wording of Donnan and Guggenheim for the Gibbs formulations. But the derivations used here can be traced directly to Gibbs. These exceed all later proofs in rigour and generality. We have abbreviated the process in that we have assembled together, in a single line of argument, all individual cases (normal heterogeneous equilibrium, electrochemical equilibrium, osmotic equilibrium, and Donnan equilibrium).

Let us state the equilibrium conditions for systems with *interfacial phenomena*[16c] without proof. For the layout sketched in Fig. 1, we have (with the assumption of N uncharged species, for which the phase boundary is permeable):

$$T' = T'' = T''', \tag{1–20.23}$$

$$P' - P'' = \sigma\left(\frac{1}{r_1} + \frac{1}{r_2}\right), \tag{1–20.24}$$

$$\mu_k' = \mu_k'' = \mu_k''' \quad (k = 1, 2, \ldots, N). \tag{1–20.25}$$

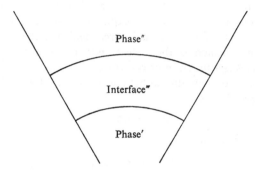

Fig. 1. Two normal fluid phases ′ and ″ (bulk phases) with an interface ″ separating them (interfacial phase).

Here σ is the interfacial tension, r_1 and r_2 are the principal radii of curvature of the interface. Obviously, the conditions for thermal and distribution equilibrium (cf. Eqs. 11 and 14) are unaffected by the presence of the interfacial layer, while the condition for mechanical equilibrium (cf. Eq. 12) is modified except in the case of plane interfaces ($r_1 \to \infty$, $r_2 \to \infty$, that is, $P' = P''$). Furthermore, Eqs. (23) and (25) state that for equilibrium the temperature T''' of the interface and the chemical potential μ_k''' of species k in the interface have the same value as they have in the adjoining bulk phases. Equation (24) is known as "Laplace's formula."

In the above equilibrium relations, the difficult concept of "pressure in the interfacial phase" is not considered. On the other hand, the negative "pressure" $-P'''$ appears formally as a work-coefficient in the fundamental equations which relate to the interface itself and which have been chosen as the starting point for the thermodynamics of interfacial phenomena, so that we do not have a clear definition of the quantity P'''. But the interfacial tension σ can only be defined for an interface whose thickness is small compared with its curvature. From this it can be shown[4d] that the pressure P''' may denote either P' or P'' or any intermediate pressure.

1–21 EQUILIBRIUM IN CONTINUOUS SYSTEMS

Matter in a gravitational, centrifugal, or electromagnetic field constitutes, in general, a continuous system in which quantities such as pressure, concentrations, and chemical potentials depend—even in equilibrium as we will see—on the space-coordinates in a continuous way if we exclude phase boundaries and do not consider exactly homogeneous fields. We restrict our derivations to the case of time-invariant (stationary) conservative force fields, as represented by the earth's gravitational field, the centrifugal field at constant angular speed, or an electrostatic field. We assume isotropic media and exclude the polarization of matter. On the other hand, electrically charged species and chemical reactions will be explicitly included in the presentation.

Let the potential of the external force field, i.e. the specific potential energy of a volume element of the system under consideration, be denoted by ψ. The potential ψ is a function of position.

In the case of the gravitational field, ψ is connected with the acceleration due to gravity $\mathbf{g}$ as follows:[†]

$$\mathbf{g} = -\operatorname{grad} \psi. \qquad (1\text{–}21.1)$$

For a centrifugal field, the relation

$$\Omega^2 \mathbf{r} = -\operatorname{grad} \psi \qquad (1\text{–}21.2)$$

is valid, where Ω denotes the angular velocity and $\mathbf{r}$ the distance from the axis of rotation. For an electrostatic field, the electric potential φ and the electric field strength $\mathfrak{E}$ are so defined that we obtain the following relations:

$$M_k \psi = z_k F \varphi, \qquad \mathfrak{E} = -\operatorname{grad} \varphi. \qquad (1\text{–}21.3)$$

Here M_k is the molar mass of species k, z_k its charge number, F the Faraday constant.

In the most general case, the potential therefore contains three terms which relate to the gravitational field, the centrifugal field, and the electrostatic field. It follows from Eqs. (1), (2), and (3) that

$$\mathbf{K}_k = -M_k \operatorname{grad} \psi = M_k(\mathbf{g} + \Omega^2 \mathbf{r}) + z_k F \mathfrak{E}. \qquad (1\text{–}21.4)$$

Here $\mathbf{K}_k$ is the molar external force acting on species k. The Coriolis force in rotating systems is excluded here, since this does not contribute to the potential energy of the system.[‡]

[†] The quantity grad ψ is, like $\mathbf{g}$ or $\mathbf{r}$, a vector and will be called "gradient of ψ." Assuming rectangular coordinates (x, y, z), the three components of the vector are

$$\frac{\partial \psi}{\partial x}, \quad \frac{\partial \psi}{\partial y}, \quad \frac{\partial \psi}{\partial z}.$$

[‡] Further information is to be found in §4–33.

We consider a volume element of the continuous system. Then the Gibbs equation in the form (1–18.8) holds:

$$T \, \delta S_V = \delta U_V - \sum_{k=1}^{N} \mu_k \, \delta c_k. \tag{1-21.5}$$

Here the symbol δ denotes a variation of the first order, S_V the entropy density, U_V the density of the internal energy, c_k the molarity of species k, T the absolute temperature, μ_k the chemical potential of species k, and N the number of species. Further, we write (cf. Eq. 1–19.2)

$$\delta c_k = \delta' c_k + \sum_{r=1}^{R} \nu_{kr} \, \delta \zeta_r \quad (k = 1, 2, \ldots, N). \tag{1-21.6}$$

Here $\delta' c_k$ is the virtual increase of c_k due to an exchange of matter with neighboring volume elements, ν_{kr} the stoichiometric number of species k in the chemical reaction r, ζ_r the reaction variable[12] of the chemical reaction r, and R the number of independent chemical reactions.

If we denote a volume element by $d\tau$ then we find for the entropy S and the internal energy U of the total system (cf. Eq. 1–18.10):

$$S = \int S_V \, d\tau, \tag{1-21.7}$$

$$U = \int U_V \, d\tau. \tag{1-21.8}$$

The integration is over the entire space occupied by the system in question. An analogous expression results for the total potential energy E_{pot} of the system:

$$E_{\text{pot}} = \int \sum_{k=1}^{N} M_k c_k \psi \, d\tau, \tag{1-21.9}$$

where

$$\sum_{k=1}^{N} M_k c_k = \rho$$

denotes the density of a volume element.

The virtual displacements from equilibrium which we must consider according to §1–18 in order to find the equilibrium conditions consist of virtual heat transfer, mass and electricity transfer between neighboring volume elements as well as of virtual local chemical reactions. We can, accordingly, regard all variations as two-directional and, using Eq. (1–18.1) and considering Eqs. (5), (6), and (7), we can write:

$$\delta S = \delta \int S_V \, d\tau = \int \delta S_V \, d\tau$$

$$= \int \frac{1}{T} \left(\delta U_V - \sum_{k=1}^{N} \mu_k \, \delta' c_k - \sum_{k=1}^{N} \sum_{r=1}^{R} \nu_{kr} \mu_k \, \delta \zeta_r \right) d\tau = 0. \quad (1\text{–}21.10)$$

Here the integration sign can be interchanged with the variation sign, because each volume element remains at its position during the virtual displacements which are of interest here. Therefore ψ in Eq. (9) can also be considered as constant.

The subsidiary conditions on Eq. (10) state, according to Eq. (1–18.2), that the sum of the internal and potential energies must remain constant for the total system and that no matter can flow through the boundary surfaces of the total system (cf. §1–18). We obtain, thus, by combining the conditions

$$\psi = \text{const},$$

$$\sum_{k=1}^{N} M_k \nu_{kr} = 0 \quad (r = 1, 2, \ldots, R) \quad \begin{array}{l} \text{(conservation of mass} \\ \text{in chemical reactions)} \end{array}$$

with Eqs. (6), (8), and (9):

$$\delta U + \delta E_{\text{pot}} = \int \left(\delta U_V + \sum_{k=1}^{N} M_k \psi \, \delta' c_k \right) d\tau = 0, \quad (1\text{–}21.11)$$

$$\int \delta' c_k \, d\tau = 0 \quad (k = 1, 2, \ldots, N). \quad (1\text{–}21.12)$$

We now multiply Eq. (11) by an arbitrary constant λ, the first, second, and $\ldots$ Nth equation in (12) by the arbitrary constants $\lambda_1, \lambda_2, \ldots, \lambda_N$, and subtract the equations so obtained from Eq. (10) according to the method of Lagrange. In this way, we find an equation freed from all subsidiary conditions

$$\int \left[\left(\frac{1}{T} - \lambda \right) \delta U_V - \sum_{k=1}^{N} \left(\frac{\mu_k}{T} + \lambda M_k \psi + \lambda_k \right) \delta' c_k \right.$$

$$\left. - \frac{1}{T} \sum_{k=1}^{N} \sum_{r=1}^{R} \nu_{kr} \mu_k \, \delta \zeta_r \right] d\tau = 0. \quad (1\text{–}21.13)$$

Now, since all variations δU_V, $\delta' c_1$, $\delta' c_2$, $\ldots$, $\delta' c_N$, $\delta \zeta_1$, $\delta \zeta_2$, $\ldots$, $\delta \zeta_R$ are independent of each other and since this must be valid for each arbitrary volume element, the integral disappears, if and only if the coefficients of the above variations disappear individually. We thus obtain

$$\frac{1}{T} = \lambda, \tag{1–21.14}$$

$$\mu_k + M_k \psi = -\frac{\lambda_k}{\lambda} \quad (k = 1, 2, \ldots, N), \tag{1–21.15}$$

$$\sum_{k=1}^{N} \nu_{kr}\mu_k = 0 \quad (r = 1, 2, \ldots, R). \tag{1–21.16}$$

Equation (14) signifies constancy in temperature, that is, thermal equilibrium:

$$T = \text{const.} \tag{1–21.17}$$

Equation (15) states that for each species k the sum of the chemical potential μ_k and the molar potential energy due to the external field $M_k\psi$ has the same value everywhere:

$$\mu_k + M_k\psi = \text{const} \quad (k = 1, 2, \ldots, N). \tag{1–21.18}$$

From Eq. (16) it follows, finally, upon consideration of Eq. (1–19.6) that the affinity A_r of each independent reaction vanishes

$$A_r = 0 \quad (r = 1, 2, \ldots, R), \tag{1–21.19}$$

by analogy with Eq. (1–19.5), valid for homogeneous systems.

In the case of an electric field, Eqs. (18) are reduced because of Eq. (3) to the conditions

$$\mu_i + z_i F\varphi = \text{const}, \tag{1–21.20}$$

$$\mu_j = \text{const}, \tag{1–21.21}$$

where the inferior i refers to all charged species (ions or electrons) and the inferior j to all neutral species. Equation (20) can be reduced to the form

$$\text{grad } \mu_i = z_i F \mathfrak{E} \tag{1–21.22}$$

with the help of Eq. (3).

If we consider Eqs. (1) and (2) then we can transform Eq. (18) into the relation

$$\text{grad } \mu_k = M_k(\mathbf{g} + \Omega^2 \mathbf{r}) \quad (k = 1, 2, \ldots, N), \tag{1–21.23}$$

which is valid for gravitational and centrifugal fields.

Due to Eqs. (4) and (18) the general equilibrium condition for the distribution of matter is

$$\text{grad } \mu_k = \mathbf{K}_k = M_k(\mathbf{g} + \Omega^2 \mathbf{r}) + z_k F \mathfrak{E} \quad (k = 1, 2, \ldots, N), \tag{1–21.24}$$

in which Eqs. (22) and (23) are contained as special cases.

The equilibrium conditions (17) to (19) remain intact even for electric polarization† and the existence of phase boundaries[16d]. In the first case, it should be pointed out that the chemical potentials μ_k depend not only on temperature, pressure, and composition but also on the electric field strength. In the second case, we must again, as with heterogeneous equilibria (§1–20), distinguish between permeable and semipermeable phase boundaries, where this can encompass either natural boundary surfaces or membranes. For semipermeable phase boundaries, Eq. (18) is valid only for those species which can pass through the boundary. A *rigid* semipermeable membrane simply prevents the establishment of mechanical equilibrium as described in §1–20.

We can summarize the equilibrium conditions derived in this and in the previous sections as follows: the relations (17), (18), and (19) *always* hold for isotropic media in stationary potential fields if the inferior k in Eq. (18) is restricted to those species which can pass the phase boundaries on hand. Only the conditions for mechanical equilibrium—in the event that this can be achieved at all—are different from case to case.

The conditions for mechanical equilibrium in continuous systems can be derived from Eqs. (17) and (18). We show this for a system without electric polarization. According to Eqs. (17) and (24), we can write

$$\text{grad } T = 0, \tag{1–21.25}$$

$$\text{grad } \mu_k = V_k \text{ grad } P + (\text{grad } \mu_k)_{T,P} = \mathbf{K}_k \quad (k = 1, 2, \ldots, N), \tag{1–21.26}$$

where the partial molar volumes V_k of species k have been introduced with the help of Eq. (1–12.12), and the term $(\text{grad } \mu_k)_{T,P}$ signifies the gradient at constant temperature T and constant pressure P and thus refers only to the change in position of μ_k due to changes in composition.

We now multiply each of the equations (26) by c_k and, taking into consideration Eqs. (1–6.11) and (1–13.6), we obtain the condition for mechanical equilibrium:

$$\text{grad } P = \sum_{k=1}^{N} c_k \mathbf{K}_k. \tag{1–21.27}$$

Equivalent to Eq. (27) is a relation which results from Eqs. (4) and (27) on introducing the density

$$\rho = \sum_{k=1}^{N} c_k M_k,$$

$$\text{grad } P = \rho(\mathbf{g} + \Omega^2 \mathbf{r}) + \sum_{k=1}^{N} z_k c_k F\mathfrak{E}. \tag{1–21.28}$$

---

† Cf. §4–34 (p. 441).

In a system without electrically charged particles or in an electrically conducting medium, the following condition is valid:

$$\sum_{k=1}^{N} z_k c_k = 0, \qquad (1\text{--}21.29)$$

which mathematically describes the disappearance of the space charge. Only for insulators which carry electric charges is the relation (29) not generally satisfied.[13] From Eqs. (28) and (29) it follows that

$$\text{grad } P = \rho(\mathbf{g} + \Omega^2 \mathbf{r}). \qquad (1\text{--}21.30)$$

Equation (1–17.1) is contained in this as a special case.

With Eqs. (3), (24), (26), and (30) and considering Eq. (1–20.15) it is found for media without space charge that

$$(M_k - \rho V_k)(\mathbf{g} + \Omega^2 \mathbf{r}) = (\text{grad } \mu_k)_{T,P} + z_k F \text{ grad } \varphi = (\text{grad } \eta_k)_{T,P}, \qquad (1\text{--}21.31)$$

where η_k represents the electrochemical potential of species k. If no external electrostatic field is present then we speak of a *sedimentation equilibrium*. The quantity φ then denotes the *sedimentation potential* in the equilibrium case. The "barometric height formula" and other known formulas for the concentration distribution of the components of an ideal gas mixture, of an ideal dilute solution, or of an ideal mixture in the gravitational field or in a centrifuge evolve from Eq. (31) by integration under the corresponding special assumptions (cf. Eq. 4–20.22, p. 309).

The equilibrium conditions for a system in the gravitational field (also regarding interfacial phenomena) were first derived by Gibbs. Later, van der Waals[23] generalized these considerations to arbitrary stationary fields. We have once more generalized the derivations here by inclusion of chemical reactions and with explicit regard to charged species[24].

1-22 STABILITY AND CRITICAL PHENOMENA

(a) General

The equilibrium conditions discussed thus far describe equilibrium with reference to the originally existing regions. If now the formation of new regions, perhaps "nuclei" of new phases (for example, droplets in a supersaturated vapor or crystallization from a supersaturated solution), is possible inside the original phases or volume elements then we have to consider a still further class of virtual state changes: the formation of new regions. The additional equilibrium conditions resulting therefrom are called *stability conditions*. It is possible, therefore, for there to be equilibrium in a system with reference to all possible changes in the original phases or volume elements while still permitting a transition of the given system into a qualitatively different system built up from other regions. In this case, we are dealing with an *unstable* equilibrium or system. If a transition of the above kind is impossible, the equilibrium or the system is called *stable*. An equilibrium or a system is called "relatively unstable" or "metastable" if the transition in question can only occur through discontinuous state changes. If the transition is possible through any small state changes, i.e. by the formation of infinitesimally neighboring regions, then the equilibrium or the system is labeled as "absolutely unstable" or shortly "unstable." Metastable systems can exist, unstable systems on the other hand—apart from a few exceptions involving crystal mixtures—cannot.

The general equilibrium criterion (1-18.1) again serves as a basis for denoting stability conditions. Now, however, all inequality signs must be retained, since one-directional changes are also found among the virtual displacements (namely, those state changes connected with the virtual formation of new regions). Mathematically this is expressed by the presence of inequalities among the subsidiary conditions.

In the following, we do not want to reproduce tedious derivations which go back to Gibbs but will satisfy ourselves with the statement of a few results. For details of the methods and other particulars see Haase[1h].

We consider the conditions necessary so that a single phase (or a volume element) is not unstable. These "stability conditions in the narrower sense" are, in the framework of our presentation of the thermodynamics of irreversible processes, the only interesting stability conditions. They define the circumstances in which a phase (or a volume element) is stable or metastable, and can be formulated as statements on the signs of the second and higher derivatives of the characteristic functions with respect to the appropriate variables ("equilibrium conditions of the second and higher orders").

If we denote the internal energy of the region by U, the entropy by S, the volume by V, the amount of substance k (component or species k) by n_k, the enthalpy by H, the pressure by P, the Helmholtz function by F, the absolute temperature by T, and the Gibbs function by G then the above-mentioned stability conditions are, in their general form (with the restriction to isotropic regions without electrification or magnetization),

$$\Delta^2 U(S, V, n_k) \geqslant 0, \tag{1-22.1}$$

$$[\Delta^2 H(P)]_{S,n_k} < 0, \tag{1-22.2}$$

$$[\Delta^2 H(S, n_k)]_P \geqslant 0, \tag{1-22.3}$$

$$[\Delta^2 F(T)]_{V,n_k} < 0, \tag{1-22.4}$$

$$[\Delta^2 F(V, n_k)]_T \geqslant 0, \tag{1-22.5}$$

$$[\Delta^2 G(T, P)]_{n_k} < 0, \tag{1-22.6}$$

$$[\Delta^2 G(n_k)]_{T,P} \geqslant 0. \tag{1-22.7}$$

Here the symbol $\Delta^2 Z$ denotes a variation of the function Z (referring to the given variables as the case may be) of second and higher order. For example, in the case of (2) we have

$$[\Delta^2 H(P)]_{S,n_k} = \frac{1}{2!}\left(\frac{\partial^2 H}{\partial P^2}\right)_{S,n_k}(\Delta P)^2 + \frac{1}{3!}\left(\frac{\partial^3 H}{\partial P^3}\right)_{S,n_k}(\Delta P)^3 + \cdots < 0,$$

where ΔP denotes a variation of the pressure.

The equality sign in (1), (3), (5), and (7) is only valid for the case in which the variation consists purely of a change in mass of the system, where all the amounts of substance are changed in the same proportion, so that the nature of the region is not changed.† Those superfluous variations can be eliminated if the extensive properties are replaced by specific or molar quantities or by generalized densities. We find then, e.g. from (7),

$$[\Delta^2 \bar{G}(x_k)]_{T,P} > 0, \tag{1-22.8}$$

where $\bar{G}$ is the molar Gibbs function and x_k the mole fraction of component k.

Even if changes in the amounts of substance are assumed to arise only through chemical reactions, the equality sign in (1), (3), (5), and (7) disappears. In place of n_k, the extents of the independent chemical reactions ξ_r must be introduced as independent variables. We obtain in this case, e.g. from (7),

$$[\Delta^2 G(\xi_r)]_{T,P} > 0. \tag{1-22.9}$$

(b) Mechanical and Thermal Stability

If we neglect stability limits and critical phenomena, it is sufficient for us to consider variations up to the second order only. Then, from the inequalities above, we can reach some conclusions as to the sign of the second derivatives of the characteristic functions. In the following, we derive a few important examples of such statements, beginning with the simplest case: constant amount and composition of the region.

Under the conditions mentioned above we obtain from (2)

$$\left(\frac{\partial^2 H}{\partial P^2}\right)_{S,n_k} < 0.$$

From this with Eq. (1–12.15) follows the condition for "mechanical stability":

$$\left(\frac{\partial V}{\partial P}\right)_{S,n_k} < 0.$$

---

† The possibility of the "coexistence of neighboring phases" in the critical region, suggested briefly by Gibbs and arising again within the framework of certain modern molecular statistical theories, is indicated by the equality signs in the above relations. Here we exclude, however, this possible complication because on the one hand it has not been definitely proved and on the other hand it is without significance to the final results[1i].

For the quantity denoted by "adiabatic compressibility"

$$\kappa_S \equiv -\frac{1}{V}\left(\frac{\partial V}{\partial P}\right)_{S,n_k} \qquad (1\text{–}22.10)$$

we have

$$\kappa_S > 0. \qquad (1\text{–}22.11)$$

From (5) we find that

$$\left(\frac{\partial^2 F}{\partial V^2}\right)_{T,n_k} > 0,$$

from which together with Eq. (1–12.16) we derive

$$\left(\frac{\partial P}{\partial V}\right)_{T,n_k} < 0.$$

This is again an expression for the "mechanical stability." For the "isothermal compressibility"

$$\kappa_T \equiv -\frac{1}{V}\left(\frac{\partial V}{\partial P}\right)_{T,n_k}, \qquad (1\text{–}22.12)$$

we thus conclude that

$$\kappa_T > 0. \qquad (1\text{–}22.13)$$

From (3) we obtain

$$\left(\frac{\partial^2 H}{\partial S^2}\right)_{P,n_k} > 0.$$

With Eq. (1–12.15) we gain the condition for "thermal stability"

$$\left(\frac{\partial S}{\partial T}\right)_{P,n_k} > 0.$$

For the "heat capacity at constant pressure" (see Eq. 1–15.5)

$$C_P = T\left(\frac{\partial S}{\partial T}\right)_{P,n_k}, \qquad (1\text{–}22.14)$$

we find, thus, that

$$C_P > 0. \qquad (1\text{–}22.15)$$

From (4) we obtain

$$\left(\frac{\partial^2 F}{\partial T^2}\right)_{V,n_k} < 0,$$

from which with Eq. (1–12.16) it follows that

$$\left(\frac{\partial S}{\partial T}\right)_{V,n_k} > 0,$$

another expression for the "thermal stability." For the "heat capacity at constant volume" (see Eq. 1–15.4)

$$C_V = T\left(\frac{\partial S}{\partial T}\right)_{V,n_k},$$ (1–22.16)

we find that

$$C_V > 0.$$ (1–22.17)

With the help of the mathematical identities

$$\left(\frac{\partial V}{\partial P}\right)_{S,n_k}\left(\frac{\partial S}{\partial T}\right)_{P,n_k} = \left(\frac{\partial V}{\partial P}\right)_{T,n_k}\left(\frac{\partial S}{\partial T}\right)_{V,n_k}$$

and

$$\left(\frac{\partial P}{\partial T}\right)_{V,n_k}\left(\frac{\partial V}{\partial P}\right)_{T,n_k} = -\left(\frac{\partial V}{\partial T}\right)_{P,n_k}$$

we derive from Eqs. (10), (12), (14), and (16) on consideration of Eq. (1–15.11):

$$\frac{\kappa_S}{\kappa_T} = \frac{C_V}{C_P},$$ (1–22.18)

$$C_P - C_V = \frac{TV\beta^2}{\kappa_T},$$ (1–22.19)

where

$$\beta \equiv \frac{1}{V}\left(\frac{\partial V}{\partial T}\right)_{P,n_k}$$ (1–22.20)

is the thermal expansivity. If we exclude the singular case $\beta = 0$ (which, for example, occurs in water at 4°C) then we find from (13), (17), (18), and (19) the general inequalities

$$C_P - C_V > 0,$$ (1–22.21)

$$\kappa_T - \kappa_S > 0.$$ (1–22.22)

(c) Chemical Stability

If we introduce the extent of reaction ξ_r, according to the model of (9), into (1), (3), (5), and (7) and consider only variations up to the second order in the ξ_r, we obtain from Eqs. (1–14.10) to (1–14.13) the following conditions for "chemical stability," valid for a closed region with R independent chemical reactions (A_r = affinity of the reaction r):

$$\frac{\partial A_r}{\partial \xi_r} < 0 \quad (r = 1, 2, \ldots, R),$$ (1–22.23)

$$\frac{\partial A_r}{\partial \xi_r}\frac{\partial A_s}{\partial \xi_s} - \left(\frac{\partial A_r}{\partial \xi_s}\right)^2 > 0 \quad \text{with} \quad \frac{\partial A_r}{\partial \xi_s} = \frac{\partial A_s}{\partial \xi_r} \quad (r, s = 1, 2, \ldots, R; r \neq s),$$

etc. (1–22.24)

The determinant

$$\left|\frac{\partial A_r}{\partial \xi_s}\right|$$

is positive or negative if it is of even or odd order respectively. As an equivalent statement we have: the quadratic form

$$\sum_{r=1}^{R} \sum_{s=1}^{R} a_r a_s \frac{\partial A_r}{\partial \xi_s} < 0 \qquad (1\text{–}22.25)$$

constructed of arbitrary variables a_r and a_s is always negative.

In the partial differentiations, one of the variable pairs

$$S, V$$
$$S, P$$
$$T, V$$
$$T, P$$

must be chosen to be constant. In the case of a single chemical reaction (extent of reaction ξ, affinity A), we have thus:

$$\left(\frac{\partial A}{\partial \xi}\right)_{S,V} < 0, \quad \left(\frac{\partial A}{\partial \xi}\right)_{S,P} < 0, \quad \left(\frac{\partial A}{\partial \xi}\right)_{T,V} < 0, \quad \left(\frac{\partial A}{\partial \xi}\right)_{T,P} < 0. \quad (1\text{–}22.26)$$

(d) Stability Limits and Critical Phases

The concepts "stability limit" and "critical phase" will be clarified by considering a binary system of components 1 and 2. If we choose the mole fraction x_1 of substance 1 as the independent variable for the description of the composition of the phase then according to (8) one of the conditions for the phase not to be unstable is

$$[\Delta^2 \overline{G}(x_1)]_{T,P} = \frac{1}{2!} \left(\frac{\partial^2 \overline{G}}{\partial x_1^2}\right)_{T,P} (\Delta x_1)^2 + \frac{1}{3!} \left(\frac{\partial^3 \overline{G}}{\partial x_1^3}\right)_{T,P} (\Delta x_1)^3 + \cdots > 0,$$

$$(1\text{–}22.27)$$

where Δx_1 denotes a variation of the mole fraction of component 1.

The *stability limit*, i.e. the limit between the stable or metastable and the unstable regions is given by the equation

$$\left(\frac{\partial^2 \overline{G}}{\partial x_1^2}\right)_{T,P} = 0, \qquad (1\text{–}22.28)$$

while for the *unstable* regions we have

$$\left(\frac{\partial^2 \overline{G}}{\partial x_1^2}\right)_{T,P} < 0, \qquad (1\text{–}22.29)$$

so that here $\Delta^2\bar{G}(x_1)$ is negative according to Eq. (27). *Inside* the stable and metastable region surrounded by the stability limit, we have the stability condition

$$\left(\frac{\partial^2\bar{G}}{\partial x_1^2}\right)_{T,P} > 0. \tag{1–22.30}$$

A *critical phase* arises from the coincidence of two coexisting phases. It lies on the one hand at the stability limit, on the other hand at the coexistence curve. Thus two equations are required for the characterization of the critical state. They are in the present case†

$$\left(\frac{\partial^2\bar{G}}{\partial x_1^2}\right)_{T,P} = 0, \qquad \left(\frac{\partial^3\bar{G}}{\partial x_1^3}\right)_{T,P} = 0. \tag{1–22.31}$$

Since the critical phase is not unstable we have, on the basis of (27),

$$\left(\frac{\partial^4\bar{G}}{\partial x_1^4}\right)_{T,P} > 0. \tag{1–22.32}$$

Should the derivative in (32) disappear then the following higher, non-disappearing derivative of *even* order of $\bar{G}$ with respect to x_1 ($T = $ const, $P = $ const) must be positive.

From Eqs. (1–6.4) and (1–11.6) there results for a two-component system:

$$\bar{G} = x_1\mu_1 + x_2\mu_2 = x_1\mu_1 + (1 - x_1)\mu_2, \tag{1–22.33}$$

where x_i and μ_i are the mole fraction and the chemical potential of component i respectively ($i = 1, 2$). Moreover, it follows from the Gibbs–Duhem relation (1–13.5) for a binary mixture that

$$x_1\left(\frac{\partial\mu_1}{\partial x_1}\right)_{T,P} + (1 - x_1)\left(\frac{\partial\mu_2}{\partial x_1}\right)_{T,P} = 0. \tag{1–22.34}$$

From the two equations we derive (cf. Eq. 1–12.23):

$$\left(\frac{\partial\bar{G}}{\partial x_1}\right)_{T,P} = \mu_1 - \mu_2, \tag{1–22.35}$$

$$\left(\frac{\partial^2\bar{G}}{\partial x_1^2}\right)_{T,P} = \frac{1}{1 - x_1}\left(\frac{\partial\mu_1}{\partial x_1}\right)_{T,P}, \tag{1–22.36}$$

$$\left(\frac{\partial^3\bar{G}}{\partial x_1^3}\right)_{T,P} = \frac{1}{(1 - x_1)^2}\left(\frac{\partial\mu_1}{\partial x_1}\right)_{T,P} + \frac{1}{1 - x_1}\left(\frac{\partial^2\mu_1}{\partial x_1^2}\right)_{T,P}. \tag{1–22.37}$$

† The most general equations for a critical phase, which are valid both for one-component and many-component systems, are found in Haase[1j].

With Eqs. (36) and (37) we can write the statements (28) to (31) in the following form:

$$\left(\frac{\partial \mu_1}{\partial x_1}\right)_{T,P} > 0 \quad \text{(stable or metastable region inside stability limit)}, \quad (1\text{–}22.38)$$

$$\left(\frac{\partial \mu_1}{\partial x_1}\right)_{T,P} = 0 \quad \text{(stability limit)}, \quad (1\text{–}22.39)$$

$$\left(\frac{\partial \mu_1}{\partial x_1}\right)_{T,P} < 0 \quad \text{(unstable region)}, \quad (1\text{–}22.40)$$

$$\left(\frac{\partial \mu_1}{\partial x_1}\right)_{T,P} = 0, \quad \left(\frac{\partial^2 \mu_1}{\partial x_1^2}\right)_{T,P} = 0 \quad \text{(critical state)}. \quad (1\text{–}22.41)$$

These formulations are important for later applications to systems of two components.

1–23 THERMODYNAMIC FUNCTIONS FOR NONEQUILIBRIUM STATES

In the derivation of the different equilibrium conditions from the general equilibrium criterion (§1–18), we have always assumed that for a virtual transition of the system from equilibrium into a neighboring nonequilibrium state the generalized Gibbs equation (1–9.4)

$$T\,dS = dU - \sum_i L_i\,dl_i - \sum_k \mu_k\,dn_k \quad (1\text{–}23.1)$$

or, in the special case of an isotropic medium without electrification and magnetization, the classical Gibbs equation (1–9.5)

$$T\,dS = dU + P\,dV - \sum_k \mu_k\,dn_k \quad (1\text{–}23.2)$$

holds for each region of the system. We now want to ask if this assumption is justified and furthermore to show that Eq. (1) can be retained under certain conditions even when actual (irreversible) processes occur in the region. If this were once proved then all relations which can be derived from Eq. (1), such as the fundamental equations (§1–12) and the Gibbs–Duhem relation (§1–13), remain valid for the processes in question.

Equation (1) contains only differentials of state variables. If, therefore, the internal state of a region can be described completely at that particular moment by quantities such as the internal energy U (or the absolute temperature T), the work-coordinates l_i (or the work-coefficients L_i), and the amounts of substance n_k then there also exists according to the second law (§1–8) the entropy S as a function of these quantities and the differential dS is given by

Eq. (1) according to §1–9. Even if undefined intermediate states are involved in a finite state change, Eq. (1) continues to hold—this time in integrated form—provided that initial and final states can be characterized by the above-mentioned set of variables. From this it then follows that the validity of the (generalized) Gibbs equation is in no way restricted to reversible changes.

In the derivation of the equilibrium conditions, the equilibrium state in question must now be compared to those neighboring nonequilibrium states which can themselves be described by the same number of variables as the equilibrium state. This is best clarified by considering heterogeneous equilibrium (§1–20), for here we know from the start that the different homogeneous bodies (phases) of the heterogeneous system can always be characterized by variables such as U, l_i (particularly volume V), and n_k with respect to their internal states, and that in the derivation of the equilibrium conditions the question is only to discover conditions for the "correct" distribution of the internal energies, volumes, amounts, etc., to the different phases of the system. Similar considerations can be posed for other kinds of equilibria. Thus Eq. (1) is always applicable to virtual displacements from equilibrium.

We can prove the last statement still more rigorously. For this we consider a nonequilibrium state in a region, the entropy of which depends on the additional variables $\lambda_1, \lambda_2, \ldots$ apart from the variables U, l_i, n_k. (The generalized definition of the entropy in the sense of statistical mechanics can, for example, result in a dependence on quantities such as gradients of temperature and of concentrations.) We find then for the entropy S of the region:

$$dS = \frac{\partial S}{\partial U}\,dU + \sum_i \frac{\partial S}{\partial l_i}\,dl_i + \sum_k \frac{\partial S}{\partial n_k}\,dn_k + \frac{\partial S}{\partial \lambda_1}\,d\lambda_1 + \frac{\partial S}{\partial \lambda_2}\,d\lambda_2 + \cdots.$$

For a virtual displacement from equilibrium, we have to consider a variation of the first order departing from the equilibrium state, so that the derivatives above refer to the equilibrium. Since the generalized Gibbs equation is itself satisfied for the equilibrium, there results from Eq. (1)

$$\frac{\partial S}{\partial U} = \frac{1}{T}, \quad \frac{\partial S}{\partial l_i} = -\frac{L_i}{T}, \quad \frac{\partial S}{\partial n_k} = -\frac{\mu_k}{T}, \quad \frac{\partial S}{\partial \lambda_1} = 0, \quad \frac{\partial S}{\partial \lambda_2} = 0, \ldots.$$

Here T, L_i, and μ_k denote the equilibrium values of the absolute temperature, of the work-coefficients (in particular, the pressure P), and of the chemical potentials. For a virtual displacement from equilibrium, therefore, the relation

$$T\,\delta S = \delta U - \sum_i L_i\,\delta l_i - \sum_k \mu_k\,\delta n_k$$

holds, as asserted above.

If we now consider a region at an arbitrary moment while actual (irreversible) processes are occurring, the decision as to the validity of Eq. (1) will be more difficult because it is not immediately certain whether the above-mentioned set of variables is sufficient to describe the internal state of the system and whether it is at all permissible to talk about an "internal state" in the thermodynamic sense. We consider next a simple example in order to clarify this situation.

Let an isotropic phase without electrification and magnetization change from state I (characterized by temperature T_I and volume V_I) into state II (distinguished by temperature T_{II} and volume V_{II}), the mass and composition having fixed values. Then the integrated form of the differential relation

$$T\,dS = dU + P\,dV, \tag{1–23.3}$$

which follows from Eq. (2) when there is no change in the amounts of substance, can, in each case, be applied to the process (I → II). We thus obtain:

$$\Delta S = S_{II} - S_I = \int_I^{II} \frac{dU}{T} + \int_I^{II} \frac{P}{T}\,dV. \tag{1–23.4}$$

The integrals can be evaluated as soon as the functions $U(T, V)$ and $P(T, V)$ are known. Since the entropy is a state function, the entropy change ΔS must be independent of the path, i.e. it can be calculated from Eq. (4) even if Eq. (3) is not valid at each instant during the state change.

The applicability of Eq. (3) at any moment during the course of a state change is likewise not restricted to reversible processes. Consider a liquid which is being slowly stirred or a homogeneous wire through which a constant electric current flows. Here the phase in question is at each moment practically homogeneous and its internal state is determined by the two variables U and V (or T and V). Thus Eq. (3) is satisfied at any arbitrary instant although the state change is irreversible. Here, however, the term

$$-P\,dV$$

does not denote the work done on the system and accordingly the expression

$$dU + P\,dV$$

does not denote the heat transferred to the system.

This state of affairs will become clearer if we further specify that the phase is an *ideal gas*, so that the equation of state

$$PV = nRT \quad (n = \text{amount of substance}, \ R = \text{gas constant}) \tag{1–23.5}$$

is valid, from which there follows with Eq. (1–15.19):

$$\left(\frac{\partial U}{\partial V}\right)_T = 0. \tag{1–23.6}$$

The state change considered in Eq. (4) is now an *isothermal expansion* ($T_I = T_{II} = T$, $V_{II} > V_I$). Then we have from Eqs. (4), (5), and (6):

$$\Delta U = U_{II} - U_I = 0, \quad \Delta S = S_{II} - S_I = nR \ln \frac{V_{II}}{V_I} > 0. \quad (1\text{-}23.7)$$

These relations are independent of the particular path of the state change and are satisfied for each isothermal expansion of an ideal gas. We consider three paths of such a process:

1. *Reversible Expansion in a Thermostat.* Following Eq. (1-3.2) and Eq. (5) for the work done on the gas during an infinitesimal change at an arbitrary instant we have

$$dW = -P \, dV = -nRT \frac{dV}{V}.$$

According to Eq. (1-4.8) and Eq. (6) for the heat supplied to the gas from the thermostat, there results

$$dQ = dU + P \, dV = P \, dV = -dW = nRT \frac{dV}{V}. \quad (1\text{-}23.8)$$

Moreover, Eq. (3) holds at each instant. We find therefore that

$$T \, dS = dU + P \, dV = dQ, \quad (1\text{-}23.9)$$

from which there follows for the total change I → II

$$\Delta S = \frac{Q}{T}, \quad (1\text{-}23.10)$$

a relation which, according to Eq. (1-10.12), must be satisfied for each reversible and isothermal change in a closed system. From Eqs. (8) and (10) we again obtain Eq. (7).

2. *Experiment of Joule and Thomson.* Here the gas is passed through a throttling valve and is thermally insulated. It has volume V_I and pressure P_I before passing through the valve (state I) and volume V_{II} and pressure P_{II} after passing through the valve (state II). Since we are dealing with an adiabatic process, the work W which is done on the gas is equal to the increase of the internal energy ΔU and thus is independent of the path. It can be calculated, therefore, on the assumption of reversible compression followed by reversible expansion. We thus obtain:

$$Q = 0, \quad (1\text{-}23.11)$$

$$W = P_I V_I - P_{II} V_{II} = \Delta U = U_{II} - U_I. \quad (1\text{-}23.12)$$

On introduction of the enthalpy (cf. Eq. 1-5.3)

$$H \equiv U + PV \quad (1\text{-}23.13)$$

there results:

$$\Delta H = H_{\text{II}} - H_{\text{I}} = 0. \tag{1-23.14}$$

Now we have from Eq. (5) with Eq. (1-15.20) for an ideal gas:

$$\left(\frac{\partial H}{\partial P}\right)_T = 0. \tag{1-23.15}$$

We can immediately derive from Eqs. (14) and (15) that the temperature of the gas during the throttling process cannot change. This is in agreement with our assumption according to which the state change I → II should consist wholly of an isothermal expansion. Equation (7) is therefore again valid. Equation (10) on the contrary ceases to hold (see Eq. 11), since an irreversible process is involved. On the other hand, Eq. (3) remains valid at each instant (for sufficiently slow flows) in each homogeneous gas part, because the state of the gas can at any one instant during the state change be described by the variables U and V (or T and V).

3. *Experiment of Gay–Lussac and Joule.* Here the gas is expanded adiabatically by flowing out into a vacuum from volume V_{I} to volume V_{II}, where no work is done on the system because it is enclosed by rigid walls. We thus obtain:

$$Q = 0, \quad W = 0, \quad \Delta U = 0, \tag{1-23.16}$$

in sharp contrast to Eq. (8) or to Eq. (12). From Eqs. (6) and (16) it then follows that the process is isothermal corresponding to our fundamental assumption. Accordingly, Eq. (7) remains valid while Eq. (10) is obviously incorrect, as is to be expected, since we are dealing with an irreversible process as in the previous example. In the present case, however, Eq. (3) cannot be applied to an arbitrary instant in the course of the irreversible state change, for, while the gas is streaming out into the vacuum, turbulent phenomena appear which not only weaken the homogeneity and our ability to describe the state of the gas by two variables but also cast doubt on the concept of the thermodynamic "state."

The case lying between the two extremes (reversible expansion and irreversible flow into a vacuum) that is, the slow flow through a throttling valve, is representative for all those cases which can be dealt with by the methods of the thermodynamics of irreversible processes: the process is irreversible, but the (generalized) Gibbs equation remains applicable to each region of the system at each instant during the state change.

If we consider a heterogeneous system consisting of any number of phases we can generalize the statements above: the (generalized) Gibbs equation holds for each phase at each moment during an irreversible state change if the processes occur so slowly that the homogeneity of the phases is not destroyed. Each phase of the heterogeneous system at any instant must be in internal thermal and mechanical equilibrium, the concentrations remaining uniform within the phase, though temperature, pressure, and compositions may vary in time. However, the conditions for chemical equilibrium or for mechanical, thermal, and distribution equilibrium between the different phases need not be fulfilled. We will be able to recognize from the kind of experiment performed whether the (generalized) Gibbs equation is applicable and if necessary be able to arrange the experimental conditions in such a way (e.g. by stirring the phases) that the above equation holds.

The question of the applicability of the generalized Gibbs equation to a volume element of a continuous system in which irreversible processes occur cannot be answered so easily. First, Eq. (1) is converted to a form which contains only intensive quantities because these can be defined for every point in space. We can use, for example, Eq. (1–9.9):

$$T\,dS_V = dU_V - \sum_i L_i\,dl_{iV} - \sum_k \mu_k\,dc_k, \qquad (1\text{–}23.17)$$

which, in the special case of an isotropic region without electrification and magnetization, is reduced to the relation (cf. Eqs. 1–9.12 and 1–21.5):

$$T\,dS_V = dU_V - \sum_k \mu_k\,dc_k. \qquad (1\text{–}23.18)$$

Here S_V is the entropy density, U_V the density of the internal energy, l_{iV} a work-coordinate referred to unit volume (such as, say, the electric polarization), and c_k the molarity of species k. Using the generalized Gibbs equation means, therefore, that in each volume element the entropy density S_V, even outside equilibrium, depends explicitly only on variables such as U_V, l_{iV}, and c_k, where, naturally, we assume that it is still reasonable to speak of the "entropy" of the region.

A statistical theory of irreversible processes, general enough to facilitate an answer to the questions posed here, does not exist. We can, however, discuss the problem kinetically in the case of chemical reactions and transport phenomena in ideal gases[25–29]. We find, then, that the Gibbs equation (18) remains applicable to a large region in the total domain of irreversible processes. This region includes all chemical reactions which occur so slowly that the Maxwellian velocity distribution for the single reacting species is not significantly disturbed (which is assumed to be obvious in most studies in reaction kinetics) and for all transport processes such as heat conduction, diffusion, and thermal diffusion, which, following Chapman and Enskog, can be described by a power series expansion of the statistical distribution functions up to the second approximation. In the framework of our macroscopic theory, this involves the linear "phenomenological laws" which are discussed later (§1–25). Once convinced of the consistency of the methods of our treatment, starting from the generalized Gibbs equation, in the special case of the ideal gas, we will not hesitate to apply the theory in other cases, to phenomena such as "slow" chemical reactions, heat conduction, diffusion, thermal diffusion, etc. Indeed, agreement with experiments always results, if we exclude extreme cases such as transfer in turbulent flows (cf. above) or large gradients of concentrations and temperature, for which the meaning of the macroscopic concepts "temperature," "entropy," etc., is doubtful anyhow.

In short, to summarize: thermodynamic functions such as temperature, entropy, etc., and the generalized Gibbs equation remain meaningful even in nonequilibrium states, if the processes in question do not occur too violently. Then, if necessary, a more precise analysis of the experimental conditions or a kinetic discussion of an analogous case which can be calculated must specify the meaning of the expression "not too violent."

1–24 ENTROPY FLOW AND ENTROPY PRODUCTION

Let us consider an actual (irreversible) process or—as a limiting case—a reversible process in an arbitrary system and let us assume that the actual process does not occur "too violently" in the sense discussed in §1–23.

First, we have for each region α of the system at an arbitrary instant the generalized Gibbs equation (1–9.4) or (1–23.1):

$$T^\alpha \, dS^\alpha = dU^\alpha - \sum_i L_i^\alpha \, dl_i^\alpha - \sum_k \mu_k^\alpha \, dn_k^\alpha, \qquad (1\text{–}24.1)$$

where T^α is the absolute temperature, S^α the entropy, U^α the internal energy, the L_i^α and l_i^α the work-coefficients and work-coordinates respectively, and the μ_k^α and n_k^α the chemical potentials and the amounts of substance respectively in the region considered. The total entropy S of the system is given, according to Eq. (1–8.2), by the relation

$$S = \sum_\alpha S^\alpha. \qquad (1\text{–}24.2)$$

Furthermore, expressions (1–8.3), (1–8.4), and (1–8.5) must be satisfied when the states under consideration [I and II in Eq. 1–8.3] are arbitrarily close together. If, therefore, the entropy of the system has the value S at time t and the value $S + dS$ at time $t + dt$ then we obtain the *entropy balance equation*

$$\frac{dS}{dt} = \frac{d_a S}{dt} + \frac{d_i S}{dt} \qquad (1\text{–}24.3)$$

and the relation

$$\frac{d_i S}{dt} \geq 0, \qquad (1\text{–}24.4)$$

where the inequality is valid for the actual (irreversible) path of the process and the equality for the reversible limiting case. Separation of the rate of increase of the entropy of the system dS/dt into two terms according to Eq. (3) is clearly established by the condition (1–8.7)

$$\frac{d_a S}{dt} = 0 \quad \text{for the case of thermal insulation of the system.} \quad (1\text{--}24.5)$$

Thus $d_a S/dt$ denotes the rate of increase of the entropy of the system by exchange of heat and matter with the surroundings, and is called the *entropy flow* (or "entropy flux"). The expression $d_i S/dt$ is the rate of increase of the entropy of the system due to processes which occur inside the system and will be described as *entropy production*.

The total rate of increase of the entropy of the system dS/dt and the entropy flow $d_a S/dt$ can be positive, negative, or zero, according to the direction and quantity of the heat and mass fluxes through which the system is connected to the surroundings. The entropy production $d_i S/dt$ on the other hand is never negative and disappears only in the reversible limiting case.

The quantitative evaluation of the relations (3) and (4) follows with the help of the generalized Gibbs equation (1) on considering Eq. (2) with the mass and energy balances (and also with Newton's second law for continuous systems). The setting up of the detailed "entropy balance" can be quite complicated. We discuss below only the simple case of the homogeneous system. For the more complicated cases (discontinuous and continuous systems), we refer the reader to the introductions to Chapters 3 and 4. It should be particularly noticed here that for continuous systems a "local entropy balance" is set up instead of Eq. (3). This balance is valid for each volume element. In it the divergence of an "entropy current density", the analog of the expression $d_a S/dt$, and the "local entropy production,"analogous to $d_i S/dt$ but referred to unit volume, both appear.

We use the abbreviation

$$\Theta \equiv \frac{d_i S}{dt} \tag{1--24.6}$$

for the entropy production in homogeneous and heterogeneous (discontinuous) systems and the symbol ϑ for the local entropy production in continuous systems. The product

$$T\Theta \equiv \Psi \tag{1--24.7}$$

or

$$T\vartheta \equiv \Psi \tag{1--24.8}$$

will be called, for the sake of brevity, by the common name *dissipation function*, where T denotes the absolute temperature of the region or the average absolute temperature in the case of a heterogeneous (discontinuous) system. In fact, for homogeneous systems, the quantity (7) contains the "dissipative effects" (see below) and the expression (8) will be identical with the "Rayleigh Dissipation Function" of hydrodynamic processes in continuous systems for the case where, apart from viscous flow, no other irreversible processes occur (see p. 411).

We consider as the simplest example a homogeneous system in which irreversible processes can take place. This should be an open phase which is isotropic and without electrification or magnetization. Then irreversible processes in the form of "dissipative effects" (passage of electricity, friction, etc.) and of chemical reactions are possible inside this homogeneous system. Equation (2) does not apply and Eq. (1) is reduced to the classical Gibbs equation (1–9.5) or (1–23.2):

$$T \, dS = dU + P \, dV - \sum_k \mu_k \, dn_k, \qquad (1\text{–}24.9)$$

where P is the pressure and V the volume. The definition of the "heat" dQ, which is supplied to an open phase during an infinitesimal change, as chosen in §1–7, results in the following expression for the "energy balance," according to Eqs. (1–7.3) and (1–7.5):

$$dQ = dU + P \, dV - dW_{\text{diss}} - \sum_k H_k \, d_a n_k. \qquad (1\text{–}24.10)$$

Here dW_{diss} denotes the work connected with the dissipative effects, H_k the partial molar enthalpy of species k, and $d_a n_k$ the increase of the amount of substance k by material exchange with the surroundings. Moreover, the "mass balance" is still to be considered (cf. Eqs. 1–7.5, 1–7.6, and 1–14.1):

$$dn_k = d_a n_k + \sum_r \nu_{kr} \, d\xi_r. \qquad (1\text{–}24.11)$$

Here ν_{kr} is the stoichiometric number of species k in the chemical reaction r and ξ_r the extent of the reaction r. Finally, if we introduce the affinity of the reaction r defined in Eq. (1–14.2) by

$$A_r = - \sum_k \nu_{kr} \mu_k$$

and note Eq. (1–12.6), which is

$$u_k = H_k - TS_k \quad (S_k = \text{partial molar entropy of species } k),$$

then from Eqs. (9), (10), and (11) we obtain

$$T \, dS = dQ + T \sum_k S_k \, d_a n_k + dW_{\text{diss}} + \sum_r A_r \, d\xi_r \qquad (1\text{–}24.12)$$

in which the first and second terms of the right-hand side relate to the heat and matter exchange respectively with the surroundings, the third term to the dissipative effects, and the fourth term to the chemical reactions inside the phase. If we rewrite Eq. (12) in the form of Eq. (3) and recall that $d_a S/dt = 0$ for thermally insulated systems ($dQ/dt = 0$, $d_a n_k/dt = 0$) then the "entropy balance" must appear in the following way:

$$\frac{dS}{dt} = \frac{d_a S}{dt} + \frac{d_i S}{dt}$$

and the "entropy flow" becomes:

$$\frac{d_a S}{dt} = \frac{1}{T}\frac{dQ}{dt} + \sum_k S_k \frac{d_a n_k}{dt} \qquad (1\text{–}24.13)$$

and the "entropy production" is

$$\frac{d_i S}{dt} = \frac{1}{T}\frac{dW_{\text{diss}}}{dt} + \frac{1}{T}\sum_r A_r w_r, \qquad (1\text{–}24.14)$$

wherein the quantity

$$w_r \equiv \frac{d\xi_r}{dt} \qquad (1\text{–}24.15)$$

was introduced as the "rate of reaction" of the chemical reaction r. Furthermore, with Eqs. (4), (6), and (7) we find:

$$\Psi = T\Theta = T\frac{d_i S}{dt} = \frac{dW_{\text{diss}}}{dt} + \sum_r A_r w_r \geqslant 0. \qquad (1\text{–}24.16)$$

This plausible form of the entropy flux (13) is the result of the definition of heat chosen in §1–7. In the entropy production (14) and in the dissipation function (16), we find, in addition to a sum of the products of the rates of reaction and the appropriate affinities, the work dissipated per unit time ("dissipated output"), which is related to such "dissipative effects" as result from passage of electricity and friction. In homogeneous systems, these appear as influences from the outside.

The entropy flux has a form analogous to Eq. (13) even for heterogeneous (discontinuous) systems (cf. Eq. 1–10.8), while the entropy production appears more complicated, since this still contains terms which refer to heat, electricity, and matter transport between the individual phases.

In continuous systems, the "entropy flux density" is analogous to Eq. (13), while in the "local entropy production" terms still appear which describe irreversible processes such as heat conduction, electric conduction, diffusion, etc.

We again consider a homogeneous system, and stipulate a specific case of dissipative effects, namely, the passage of an electric current of strength I

with an externally applied voltage Φ, without decomposition of matter. Then, using

$$\frac{dW_{\text{diss}}}{dt} = I\Phi$$

from Eq. (16),† we have for the dissipation function:

$$\Psi = T\Theta = I\Phi + \sum_r w_r A_r \geqslant 0. \qquad (1\text{-}24.17)$$

Since a passage of current without decomposition of matter through a body, assumed to be homogeneous, must be related to conduction of electrons, both kinds of irreversible processes (electric conduction and chemical reactions) are independent of each other. In practice, only the one or the other need be considered. We write, therefore, for the electric conduction:

$$\Psi = I\Phi \geqslant 0 \qquad (1\text{-}24.18)$$

---

† The expression

$$dW_{\text{diss}} = I\Phi \, dt$$

is usually called the "Joule heat" (for the time element dt). It can be seen, however, on introducing this expression into Eq. (10), that the relation

$$dQ = -I\Phi \, dt$$

only holds for the special case

$$dU + P \, dV = \sum_k H_k \, d_a n_k.$$

This condition is, for example, satisfied, if the temperature T and the pressure P are constant throughout the phase and if no chemical reactions occur. It then follows from Eq. (11) with Eqs. (1-5.3) and (1-6.3) that the enthalpy H is given by

$$dH = dU + P \, dV = \sum_k H_k \, dn_k = \sum_k H_k \, d_a n_k.$$

If the phase being considered is closed, we have

$$d_a n_k = 0, \qquad dU + P \, dV = 0.$$

This case is realized when an electric current flows through a metal wire of a given mass and composition, for in this case a stationary state will be obtained corresponding to a time invariance of T and P.

Strictly speaking, a conductor to which an electric potential difference is applied is not a homogeneous but a continuous system. If, however, we neglect the concentration difference (chemically indeterminate) of the electrons, the state of the media discussed here is homogeneous with reference to temperature, pressure, and composition. We can thus, as we are not interested in electric conduction at present, relinquish the greater requirement for the treatment of irreversible processes in continuous systems (Chapter 4).

and for the chemical reactions:

$$\Psi = \sum_r w_r A_r \geqslant 0. \tag{1-24.19}$$

The latter relation is the inequality of de Donder[18].

In Eq. (18), the equality denotes the reversible limiting case, i.e. no flow of current ($I = 0$, $\Phi = 0$). In Eq. (19), the equality sign corresponds, if the abstract reversibility concepts of traditional thermodynamics are excluded, either to frozen reactions ($w_r = 0$, $A_r \neq 0$) or to chemical equilibrium ($w_r = 0$, $A_r = 0$) (cf. §1–19). For the actual (irreversible) course of the processes, there consequently results

$$\Psi = I\Phi > 0 \tag{1-24.20}$$

and

$$\Psi = \sum_r w_r A_r > 0. \tag{1-24.21}$$

We will come back to relation (20) in §1–25. The inequality in (19) and (21) will be dealt with in detail in Chapter 2.

As will be shown in the discussion of individual problems in the following chapters, the dissipation function Ψ always has the form

$$\Psi = \sum_i J_i X_i \geqslant 0. \tag{1-24.22}$$

Here the J_i are the *generalized fluxes* (in the above example, the electric current and rates of reaction) and the X_i the appropriate (conjugate) *generalized forces* (above, the electric potential difference and the affinities), where these quantities can be either scalars, vectors, or tensors. With a suitable choice of the "fluxes" and "forces," all disappear individually at equilibrium, as is the case with the rates of reaction and affinities of all (linearly independent) chemical reactions (cf. §1–19). If we omit the equality sign in (22), which refers to the reversible limiting case and thus also to equilibrium, then the following expression is found for the actual (irreversible) course of the process:

$$\Psi = \sum_i J_i X_i > 0. \tag{1-24.23}$$

As we will see in detail later, we have omitted the cases $\Psi = 0, J_i = 0, X_i = 0$ (equilibrium) and $\Psi = 0$, $J_i = 0$, $X_i \neq 0$ (restrained processes as defined above and in §2–3 or, for dependent fluxes, reversible movements as discussed in §1–27 and §4–12) in the transition from (22) to (23).

From the above the entropy flow is seen to contain the heat flux and the material fluxes which occur between a system and its surroundings; the dissipation function, however, contains quantities like the rate of reactions.

From this it is immediately obvious that entropy flux and (local) entropy production—in contrast to the total entropy of a system or to the entropy of a volume element—do not represent state functions; they depend on the particular paths of the changes.

The statements sketched in this section, and elaborated further in the introductory sections of each chapter, form the contents of the thermodynamic part of the "thermodynamic–phenomenological theory of irreversible processes." Thereby, as is obvious, use is made only of a generalization and a precise definition of the concepts of classical thermodynamics. Accordingly, this thermodynamic part of the theory was already developed in principle at a time when the "phenomenological" part of the theory was not yet known (§1–25 and §1–26). Thus Jaumann[30, 31] and Lohr[32, 33] have already completed calculations relating to this. The works of these authors, however, have passed into oblivion. Later publications of de Donder (cf. above and §1–25) are independent of the above authors. The theory gained new impetus through Onsager (cf. §1–25 and §1–26). The extension of thermodynamics of irreversible processes into a systematic theory is due to such authors as Eckart[34], Meixner[35, 36a], Prigogine[10, 37], and de Groot[38, 39].

The introduction and application of an arbitrary number of work-coordinates and work-coefficients (§1–3), the critical discussion of the concept of heat (§1–4 and §1–7), and the generalized formulation of the second law (§1–8 to §1–10) can be taken to be, among others, the special features of this present work. With this an attempt has been made to achieve a general, modern, and logical presentation of the thermodynamics of irreversible processes.† [14]

1–25 PHENOMENOLOGICAL EQUATIONS

Since at thermodynamic equilibrium the "fluxes" J_i and the corresponding "driving forces" X_i disappear—suitable choice of these quantities having been assumed (cf. §1–24)—it is natural to describe the deviations from equilibrium and, thus, the irreversible processes relating to them as a first approximation by a *linear* dependence of the generalized fluxes on the generalized forces. According to Onsager[44a,b], the following equations are then used as the most general expression of a linear dependence of the "fluxes" $J_1, J_2, \ldots, J_m$ on the "forces" $X_1, X_2, \ldots, X_m$:

$$J_i = \sum_{j=1}^{m} \alpha_{ij} X_j \quad (i = 1, 2, \ldots, m). \tag{1-25.1}$$

These homogeneous linear relations are described as *phenomenological equations*. The quantities α_{ij} $(i, j = 1, 2, \ldots, m)$ are called *phenomenological coefficients*. They are not thermodynamic functions, but kinetic quantities,

† There are other monographs and comprehensive articles[40–43].

the significance of which is shown in discussions of particular problems. They can be arbitrary functions of state variables (temperature, pressure, composition, etc.), but supposedly do not depend on the J_i and the X_i.

In the discussion of the individual phenomena, it will become obvious to what extent the laws (1), which require that the system be not far removed from equilibrium, relate to reality. Here let it be stated in advance that the relations (1) are of restricted applicability for chemical reactions, since most reactions which actually occur are so far from equilibrium that a linear dependence of the reaction rates on the affinities does not arise. For the majority of transport phenomena, however, Eq. (1) is completely valid.

At this point consider, as one example, the problem of pure electric conduction (§1–24).

We find, for the passage of current through a homogeneous body (without chemical reactions), that the dissipation function becomes, according to Eq. (1–24.20),

$$\Psi = I\Phi > 0, \tag{1–25.2}$$

where I denotes the electric current and Φ the electric potential difference applied to the phase. The phenomenological laws state in this simple case:

$$I = \alpha\Phi, \tag{1–25.3}$$

where the phenomenological coefficient α can be a function of the temperature, of the pressure, and of the composition of the phase, but cannot depend on I or on Φ. Obviously, Eq. (3) is none other than Ohm's law, where α denotes the electric conductivity (the reciprocal of resistance).[15] From the inequality (2) there follows with (3):

$$\alpha > 0, \tag{1–25.4}$$

so that the positive sign of the entropy production leads to a relatively trivial statement. We recognize furthermore that deviations from the linear law (3) would demand invalidity of Ohm's law, for example, at extremely high electric field strength.

We will see later that other known linear relations for irreversible processes apart from Ohm's law for electric conduction are also contained in the phenomenological equations (1). Fourier's law of heat conduction, a generalized form of Fick's law for diffusion, and correspondingly the linear laws for the cross effects (thermoelectric effects, thermal diffusion, diffusion thermal effect, etc.) as well as the linear relations between reaction rates and affinities in chemical reactions holding for states near equilibrium which de Donder[18, 45] has presented are all of this class. Here the coefficients α_{ij} for $i = j$ describe the "direct" effects (such as electric conduction, heat conduction, and diffusion in isotropic media with at most two species) and for $i \neq j$ they describe the cross effects.

From the inequality (1–24.22) the general expression for the dissipation function follows with the phenomenological laws (1)

$$\Psi = \sum_{i=1}^{m} \sum_{j=1}^{m} \alpha_{ij} X_i X_j \geqslant 0. \qquad (1\text{–}25.5)$$

From this results a series of inequalities for the phenomenological coefficients. We obtain, for example, for $m = 2$:

$$\alpha_{11}\alpha_{22} - \tfrac{1}{4}(\alpha_{12} + \alpha_{21})^2 \geqslant 0, \quad \alpha_{11} \geqslant 0, \quad \alpha_{22} \geqslant 0. \qquad (1\text{–}25.6)$$

If all "fluxes" and "forces" and thus all phenomenological coefficients are independent of each other and if no restrained process is present (cf. §1–24) then the equality signs in (5) and (6) must be excluded, for then the equilibrium conditions $J_i = 0$, $X_i = 0$ (for all i) refer precisely to the condition $\Psi = 0$ in accordance with Eq. (1). If, therefore, the α_{ij} are independent and if only the actual course of processes is considered then relations (5) and (6) hold with the inequality sign. On the other hand, for restrained processes and for dependent fluxes or forces, individual phenomenological coefficients or certain combinations of these quantities (cf. 6) can disappear (cf. also §1–27).

1–26 ONSAGER'S RECIPROCITY RELATIONS

As both the classical and the quantum mechanical equations of motion indicate, all microphysical laws which determine the motions of single particles of a system are symmetric with reference to past and future, i.e. invariant to a transformation in time ($t \to -t$) ("Principle of Microscopic Reversibility"). It is assumed that no forces which are uneven functions of the speed act on the particles. Examples of these include the Coriolis force for rotating systems or the Lorentz force in magnetic fields, both of which depend linearly on the velocities.

If the above "principle of microscopic reversibility" is built into the framework of a general molecular statistical theory of irreversible processes then we obtain the result indicated by Onsager[44a,b] in his two brilliant papers. Thus, in the simplest case, the following expression for the pheno-menological coefficients in the equations (1–25.1) results:

$$\alpha_{ij} = \alpha_{ji} \quad (i, j = 1, 2, \ldots, m). \qquad (1\text{–}26.1)$$

The coefficient matrix is, therefore, symmetric.

The addition "in the simplest case" should mean that Eq. (1) represents a special case of a more general statement. We must, as Casimir[46] has shown, distinguish between two different types of generalized forces which are called "forces of the α-type" and "forces of the β-type." The "forces of the α-type" are even with reference to time reversal while "forces of the

β-type" are uneven. All generalized forces appearing in the framework of our discussion of processes in homogeneous systems (Chapter 2) and in discontinuous systems (Chapter 3) as well as the majority of the "forces" appearing in continuous systems (Chapter 4) belong to the α-type. The first example for forces of the β-type which we will encounter relates to continuous systems *with viscous flow* (cf. §4–12 and §4–32): it is the velocity gradients (or linear combinations of these quantities) which are conjugate to the viscous pressures.

Thus we formulate a generalization of the symmetry expression (1):

$$\alpha_{ij} = \varepsilon_i \varepsilon_j \alpha_{ji} \quad (i, j = 1, 2, \ldots, m). \tag{1–26.2}$$

Here

$$\varepsilon_k = 1 \qquad \text{if } X_k \text{ is a force of } \alpha\text{-type},$$

$$\varepsilon_k = -1 \quad \text{if } X_k \text{ is a force of } \beta\text{-type},$$

where the inferior k stands for i or for j. If X_i and X_j are of the same type (in particular, as in the majority of cases, both of the α-type) then Eq. (2) reduces to Eq. (1). If X_i and X_j belong to different types of forces then there follows from Eq. (2):

$$\alpha_{ij} = -\alpha_{ji}. \tag{1–26.3}$$

When Eq. (1) holds, the phenomenological coefficients α_{ij} and α_{ji} are referred to as "Onsager coefficients" and, when Eq. (3) is valid, as "Casimir coefficients."

If Coriolis forces or magnetic fields appear then the above relations must be modified, since here "microscopic reversibility" is satisfied only if the direction of the force fields is simultaneously reversed with all particle velocities[44b]. If we designate the angular velocity of rotational motion and the magnetic induction by the common symbol **b**, we arrive at the following expression[46] in place of Eq. (2):

$$\alpha_{ij}(\mathbf{b}) = \varepsilon_i \varepsilon_j \alpha_{ji}(-\mathbf{b}). \tag{1–26.4}$$

More on this is found in §4–35.

The equations (1) are denoted as the *Onsager reciprocity relations*, the equations (2) and (4) as the *Onsager–Casimir reciprocity relations*. They are a real supplement to the general statements of macroscopic physics, since they cannot be derived from structural symmetry properties of matter (such as from the isotropy in fluid media or from symmetry relations in crystals). The greater part of the general statements of the thermodynamics of irreversible processes depends on the reciprocity relations as is shown in detail later. These relations connect the cross effects (e.g. thermal diffusion) and the appropriate reciprocal effects (e.g. diffusion thermal effect) in a completely general way. The significance and the historical origin of the reciprocity laws can be most simply appreciated by the examples of the "coupling" of two chemical reactions (§2–7) or of heat conduction in anisotropic media

(§4–39). More recent investigations have led to a generalization of the statistical interpretation of the reciprocity relations[36b, 46a–51a].

If we exclude case (4), we can make some general statements about the directions of irreversible processes not far from equilibrium. If we introduce the reciprocity relations (2) into the inequality (1–25.5) for the dissipation function then according to (3) all terms with Casimir coefficients vanish. There remains a positive-definite quadratic form assuming that all phenomenological coefficients are independent (cf. §1–25). Then the determinant of these coefficients together with all chief minors are positive:

$$\alpha_{ii} > 0, \quad \alpha_{ii}\alpha_{jj} - \alpha_{ij}^2 > 0, \quad \ldots \quad (i, j = 1, 2, \ldots, m; i \neq j), \quad (1\text{–}26.5)$$

$$\begin{vmatrix} \alpha_{11} & \alpha_{12} & \cdots & \alpha_{1m} \\ \alpha_{21} & \alpha_{22} & \cdots & \alpha_{2m} \\ \cdot & \cdot & \cdot & \cdot \\ \alpha_{m1} & \alpha_{m2} & \cdots & \alpha_{mm} \end{vmatrix} > 0. \qquad (1\text{–}26.6)$$

For $m = 2$, we have therefore the expressions:

$$\alpha_{11} > 0, \quad \alpha_{22} > 0, \quad \alpha_{11}\alpha_{22} - \alpha_{12}^2 > 0. \qquad (1\text{–}26.7)$$

These can also be derived directly from (1–25.6) with (1). We will mention many examples for the inequalities (7) in the course of the discussion of particular processes.

1–27 TRANSFORMATIONS OF THE GENERALIZED FLUXES AND FORCES

It was noted in §1–7 that, besides our chosen definition for open systems, there are still other concepts of "heat" in the literature. We will show for continuous systems (§4–7) in detail how the different "heat flows" are related to each other. Also, in continuous systems, different definitions of "diffusion fluxes" depending on different reference velocities are used (§4–3). Finally, a sequence of chemical reactions can be described by different, macroscopically equivalent formulations (cf. below). This again corresponds to a certain freedom in the choice of the "generalized fluxes."

In all these and similar cases, we start from a given set of "generalized fluxes" J_i and "generalized forces" X_i and introduce new "fluxes" J_i', which are homogeneous linear functions of the old "fluxes" J_i. Then the new "forces" X_i' also depend homogeneously and linearly on the old "forces" X_i. Since the dissipation function Ψ must, on physical grounds, be independent of the particular choice of the fluxes and forces then, according to Eq. (1–24.22), the following relation is obtained:

$$\Psi = \sum_i J_i X_i = \sum J_i' X_i'. \qquad (1\text{–}27.1)$$

Here the forces X_i' belonging to the new fluxes J_i' are determined by Eq. (1), as are also the conjugate fluxes J_i' for newly chosen forces X_i'. It is assumed first that the fluxes and forces in each of the two summations are independent of each other. Thus both summations in (1) contain the same number of terms. We discuss the case of dependent fluxes and forces below.

Before we develop a general theory for the above-mentioned transformations of fluxes and forces and for the important question of the invariance of the Onsager reciprocity relations during such transformations, we want to mention a simple example.

Consider the reaction scheme

$$L \longrightarrow M \quad (I),$$
$$M \longrightarrow N \quad (II),$$

which takes place in a homogeneous closed system. If the amounts of species L, M, and N are denoted by n_L, n_M, and n_N and the reaction rates of the chemical reactions I and II by w_I and w_{II}, Eqs. (1–24.11) and (1–24.15) indicate that

$$\frac{dn_L}{dt} = -w_I, \quad \frac{dn_M}{dt} = w_I - w_{II}, \quad \frac{dn_N}{dt} = w_{II}, \tag{1–27.2}$$

where t denotes the time. The affinities A_I and A_{II} of the two reactions I and II are given by Eq. (1–14.2):

$$A_I = \mu_L - \mu_M, \quad A_{II} = \mu_M - \mu_N. \tag{1–27.3}$$

Here μ_k is the chemical potential of species k in the phase considered.

Now, the reaction system under consideration can be described macroscopically by the reaction equations:

$$L \longrightarrow N \quad (I'),$$
$$M \longrightarrow N \quad (II').$$

Formally, Eqs. (I') and (II') arise through a linear combination of Eqs. (I) and (II). The reaction rates w_I' and w_{II}' and the corresponding affinities A_I' and A_{II}' in the new formulation result again from Eqs. (1–24.11) and (1–24.15) respectively and from Eq. (1–14.2):

$$\frac{dn_L}{dt} = -w_I', \quad \frac{dn_M}{dt} = -w_{II}', \quad \frac{dn_N}{dt} = w_I' + w_{II}', \tag{1–27.4}$$

$$A_I' = \mu_L - \mu_N, \quad A_{II}' = \mu_M - \mu_N. \tag{1–27.5}$$

The comparison of Eq. (2) with Eq. (4) and of Eq. (3) with Eq. (5) gives us the relation between the old and the new quantities:

$$w_I' = w_I, \quad w_{II}' = w_{II} - w_I, \tag{1–27.6}$$

$$A_I' = A_I + A_{II}, \quad A_{II}' = A_{II}. \tag{1–27.7}$$

The new "fluxes" w'_I and w'_{II} and the new "forces" A'_I and A'_{II} are, therefore, homogeneous linear functions of the old "fluxes" w_I and w_{II} and the old "forces" A_1 and A_2 respectively. Here the affinities are transformed according to Eq. (7) as are the reaction equations belonging to them, while the reaction rates undergo a complementary transformation according to Eq. (6). As can be derived from Eq. (1–24.19) as well as Eqs. (6) and (7), the dissipation function Ψ remains invariant to this linear transformation of the fluxes and the forces:

$$\Psi = w_I A_I + w_{II} A_{II} = w'_I A'_I + w'_{II} A'_{II}. \qquad (1\text{–}27.8)$$

This is in agreement with the general relation (1). Analogous conclusions are valid for any chemical reactions[52].

In Eq. (8), both, w_I and w_{II} or A_I and A_{II} and w'_I and w'_{II} or A'_I and A'_{II} are independent of each other, corresponding therefore to the assumption made in Eq. (1) that in each summation both the fluxes and the forces are individually independent of each other.

We consider now an arbitrary linear transformation from the m independent fluxes J_j to the m independent fluxes J'_i:

$$J'_i = \sum_{j=1}^{m} \beta_{ij} J_j \quad (i = 1, 2, \ldots, m), \qquad (1\text{–}27.9)$$

where the β_{ij} are unspecified coefficients. Then from Eq. (1):

$$\sum_{i=1}^{m} J'_i X'_i = \sum_{j=1}^{m} J_j \sum_{i=1}^{m} \beta_{ij} X'_i = \sum_{j=1}^{m} J_j X_j.$$

From this the relation between the old forces X_j and the new forces X'_i results:

$$X_j = \sum_{i=1}^{m} \beta_{ij} X'_i \quad (j = 1, 2, \ldots, m). \qquad (1\text{–}27.10)$$

As an example of the conversion relations (9) and (10), we will consider the transition from reaction equations (I) and (II) to the reaction equations (I′), (II′). With the relations following from Eq. (9) for $m = 2$

$$J'_1 = \beta_{11} J_1 + \beta_{12} J_2,$$
$$J'_2 = \beta_{21} J_1 + \beta_{22} J_2$$

we obtain from Eq. (6):

$$\beta_{11} = 1, \quad \beta_{12} = 0, \quad \beta_{21} = -1, \quad \beta_{22} = 1$$

if we identify w'_I, w'_{II}, w_I, w_{II} with J'_1, J'_2, J_1, J_2 in that order. According to Eq. (10), we obtain:

$$X_1 = \beta_{11} X'_1 + \beta_{21} X'_2,$$
$$X_2 = \beta_{12} X'_1 + \beta_{22} X'_2.$$

Hence there results:

$$X_1 = X_1' - X_2', \quad X_2 = X_2',$$

which agrees with Eq. (7) if we put $X_1 = A_{\mathrm{I}}$, $X_2 = A_{\mathrm{II}}$, $X_1' = A_{\mathrm{I}}'$, $X_2' = A_{\mathrm{II}}'$.

We now set up the phenomenological equations and examine the transformation properties of the phenomenological coefficients[53–56]. We have, according to Eq. (1–25.1),

$$J_j = \sum_{k=1}^{m} \alpha_{jk} X_k \quad (j = 1, 2, \ldots, m), \tag{1–27.11}$$

where the J_j are the old fluxes, the X_k the old forces, and the α_{jk} the old phenomenological coefficients. The corresponding linear laws for the new fluxes and forces (J_i' and X_i') with new phenomenological coefficients α_{il}' are:

$$J_i' = \sum_{l=1}^{m} \alpha_{il}' X_l' \quad (i = 1, 2, \ldots, m). \tag{1–27.12}$$

The relations between the two sets of coefficients can be found, if we write Eq. (10) with changed inferior labeling:

$$X_k = \sum_{l=1}^{m} \beta_{lk} X_l' \quad (k = 1, 2, \ldots, m). \tag{1–27.13}$$

By inserting Eq. (13) into Eq. (11) and considering Eq. (9) we obtain

$$J_i' = \sum_{j=1}^{m} \sum_{k=1}^{m} \sum_{l=1}^{m} \alpha_{jk} \beta_{ij} \beta_{lk} X_l' \quad (i = 1, 2, \ldots, m). \tag{1–27.14}$$

Comparison of Eq. (12) with Eq. (14) yields:

$$\alpha_{il}' = \sum_{j=1}^{m} \sum_{k=1}^{m} \alpha_{jk} \beta_{ij} \beta_{lk} \quad (i, l = 1, 2, \ldots, m). \tag{1–27.15}$$

These are the relations between the old and the new phenomenological coefficients.

We restrict the following discussion to the case of the "Onsager coefficients" (§1–26). The Onsager reciprocity law in the form (1–26.1) is valid for each system of phenomenological equations obtained from the dissipation function if the fluxes and forces are independent. We could therefore anticipate the validity of the reciprocity relations immediately, both in the form (11) and in the form (12).

But we can show this explicitly, too. It follows from the symmetry relation that:

$$\alpha_{jk} = \alpha_{kj} \quad (j, k = 1, 2, \ldots, m). \tag{1–27.16}$$

With the help of Eq. (15) we find immediately:

$$\alpha'_{il} = \alpha'_{li} \quad (i, l = 1, 2, \ldots, m) \tag{1-27.17}$$

and vice versa.†

If there is a linear dependence between the fluxes or between the forces, examples of which will be mentioned in §2–7 and §4–2, then the validity of the Onsager reciprocity law is no longer obvious. The case of linearly dependent fluxes and forces requires, therefore, a special examination[53].

First, *dependent fluxes* with independent forces will be assumed. Thus, considering the entropy balance for continuous systems, we obtain an expression for the dissipation function in which forces X_i appear which—at least in the most general case—are independent, while for the appropriate fluxes J_i there exists a linear dependence in the general form (cf. §4–12):

$$\sum_{i=1}^{m} a_i J_i = 0, \tag{1-27.18}$$

where the a_i are constants. If we eliminate the flux J_m, which is chosen so that the coefficient a_m does not disappear, then we find that

$$J_m = -\sum_{i=1}^{m-1} \frac{a_i}{a_m} J_i. \tag{1-27.19}$$

The dissipation function Ψ can be written in its original form

$$\Psi = \sum_{i=1}^{m} J_i X_i \tag{1-27.20}$$

but also in the abbreviated form obtained from Eq. (19):

$$\Psi = \sum_{i=1}^{m-1} J_i \left(X_i - \frac{a_i}{a_m} X_m \right), \tag{1-27.21}$$

in which only $m - 1$ independent flows J_i and $m - 1$ independent forces $X_i - (a_i/a_m)X_m$ are contained (cf. §4–12).

We now formulate the phenomenological laws from expressions (20) and (21):

$$J_i = \sum_{k=1}^{m} \alpha'_{ik} X_k \quad (i = 1, 2, \ldots, m), \tag{1-27.22}$$

$$J_i = \sum_{j=1}^{m-1} \alpha_{ij} \left(X_j - \frac{a_j}{a_m} X_m \right) \quad (i = 1, 2, \ldots, m-1). \tag{1-27.23}$$

† A very general discussion of the invariance of Onsager's reciprocity relations is given by Coleman and Truesdell[57].

Here the α'_{ik} and α_{ij} are the phenomenological coefficients. On the basis of Eq. (18) there immediately exist between the α'_{ik} the following linear relations:

$$\sum_{i=1}^{m} a_i \alpha'_{ik} = 0 \quad (k = 1, 2, \ldots, m). \tag{1-27.24}$$

The connection between the two sets of phenomenological coefficients results from a comparison of Eq. (22) with Eq. (23) on consideration of Eq. (19):

$$\alpha'_{ij} = \alpha_{ij} \quad (i, j = 1, 2, \ldots, m - 1), \tag{1-27.25}$$

$$\alpha'_{im} = -\sum_{j=1}^{m-1} \frac{a_j}{a_m} \alpha_{ij} \quad (i = 1, 2, \ldots, m - 1), \tag{1-27.26}$$

$$\alpha'_{mi} = -\sum_{j=1}^{m-1} \frac{a_j}{a_m} \alpha_{ji} \quad (i = 1, 2, \ldots, m - 1), \tag{1-27.27}$$

$$\alpha'_{mm} = \sum_{i=1}^{m-1} \sum_{j=1}^{m-1} \frac{a_i a_j}{a_m^2} \alpha_{ij}. \tag{1-27.28}$$

The relations (24) are also contained in these equations.

Since only independent fluxes and forces are present in Eq. (23), the Onsager reciprocity law is valid:

$$\alpha_{ij} = \alpha_{ij} \quad (i, j = 1, 2, \ldots, m - 1). \tag{1-27.29}$$

There then follows from Eqs. (25) to (27):

$$\alpha'_{ij} = \alpha'_{ji}, \quad \alpha'_{im} = \alpha'_{mi} \quad (i, j = 1, 2, \ldots, m - 1) \tag{1-27.30}$$

and the reciprocity relations remain applicable even for the (linearly dependent) phenomenological coefficients. Since the dissipation function Ψ disappears when

$$X_i = \frac{a_i}{a_m} X_m \quad (i = 1, 2, \ldots, m - 1),$$

according to Eq. (21) it no longer represents a positive-definite quadratic of the form (20), as it does in the case of independent fluxes and forces (cf. §1-26), but is positive or zero (positive-semidefinite). This situation occurring in continuous systems for $\Psi = 0$, $X_i \neq 0$, $X_m \neq 0$ (at least for some forces) is that of "reversible motions" (cf. §4-12).

As a simple example let us consider two dependent fluxes (J_1 and J_2) and two independent forces (X_1 and X_2). The linear relations therein (18) should have the form

$$J_1 + J_2 = 0. \tag{1-27.31}$$

The dissipation function can be written either in the form (20)

$$\Psi = J_1 X_1 + J_2 X_2 \qquad (1\text{–}27.32)$$

or in the form (21)

$$\Psi = J_1(X_1 - X_2). \qquad (1\text{–}27.33)$$

The phenomenological laws are given by Eqs. (22) and (23)

$$\left.\begin{aligned} J_1 &= \alpha'_{11} X_1 + \alpha'_{12} X_2, \\ J_2 &= \alpha'_{21} X_1 + \alpha'_{22} X_2, \end{aligned}\right\} \qquad (1\text{–}27.34)$$

$$J_1 = \alpha(X_1 - X_2), \qquad (1\text{–}27.35)$$

where $\alpha \ (= \alpha_{11})$ is the only phenomenological coefficient obtained by taking the independent flow J_1 and the conjugate force $X_1 - X_2$ as a basis. If condition (31) is introduced into the relations (34) then we obtain—as a special case of Eq. (24)—the following linear relation between the four phenomenological coefficients of Eqs. (34):

$$\alpha'_{11} + \alpha'_{21} = 0, \quad \alpha'_{12} + \alpha'_{22} = 0. \qquad (1\text{–}27.36)$$

From this there results with Eqs. (31) and (34)

$$J_1 = -J_2 = \alpha'_{11} X_1 - \alpha'_{22} X_2. \qquad (1\text{–}27.37)$$

In order that this expression agrees with Eq. (35), we must have

$$\alpha'_{11} = \alpha'_{22} = \alpha. \qquad (1\text{–}27.38)$$

Thus, according to Eq. (36)

$$\alpha'_{12} = \alpha'_{21}, \qquad (1\text{–}27.39)$$

the Onsager reciprocity law has been formally satisfied in the statement (34).

The situation becomes more complicated if we have *dependent forces* (cf. §2–7). We will not examine the most general case here[58] but restrict the discussion to a few simple examples.

We consider first two independent fluxes (J_1 and J_2) and two dependent forces (X_1 and X_2). The linear relation between the forces shall have the form

$$X_1 + X_2 = 0. \qquad (1\text{–}27.40)$$

The dissipation function can then be written

$$\Psi = J_1 X_1 + J_2 X_2 \qquad (1\text{–}27.41)$$

or

$$\Psi = (J_1 - J_2) X_1. \qquad (1\text{–}27.42)$$

The corresponding phenomenological laws are (on exchanging the roles of the fluxes and forces)

$$\left.\begin{aligned} X_1 &= \alpha_{11} J_1 + \alpha_{12} J_2, \\ X_2 &= \alpha_{21} J_1 + \alpha_{22} J_2 \end{aligned}\right\} \qquad (1\text{–}27.43)$$

or

$$X_1 = \alpha(J_1 - J_2). \tag{1-27.44}$$

On comparison of Eq. (43) with Eq. (44) there results from Eq. (40)

$$\alpha_{11} = \alpha_{22} = -\alpha_{12} = -\alpha_{21} = \alpha. \tag{1-27.45}$$

Here, also, the reciprocity relation in Eq. (43) is formally satisfied.

As a second example, we deal with two dependent fluxes (J_1 and J_2) and two dependent forces (X_1 and X_2). Let the linear relations

$$X_1 + X_2 = 0, \tag{1-27.46}$$

$$2J_1 + J_2 = 0 \tag{1-27.47}$$

hold here. Then the dissipation function has the form

$$\Psi = J_1 X_1 + J_2 X_2 \tag{1-27.48}$$

or

$$\Psi = 3J_1 X_1. \tag{1-27.49}$$

The corresponding phenomenological equations are (cf. Eq. 34)

$$\left.\begin{array}{l} J_1 = \alpha_{11} X_1 + \alpha_{12} X_2, \\ J_2 = \alpha_{21} X_1 + \alpha_{22} X_2 \end{array}\right\} \tag{1-27.50}$$

or

$$J_1 = \alpha X, \tag{1-27.51}$$

where

$$X \equiv 3X_1 = -3X_2. \tag{1-27.52}$$

From Eqs. (47), (51), and (52) it follows that

$$\left.\begin{array}{l} J_1 = 3\alpha X_1, \\ J_2 = -6\alpha X_1. \end{array}\right\} \tag{1-27.53}$$

From Eqs. (46) and (50) there results:

$$\left.\begin{array}{l} J_1 = (\alpha_{11} - \alpha_{12})X_1, \\ J_2 = (\alpha_{21} - \alpha_{22})X_1. \end{array}\right\} \tag{1-27.54}$$

Here we only obtain the conditions

$$\left.\begin{array}{l} \alpha_{11} - \alpha_{12} = 3\alpha, \\ \alpha_{21} - \alpha_{22} = -6\alpha. \end{array}\right\} \tag{1-27.55}$$

Thus in Eq. (50) *two coefficients can be freely chosen.* We can, therefore, arbitrarily prescribe two condition equations. If we choose

$$\alpha_{12} = \alpha_{21}, \tag{1-27.56}$$

$$\alpha_{22} = 5\alpha \tag{1-27.57}$$

then, according to Eq. (55), all coefficients are determined:

$$\alpha_{11} = 2\alpha, \quad \alpha_{12} = \alpha_{21} = -\alpha, \quad \alpha_{22} = 5\alpha. \qquad (1\text{-}27.58)$$

Now Onsager's reciprocity law is satisfied, a condition which would not necessarily be true for another choice of coefficients. Moreover, for $\alpha > 0$ the dissipation function is always positive. It follows, indeed, from Eq. (58):

$$\begin{vmatrix} \alpha_{11} & \alpha_{12} \\ \alpha_{21} & \alpha_{22} \end{vmatrix} = \begin{vmatrix} 2\alpha & -\alpha \\ -\alpha & 5\alpha \end{vmatrix} = 9\alpha^2 > 0, \qquad (1\text{-}27.59)$$

from which there results, with Eqs. (48) and (50), using the condition that $\alpha > 0$, the fact that:

$$\Psi > 0. \qquad (1\text{-}27.60)$$

This result is also obtained directly from (49) and (53) with $\alpha > 0$.

The validity of the Onsager reciprocity law is generally guaranteed for dependent fluxes *and* dependent forces only for a *prescribed choice of coefficients*.

According to the above, the safest and clearest procedure is to choose both the fluxes and the forces so that both sets of variables are independent. In doing so, care must be taken at the same time to ensure that all fluxes and forces disappear at equilibrium, for only then is it permissible to formulate the phenomenological laws. This was assumed as obvious in the preceding equations; there are, however, cases in which we must be convinced of this further property of the fluxes and forces before proceeding to set up the phenomenological laws (cf. §2–7 and §3–5).

To sum up, we proceed with the application of the theory to special cases as follows: the entropy balance is set up, an explicit expression for the dissipation function derived from it, then the fluxes and the forces determined so that all these quantities are independent and disappear at equilibrium; with this the phenomenological laws are formulated, and finally Onsager's reciprocity law is applied.

1–28 IRREVERSIBLE PROCESSES AND EQUILIBRIUM

Irreversible processes in a given system will occur continuously until either a stationary state or equilibrium has been achieved (cf. §1–17). (Periodic processes are excluded here.) If several irreversible processes are superimposed and if the final state achieved corresponds to an equilibrium state then we can, in certain cases, make general statements about the coefficients which describe the irreversible phenomena without the thermodynamics of irreversible processes.

As a simple example, let us consider the homogeneous reaction

$$L \rightleftharpoons M.$$

If the reaction takes place in an ideal gas or in an ideal dilute solution then we know from experience that the rate of reaction w is given by the equation (cf. §2–5)

$$w = \kappa c_L - \kappa' c_M. \tag{1-28.1}$$

Here c_L and c_M are the molarities of the species L and M respectively, and κ and κ' are the rate constants for the reaction from left to right and from right to left respectively. Equilibrium is achieved (homogeneous chemical equilibrium) if the values of the rates of both the reactions are equal; the reaction rate of the total reaction thus disappears:

$$w = 0 \quad \text{(equilibrium).} \tag{1-28.2}$$

Now the equilibrium conditions of classical thermodynamics for the above case (cf. §2–5) state:

$$\frac{c_M}{c_L} = K \quad \text{(equilibrium),} \tag{1-28.3}$$

where K represents the equilibrium constant. The quantity K is, like κ and κ', independent of c_L and c_M. From Eqs. (1), (2), and (3) there follows immediately:

$$\frac{\kappa}{\kappa'} = K. \tag{1-28.4}$$

We obtain thus from Eq. (1) (which is taken directly from experience) and from Eqs. (2) and (3) (which stem from classical thermodynamics) a relation between coefficients (κ and κ') describing irreversible processes and a quantity relating to equilibrium (K).

Another example concerns a less familiar problem: the relation between diffusion and sedimentation (cf. §4–20). In a continuous system with two independently moving substances in which concentration gradients and a stationary external force field (gravitational or centrifugal field) are present, diffusion (mass transport due to concentration gradients) and sedimentation (flow of matter due to external fields) in a volume element under normal conditions can, according to experience, be described by the following law:

$$_w\mathbf{J}_2 = - D \operatorname{grad} c_2 + c_2 s \mathbf{g}. \tag{1-28.5}$$

Here $_w\mathbf{J}_2$ is the diffusion flux vector of species 2, c_2 is the molarity of substance 2, $\mathbf{g}$ the vector of gravitational or centrifugal acceleration, D the diffusion coefficient, and s the sedimentation coefficient. The last two quantities are independent of $\operatorname{grad} c_2$ and $\mathbf{g}$. At equilibrium (sedimentation equilibrium) the diffusion and sedimentation must obviously compensate in each volume element. We have therefore:

$$_w\mathbf{J}_2 = 0 \quad \text{(equilibrium).} \tag{1-28.6}$$

Now, however, the equilibrium conditions of classical thermodynamics state that, for this case,

$$(M_2 - \rho V_2)\mathbf{g} = \left(\frac{\partial \mu_2}{\partial c_2}\right)_{T,P} \text{grad } c_2 \quad \text{(equilibrium)}. \qquad (1\text{–}28.7)$$

This can be derived from Eq. (1–21.31) (cf. also §4–20). Here M_2 is the molar mass, V_2 the partial molar volume, and μ_2 the chemical potential of substance 2; ρ denotes the density, T the temperature, and P the pressure of the volume element. Comparison of Eqs. (5), (6), and (7) results in:

$$\frac{D}{s} = \frac{c_2}{M_2 - V_2\rho} \left(\frac{\partial \mu_2}{\partial c_2}\right)_{T,P}. \qquad (1\text{–}28.8)$$

We obtain, thus, a relation between two transport coefficients (D and s) and equilibrium quantities. This results from the empirical law (5) and from Eqs. (6) and (7) deriving from classical thermodynamics. The generalization of this relation to systems with any number of substances will be derived in an analogous way in §4–20.

In §3–12, we will obtain a general relation between permeation and osmosis for binary membrane systems in a similar manner.

If we seek a general criterion relating to transport processes which allows such relations to be derived in this way, we come to the following conclusion. If several transport processes in a system overlap in such a way that the final state of the system corresponds to an *equilibrium in a nonhomogeneous system* then *classical thermodynamics* will provide us with the deduction of a general relation between the different transport coefficients. This general statement[24] makes the discovery of relations between transport quantities much easier, since we need not apply the more cumbersome methods of nonequilibrium thermodynamics.

The method described cannot be applied if the superposition of several transport processes leads to equilibrium in a homogeneous system; for now all gradients which appear in the transport equations will disappear at equilibrium, so that no statement on the transport coefficients can be obtained with the help of the equilibrium conditions.

At first sight we could object to the above considerations; the relations for the rate constants and transport coefficients are referred, according to the derivation, only to the equilibrium state, while they must in fact be valid for nonequilibrium states. This objection can easily be overruled. The above coefficients are indeed only functions of the local state variables (temperature, pressure, concentrations, etc.) and not explicit functions of the distance from equilibrium. We can, therefore, at given temperature and at given pressure, prescribe arbitrary concentrations and allow the appropriate irreversible processes to occur until equilibrium is reached. Since the coefficients which are now dependent on the concentrations do not "know" whether the concentrations represent equilibrium or nonequilibrium states, the argument must be quite general.

REFERENCES

1. R. HAASE, *Thermodynamik der Mischphasen*, Berlin–Goettingen–Heidelberg (1956), (a) p. 57; (b) p. 66; (c) p. 48; (d) p. 183; (e) p. 118; (f) p. 105; (g) p. 111; (h) p. 135; (i) p. 179; (j) p. 177.
2. A. SOMMERFELD, *Vorlesungen ueber Theoretische Physik*, Vol. II: *Mechanik der deformierbaren Medien*, 4th ed., Leipzig (1957).
3. R. DEFAY, *Etude thermodynamique de la Tension superficielle*, Paris (1934).
4. E. A. GUGGENHEIM, *Thermodynamics*, 3rd ed., Amsterdam (1957), (a) p. 47; (b) p. 438; (c) p. 50; (d) p. 53.
5. M. W. ZEMANSKY, *Heat and Thermodynamics*, London (1951), pp. 51, 292.
6. H. S. FRANK, *J. Chem. Phys.* **23**, 2023 (1955).
7. J. C. M. LI and T. W. TING, *J. Chem. Phys.* **27**, 693 (1957).
8. R. DEFAY, *Bull. Classe Sci., Acad. Roy. Belg.* **15**, 678 (1929).
9. L. J. GILLESPIE and J. R. COE, JR., *J. Chem. Phys.* **1**, 103 (1933).
10. I. PRIGOGINE, *Etude thermodynamique des Phénomènes irréversibles*, Paris–Liége (1947).
11. H. A. TOLHOEK and S. R. DE GROOT, *Physica* **18**, 780 (1952).
12. R. HAASE, *Z. Naturforsch.* **8a**, 729 (1953).
13. R. HAASE, *Z. Physik. Chem. (Frankfurt)* **12**, 1 (1957).
14. R. HAASE, *Z. Physik. Chem. (Frankfurt)* **9**, 355 (1956).
15. R. HAASE, *Med. Grundlagenforsch.* **II**, 717 (1959).
16. J. W. GIBBS, *The Collected Works*, Vol. I: *Thermodynamics*, New Haven (1948), (a) p. 55; (b) p. 56; (c) p. 219; (d) p. 276.
17. W. KUHN, I. TÓTH, and H. J. KUHN, *Helv. Chim. Acta* **45**, 2325 (1962).
18. TH. DE DONDER, *L'Affinité*, Paris (1927–1936).
19. M. PLANCK, *Vorlesungen ueber Thermodynamik*, Berlin (1897–1954).
20. R. HAASE, *Z. Physik. Chem. (Frankfurt)* **39**, 360 (1963).
21. R. HAASE, *Angew. Chem. Intern. Ed. Engl.* **4**, 485 (1965).
22. E. LANGE, *Handbuch d. Experimentalphysik*, Vol. 12/2, Leipzig (1933), p. 305.
23. J. D. VAN DER WAALS, *Lehrbuch der Thermodynamik*, Part 2, Leipzig (1912), p. 560.
24. R. HAASE, *Z. Physik. Chem. (Frankfurt)* **25**, 26 (1960).
25. J. MEIXNER, *Ann. Physik* (5) **39**, 333 (1941).
26. J. MEIXNER, *Z. Physik. Chem. (Leipzig)* (B)**53**, 235 (1943).
27. I. PRIGOGINE, *Physica* **15**, 272 (1949).
28. I. PRIGOGINE, *J. Phys. Colloid Chem.* **55**, 765 (1951).
29. H. G. REIK, *Z. Physik* **148**, 156, 333 (1957).
30. G. JAUMANN, *Sitzber. Akad. Wiss. Wien, Math.-Naturw. Kl.* **120**, 385 (1911).
31. G. JAUMANN, *Denkschr. Akad. Wiss. Wien, Math.-Naturw. Kl.* **95**, 461 (1918).
32. E. LOHR, *Denkschr. Akad. Wiss. Wien, Math.-Naturw. Kl.* **93**, 339 (1916).
33. E. LOHR, *Festschr. Dtsch. T.H. in Bruenn*, Bruenn (1924), p. 176.
34. C. ECKART, *Phys. Rev.* **58**, 267, 269, 919 (1940).
35. J. MEIXNER, *Thermodynamik der irreversiblen Prozesse*, Aachen (1954) (photomechanical reproduction of original papers).
36. J. MEIXNER and H. G. REIK, *Handbuch der Physik*, Vol. III/2, Berlin–Goettingen–Heidelberg (1959), (a) p. 413; (b) p. 505.
37. I. PRIGOGINE, *Introduction to Thermodynamics of Irreversible Processes*, Springfield, Illinois (1955).
38. S. R. DE GROOT, *Thermodynamics of Irreversible Processes*, Amsterdam (1951).

39. S. R. DE GROOT and P. MAZUR, *Non-Equilibrium Thermodynamics*, Amsterdam (1962).
40. K. G. DENBIGH, *The Thermodynamics of the Steady State*, London–New York (1951).
41. D. D. FITTS, *Nonequilibrium Thermodynamics*, New York–San Francisco–Toronto–London (1962).
42. R. HAASE, *Ergeb. Exakt. Naturw.* **26**, 56 (1952).
43. H. KNOF, *Thermodynamics of Irreversible Processes in Liquid Metals*, Braunschweig (1966).
44. L. ONSAGER, *Phys. Rev.* (a) **37**, 405 (1931); (b) **38**, 2265 (1931).
45. TH. DE DONDER, *Bull. Classe Sci., Acad. Roy. Belg.* **24**, 15 (1938).
46. H. B. G. CASIMIR, *Rev. Mod. Phys.* **17**, 343 (1945); (a) R. T. COX, *Rev. Mod. Phys.* **22**, 238 (1950).
47. L. ONSAGER and S. MACHLUP, *Phys. Rev.* **91**, 1505, 1512 (1953).
48. S. R. DE GROOT and P. MAZUR, *Physica* **23**, 73 (1957).
49. J. G. KIRKWOOD and D. D. FITTS, *J. Chem. Phys.* **33**, 317 (1960).
50. R. ZWANZIG, *Phys. Rev.* **124**, 983 (1961).
51. R. E. NETTLETON, *J. Chem. Phys.* **40**, 112 (1964); (a) F. SCHLÖGL, *Z. Physik* **193**, 163 (1966).
52. TH. DE DONDER, *Bull. Classe Sci., Acad. Roy. Belg.* **23**, 936 (1937).
53. J. MEIXNER, *Ann. Physik* (5) **43**, 244 (1943).
54. I. PRIGOGINE, *Bull. Classe Sci., Acad. Roy. Belg.* (5) **32**, 30 (1946).
55. G. J. HOOYMAN and S. R. DE GROOT, *Physica* **21**, 73 (1955).
56. G. J. HOOYMAN, S. R. DE GROOT, and P. MAZUR, *Physica* **21**, 360 (1955).
57. B. D. COLEMAN and C. TRUESDELL, *J. Chem. Phys.* **33**, 28 (1960).
58. J. K. RASTAS (unpublished).

CHAPTER 2

PROCESSES IN HOMOGENEOUS SYSTEMS

2-1 INTRODUCTION

We restrict the discussion of "processes in homogeneous systems" to the most important special case, namely, to chemical reactions and relaxation phenomena (chemical conversions, excitation processes, internal changes, etc.), in a homogeneous, isotropic medium without electrification and magnetization. Thus our formulas are considerably simplified, without the important features of the thermodynamic–phenomenological theory being lost.

Due to these restrictions all terms which relate to dissipative effects and changes in the external coordinates in the general equations of the previous chapter drop out, and the volume and the negative pressure represent the only work-coordinate and the only work-coefficient respectively. In particular, the energy equation is valid in the form (1–7.4) and the Gibbs equation in the form (1–9.5).

We allow arbitrary exchange of matter with the environment, since this requires no real complications. The homogeneous systems dealt with are thus "open phases".

In order that the methods of thermodynamics of irreversible processes be applicable to homogeneous systems, in particular so that the Gibbs equation remains valid for any arbitrary instant during the course of irreversible processes, it must be assumed that the chemical reactions and relaxation phenomena occur sufficiently slowly so that the homogeneity of the phase in question is not disturbed. Then the phase is in internal thermal and mechanical equilibrium at each moment and has uniform composition although temperature, pressure, and concentrations change in the course of time (cf. §1–23). It will be shown later (Chapter 4) how chemical reactions, in the presence of gradients of temperature, pressure, and concentrations, i.e. with the simultaneous occurrence of other irreversible processes (heat conduction, diffusion, etc.), are to be dealt with. The restriction to sufficiently slow progress, which is implicit in the elementary kinetic theories, does not demand a near-equilibrium state. This will first be postulated with the linear phenomenological laws (§2–4).

2–2 ENTROPY BALANCE

Let n_k be the amount of substance of species k, ν_{kr} the stoichiometric number of species k in the reaction r, and ξ_r the extent of the reaction r. In relaxation phenomena, n_k denotes the amount of a particular species which is in a definite molecular state (excited, vibrational, rotational state, etc.), so that ξ_r becomes an "internal parameter" assigned to the rth relaxation mechanism. The "mass balance" is then (cf. Eq. 1–24.11)

$$dn_k = \sum_r \nu_{kr} \, d\xi_r + d_a n_k, \qquad (2\text{–}2.1)$$

where the summation is over all chemical reactions (or relaxation mechanisms) and $d_a n_k$ denotes the infinitesimal increase in amount of species k from the outside. If we consider the time element dt and introduce† the "reaction rate" of reaction r

$$w_r \equiv \frac{d\xi_r}{dt} \qquad (2\text{–}2.2)$$

then we obtain from Eq. (1) for the rate of increase of the amount of species k:

$$\frac{dn_k}{dt} = \sum_r \nu_{kr} w_r + \frac{d_a n_k}{dt}, \qquad (2\text{–}2.3)$$

where the last term is the "flow of matter" of species k from the surroundings.

The "energy balance" follows from Eq. (1–7.4) which, strictly speaking, represents a definition of the "heat" dQ for an infinitesimal internal state change in an open isotropic phase without electrification and magnetization and without dissipative effects:

$$dU = dQ - P \, dV + \sum_k H_k \, d_a n_k. \qquad (2\text{–}2.4)$$

Here U denotes the internal energy, P the pressure, V the volume, and H_k the partial molar enthalpy of species k. The summation is over all species. The quantity dQ/dt is the "heat flow" from the surroundings.

---

† If we denote the "reaction rate" used in reaction kinetics by b_r, then

$$b_r = \frac{d\zeta_r}{dt} = \frac{1}{V} \frac{d\xi_r}{dt} = \frac{w_r}{V}.$$

ζ_r is the reaction variable introduced on p. 65 and V is the volume of the phase.

Furthermore, the Gibbs equation (1–9.5) is valid under the assumptions discussed above (§2–1). We obtain therefore, for the differential of the entropy S of the phase,

$$T \, dS = dU + P \, dV - \sum_k \mu_k \, dn_k. \tag{2-2.5}$$

Here T is the absolute temperature and μ_k the chemical potential of species k. If we introduce the affinity of the reaction r by means of Eq. (1–14.2)

$$A_r = - \sum_k \nu_{kr} \mu_k, \tag{2-2.6}$$

we find from Eqs. (1) and (5) that

$$T \, dS = dU + P \, dV + \sum_r A_r \, d\xi_r - \sum_k \mu_k \, d_a n_k. \tag{2-2.7}$$

Moreover, we note Eq. (1–12.6)

$$TS_k = H_k - \mu_k, \tag{2-2.8}$$

where S_k is the partial molar entropy of species k. From Eqs. (4), (7), and (8) we derive:

$$T \, dS = dQ + T \sum_k S_k \, d_a n_k + \sum_r A_r \, d\xi_r. \tag{2-2.9}$$

This represents a special case of Eq. (1–24.12).

If we introduce the time t and consider Eq. (2), we follow the general model of Eqs. (1–24.3), (1–24.4), and (1–24.5) to obtain the *entropy balance*

$$\frac{dS}{dt} = \frac{d_a S}{dt} + \frac{d_i S}{dt} \tag{2-2.10}$$

with the *entropy flow* (cf. Eq. 1–24.13)

$$\frac{d_a S}{dt} = \frac{1}{T} \frac{dQ}{dt} + \sum_k S_k \frac{d_a n_k}{dt} \tag{2-2.11}$$

and the *entropy production* (cf. Eq. 1–24.6)

$$\Theta = \frac{d_i S}{dt} = \frac{1}{T} \sum_r A_r w_r \geqslant 0. \tag{2-2.12}$$

The inequality in (12) is valid for an actual (irreversible) path of the reaction (or relaxation mechanism), the equality for the reversible limiting case. The relation (12) is a special case of Eqs. (1–24.14) and (1–24.16).

The entropy flow (11) includes the "heat flow" dQ/dt and the "flows of matter" $d_a n_k/dt$, which pass from the surroundings into the homogeneous system under consideration. It can be positive, negative, or zero (cf. §1–24). In closed systems, the terms with the mass flows drop out. The plausible form of Eq. (11) for the case of open systems is a result of the definition of heat chosen in §1–7 leading to Eq. (4).

The entropy production (12) relates to the chemical reactions (or relaxation phenomena) which occur inside the phase. It is never negative but is always positive for actual completion of the processes.

The *dissipation function* Ψ has, according to Eq. (1–24.7) and Eq. (12), the form

$$\Psi = T\Theta = \sum_r w_r A_r \geqslant 0, \qquad (2\text{-}2.13)$$

and represents thus a special case of the relation (1–24.16) or the still more general formula (1–24.22). The reaction rates w_r are the "generalized fluxes" and the affinities A_r are the appropriate "generalized forces" (cf. §1–24).

In this chapter, we will proceed from the relation (13), which is identical with "de Donder's inequality" (cf. Eq. 1–24.19).

2-3 REACTION RATES AND AFFINITIES

Let there be R independent chemical reactions (or relaxation phenomena) possible in the phase under consideration. Then the inequality of de Donder (2–2.13) states for the reaction rates $w_1, w_2, \ldots, w_R$ and the affinities $A_1, A_2, \ldots, A_R$:

$$\sum_{r=1}^{R} w_r A_r \geqslant 0, \qquad (2\text{-}3.1)$$

where the inequality is valid for an actual (irreversible) path of the processes and the equality for the reversible limiting case.

From the definition of equilibrium (§1–17) and the general equilibrium conditions for homogeneous systems (§1–19) all reaction rates and affinities of the independent reactions must disappear at *chemical equilibrium*:

$$w_r = 0, \quad A_r = 0 \quad (r = 1, 2, \ldots, R). \qquad (2\text{-}3.2)$$

In this case, the equality sign is valid in Eq. (1).

A "reversible limiting case" would, however, also be present in a formal sense if in (1) the conditions

$$w_r = 0, \quad A_r \neq 0$$

or

$$w_r \neq 0, \quad A_r = 0$$

were satisfied. The first possibility corresponds to "inhibited reactions" or "false equilibria", as they are often observed in the absence of catalysts.

The second possibility corresponds to certain abstract experiments in traditional thermodynamics in which changes at "chemical equilibrium" are considered. Such changes cannot be realized for, at disappearing affinities, the reaction rates cannot be different from zero, just as the "infinitely slow processes" have no place at all in our considerations. The only case of interest to us, for which the equality in (1) is valid, is that of chemical equilibrium, described by Eq. (2).

If the inequality is valid in (1) then we have to deal with an *actual path* of the reactions:

$$w_r \neq 0, \quad A_r \neq 0 \quad (r = 1, 2, \ldots, R), \quad \sum_{r=1}^{R} w_r A_r > 0. \qquad (2\text{–}3.3)$$

For a single reaction, in accordance with (3), the reaction rate w and affinity A always have the same sign:

$$wA > 0. \qquad (2\text{–}3.4)$$

If two independent reactions occur simultaneously then the inequality (3) states:

$$w_1 A_1 + w_2 A_2 > 0. \qquad (2\text{–}3.5)$$

The case for which

$$w_1 A_1 < 0 \qquad (2\text{–}3.6)$$

can appear and, naturally, according to (5), the conditions

$$w_2 A_2 > 0, \qquad (2\text{–}3.7)$$

$$w_2 A_2 > |w_1 A_1| \qquad (2\text{–}3.8)$$

must be satisfied.

In §2–4, it will be proved—for near-equilibrium situations—that case (6) can appear if the two reactions are "coupled". We will clarify the concept of "coupling" of two reactions with a simple reaction sequence (cf. §2–5 and §2–7), i.e.

$$L \rightleftharpoons M, \qquad (2\text{–}3.9)$$

$$M \rightleftharpoons N. \qquad (2\text{–}3.10)$$

If the direct transition between the molecular species L and N, that is, the conversion reaction

$$L \rightleftharpoons N, \qquad (2\text{–}3.11)$$

is impossible then the number of the elementary reactions is the same as the number of the linearly independent reaction equations, so that the reactions (9) and (10) are not, by definition, coupled to one another although the molecules of type M appear in both reactions. If, however, the conversion is possible and the number of elementary reactions (three) surpasses those of the linearly independent reaction equations (two) then, by definition, "coupling" of the two reactions (9) and (10) is present. Further treatment of coupled reactions follows in §2–7 and §2–8.

2-4 PHENOMENOLOGICAL EQUATIONS AND ONSAGER'S RECIPROCITY LAW

Since, according to Eq. (2–3.2), all reaction rates $w_1, w_2, \ldots, w_R$ and all affinities $A_1, A_2, \ldots, A_R$ for R independent reactions in an arbitrary phase disappear at equilibrium, it is reasonable to establish homogeneous linear relations between reaction rates and affinities for the actual course of the reaction *not far from equilibrium*[1]:

$$w_r = \sum_{s=1}^{R} a_{rs} A_s \quad (r = 1, 2, \ldots, R). \tag{2-4.1}$$

These are the *phenomenological equations* in the sense of Eq. (1–25.1) with the *phenomenological coefficients* $a_{11}, a_{12}, a_{21}, \ldots, a_{RR}$, the meaning of which will be explained by examples presented in the following sections.

For $r \neq s$ the statement $a_{rs} \neq 0$ signifies a "coupling" of the reactions r and s (cf. §2–7 and §2–8). For such coupled reactions, the *reciprocity law of Onsager* (1–26.1) is valid:

$$a_{rs} = a_{sr} \quad (r, s = 1, 2, \ldots, R). \tag{2-4.2}$$

These reciprocity relations will be discussed in detail.

The dissipation function Ψ, that is, the product of the absolute temperature T and the entropy production Θ, results from Eq. (2–2.13) and Eq. (1) for the actual path of the reactions in the neighborhood of equilibrium (cf. Eq. 2–3.3):

$$\Psi = T\Theta = \sum_{r=1}^{R} \sum_{s=1}^{R} a_{rs} A_r A_s > 0. \tag{2-4.3}$$

This expression is a symmetric, positive quadratic due to the reciprocity relations (2). It follows then (cf. Eqs. 1–26.5 and 1–26.6):

$$a_{rr} > 0, \quad a_{rr} a_{ss} - a_{rs}^2 > 0, \quad \ldots (r, s = 1, 2, \ldots, R, r \neq s) \tag{2-4.4}$$

$$\begin{vmatrix} a_{11} & a_{12} & \cdots & a_{1R} \\ a_{21} & a_{22} & \cdots & a_{2R} \\ \cdot & \cdot & \cdot & \cdot \\ a_{R1} & a_{R2} & \cdots & a_{RR} \end{vmatrix} > 0. \tag{2-4.5}$$

Thus the determinant of the phenomenological coefficients with the corresponding principal cofactors is positive.

For two independent reactions $R = 2$, we obtain from Eqs. (1) to (5):

$$\left. \begin{array}{l} w_1 = a_{11}A_1 + a_{12}A_2, \\ w_2 = a_{21}A_1 + a_{22}A_2, \end{array} \right\} \tag{2-4.6}$$

$$a_{12} = a_{21}, \tag{2-4.7}$$

$$\Psi = a_{11}A_1^2 + 2a_{12}A_1A_2 + a_{22}A_2^2 > 0, \tag{2-4.8}$$

$$a_{11} > 0, \quad a_{22} > 0, \quad a_{11}a_{22} - a_{12}^2 > 0. \tag{2-4.9}$$

The last equations are a special case of relations (1–26.7).

For $a_{12} = a_{21} = 0$ (no coupling) it follows from (6) and (9) that

$$w_1A_1 = a_{11}A_1^2 > 0, \qquad w_2A_2 = a_{22}A_2^2 > 0,$$

so that case (2–3.6) is impossible. For $a_{12} = a_{21} \neq 0$ (coupling of the two reactions) we find from (6), (7), and (9):

$$w_1A_1 = a_{11}A_1^2 + a_{12}A_1A_2,$$
$$w_2A_2 = a_{12}A_1A_2 + a_{22}A_2^2.$$

Here, therefore, case (2–3.6) is possible.

2-5 REGION OF VALIDITY OF THE PHENOMENOLOGICAL EQUATIONS

To specify the region of validity of the phenomenological equations we consider the simplest case: a single chemical reaction.[16]

From Eq. (2–4.1) holding for states near equilibrium we obtain a phenomenological equation which relates the reaction rate w to the affinity A in the form of a linear relation:

$$w = aA, \tag{2-5.1}$$

where the phenomenological coefficient must be

$$a > 0 \tag{2-5.2}$$

in accordance with (2–4.4). We now show, using general considerations and particular examples, under which conditions Eq. (1) is valid and the significance of the coefficient a.

We begin with a general consideration[2]. The reaction rate w of the overall reaction, the mechanism of which need not be known, will depend on several macroscopic variables, e.g. temperature, pressure, concentrations of the reacting species, amounts and compositions of catalysts. If we denote these variables by $x_1, x_2, \ldots, x_n, x_{n+1}$ then:

$$w = w(x_1, x_2, \ldots, x_n, x_{n+1}).$$

On the other hand, the affinity A of the reaction is also dependent on some of these variables.† We thus write:

$$A = A(x_1, \ldots, x_n, x_{n+1}).$$

By elimination of x_{n+1} it follows that

$$w = f(x_1, x_2, \ldots, x_n, A). \tag{2–5.3}$$

If $\bar{x}_i$ is the equilibrium value of x_i then, according to Eq. (2–3.2), the general equilibrium condition

$$w = f(A, x_1, x_2, \ldots, x_n) = 0 \quad \text{for } A = 0, \; x_i = \bar{x}_i \quad (i = 1, 2, \ldots, n) \tag{2–5.4}$$

must be satisfied. Since the function A is already a measure of the distance from equilibrium, we will set up the following series expression

$$w = \left(\frac{\partial f}{\partial A}\right)_{x_i = \bar{x}_i} A + \cdots \quad (i = 1, 2, \ldots, n). \tag{2–5.5}$$

"Near to equilibrium", that is, for small values of A, we may ignore the terms in the power series after the linear term in A. We thus obtain a phenomenological law of the form (1)

$$w = aA$$

with

$$a \equiv \left(\frac{\partial f}{\partial A}\right)_{x_i = \bar{x}_i}. \tag{2–5.6}$$

Thus the phenomenological coefficient a depends first on the n quantities $\bar{x}_1, \ldots, \bar{x}_n$. Now, however, the equilibrium condition $A = 0$ results in a relation between the temperature, pressure, and equilibrium concentrations. Therefore a is a function of $n - 1$ variables. If, during a series of tests, at least $n - 1$ of the quantities $x_1, x_2, \ldots, x_n, x_{n+1}$ are held constant, these originally appearing as macroscopic variables, then at most two quantities can be varied and the relation $w = aA$ with constant coefficient a holds in the vicinity of equilibrium.

The exact meaning of "near to equilibrium" and the physical meaning of the phenomenological coefficient a for a given overall reaction will become clearer if the pertinent elementary reactions are formulated—where,

---

† In the most general case, A is a function of the temperature, pressure, and of the concentrations of the species present. In the special case of an ideal gas mixture or an ideal dilute solution, A depends on the temperature, the pressure (for ideal dilute solutions), and the concentrations of the reacting substances (see Eq. 2–5.12).

to begin with, we allow only *one* elementary reaction—and if the considerations are restricted to *ideal gas mixtures* or *ideal dilute solutions*. The latter assumption permits application of a particular reaction kinetic law, "the kinetic law of mass action", which is valid for any arbitrary distance from equilibrium.[17] Also, the chemical potential μ_k of species k assumes an analytically simple form:

$$\mu_k = \mu_k^{\ominus} + RT \ln \frac{c_k}{c^{\dagger}}. \tag{2-5.7}$$

Here $\mu_k^{\ominus}$ is a standard value of the chemical potential, R the gas constant, T the absolute temperature, c_k the molarity of species k, and $c^{\dagger}$ a standard concentration ($c^{\dagger} = 1 \text{ mol l}^{-1}$). The quantity $\mu_k^{\ominus}$ depends, for ideal gas mixtures, only on the temperature, while for ideal dilute solutions also on the pressure. In the latter case, Eq. (7) is valid for all solute species (in what follows we exclude solution reactions in which the solvent plays a part).†

We consider an elementary reaction of the general form

$$\sum_i \nu_i[\text{M}_i] \underset{\kappa'}{\overset{\kappa}{\rightleftharpoons}} \sum_j \nu_j[\text{M}_j], \tag{2-5.8}$$

where the symbols $[\text{M}_i]$ and $[\text{M}_j]$ denote the various reacting species. The numbers ν_i and ν_j are equal to the negative and positive respectively of the stoichiometric numbers ν_k which, by definition, are to be counted negative or positive according to whether they appear on the left or the right side of the reaction equation respectively. We find, then, the relation:

$$\prod_k c_k^{\nu_k} = \prod_j c_j^{\nu_j} \Big/ \prod_i c_i^{\nu_i}. \tag{2-5.9}$$

The symbol κ in Eq. (8) denotes the "rate constant" for the "forward reaction," the symbol κ' denotes the "rate constant" for the "reverse reaction". Both quantities are positive, depending only on temperature for ideal gas mixtures, but on temperature and pressure for ideal dilute solutions, in addition to the amounts and concentrations of catalysts when they are present.

If we regard the volume V of the system as given and if we consider the constant factor V included in κ and κ' then we may use w instead of the conventional "reaction rate" w/V in classical reaction kinetics (see footnote to p. 106). We can thus write the "kinetic law of mass action" for the reaction

---

† In §1–19 we introduced the standard value $\mu_k^{\ominus}$ valid for the molality scale. Here we use the symbol $\mu_k^{\ominus}$ to denote a standard value in the molarity scale.

rate w of the chemical reaction (8), considering Eq. (9), in the following form:

$$w = \kappa \prod_i c_i^{\nu_i} - \kappa' \prod_j c_j^{\nu_j} = \omega\left(1 - \lambda \prod_k c_k^{\nu_k}\right) \qquad (2\text{-}5.10)$$

with

$$\omega \equiv \kappa \prod_i c_i^{\nu_i}, \qquad \lambda \equiv \frac{\kappa'}{\kappa}. \qquad (2\text{-}5.11)$$

The term ω is the rate of the forward reaction in (8).

The affinity A of the reaction (8) can be calculated from Eq. (2-2.6) and Eq. (7):

$$A = -\sum_k \nu_k \mu_k = RT \ln \frac{K_c(c^\dagger)^\nu}{\prod_k c_k^{\nu_k}}, \qquad (2\text{-}5.12)$$

with

$$\nu \equiv \sum_k \nu_k, \qquad RT \ln K_c \equiv -\sum_k \nu_k \mu_k^\square. \qquad (2\text{-}5.13)$$

Insertion of the equilibrium condition $A = 0$ into Eq. (12) produces the "classical law of mass action" for chemical equilibrium:

$$\prod_k \bar{c}_k^{\nu_k}(c^\dagger)^{-\nu} = K_c, \qquad (2\text{-}5.14)$$

where $\bar{c}_k$ is the equilibrium molarity of the species k, and K_c the "equilibrium constant."† For ideal gas mixtures, K_c depends only on the temperature, while for ideal dilute solutions it depends on the temperature and on the pressure according to Eq. (13).

---

†For *ideal gas mixtures*, we have:

$$p_k = c_k RT = Px_k, \qquad (2\text{-}5.14a)$$

where p_k and x_k are the partial pressure and mole fraction respectively of species k, and P the total pressure. With the definitions

$$K_c\left(\frac{c^\dagger RT}{P^\dagger}\right)^\nu \equiv K_p, \qquad K_c\left(\frac{c^\dagger RT}{P}\right)^\nu \equiv K_x, \qquad (2\text{-}5.14b)$$

where $P^\dagger$ denotes a standard pressure ($P^\dagger = 1$ atm), there follows from Eqs. (14) and (14a):

$$\prod_k \bar{p}_k^{\nu_k}(P^\dagger)^{-\nu} = K_p, \qquad \prod_k \bar{x}_k^{\nu_k} = K_x. \qquad (2\text{-}5.14c)$$

Here the $\bar{p}_k$ and $\bar{x}_k$ are the equilibrium values of p_k and x_k respectively. The quantities K_p and K_x are similarly called "equilibrium constants." All three constants K_c, K_p, K_x are dimensionless. While K_c and K_p depend only on the temperature T, K_x is a function of the temperature T and the pressure P.

By combination of Eq. (10) with Eq. (12) we obtain the general relation between the rate of reaction and the affinity for ideal gas mixtures and ideal dilute solutions:

$$w = \omega \left[1 - \lambda K_c(c^\dagger)^\nu \exp\left(-\frac{A}{RT} \right) \right].$$
(2–5.15)

This holds for any distance from equilibrium.

With the equilibrium condition $w = 0$ it follows from Eqs. (10), (11), and (14) that

$$\frac{\kappa}{\kappa'} = \frac{1}{\lambda} = K_c(c^\dagger)^\nu.$$
(2–5.16)

In the case of a single elementary reaction, this relation between the rate constants and the equilibrium constant is a consequence of the kinetic law of mass action and the thermodynamic conditions of equilibrium (cf. §1–28).

From Eqs. (15) and (16) we find the simple expression:

$$w = \omega \left[1 - \exp\left(-\frac{A}{RT} \right) \right],$$
(2–5.17)

which, like Eq. (15), is valid for any distance from equilibrium.

In Eq. (17), we now require *proximity to equilibrium*, which we define by the condition:

$$\left| \frac{A}{RT} \right| \ll 1.$$
(2–5.18)

Then, on expansion of the exponential function:

$$1 - e^{-A/RT} = \frac{A}{RT} - \frac{1}{2!}\left(\frac{A}{RT} \right)^2 + \frac{1}{3!}\left(\frac{A}{RT} \right)^3 \cdots$$

In the case of *ideal dilute solutions*, for each species k which belongs to the solutes we find that

$$c_k = \frac{x_k}{V_{01}} = \frac{M_1}{V_{01}} m_k = \rho_{01} m_k.$$
(2–5.14d)

Here m_k is the molality of species k (equilibrium value: $\bar{m}_k$), V_{01} the molar volume, M_1 the molar mass, and ρ_{01} the density of pure solvent (cf. Eq. 1–19.12). Using the definitions

$$K_c(c^\dagger V_{01})^\nu \equiv K_x, \qquad K_c\left(\frac{c^\dagger}{m^\dagger \rho_{01}} \right)^\nu \equiv K_m,$$
(2–5.14e)

we obtain from Eqs. (14) and (14d)

$$\prod \bar{x}_k^{\nu_k} = K_x, \qquad \prod \bar{m}_k^{\nu_k}(m^!)^{-\nu} = K_m.$$
(2–5.14f)

$m^\dagger$ denotes a standard molality (1 mol kg^{-1}); K_m is the "equilibrium constant in the m-scale." It is used particularly for electrolyte solutions (cf. p. 57). The three dimensionless quantities (K_c, K_x, and K_m) here depend on T and P.

we can ignore the terms in the series after the first term and write Eq. (17) in the form

$$w = \omega \frac{A}{RT}.$$

Since this corresponds to a truncation of a Taylor expansion for the function $w(A)$ after the first term, the coefficient ω refers to the point $w = 0$, $A = 0$ (equilibrium) and will accordingly be denoted by $\bar{\omega}$. According to Eq. (11):

$$\bar{\omega} = \kappa \prod_i \bar{c}_i{}^{\nu_i}. \tag{2-5.19}$$

We obtain therefore in place of Eq. (17):

$$w = \frac{\bar{\omega}}{RT} A. \tag{2-5.20}$$

This formula is only valid for reactions in ideal gas mixtures and ideal dilute solutions close to equilibrium as established by (18). It represents a phenomenological equation in the sense of Eq. (1), the phenomenological coefficient being given by

$$a = \frac{\bar{\omega}}{RT}. \tag{2-5.21}$$

The phenomenological coefficient a is always positive according to Eqs. (19) and (21) and in agreement with Eq. (2). Further, a depends, as has already been shown by Eq. (6), on a smaller number of variables than the reaction rate w. According to Eqs. (19) and (21), a is a function of the temperature T, of the pressure (for ideal dilute solutions), of the amounts and compositions of the catalysts if they are present, as well as of the equilibrium concentrations $\bar{c}_i$ of the species on the left side of the reaction equation (8), while w, according to Eqs. (10), (12), and (20), depends on the *instantaneous* concentrations c_k of *all* substances instead of on $\bar{c}_i$.

For *several reactions*, which take place simultaneously in a phase, we must distinguish between two cases.

In the first case, the number of elementary reactions is equal to those of the linearly independent reactions. Then we can write the relations (10) to (21) for each single elementary reaction where only the quantities w, κ, κ', A, K, etc., are to be distinguished by inferior labeling. Here we are dealing with reactions without "coupling" (cf. §2–3).

In the second case, the number of elementary reactions exceeds those of the linearly independent reactions, so that formally certain reaction equations can be derived from a linear combination of other reaction equations (cf. §2–3). Then the relations (16) and (17) cannot be obtained from the kinetic law of mass action and the thermodynamic equilibrium conditions only, and we are dealing with reactions with "coupling." We exclude these from our considerations in this section.

We now discuss some *reaction sequences* which can be represented by linearly independent reaction equations and thus are uncoupled.

The simplest example is the reaction sequence discussed in §2–3 (2–3.9 and 2–3.10)

$$L \rightleftharpoons M \text{ (reaction 1)},$$
$$M \rightleftharpoons N \text{ (reaction 2)},$$

$$(2-5.22)$$

which corresponds to two rearrangement reactions occurring simultaneously.†
Under the conditions (cf. Eq. 18)

$$\left|\frac{A_1}{RT}\right| \ll 1, \quad \left|\frac{A_2}{RT}\right| \ll 1 \tag{2-5.23}$$

the phenomenological statements (cf. Eqs. 20 and 21)

$$w_1 = a_{11}A_1, \quad w_2 = a_{22}A_2 \tag{2-5.24}$$

hold with the phenomenological coefficients

$$a_{11} = \frac{\bar{\omega}_1}{RT}, \quad a_{22} = \frac{\bar{\omega}_2}{RT}, \tag{2-5.25}$$

where $\bar{\omega}_1$ and $\bar{\omega}_2$ are the equilibrium values of the reaction rates of the reaction steps $L \rightarrow M$ and $M \rightarrow N$ respectively. By comparing Eq. (24) with Eq. (2–4.6) we recognize that here there is no "coupling" of the two reactions present ($a_{12} = a_{21} = 0$). The situation immediately becomes more complicated if we allow the conversion ($L \rightleftharpoons N$) which is forbidden in (22). This will be discussed in §2–7.

---

† The isomerization reactions

$$o\text{-xylene} \underset{\kappa_1'}{\overset{\kappa_1}{\rightleftharpoons}} m\text{-xylene},$$

$$m\text{-xylene} \underset{\kappa_2'}{\overset{\kappa_2}{\rightleftharpoons}} p\text{-xylene},$$

representing an example of the consecutive reactions (22), were quantitatively examined for dilute solutions of xylene in toluene at 50°C (with aluminum chloride as a catalyst in the presence of HCl) by Allen and Yats[3]. There it was shown that the direct conversion o-xylene $\rightleftharpoons$ p-xylene does not take place and the rate constants have the following relative values:

$$\frac{\kappa_1}{\kappa_1'} = 3.6,$$

$$\frac{\kappa_2}{\kappa_1'} = 2.1, \quad \frac{\kappa_2'}{\kappa_1'} = 6.0.$$

In a similar way, consider the reaction sequence [4]

$$\left.\begin{array}{l} L \rightleftharpoons M \text{ (reaction 1),} \\ M \rightleftharpoons N \text{ (reaction 2),} \\ \cdot \quad \cdot \quad \cdot \quad \cdot \quad \cdot \quad \cdot \quad \cdot \\ Y \rightleftharpoons Z \text{ (reaction } R) \end{array}\right\} \tag{2-5.26}$$

with the assumption of conditions

$$\left|\frac{A_r}{RT}\right| \ll 1 \quad (r = 1, 2, \ldots, R) \tag{2-5.27}$$

and the phenomenological equations:

$$w_r = a_{rr}A_r \quad (r = 1, 2, \ldots, R), \tag{2-5.28}$$

corresponding to the relations (2–4.1) with $a_{rs} = 0$ for $r \neq s$ (no coupling). If the single steps in (26) are not known then it is expedient to introduce for the overall reaction

$$L \rightleftharpoons Z \tag{2-5.29}$$

an affinity A and a reaction rate w. From the definition of affinities in Eq. (2–2.6) it follows directly that

$$A = A_1 + A_2 + \cdots + A_R. \tag{2-5.30}$$

Further, if the intermediate products M, N, ..., Y are unstable then, following standard practice in reaction kinetics, we assume a quasistationary state:

$$w_1 = w_2 = \cdots = w_R = w. \tag{2-5.31}$$

From Eqs. (28) and (31) there results:

$$w = a_{11}A_1 = a_{22}A_2 = \cdots = a_{RR}A_R. \tag{2-5.32}$$

From Eqs. (30) and (32) we obtain:

$$w = aA \tag{2-5.33}$$

with

$$\frac{1}{a} \equiv \sum_{r=1}^{R} \frac{1}{a_{rr}}. \tag{2-5.34}$$

According to Eqs. (30) and (33), the measured reaction rate w can be proportional to the affinity A of the total reaction (29) without the condition

$$\left|\frac{A}{RT}\right| \ll 1 \tag{2-5.35}$$

having to be fulfilled, provided the assumptions (27) are valid.

2–6 EXPERIMENTAL EXAMPLE

The gas phase reaction

$$C_6H_{12} \rightleftharpoons C_6H_6 + 3H_2 \tag{2–6.1}$$

which is the dehydration of cyclohexane or the hydration of benzene in the gas phase has been experimentally and theoretically examined by Prigogine, Outer, and Herbo[2].

The reaction (1) can be assumed to take place in an ideal gas phase. The reaction rate w was measured by the authors using a flow reactor in terms of the rate of formation of benzene (M_2w, M_2 = molar mass of benzene). The affinity A was calculated from the experimentally determined equilibrium constant following Eqs. (2–5.12), (2–5.14a), and (2–5.14b) as:

$$\frac{A}{RT} = \ln \frac{K_c(c^\dagger)^3 c_1}{c_2 c_3^3} = \ln \frac{K_p(P^\dagger)^3 p_1}{p_2 p_3^3}. \tag{2–6.2}$$

Here R is the gas constant, T the absolute temperature, $P^\dagger$ the standard pressure, $c^\dagger$ the standard molarity, c_k the molarity of species k,

$$p_k = c_k RT \quad (k = 1, 2, 3) \tag{2–6.3}$$

the partial pressure of k. For the equilibrium constants in the c-scale and p-scale, K_c and K_p respectively, the relation

$$K_p = K_c \left(\frac{c^\dagger RT}{P^\dagger} \right)^3 \tag{2–6.4}$$

holds. The inferior 1, 2, or 3 refers to cyclohexane, benzene, or hydrogen respectively.

As is shown in Table 2 and in Figs. 2 and 3, the phenomenological equation

$$w = aA \tag{2–6.5}$$

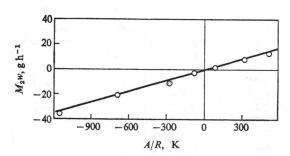

Fig. 2. See text.

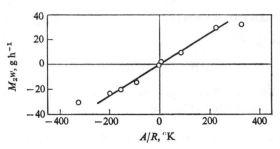

Fig. 3. See text above.

is valid with constant phenomenological coefficient a, without the condition

$$\left|\frac{A}{RT}\right| \ll 1 \tag{2-6.6}$$

being satisfied in each case. Noting the comment at the end of §2–5, this result corresponds to a reaction sequence of the scheme (2-5.26) under

Table 2

Reaction Rate (w) and Affinity (A) of the Reaction (1)

1. Catalyst: 50% Ni, 25% ZnO, 23% Cr_2O_3 (mole percent)
   Amount: 200 mg   Graphical presentation in Fig. 2.

p_1 atm	p_2 atm	p_3 atm	T K	A/R K	M_2w g h^{-1}
0.0262	0.0977	0.925	500	−1150	−36
			524	−686	−21
			545	−271	−11.5
			554	−85	−3
			561	+84	+1
			571	+323	+8
			582	+526	+13

2. Catalyst 50% Ni, 10% ZnO, 40% Cr_2O_3 (mole percent)
   Amount: 175 mg   Graphical presentation in Fig. 3

0.073	0.179	0.792	548	−197	−24
0.048	0.204	0.790		−323	−31
0.195	0.051	0.799		+328	+32
0.145	0.107	0.796		+86	+9
0.097	0.154	0.792		−92	−15
0.121	0.128	0.797		0	−1
0.017	0.077	0.786		+227	+29
0.073	0.179	0.781		−154	−21
0.121	0.128	0.782		+1	+1

conditions (2–5.27) and (2–5.31). In the series of experiments appertaining to Fig. 2, only the temperature of all the macroscopic variables was changed (see Table 2). In Fig. 3, on the other hand, a set of experiments is presented in which three quantities, namely p_1, p_2, and p_3, are varied (see Table 2). The coefficient a in Eq. (5) is constant in accordance with the statements on p. 112, if only the temperature or, at given temperature, only two concentrations are changed. These conditions are satisfied strictly in the data of Fig. 2 and approximately in those of Fig. 3 since p_3 varies very little.

2–7 COUPLING OF TWO REACTIONS

Consider the reciprocal conversion of three substances L, M, and N:

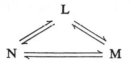

This "unimolecular triangular reaction" can be described in two different ways: either by the two *linearly independent reactions*

$$\left. \begin{array}{l} \text{L} \rightleftharpoons \text{M} \quad \text{(reaction 1),} \\ \text{M} \rightleftharpoons \text{N} \quad \text{(reaction 2)} \end{array} \right\} \qquad (2\text{–}7.1)$$

or by the three linearly dependent *elementary reactions*

$$\left. \begin{array}{l} \text{L} \underset{\kappa_{\text{I}}'}{\overset{\kappa_{\text{I}}}{\rightleftharpoons}} \text{M} \quad \text{(reaction I),} \\[2mm] \text{M} \underset{\kappa_{\text{II}}'}{\overset{\kappa_{\text{II}}}{\rightleftharpoons}} \text{N} \quad \text{(reaction II),} \\[2mm] \text{N} \underset{\kappa_{\text{III}}'}{\overset{\kappa_{\text{III}}}{\rightleftharpoons}} \text{L} \quad \text{(reaction III).} \end{array} \right\} \qquad (2\text{–}7.2)$$

The first reaction scheme corresponds to a purely macroscopic description in which only the overall process is described by reaction equations which are mathematically independent of each other. The second reaction scheme (in which the rate constants κ_{I}, κ_{I}', etc., are introduced for the individual reaction steps, cf. §2–5) depends on the reaction mechanism, i.e. the actual course of the chemical conversion. Here we deal with the *coupling* of two reactions.

We next treat this reaction sequence exactly as we did for reactions without coupling, following §2–5: assuming ideal gas mixtures or ideal dilute solutions and applying both the equilibrium conditions of classical thermodynamics and the statements of classical reaction kinetics. For simplicity, we assume a homogeneous closed system of constant volume. We begin our discussion with a few general considerations.

Let n_k be the amount of species k, t the time, w_1 and w_2 reaction rates of reactions 1 and 2 respectively in scheme (1), w_I, w_{II}, and w_{III} reaction rates of the elementary reactions I, II, and III respectively in scheme (2). Then, in accordance with Eq. (2–2.3), for our closed system we write:

$$\left.\begin{aligned}
-\frac{dn_L}{dt} &= w_1 = w_I - w_{III}, \\
-\frac{dn_M}{dt} &= w_2 - w_1 = w_{II} - w_I, \\
-\frac{dn_N}{dt} &= -w_2 = w_{III} - w_{II}.
\end{aligned}\right\} \tag{2–7.3}$$

From this there follows:

$$w_1 = w_I - w_{III}, \quad w_2 = w_{II} - w_{III}. \tag{2–7.4}$$

If we denote the chemical potential of species k by μ_k, the affinity of the overall reaction 1 or 2 in scheme (1) by A_1 or A_2 respectively, and the affinities of the elementary reactions I, II, and III in scheme (2) by A_I, A_{II}, and A_{III} respectively, then we derive from Eq. (2–2.6)

$$\left.\begin{aligned}
\mu_L - \mu_M &= A_1 = A_I, \\
\mu_M - \mu_N &= A_2 = A_{II}, \\
\mu_N - \mu_L &= -(A_1 + A_2) = A_{III}.
\end{aligned}\right\} \tag{2–7.5}$$

With this we find:

$$A_I = A_1, \quad A_{II} = A_2, \quad A_{III} = -A_1 - A_2. \tag{2–7.6}$$

The "fluxes" ascribed to reaction scheme (1) (reaction rates) and "forces" (affinities) are the independent quantities

$$w_1, w_2 \quad \text{and} \quad A_1, A_2.$$

For reaction scheme (2), the linearly dependent fluxes and forces can be written thus:

$$w_I, w_{II}, w_{III} \quad \text{and} \quad A_I, A_{II}, A_{III}.$$

The dissipation function Ψ remains invariant for the transition from one set of fluxes and forces to another set (cf. §1–27). This results explicitly from Eq. (2–2.13) on consideration of the relations (4) and (6):

$$\Psi = w_1 A_1 + w_2 A_2 = w_I A_I + w_{II} A_{II} + w_{III} A_{III}.$$

At *chemical equilibrium* the principles of thermodynamics require that reaction rates and affinities of all linearly *independent* chemical reactions disappear (cf. §1–19, §2–3, and §2–5):

$$w_1 = 0, \quad w_2 = 0 \quad \text{(equilibrium)}, \tag{2-7.7}$$

$$A_1 = 0, \quad A_2 = 0 \quad \text{(equilibrium)}, \tag{2-7.8}$$

from which it follows, with Eqs. (4) and (6) respectively, that:

$$w_{\text{I}} = w_{\text{II}} = w_{\text{III}} \quad \text{(equilibrium)}, \tag{2-7.9}$$

$$A_{\text{I}} = 0, \quad A_{\text{II}} = 0, \quad A_{\text{III}} = 0 \quad \text{(equilibrium)}. \tag{2-7.10}$$

Conditions (7) and (9) require, according to Eq. (3), that at equilibrium the amounts n_{L}, n_{M}, and n_{N} be constant. Since we have assumed a given volume, this requires time invariance of the concentrations of the species L, M, and N. It is significant that the rates w_{I}, w_{II}, and w_{III} of the three elementary reactions do not need to disappear individually in order to satisfy the thermodynamic equilibrium conditions. Should we wish to set up phenomenological equations for the near-equilibrium case at this stage, then the choice of reaction rates or affinities of the three elementary reactions as fluxes or forces respectively would be forbidden, because both sets of variables are linearly dependent (cf. §1–27).

Up to this point our consideration of the triangular reaction in a closed homogeneous system of constant volume has been quite general. We now assume that the system represents an ideal gas mixture or an ideal dilute solution. Then both the classical law of mass action for chemical equilibrium and the kinetic law of mass action for the rates of the elementary reactions hold.

First, with the help of Eq. (2–5.7) from Eq. (5) for the affinity of the three elementary reactions (cf. Eqs. 2–5.12 and 2–5.13) we find:

$$\left.\begin{aligned}
A_{\text{I}} &= RT \ln \left(K_{\text{I}} \frac{c_{\text{L}}}{c_{\text{M}}} \right), \\[6pt]
A_{\text{II}} &= RT \ln \left(K_{\text{II}} \frac{c_{\text{M}}}{c_{\text{N}}} \right), \\[6pt]
A_{\text{III}} &= RT \ln \left(K_{\text{III}} \frac{c_{\text{N}}}{c_{\text{L}}} \right),
\end{aligned}\right\} \tag{2-7.11}$$

with the abbreviation

$$\left.\begin{aligned}
RT \ln K_{\text{I}} &\equiv \mu_{\text{L}}^{\ominus} - \mu_{\text{M}}^{\ominus}, \\
RT \ln K_{\text{II}} &\equiv \mu_{\text{M}}^{\ominus} - \mu_{\text{N}}^{\ominus}, \\
RT \ln K_{\text{III}} &\equiv \mu_{\text{N}}^{\ominus} - \mu_{\text{L}}^{\ominus}.
\end{aligned}\right\} \tag{2-7.12}$$

Here R is the gas constant, T the absolute temperature, c_k the molarity of species k, and $\mu_k^\ominus$ the standard value of the chemical potential of particle species k. The $\mu_k^\ominus$ depend, just as the K defined by Eq. (12), on the temperature (for ideal gas mixtures) or on the temperature and on the pressure (for ideal dilute solutions).

By insertion of the equilibrium conditions (10) into the relations (11) we find the *classical law of mass action* for each of the three elementary reactions (cf. Eq. 2-5.14):

$$
\left.
\begin{aligned}
\frac{\bar{c}_M}{\bar{c}_L} &= K_I, \\[2mm]
\frac{\bar{c}_N}{\bar{c}_M} &= K_{II}, \\[2mm]
\frac{\bar{c}_L}{\bar{c}_N} &= K_{III},
\end{aligned}
\right\}
\tag{2-7.13}
$$

where $\bar{c}_k$ is the equilibrium concentration of species k. We recognize from this that K_I, K_{II}, and K_{III} represent the three equilibrium constants (on the c-scale) of the three elementary reactions I, II, and III. They are not independent of one another but are connected by Eq. (13) as follows:

$$
K_I K_{II} K_{III} = 1.
\tag{2-7.14}
$$

Further, for our closed system of constant volume, due to conservation of mass we write

$$
c_L + c_M + c_N = \bar{c}_L + \bar{c}_M + \bar{c}_N.
\tag{2-7.15}
$$

With this we have three independent equations available, which determine the three equilibrium concentrations $\bar{c}_L$, $\bar{c}_M$, and $\bar{c}_N$.

Application of the *kinetic law of mass action* to the rates w_I, w_{II}, and w_{III} of the three elementary reactions I, II, and III for the reaction scheme (2) produces (cf. Eqs. 2-5.10 and 2-5.11):

$$
\left.
\begin{aligned}
w_I &= \kappa_I c_L - \kappa_I' c_M = \omega_I\left(1 - \lambda_I \frac{c_M}{c_L}\right), \\[2mm]
w_{II} &= \kappa_{II} c_M - \kappa_{II}' c_N = \omega_{II}\left(1 - \lambda_{II} \frac{c_N}{c_M}\right), \\[2mm]
w_{III} &= \kappa_{III} c_N - \kappa_{III}' c_L = \omega_{III}\left(1 - \lambda_{III} \frac{c_L}{c_N}\right)
\end{aligned}
\right\}
\tag{2-7.16}
$$

with the abbreviations

$$
\omega_I \equiv \kappa_I c_L, \quad \omega_{II} \equiv \kappa_{II} c_M, \quad \omega_{III} \equiv \kappa_{III} c_N,
\tag{2-7.17}
$$

$$
\lambda_I \equiv \frac{\kappa_I'}{\kappa_I}, \quad \lambda_{II} \equiv \frac{\kappa_{II}'}{\kappa_{II}}, \quad \lambda_{III} \equiv \frac{\kappa_{III}'}{\kappa_{III}}.
\tag{2-7.18}
$$

Here the κ and κ' are the rate constants of the reactions while the ω are the rates of the three forward reactions in scheme (2). The κ and κ' depend on temperature for ideal gas mixtures and on both temperature and pressure for ideal dilute solutions.

By inserting Eq. (11) into Eq. (16) we obtain the general relation between the reaction rates and the affinities of the three elementary reactions (cf. Eq. 2–5.15):

$$
\left.
\begin{aligned}
w_{\text{I}} &= \omega_{\text{I}}\left[1 - \lambda_{\text{I}}K_{\text{I}}\exp\left(-\frac{A_{\text{I}}}{RT}\right)\right], \\[2mm]
w_{\text{II}} &= \omega_{\text{II}}\left[1 - \lambda_{\text{II}}K_{\text{II}}\exp\left(-\frac{A_{\text{II}}}{RT}\right)\right], \\[2mm]
w_{\text{III}} &= \omega_{\text{III}}\left[1 - \lambda_{\text{III}}K_{\text{III}}\exp\left(-\frac{A_{\text{III}}}{RT}\right)\right].
\end{aligned}
\right\}
\qquad (2\text{-}7.19)
$$

These relations are valid, like the reaction kinetic equations (16), for any distance from equilibrium.

On application of (19) the equilibrium conditions (9) and (10) yield:

$$
\bar{\omega}_{\text{I}}(1 - \lambda_{\text{I}}K_{\text{I}}) = \bar{\omega}_{\text{II}}(1 - \lambda_{\text{II}}K_{\text{II}}) = \bar{\omega}_{\text{III}}(1 - \lambda_{\text{III}}K_{\text{III}}) \qquad (2\text{-}7.20)
$$

with the abbreviations (see Eq. 17)

$$
\bar{\omega}_{\text{I}} \equiv \kappa_{\text{I}}\bar{c}_{\text{L}}, \quad \bar{\omega}_{\text{II}} \equiv \kappa_{\text{II}}\bar{c}_{\text{M}}, \quad \bar{\omega}_{\text{III}} \equiv \kappa_{\text{III}}\bar{c}_{\text{N}} \qquad (2\text{-}7.21)
$$

for the equilibrium values of the rates of the three "forward reactions" in scheme (2). Due to Eqs. (18) and (21), Eq. (20) represents two independent relations between the six rate constants (κ_{I}, κ'_{I}, κ_{II}, κ'_{II}, κ_{III}, and κ'_{III}).

All statements obtainable with the help of the thermodynamic equilibrium conditions and the reaction kinetic laws are contained in the preceding formulas. As is shown by Eq. (9), the possibility of a cyclic reaction for the case of chemical equilibrium remains open. This can occur in the form

$$(2\text{-}7.22)$$

As noted by Onsager[5], such a reaction would contradict none of the relations derived above.

Now a law often considered obvious in classical reaction kinetics but a new law in the framework of our presentation states that *each* elementary reaction at equilibrium occurs exactly as often from left to right as from right to left. This *principle of detailed equilibrium* represents a special case of the "principle of microscopic reversibility" (§1–26).† It leads to the equilibrium condition $w_l = 0$ for each reaction mechanism (w_l = reaction rate of the lth elementary reaction). Thus in our case we find:

$$w_{\mathrm{I}} = 0, \quad w_{\mathrm{II}} = 0, \quad w_{\mathrm{III}} = 0 \quad \text{(equilibrium).} \tag{2–7.23}$$

The principle of detailed equilibrium thus excludes the cyclic conversion (22) and makes statement (9) more precise.

From Eqs. (19) and (23) it follows, on consideration of Eq. (10), that:

$$\lambda_{\mathrm{I}} K_{\mathrm{I}} = \lambda_{\mathrm{II}} K_{\mathrm{II}} = \lambda_{\mathrm{III}} K_{\mathrm{III}} = 1 \tag{2–7.24}$$

or with Eq. (18)

$$\frac{\kappa_{\mathrm{I}}}{\kappa_{\mathrm{I}}'} = K_{\mathrm{I}}, \quad \frac{\kappa_{\mathrm{II}}}{\kappa_{\mathrm{II}}'} = K_{\mathrm{II}}, \quad \frac{\kappa_{\mathrm{III}}}{\kappa_{\mathrm{III}}'} = K_{\mathrm{III}}. \tag{2–7.25}$$

These formulas satisfy Eq. (20) identically and represent three independent relations between the six rate constants.

It is noteworthy that in the case of *coupled* reactions relations of the form (25) can only be obtained with the help of the principle of detailed equilibrium, while they are otherwise a consequence of the thermodynamic equilibrium conditions and of the reaction kinetic statements (cf. p. 115).

From Eqs. (19) and (24) we derive (cf. Eq. 2–5.17):

$$\left. \begin{aligned} w_{\mathrm{I}} &= \omega_{\mathrm{I}}\left[1 - \exp\left(-\frac{A_{\mathrm{I}}}{RT}\right)\right], \\[2mm] w_{\mathrm{II}} &= \omega_{\mathrm{II}}\left[1 - \exp\left(-\frac{A_{\mathrm{II}}}{RT}\right)\right], \\[2mm] w_{\mathrm{III}} &= \omega_{\mathrm{III}}\left[1 - \exp\left(-\frac{A_{\mathrm{III}}}{RT}\right)\right], \end{aligned} \right\} \tag{2–7.26}$$

This relation contains the principle of detailed equilibrium and is again true for any distance from equilibrium.

We now require *proximity to equilibrium* by conditions (cf. Eqs. 2–5.18, 2–5.23, and 2–5.27)

$$\left|\frac{A_{\mathrm{I}}}{RT}\right| \ll 1, \quad \left|\frac{A_{\mathrm{II}}}{RT}\right| \ll 1, \quad \left|\frac{A_{\mathrm{III}}}{RT}\right| \ll 1. \tag{2–7.27}$$

Accordingly, we expand the exponential functions in (26) to the linear terms and notice that Eqs. (10) and (23) require that the reaction rates w_{I}, w_{II}, w_{III} and the affinities A_{I}, A_{II}, A_{III} disappear at equilibrium. In a Taylor expansion

---

† The name "principle of detailed equilibrium" is due to Fowler[6], the term "principle of microscopic reversibility" is due to Tolman[7, 8]. For an application of these principles to complex reactions see Burwell and Pearson[9].

of the functions $w(A)$ about equilibrium, the coefficients ω assume their equilibrium values $\bar{\omega}$ given by Eq. (21). We thus obtain

$$
\left.
\begin{aligned}
w_{\mathrm{I}} &= \frac{\bar{\omega}_{\mathrm{I}}}{RT} A_{\mathrm{I}}, \\[2mm]
w_{\mathrm{II}} &= \frac{\bar{\omega}_{\mathrm{II}}}{RT} A_{\mathrm{II}}, \\[2mm]
w_{\mathrm{III}} &= \frac{\bar{\omega}_{\mathrm{III}}}{RT} A_{\mathrm{III}},
\end{aligned}
\right\}
\tag{2–7.28}
$$

by analogy with Eqs. (2–5.20) and (2–5.24).

Using (28) with the help of Eqs. (4) and (6) for the reaction rates w_1, w_2 and affinities A_1, A_2 of the linearly independent reactions 1 and 2 (overall reactions) in the scheme (1), it follows that:

$$
\left.
\begin{aligned}
w_1 &= \frac{\bar{\omega}_{\mathrm{I}} + \bar{\omega}_{\mathrm{III}}}{RT} A_1 + \frac{\bar{\omega}_{\mathrm{III}}}{RT} A_2, \\[2mm]
w_2 &= \frac{\bar{\omega}_{\mathrm{III}}}{RT} A_1 + \frac{\bar{\omega}_{\mathrm{II}} + \bar{\omega}_{\mathrm{III}}}{RT} A_2.
\end{aligned}
\right\}
\tag{2–7.29}
$$

These expressions have the general form of the phenomenological equations

$$
\left.
\begin{aligned}
w_1 &= a_{11}A_1 + a_{12}A_2, \\
w_2 &= a_{21}A_1 + a_{22}A_2,
\end{aligned}
\right\}
\tag{2–7.30}
$$

where the Onsager reciprocity law (2–4.7)

$$
a_{12} = a_{21}
\tag{2–7.31}
$$

is satisfied. The meaning of the phenomenological coefficients a_{ik} ($i, k = 1, 2$) is thus established by comparison of Eq. (29) and Eq. (30):

$$
\left.
\begin{aligned}
a_{11} &= \frac{\bar{\omega}_{\mathrm{I}} + \bar{\omega}_{\mathrm{III}}}{RT}, \\[2mm]
a_{12} = a_{21} &= \frac{\bar{\omega}_{\mathrm{III}}}{RT}, \\[2mm]
a_{22} &= \frac{\bar{\omega}_{\mathrm{II}} + \bar{\omega}_{\mathrm{III}}}{RT}.
\end{aligned}
\right\}
\tag{2–7.32}
$$

We recognize that the reciprocity relation (31) has its origin in the principle of detailed equilibrium (23). This important discovery is due to Onsager[5].

If we introduce conditions (27) for proximity to equilibrium into Eqs. (19) which do not yet contain the principle of detailed equilibrium, and if we use Eq. (20), then we would *not* find the phenomenological equations with

Onsager's reciprocity law. In this case, we may not set $\omega = \bar{\omega}$ in the expression following from (19) and (27) because we would then neglect terms of the same order of magnitude as A/RT. (The argument that we have used in Eq. (28) for the substitution of the ω by the $\bar{\omega}$ is not valid, for without the principle (23) we cannot draw conclusions about the simultaneous disappearance of the w and A.) The resulting equations are so constructed that phenomenological equations (30) with the reciprocity relation (31) result on the introduction of Eq. (24).

Since the quantities w_{I}, w_{II}, w_{III} and A_{I}, A_{II}, A_{III} represent "fluxes" and "forces" which disappear at equilibrium, the linear relations (28) could be interpreted as phenomenological equations which satisfy Onsager's reciprocity law in a trivial way in that all "coupling coefficients" disappear. Meanwhile, such a statement is not clear because both the "fluxes" and "forces" in (28) are linearly dependent, so that there are an infinite number of possibilities for homogeneous linear relations between the w and the A (cf. §1–27). If we add to the right-hand side of the first equation in (28) an expression which identically disappears (according to Eq. 6), that is,

$$\mathrm{const}\,(A_{\mathrm{I}} + A_{\mathrm{II}} + A_{\mathrm{III}}) = 0,$$

then the Onsager reciprocity law in the form of Eq. (28) is no longer valid.†

We see thus, that the invariant form (29) or (30) refers to the real physical content for two reactions near equilibrium and that the condition $a_{12} = a_{21} \neq 0$ actually denotes a coupling of the reactions (cf. Eqs. 2–5.24 and 2–5.25).

2–8 COUPLING OF ANY NUMBER OF REACTIONS

The considerations of the preceding section will now be extended to any number of coupled reactions in ideal gas mixtures or ideal dilute solutions [10a]. [18]

If we express the entropy production and the reaction rates per unit volume, as in continuous systems (cf. §4–13), then the assumption of a closed system of constant volume made in §2–7 is unnecessary. The following calculations are thus valid in principle for open phases at variable volumes.

Let there be L elementary reactions possible in our phase. To this let there belong R linearly independent reactions. Generally we have: $L \geqslant R$. In the case $L > R$, coupling between the independent reactions is present. For the example considered in §2–7, we have $L = 3$, $R = 2$.

---

† Such a mathematical ambiguity is found also in the original method of presentation of Onsager[5], although the physical root of the matter was first worked out by him.

The reaction equations for the elementary reactions $1, 2, \ldots, L$ are written in the following general form:

$$\sum_i \nu_{il}[M_i] \underset{\kappa_l'}{\overset{\kappa_l}{\rightleftharpoons}} \sum_j \nu_{jl}[M_j] \quad (l = 1, 2, \ldots, L). \qquad (2\text{-}8.1)$$

Here:

$$\prod_k c_k^{\nu_{kl}} = \prod_j c_j^{\nu_{jl}} \Big/ \prod_i c_i^{\nu_{il}}. \qquad (2\text{-}8.2)$$

The numbers ν_{il} or ν_{jl} are equal to the negative or positive values respectively of the stoichiometric numbers ν_{kl}. The symbol $[M_i]$ or $[M_j]$ denotes the different reacting species. The c_i, c_j, and c_k are the molarities of the species present. Finally, the symbols κ_l and κ_l' denote the rate constants of the individual reactions (temperature-dependent for ideal gas mixtures, temperature- and pressure-dependent for ideal dilute solutions). Equations (1) and (2) represent generalizations of Eqs. (2–5.8) and (2–5.9).

The kinetic equations for the conversions (1) can be formulated by considering Eq. (2) in the following way:

$$w_l = \kappa_l \prod_i c_i^{\nu_{il}} - \kappa_l' \prod_j c_j^{\nu_{jl}} = \omega_l \left(1 - \lambda_l \prod_k c_k^{\nu_{kl}} \right)$$

$$(l = 1, 2, \ldots, L) \qquad (2\text{-}8.3)$$

with

$$\omega_l \equiv \kappa_l \prod_i c_i^{\nu_{il}}, \quad \lambda_l \equiv \frac{\kappa_l'}{\kappa_l} \quad (l = 1, 2, \ldots, L). \qquad (2\text{-}8.4)$$

Here w_l is the rate of the lth elementary reaction in (1) and ω_l the rate of the lth "forward reaction" in (1), i.e. the reaction step from left to right for the lth mechanism in (1). Equation (3) represents the most general form of the *kinetic law of mass action*.

It follows from Eqs. (2–2.6) and (2–5.7) for the affinities A_l of the elementary reactions that

$$A_l = -\sum_k \nu_{kl}\mu_k = RT \ln \frac{K_l(c^\dagger)^{\nu_l}}{\prod_k c_k^{\nu_{kl}}} \quad (l = 1, 2, \ldots, L) \qquad (2\text{-}8.5)$$

with

$$\nu_l \equiv \sum_k \nu_{kl}, \quad RT \ln K_l \equiv -\sum_k \nu_{kl}\mu_k^\ominus \quad (l = 1, 2, \ldots, L), \qquad (2\text{-}8.6)$$

where μ_k is the chemical potential of species k, $\mu_k^\ominus$ a standard value of μ_k, $c^\dagger$ a standard state concentration (1 mol l^{-1}), R the gas constant, and T the absolute temperature. The quantities $\mu_k^\ominus$ and K_l are only functions of the temperature or the temperature and the pressure for ideal gas mixtures and ideal dilute solutions respectively.

Inserting Eq. (5) into Eq. (3) produces the relations between the reaction rates and the affinities of the elementary reactions for ideal gas mixtures and ideal dilute solutions:

$$w_l = \omega_l\left[1 - \lambda_l K_l (c^\dagger)^{\nu_l} \exp\left(-\frac{A_l}{RT}\right)\right] \quad (l = 1, 2, \ldots, L). \quad (2\text{–}8.7)$$

These relations contain only the kinetic equations and the expressions for the affinities. They are thus valid for any departure from equilibrium.

The thermodynamic equilibrium conditions require only (cf. §2–3 and §2–7) that the reaction rates w_r ($r = 1, 2, \ldots, R$) and affinities A_r ($r = 1, 2, \ldots, R$) of the linearly *independent* reactions $1, 2, \ldots, R$ disappear in the case of chemical equilibrium

$$w_r = 0, \quad A_r = 0 \quad (r = 1, 2, \ldots, R) \quad \text{(equilibrium)}. \quad (2\text{–}8.8)$$

Since the affinities of the individual reactions can be added like the reaction equations, homogeneous linear relations of the form

$$A_l = \sum_{r=1}^{R} b_{rl} A_r \quad (l = 1, 2, \ldots, L) \quad (2\text{–}8.9)$$

must exist between the affinities A_l of the elementary reactions and the affinities A_r of the linearly independent reactions, where the b_{rl} are constants. Equations (2–7.6) represent an example for the relations (9).

We find the reaction rates w_r for the affinities A_r from the invariance condition on the dissipation function (2–2.13):

$$\sum_{l=1}^{L} w_l A_l = \sum_{r=1}^{R} w_r A_r.$$

From this there results, with Eq. (9),

$$\sum_{r=1}^{R} \sum_{l=1}^{L} b_{rl} w_l A_r = \sum_{r=1}^{R} w_r A_r,$$

from which it follows that

$$w_r = \sum_{l=1}^{L} b_{rl} w_l \quad (r = 1, 2, \ldots, R). \quad (2\text{–}8.10)$$

These general relations between the rates w_r of the linearly independent reactions and the rates w_l of the elementary reactions are verified by Eqs. (2–7.4) for the special case $R = 2$, $L = 3$.

For chemical equilibrium, we now obtain from Eqs. (8), (9), and (10):

$$\sum_{l=1}^{L} b_{rl}w_l = 0 \quad (r = 1, 2, \ldots, R) \quad \text{(equilibrium)}, \tag{2-8.11}$$

$$A_l = 0 \quad (l = 1, 2, \ldots, L) \quad \text{(equilibrium)}. \tag{2-8.12}$$

We could thus, as in Eqs. (2–7.9) and (2–7.10), decide that certain linear combinations of the w_l become zero while all A_l disappear. This conclusion is valid for any reaction.

We thus find the *classical law of mass action* for each elementary reaction at chemical equilibrium (Eqs. 5 and 12):

$$\prod_k \bar{c}_k^{\nu_{kl}}(c^\dagger)^{\nu_l} = K_l \quad (l = 1, 2, \ldots, L), \tag{2-8.13}$$

where $\bar{c}_k$ denotes the equilibrium concentration of species k. The dimensionless quantity K_l is, accordingly, the equilibrium constant on the c-scale of the elementary reaction l. Equation (13) is again only valid for ideal gas mixtures and ideal dilute solutions.

We now introduce an expression, as in §2–7, which is not derived from the requirements of thermodynamic equilibrium and the kinetic equations: the *principle of detailed equilibrium*

$$w_l = 0 \quad (l = 1, 2, \ldots, L) \quad \text{(equilibrium)}. \tag{2-8.14}$$

With this Eq. (11) is made more precise and any cyclic reaction mechanism for equilibrium, such as the conversion (2–7.22), is excluded.

With the principle of detailed equilibrium (14) there results from Eqs. (7) and (12):

$$\lambda_l K_l (c^\dagger)^{\nu_l} = 1 \quad (l = 1, 2, \ldots, L), \tag{2-8.15}$$

from which with Eq. (4) the general relation between the rate constants for the forward and reverse reaction (κ_l and κ_l') and the equilibrium constant K_l follows:

$$\frac{\kappa_l}{\kappa_l'} = K_l(c^\dagger)^{\nu_l} \quad (l = 1, 2, \ldots, L). \tag{2-8.16}$$

Equations (15) and (16) generalize Eqs. (2–5.16), (2–7.24), and (2–7.25).

From Eqs. (7) and (15) we immediately derive:

$$w_l = \omega_l\left[1 - \exp\left(-\frac{A_l}{RT}\right)\right] \quad (l = 1, 2, \ldots, L). \tag{2-8.17}$$

These important relations generalize Eqs. (2–5.17) and (2–7.26). They hold for any departure from equilibrium for ideal gas mixtures and ideal dilute solutions.

For *states near to equilibrium*, defined by the conditions

$$\left|\frac{A_l}{RT}\right| \ll 1 \quad (l = 1, 2, \ldots, L), \qquad (2\text{–}8.18)$$

we can expand the exponential function in (17), truncate it after the first term, and according to Eq. (4) replace ω_l by the equilibrium value

$$\bar{\omega}_l \equiv \kappa_l \prod_i \bar{c}_i^{\gamma_{il}} \quad (l = 1, 2, \ldots, L). \qquad (2\text{–}8.19)$$

We then obtain:

$$w_l = \frac{\bar{\omega}_l}{RT} A_l \quad (l = 1, 2, \ldots, L). \qquad (2\text{–}8.20)$$

By considering Eqs. (9) and (10) there follows from this:

$$w_r = \frac{1}{RT} \sum_{l=1}^{L} b_{rl} \bar{\omega}_l A_l = \frac{1}{RT} \sum_{l=1}^{L} \sum_{s=1}^{R} b_{rl} b_{sl} \bar{\omega}_l A_s$$

$$(r = 1, 2, \ldots, R). \qquad (2\text{–}8.21)$$

These equations generalize Eqs. (2–7.28) and (2–7.29).

The homogeneous linear relation (21) can be written in the form

$$w_r = \sum_{s=1}^{R} a_{rs} A_s \quad (r = 1, 2, \ldots, R) \qquad (2\text{–}8.22)$$

with the "phenomenological coefficients"

$$a_{rs} \equiv \frac{1}{RT} \sum_{l=1}^{L} b_{rl} b_{sl} \bar{\omega}_l \quad (r, s = 1, 2, \ldots, R). \qquad (2\text{–}8.23)$$

By exchanging the inferiors r and s in Eq. (23) we recognize that the symmetry relation

$$a_{rs} = a_{sr} \quad (r, s = 1, 2, \ldots, R) \qquad (2\text{–}8.24)$$

is satisfied. In the special case $L = R$ (no coupling), Eqs. (20) and (22) become identical because then $a_{rs} = 0$ for $r \neq s$, so that Eq. (24) is fulfilled in a trivial manner.

As shown by comparison of Eq. (22) or (24) with Eq. (2–4.1) or (2–4.2) respectively, we have gained the *phenomenological equations* (22) and the *Onsager reciprocity relations* (24) by this calculation. We see that the principle of detailed equilibrium was again used in the derivation. This was to be expected because both this principle and the Onsager reciprocity law in its most general form (§1–26) follow from the "principle of microscopic reversibility."

Equations (17) are valid for any distance from equilibrium, while Eqs. (22) to (24), because of conditions (18), are correct only for near-equilibrium states. Accordingly, the kinetic equations and the principle of detailed equilibrium state more than do the phenomenological equations and the Onsager reciprocity relations, provided that the reaction mechanisms are known and that we are dealing with reactions in ideal gas mixtures or in ideal dilute solutions. Under these assumptions more can be learnt from the classical reaction kinetic treatment than from the thermodynamic–phenomenological description both for any departure from equilibrium[11] and for near-equilibrium conditions[12].

If, however, reactions with unknown mechanisms are present, which take place in any arbitrary phase, as is the case in after-effects and relaxation phenomena, then neither the kinetic equations nor the principle of detailed equilibrium can be formulated. In this case, we may, however, on the assumption of nearness to equilibrium—usually a sufficient approximation (cf. §2–11) for after-effects and relaxation phenomena—use the phenomenological relations and the Onsager reciprocity law because these can generally be derived from the principle of microscopic reversibility (cf. §1–26) without reference to a particular mechanism. Such general considerations are fundamental to our discussion in the following sections.

As will be shown in §2–10, it is possible to transform each system of independent reaction equations so that the coupling coefficients $a_{rs}(r \neq s)$ disappear. If considerations are not restricted to near-equilibrium states then it can be generally proved that it is always possible to formulate the reactions so that each individual term to be summed in the expression

$$\sum_r w_r A_r$$

(cf. §2–3) is positive[13,14]. It could be concluded from this that the concept of "coupling" in chemical reactions is of purely formal significance. As may be seen from the discussions of this and the preceding sections, the coupling coefficients play the decisive role in the relation between the elementary reactions and the "pertinent" gross reactions—as, for instance, in the transition from reaction scheme (2–7.2) to scheme (2–7.1)—and that this concept was the starting point for Onsager in the development of the reciprocity law.

2–9 RELAXATION TIME OF A REACTION

While the "reaction rate" of the rth reaction

$$w_r = \frac{d\xi_r}{dt}$$

can be defined for any system, following Eqs. (2–2.2) and (2–2.3), the extent of reaction ξ_r may be interpreted as a state variable and accordingly the reaction rate w_r as a derivative of a state quantity with respect to the time t for *closed systems* and for linearly *independent* reactions only[15]. In open systems and in linearly dependent reactions, a change in the mass or amount of a species can be brought about by several processes, thus leading to values of ξ_r depending on the path of the state change. Since we assume that ξ_r can be regarded as a function of the time t in the following discussion, we will from now on assume a closed system and consider the reaction equations so formulated that they correspond to linearly independent reactions.

We first consider a single reaction (extent of reaction ξ, affinity A) which occurs in any single closed phase under conditions near to equilibrium. Then the reaction can be formulated as a "relaxation process" in which the variable ξ decreases to the equilibrium value $\bar{\xi}$ according to an exponential law:

$$\xi = \bar{\xi} + C\,e^{-t/\tau}. \tag{2–9.1}$$

Here C is an integration constant (the initial value of $\xi - \bar{\xi}$) and τ a "relaxation time." Certain variables (e.g. temperature T and pressure P) are to be held constant. In the following, we develop this type of formulation more exactly.

We begin with the phenomenological equation following from Eqs. (2–2.2) and (2–5.1) valid for near-equilibrium conditions

$$\frac{d\xi}{dt} = aA. \tag{2–9.2}$$

Here a is a phenomenological coefficient.

Since our system is assumed to be closed, the affinity A depends on the extent of the reaction ξ and on two other variables. It is expedient to choose one of the four following combinations as the appropriate pair of variables:

temperature T and pressure P: $A = A(T, P, \xi)$,
temperature T and volume V: $A = A(T, V, \xi)$,
entropy S and pressure P: $A = A(S, P, \xi)$,
entropy S and volume V: $A = A(S, V, \xi)$.

The constancy of one of the above pairs of variables corresponds to the following physical conditions:

T, P const: isothermal–isobaric   (2–9.3)
T, V const: isothermal–isochoric   (2–9.4)
S, P const: adiabatic–isobaric   (2–9.5)
S, V const: adiabatic–isochoric   (2–9.6)

We recognize in the following consideration that, in the assumption of small deviations from equilibrium considered here, the terms "isentropic" and "adiabatic" are practically identical. In the closed, homogeneous, and isotropic system, in which, according to our assumptions, electrification, magnetization, and dissipative effects (friction, current passage, etc.) are excluded, an adiabatic–isochoric state change, according to Eq. (1–4.10) and Eq. (1–5.5), is denoted by the conditions

$$U = \text{const}, \qquad V = \text{const} \tag{2-9.7}$$

and an adiabatic–isobaric state change by the conditions

$$H = \text{const}, \qquad P = \text{const}. \tag{2-9.8}$$

Here U or H denote the internal energy or the enthalpy respectively of the system. We consider the entropy S of the system to depend on U, V, and ξ or on H, P, and ξ. We accordingly expand the function $S(U, V, \xi)$ or $S(H, P, \xi)$ about the equilibrium $\xi = \bar{\xi}$ for a thermally insulated system at constant volume or constant pressure:

$$S(\xi) - S(\bar{\xi}) = \frac{\partial S}{\partial \xi}(\xi - \bar{\xi}) + \frac{1}{2}\frac{\partial^2 S}{\partial \xi^2}(\xi - \bar{\xi})^2 + \cdots. \tag{2-9.9}$$

Here the derivatives must be evaluated at equilibrium under conditions (7) or (8). We now find from Eqs. (1–14.10) and (1–14.11) that:

$$\left(\frac{\partial S}{\partial \xi}\right)_{U,V} = \left(\frac{\partial S}{\partial \xi}\right)_{H,P} = \frac{A}{T}. \tag{2-9.10}$$

The derivative $\partial S/\partial \xi$, valid for equilibrium ($A = 0$), disappears. Thus the entropy S of the system remains constant for an adiabatic–isochoric or adiabatic–isobaric state change *for small deviations from equilibrium.*

We now expand the affinity A about equilibrium ($A = 0$, $\xi = \bar{\xi}$) under the same conditions (isothermal–isobaric, etc.) and truncate the series after the first term, corresponding to a near-equilibrium condition as assumed in Eq. (2):

$$A = \frac{\partial A}{\partial \xi}(\xi - \bar{\xi}), \tag{2-9.11}$$

where the derivative is to be formed at equilibrium under the previously assumed conditions (T, P const, etc.).

From Eqs. (2) and (11) there follows:

$$\frac{d\xi}{dt} = a\frac{\partial A}{\partial \xi}(\xi - \bar{\xi}). \tag{2-9.12}$$

We have four different equations of the type (12) corresponding to the four conditions (3), (4), (5), and (6). The quantities $\partial A/\partial \xi$ and $\bar{\xi}$, which relate to the equilibrium state, are constant under those conditions. According to our discussion on p. 116, the phenomenological coefficient a here also represents a constant. If we always proceed from the same temperature and the same pressure then the system does not depart far from the initial values of the temperature and the pressure even for an adiabatic or isochoric process. We could thus, under these circumstances, regard the coefficients a as constants valid for all cases (isothermal–isobaric, adiabatic–isochoric, etc.). The four derivatives

$$\left(\frac{\partial A}{\partial \xi}\right)_{T,P}, \quad \left(\frac{\partial A}{\partial \xi}\right)_{T,V}, \quad \left(\frac{\partial A}{\partial \xi}\right)_{S,P}, \quad \left(\frac{\partial A}{\partial \xi}\right)_{S,V} \qquad (2\text{–}9.13)$$

are, on the other hand, generally different from one another. For these quantities, the general inequality is valid in accordance with the stability condition (1–22.26):

$$\frac{\partial A}{\partial \xi} < 0. \qquad (2\text{–}9.14)$$

The phenomenological coefficient a is, according to (2–5.2), always positive:

$$a > 0. \qquad (2\text{–}9.15)$$

With these preliminary remarks we can integrate Eq. (12) immediately. For the sake of clarity, we treat the four cases separately and introduce the abbreviations

$$\tau_{TP} \equiv -\frac{1}{a}\left(\frac{\partial \xi}{\partial A}\right)_{T,P}, \qquad (2\text{–}9.16)$$

$$\tau_{TV} \equiv -\frac{1}{a}\left(\frac{\partial \xi}{\partial A}\right)_{T,V}, \qquad (2\text{–}9.17)$$

$$\tau_{SP} \equiv -\frac{1}{a}\left(\frac{\partial \xi}{\partial A}\right)_{S,P}, \qquad (2\text{–}9.18)$$

$$\tau_{SV} \equiv -\frac{1}{a}\left(\frac{\partial \xi}{\partial A}\right)_{S,V}. \qquad (2\text{–}9.19)$$

Then we find:

$$\xi(t) - \xi(T, P) = C_1 \exp\left(-\frac{t}{\tau_{TP}}\right), \qquad (2\text{–}9.20)$$

$$\xi(t) - \xi(T, V) = C_2 \exp\left(-\frac{t}{\tau_{TV}}\right), \qquad (2\text{–}9.21)$$

$$\xi(t) - \xi(S, P) = C_3 \exp\left(-\frac{t}{\tau_{SP}}\right), \qquad (2\text{–}9.22)$$

$$\xi(t) - \xi(S, V) = C_4 \exp\left(-\frac{t}{\tau_{SV}}\right), \qquad (2\text{–}9.23)$$

where C_1, C_2, C_3, C_4 are integration constants.

Equations (20) to (23) have the form of Eq. (1). The four quantities τ_{TP}, τ_{TV}, τ_{SP}, and τ_{SV} are thus the *relaxation times* for the four conditions (3) to (6). Due to (14) to (19) the general statement:

$$\tau > 0 \qquad (2\text{-}9.24)$$

holds for all τ. The relation between the four relaxation times results from Eqs. (16) to (19) with the help of thermodynamic manipulations. This will be shown in the following[16].

We have the mathematical identities

$$\left(\frac{\partial A}{\partial \xi}\right)_{T,V} = \left(\frac{\partial A}{\partial \xi}\right)_{T,P} + \left(\frac{\partial A}{\partial P}\right)_{T,\xi}\left(\frac{\partial P}{\partial \xi}\right)_{T,V}, \qquad (2\text{-}9.25)$$

$$\left(\frac{\partial V}{\partial \xi}\right)_{T,P} = -\left(\frac{\partial P}{\partial \xi}\right)_{T,V}\left(\frac{\partial V}{\partial P}\right)_{T,\xi}, \qquad (2\text{-}9.26)$$

the equation (1–14.21)

$$\left(\frac{\partial A}{\partial P}\right)_{T,\xi} = -\left(\frac{\partial V}{\partial \xi}\right)_{T,P}, \qquad (2\text{-}9.27)$$

the definition (1–14.21) of the reaction volume change

$$v_{TP} \equiv \left(\frac{\partial V}{\partial \xi}\right)_{T,P}, \qquad (2\text{-}9.28)$$

and the relations (1–22.12) and (1–22.13) for the isothermal compressibility

$$\kappa_T = -\frac{1}{V}\left(\frac{\partial V}{\partial P}\right)_{T,\xi} > 0. \qquad (2\text{-}9.29)$$

With this and (14), (16), and (17) we obtain

$$\frac{\tau_{TP}}{\tau_{TV}} = 1 - \frac{(v_{TP})^2}{\kappa_T V}\left(\frac{\partial \xi}{\partial A}\right)_{T,P} \geqslant 1, \qquad (2\text{-}9.30)$$

where the equality sign holds for the special case $v_{TP} = 0$.

Furthermore, the mathematical identities

$$\left(\frac{\partial A}{\partial \xi}\right)_{S,V} = \left(\frac{\partial A}{\partial \xi}\right)_{S,P} + \left(\frac{\partial A}{\partial P}\right)_{S,\xi}\left(\frac{\partial P}{\partial \xi}\right)_{S,V}, \qquad (2\text{-}9.31)$$

$$\left(\frac{\partial V}{\partial \xi}\right)_{S,P} = -\left(\frac{\partial P}{\partial \xi}\right)_{S,V}\left(\frac{\partial V}{\partial P}\right)_{S,\xi}, \qquad (2\text{-}9.32)$$

and the following relation from Eq. (1–14.11):

$$\left(\frac{\partial A}{\partial P}\right)_{S,\xi} = -\left(\frac{\partial V}{\partial \xi}\right)_{S,P}, \qquad (2\text{-}9.33)$$

with the definition of the "reaction volume change for an adiabatic–isobaric course"

$$v_{SP} \equiv \left(\frac{\partial V}{\partial \xi}\right)_{S,P}, \tag{2–9.34}$$

and the relations (1–22.10) and (1–22.11) for the adiabatic compressibility

$$\kappa_S = -\frac{1}{V}\left(\frac{\partial V}{\partial P}\right)_{S,\xi} > 0 \tag{2–9.35}$$

are all valid. With this and (14), (18), and (19) we find

$$\frac{\tau_{SP}}{\tau_{SV}} = 1 - \frac{(v_{SP})^2}{\kappa_S V}\left(\frac{\partial \xi}{\partial A}\right)_{S,P} \geqslant 1, \tag{2–9.36}$$

where the equality sign holds for the special case $v_{SP} = 0$.

Further, we have the mathematical identities

$$\left(\frac{\partial A}{\partial \xi}\right)_{S,V} = \left(\frac{\partial A}{\partial \xi}\right)_{T,V} + \left(\frac{\partial A}{\partial T}\right)_{V,\xi}\left(\frac{\partial T}{\partial \xi}\right)_{S,V}, \tag{2–9.37}$$

$$\left(\frac{\partial S}{\partial \xi}\right)_{T,V} = -\left(\frac{\partial T}{\partial \xi}\right)_{S,V}\left(\frac{\partial S}{\partial T}\right)_{V,\xi}, \tag{2–9.38}$$

and the following relation from (1–14.12):

$$\left(\frac{\partial A}{\partial T}\right)_{V,\xi} = \left(\frac{\partial S}{\partial \xi}\right)_{T,V}, \tag{2–9.39}$$

plus the definition of the "entropy of reaction for an isothermal–isochoric course"

$$s_{TV} \equiv \left(\frac{\partial S}{\partial \xi}\right)_{T,V}, \tag{2–9.40}$$

and the relations (1–22.16) and (1–22.17) for the heat capacity at constant volume

$$C_V = T\left(\frac{\partial S}{\partial T}\right)_{V,\xi} > 0. \tag{2–9.41}$$

From this and (14), (17), and (19) we derive

$$\frac{\tau_{TV}}{\tau_{SV}} = 1 - \frac{T(s_{TV})^2}{C_V}\left(\frac{\partial \xi}{\partial A}\right)_{T,V} \geqslant 1, \tag{2–9.42}$$

where the equality sign holds for the special case $s_{TV} = 0$.

Finally, if we consider the mathematical identities

$$\left(\frac{\partial A}{\partial \xi}\right)_{S,P} = \left(\frac{\partial A}{\partial \xi}\right)_{T,P} + \left(\frac{\partial A}{\partial T}\right)_{P,\xi} \left(\frac{\partial T}{\partial \xi}\right)_{S,P}, \tag{2-9.43}$$

$$\left(\frac{\partial S}{\partial \xi}\right)_{T,P} = - \left(\frac{\partial T}{\partial \xi}\right)_{S,P} \left(\frac{\partial S}{\partial T}\right)_{P,\xi}, \tag{2-9.44}$$

and the relation (1–14.20)

$$\left(\frac{\partial A}{\partial T}\right)_{P,\xi} = \left(\frac{\partial S}{\partial \xi}\right)_{T,P}, \tag{2-9.45}$$

with the definition (1–14.19) of the entropy of reaction

$$s_{TP} \equiv \left(\frac{\partial S}{\partial \xi}\right)_{T,P}, \tag{2-9.46}$$

and the relations (1–22.14) and (1–22.15) for the heat capacity at constant pressure

$$C_P = T \left(\frac{\partial S}{\partial T}\right)_{P,\xi} > 0, \tag{2-9.47}$$

we obtain from (14), (16), and (18):

$$\frac{\tau_{TP}}{\tau_{SP}} = 1 - \frac{T(s_{TP})^2}{C_P} \left(\frac{\partial \xi}{\partial A}\right)_{T,P} \geqslant 1, \tag{2-9.48}$$

where the equality sign is valid for the special case $s_{TP} = 0$.

The thermodynamic quantities in (30), (36), (42), and (48) always refer to equilibrium ($A = 0$, $\xi = \bar{\xi}$). From this it follows that the reaction entropy s_{TP} appearing in (48) is, according to Eq. (1–14.17), related to the reaction enthalpy h_{TP}:

$$h_{TP} = T s_{TP}. \tag{2-9.49}$$

The special case $s_{TP} = 0$ thus corresponds to a thermally neutral reaction.

By combining (30), (36), (42), and (48) the following general inequalities [16] are obtained:

$$\left. \begin{array}{l} \tau_{TP} \geqslant \tau_{TV} \geqslant \tau_{SV}, \\ \tau_{TP} \geqslant \tau_{SP} \geqslant \tau_{SV}. \end{array} \right\} \tag{2-9.50}$$

These statements call to mind the classical stability conditions (1–22.21) and (1–22.22):

$$C_P > C_V, \quad \kappa_T > \kappa_S.$$

2-10 RELAXATION TIMES OF ANY NUMBER OF REACTIONS

We now generalize the investigations of the previous section to any number of reactions in closed systems. Then let the reaction equations always be so formulated that they refer to linearly independent reactions. Furthermore, we will again assume near-equilibrium conditions. Under these circumstances the extents of the reactions ξ_r are functions of the time t (cf. p. 134), and the phenomenological equations (2–4.1)

$$\frac{d\xi_r}{dt} = \sum_{s=1}^{R} a_{rs} A_s \quad (r = 1, 2, \ldots, R) \qquad (2\text{–}10.1)$$

are valid for the reaction rates $d\xi_r/dt$ for R independent reactions with the Onsager reciprocity relations (2–4.2)

$$a_{rs} = a_{sr} \quad (r, s = 1, 2, \ldots, R). \qquad (2\text{–}10.2)$$

Here the A_s denote the affinities and the a_{rs} the phenomenological coefficients.

The affinities $A_1, A_2, \ldots, A_R$ are state functions in closed systems and thus depend on the extents of the reactions $\xi_1, \xi_2, \ldots, \xi_R$ as well as on two more variables. For these we can choose T and P, or T and V, or S and P, or S and V (T = absolute temperature, P = pressure, V = volume, S = entropy). We assume, as in Eqs. (2–9.3) to (2–9.6), that one of the above-mentioned variable pairs is constant. Then we can write for near-equilibrium states:

$$A_s = \sum_{q=1}^{R} \frac{\partial A_s}{\partial \xi_q} (\xi_q - \bar{\xi}_q) \quad (s = 1, 2, \ldots, R). \qquad (2\text{–}10.3)$$

Here $\bar{\xi}_q$ is the equilibrium value of ξ_q. The derivatives are evaluated at equilibrium ($A_r = 0$, $\xi_r = \bar{\xi}_r$, $r = 1, 2, \ldots, R$) and for T = const, P = const or T = const, V = const, etc.

Introducing Eq. (3) into Eq. (1) produces:

$$\frac{d\xi_r}{dt} = \sum_{s=1}^{R} \sum_{q=1}^{R} a_{rs} \frac{dA_s}{\partial \xi_q} (\xi_q - \bar{\xi}_q) \quad (r = 1, 2, \ldots, R). \qquad (2\text{–}10.4)$$

Besides the reciprocity relation (2) the further symmetry condition (1–22.24) holds:

$$\frac{\partial A_s}{\partial \xi_q} = \frac{\partial A_q}{\partial \xi_s} (q, s = 1, 2, \ldots, R). \qquad (2\text{–}10.5)$$

Moreover, it follows from the positive character of the entropy production (2–4.3) and from the stability conditions (1–22.23) to (1–22.25) that the determinant $|a_{rs}|$ with all principal cofactors is positive (see Eq. 2–4.4 and 2–4.5) and that in the determinant $|\partial A_s/\partial \xi_q|$ all principal cofactors of even order are positive and all principal cofactors of odd order (including also the diagonal terms) are negative.

By certain transformations the system of equations (4) can be brought to a mathematically simpler form[16,17]. If, indeed, we introduce new extents of reaction ξ_r', which are so chosen that all coupling terms in Eq. (1) disappear, i.e. all a_{rs} for $r \neq s$ become zero, then we have new "fluxes" $d\xi'_r/dt$ ($r = 1$, $2, \ldots, R$) which are homogeneous linear functions of the old "fluxes" $d\xi_r/dt$ ($r = 1, 2, \ldots, R$). For the $d\xi'_r/dt$, there are new "forces" A'_r ($r = 1$, $2, \ldots, R$) which represent homogeneous linear functions of the old "forces" A_r ($r = 1, 2, \ldots, R$) and which, according to §1–27, are to be determined from the invariance condition for the dissipation function (2–2.13):

$$\sum_{r=1}^{R} \frac{d\xi_r}{dt} A_r = \sum_{r=1}^{R} \frac{d\xi'_r}{dt} A'_r. \qquad (2\text{--}10.6)$$

The extents of reaction ξ'_r and affinities A'_r correspond to new reaction equations which are obtained by linear combinations of the old reaction equations but which do not necessarily have a simple significance with respect to the actual reaction mechanisms.

We thus obtain relations of the form

$$\xi'_r = \sum_{s=1}^{R} \beta_{rs} \xi_s \quad (r = 1, 2, \ldots, R), \qquad (2\text{--}10.7)$$

where the β_{rs} are those (real) constants which distribute to the ξ'_r the required property (diagonalization of the matrix of the phenomenological coefficients). With Eq. (6) the relations between the old and the new affinities can be derived from (7) (cf. Eqs. 1–27.9 and 1–27.10).

We find now in place of Eq. (4) the simple system of equations

$$\frac{d\xi'_r}{dt} = -\frac{1}{\tau_r}(\xi'_r - \bar{\xi}'_r) \quad (r = 1, 2, \ldots, R) \qquad (2\text{--}10.8)$$

with

$$\frac{1}{\tau_r} \equiv -a'_{rr} \frac{\partial A'_r}{\partial \xi'_r} \quad (r = 1, 2, \ldots, R), \qquad (2\text{--}10.9)$$

where the a'_{rr} are the new phenomenological coefficients.

From the inequality statements (cf. above)

$$a'_{rr} > 0, \quad \frac{\partial A'_r}{\partial \xi'_r} < 0 \qquad (2\text{--}10.10)$$

there results with Eq. (9):

$$\tau_r > 0. \qquad (2\text{--}10.11)$$

From considerations similar to those on p. 116 the positive (real) quantities τ_r can be regarded as constants for a given reaction system, in the same way as the $\bar{\xi}'_r$.

By integration of Eq. (8) we find equations analogous to the relations (2–9.20) to (2–9.23):

$$\xi'_r - \bar{\xi}'_r = C_r\, e^{-t/\tau_r} \quad (r = 1, 2, \ldots, R), \qquad (2\text{--}10.12)$$

where C_r is an integration constant (the initial value of $\xi'_r - \bar{\xi}'_r$). We thus recognize that each extent of reaction ξ'_r approaches its equilibrium value $\bar{\xi}_r$ exponentially and that τ_r has the significance of a *relaxation time*. Here we should, as was explained previously (§2–9), distinguish between different relaxation times (for isothermal–isobaric, adiabatic–isochoric courses, etc.) depending on the auxiliary conditions (T and P constant, S and V constant, etc.).

The original extents of reaction ξ_r are homogeneous linear functions of the transformed extents of reaction ξ'_r according to Eq. (7). Correspondingly, each of the quantities

$$\xi_r - \bar{\xi}_r \quad (r = 1, 2, \ldots, R) \tag{2–10.13}$$

is an aggregate of R exponentials of the form (12) with real coefficients. According to the general mathematical laws relating to aggregates of exponential functions, there are at most $R - 1$ zero points and at most $R - 1$ extreme values for each of the functions (13). From this it follows that each extent of reaction ξ_r passes at most $R - 1$ times through its equilibrium value $\bar{\xi}_r$ and assumes at most $R - 1$ extremes, so that periodic behavior of a reaction system having a finite number of reactions (i.e. for a finite value of R) is excluded. This general statement is due to Meixner[17]. Periodic behavior of reaction systems far from equilibrium is discussed by Lefever, Nicolis, and Prigogine[17a].

The same result had already been obtained by Jost[11] from reaction kinetic arguments. While the above reasoning on the one hand is applicable to arbitrary reaction orders and arbitrary phases, on the other hand it is valid only for near-equilibrium conditions; Jost's derivation is restricted to reactions of the first order and certain combinations of reactions of the first and second order occurring in ideal gas mixtures or in ideal dilute solutions, but it holds for any distance from equilibrium.

The entity of relaxation times $\tau_1, \tau_2, \ldots, \tau_R$ will be denoted as a *relaxation spectrum*. If we distinguish between the relaxation times belonging to the four variable sets T, P; T, V; S, P; S, V by the corresponding inferior labeling (cf. §2–9) and if they are arranged according to size (convention: $\tau_R > \cdots \tau_2 > \tau_1$) then general statements are found[16,17] which are analogous to the relations (2–9.50). For the sake of simplicity, let the limiting case of the equality sign be ignored and let the individual relaxation times of a spectrum be assumed different from one another. The following statement is then found for the connection between the isothermal–isobaric and the isothermal–isochoric spectrum:

$$(\tau_R)_{TP} > (\tau_R)_{TV} > \cdots > (\tau_2)_{TP} > (\tau_2)_{TV} > (\tau_1)_{TP} > (\tau_1)_{TV}. \tag{2–10.14}$$

The relaxation times for the isothermal–isobaric case ($T = $ const, $P = $ const) are accordingly separated from those for the isothermal–isochoric case

(T = const, V = const) in the sense that the largest relaxation time of the first spectrum is larger than the largest relaxation time of the second spectrum and, moreover, the two spectra are packed into one another. With Meixner we represent this state of affairs symbolically as follows:

$$\text{Sp}(T, P) \not\gtrless \text{Sp}(T, V). \qquad (2\text{--}10.15)$$

In an analogous way, it is found that

$$\text{Sp}(T, V) \not\gtrless \text{Sp}(S, V), \qquad (2\text{--}10.16)$$

$$\text{Sp}(T, P) \not\gtrless \text{Sp}(S, P), \qquad (2\text{--}10.17)$$

$$\text{Sp}(S, P) \not\gtrless \text{Sp}(S, V). \qquad (2\text{--}10.18)$$

2-11 AFTER-EFFECTS AND RELAXATION PROCESSES

The characteristic behavior of glasses (also plastics in the glass state) differs from the systems dealt with in classical thermodynamics: at given temperature and given pressure (or, more generally, given work-coefficients) the state of the phase in question, which ought to be of constant mass and constant composition, is not uniquely determined. Under these conditions quantities such as the volume (or, more generally, the work-coordinates), the refractive index, etc., can change with time. We then speak of "after-effects" and denote the systems in which these phenomena appear as "frozen phases," whereas most crystals, normal fluids, and all gases are called "phases in internal equilibrium." In particular, glasses represent "frozen supercooled liquids."

After-effects have their source in "internal transformations" or "relaxation processes." The molecular physical mechanisms of these processes are mostly unexplained in detail or at least disputed. But the general nature of the "frozen state" is doubtless of the same type as for those crystals (e.g. CO, NO, N_2O, and H_2O) in which the complete "order" postulated by the Nernst heat theorem for absolute zero is not achieved. The molecular building blocks of the frozen phases have, with respect to their arrangement or numbers of internal degrees of freedom (vibrational–rotational states, etc.), not attained the equilibrium positions corresponding to those configurations which represent the statistical equilibrium under the conditions in question (temperature, pressure, etc.). They are the configurations belonging to a higher temperature which have been "frozen." Accordingly, if the "restraints" are lifted, relaxation processes (internal transformations) take place and are observed as after-effects until the statistical equilibrium relating to the unrestrained degrees of freedom is reached and the phase is in a state of internal equilibrium.

This interpretation of after-effects suggests that the relaxation processes can be treated formally as chemical reactions with unknown mechanisms and the corresponding extents of reaction regarded as "internal parameters" which are pertinent to the internal transformations in question (cf. §1–16 and §2–2). If we also assume that the system is near to equilibrium then the statements in §2–9 and §2–10 can be applied directly. Thus the concepts "relaxation time" and "relaxation spectrum" in §2–9 and §2–10 were introduced with consideration of their significance for these phenomena. The "near-equilibrium" assumption is more justified for these processes than for true chemical reactions. It can be shown that almost all after-effects examined experimentally can be described by a theory which depends only on the *causal principle* and on Boltzmann's *principle of superposition*; for that part of the theory of after-effects which is covered by the thermodynamic theory of relaxation processes, namely, in the region of total monotonic after-effect functions, the regions of validity of both theories coincide[16].

As long as homogeneous isotropic media with a finite number of reactions are assumed—as throughout this chapter—certain after-effects and relaxation processes cannot be completely understood in the most general sense. Thus we must consider an infinite number of internal transformations for the thermodynamic–phenomenological treatment of dielectric relaxation[16], anisotropic media for that of elastic after-effects[18], and continuous media with transport phenomena for that of acoustic relaxation[10b]. We cannot investigate here all those conditions involved which might lead to a generalization of the thermodynamics of irreversible processes[19].

We restrict further considerations in this chapter to simple problems which fit into the framework established in §2–9 and §2–10: the dynamic equation of state (§2–12), the after-effect functions (§2–13), and the dispersion and absorption of sound (§2–14), where in all cases a homogeneous isotropic medium with a single internal transformation is assumed. Countless experimental examinations indicate that after-effects in glasses cannot be described by a single relaxation mechanism. The discussion in §2–12 to §2–14 should thus only serve as an illustration of the method.

2-12 THE DYNAMIC EQUATION OF STATE

We now consider a closed system, in which a single relaxation process corresponding to a single internal transformation with unknown mechanism (cf. §2–11) shall take place. In contrast to §2–9, we now allow arbitrary time dependences for those two variables (e.g. temperature T and pressure P) which were regarded as constant in §2–9. However, we assume small deviations from a fixed "reference state." Let this reference state be described by the temperature T_0, the pressure P_0, and the internal variable $\xi_0 = \xi(T_0, P_0)$. Here ξ denotes the equilibrium value of the internal parameter ξ determined by T_0 and P_0. Thus, the reference state corresponds to an internal equilibrium, so that we always find ourselves near to an equilibrium state.

Under the above-mentioned assumptions the phenomenological equation (2–9.2)

$$\frac{d\xi}{dt} = aA \tag{2–12.1}$$

holds. Here t is the time, A the affinity of the internal transformation, and a a phenomenological coefficient which depends only on T_0, P_0, and ξ_0, i.e. it represents a constant for the given system.

For the sake of simplicity, we restrict our discussion to the independent variables T and P. We thus find that the affinity:

$$A = A(T, P, \xi)$$

disappears for the reference state. Accordingly, we write for the vicinity of the reference state T_0, P_0, ξ_0:

$$A = \left(\frac{\partial A}{\partial T}\right)_{P,\xi} (T - T_0) + \left(\frac{\partial A}{\partial P}\right)_{T,\xi} (P - P_0) + \left(\frac{\partial A}{\partial \xi}\right)_{T,P} (\xi - \xi_0). \tag{2–12.2}$$

Moreover, an (unspecified) equation of state

$$V = V(T, P, \xi) \tag{2–12.3}$$

is valid, where V is the volume of the system. In the region around the reference state T_0, P_0, ξ_0, we can formulate Eq. (3) in the following manner:

$$V - V_0 = \left(\frac{\partial V}{\partial T}\right)_{P,\xi} (T - T_0) + \left(\frac{\partial V}{\partial P}\right)_{T,\xi} (P - P_0)$$

$$+ \left(\frac{\partial V}{\partial \xi}\right)_{T,P} (\xi - \xi_0). \tag{2–12.4}$$

Here $V_0 = V(T_0, P_0, \xi_0)$ is the volume of the system in the reference state.

The partial derivatives in Eqs. (2) and (4) are to be formed for the reference state and accordingly are constant for a given system. We relate them back to known quantities using the following definitions (cf. Eqs. 1–22.20, 2–9.27, 2–9.28, 2–9.29, 2–9.45, and 2–9.46):

$$\beta \equiv \frac{1}{V_0} \left(\frac{\partial V}{\partial T}\right)_{P,\xi}, \tag{2–12.5}$$

$$\kappa_T \equiv -\frac{1}{V_0} \left(\frac{\partial V}{\partial P}\right)_{T,\xi}, \tag{2–12.6}$$

$$v_{TP} \equiv \left(\frac{\partial V}{\partial \xi}\right)_{T,P} = -\left(\frac{\partial A}{\partial P}\right)_{T,\xi}, \tag{2–12.7}$$

$$s_{TP} \equiv \left(\frac{\partial S}{\partial \xi}\right)_{T,P} = \left(\frac{\partial A}{\partial T}\right)_{P,\xi}. \tag{2–12.8}$$

Here S is the entropy of the system. All partial derivatives are to be taken for the reference state; β denotes the thermal expansivity, κ_T the isothermal compressibility, v_{TP} the "reaction volume change", and s_{TP} the "reaction entropy," each for the reference state.

From Eqs. (1) and (2) as well as Eqs. (4) to (8) and denoting

$$\dot{x} = \frac{dx}{dt}$$

we find:

$$\dot{\xi} = as_{TP}(T - T_0) - av_{TP}(P - P_0) + a\left(\frac{\partial A}{\partial \xi}\right)_{T,P}(\xi - \xi_0),$$

$$v_{TP}(\xi - \xi_0) = V - V_0 - \beta V_0(T - T_0) + \kappa_T V_0(P - P_0),$$

$$v_{TP}\dot{\xi} = \dot{V} - \beta V_0\dot{T} + \kappa_T V_0\dot{P}.$$

By eliminating $\xi - \xi_0$ and $\dot{\xi}$ from this we derive

$$\dot{V} + \kappa_T V_0\dot{P} - \beta V_0\dot{T} = a\left(\frac{\partial A}{\partial \xi}\right)_{T,P}(V - V_0)$$

$$+ a\left[\kappa_T V_0\left(\frac{\partial A}{\partial \xi}\right)_{T,P} - (v_{TP})^2\right](P - P_0)$$

$$+ a\left[s_{TP}v_{TP} - \beta V_0\left(\frac{\partial A}{\partial \xi}\right)_{T,P}\right](T - T_0). \qquad (2\text{–}12.9)$$

This relation will be transformed in the following.

From the mathematical identity

$$\left(\frac{\partial A}{\partial \xi}\right)_{P,V} = \left(\frac{\partial A}{\partial \xi}\right)_{T,P} - \left(\frac{\partial A}{\partial T}\right)_{P,\xi}\frac{(\partial V/\partial \xi)_{T,P}}{(\partial V/\partial T)_{P,\xi}}$$

it follows, with Eqs. (5), (7), and (8), that

$$\left(\frac{\partial A}{\partial \xi}\right)_{P,V} = \left(\frac{\partial A}{\partial \xi}\right)_{T,P} - s_{TP}\frac{v_{TP}}{\beta V_0}. \qquad (2\text{–}12.10)$$

Furthermore, according to Eqs. (2–9.25), (2–9.26), (2–9.27), and Eqs. (6) and (7):

$$\left(\frac{\partial A}{\partial \xi}\right)_{T,V} = \left(\frac{\partial A}{\partial \xi}\right)_{T,P} - \frac{(v_{TP})^2}{\kappa_T V_0}. \qquad (2\text{–}12.11)$$

Combining Eq. (9) with Eqs. (10) and (11) we find:

$$\dot{V} + \kappa_T V_0\dot{P} - \beta V_0\dot{T} = a\left(\frac{\partial A}{\partial \xi}\right)_{T,P}(V - V_0) + a\kappa_T V_0\left(\frac{\partial A}{\partial \xi}\right)_{T,V}(P - P_0)$$

$$- a\beta V_0\left(\frac{\partial A}{\partial \xi}\right)_{P,V}(T - T_0). \qquad (2\text{–}12.12)$$

Even this equation permits still further transformations since it is impossible to ascertain quantities such as ξ and A for an unknown relaxation mechanism.

We now introduce the "relaxation time" discussed in §2–9 for isothermal–isobaric (τ_{TP}) and isothermal–isochoric (τ_{TV}) processes. According to Eqs. (2–9.16) and (2–9.17),

$$\frac{1}{\tau_{TP}} \equiv -a\left(\frac{\partial A}{\partial \xi}\right)_{T,P}, \qquad (2\text{–}12.13)$$

$$\frac{1}{\tau_{TV}} \equiv -a\left(\frac{\partial A}{\partial \xi}\right)_{T,V}. \qquad (2\text{–}12.14)$$

Furthermore, we need a "relaxation time" for the isobaric–isochoric path τ_{PV} of the internal transformation:†

$$\frac{1}{\tau_{PV}} \equiv -a\left(\frac{\partial A}{\partial \xi}\right)_{P,V}. \qquad (2\text{–}12.15)$$

With the help of these three relations we eliminate the phenomenological coefficient a in Eq. (12):

$$\dot{V} + \frac{1}{\tau_{TP}}(V - V_0) + \kappa_T V_0\left[\dot{P} + \frac{1}{\tau_{TV}}(P - P_0)\right]$$

$$-\beta V_0\left[\dot{T} + \frac{1}{\tau_{PV}}(T - T_0)\right] = 0. \qquad (2\text{–}12.16)$$

This is a differential form of the *dynamic equation of state*. Equation (16) relates the three variables T, P, V and their time derivatives $\dot{T}$, $\dot{P}$, $\dot{V}$ to one another. It contains the quantities κ_T and β accessible from equilibrium measurements, and the three relaxation times τ_{TP}, τ_{TV}, and τ_{PV}. If all these quantities are known then the time variation of the volume can be calculated, that is, the function $V(t)$ if the pressure and temperature are available as functions of the time, $P(t)$ and $T(t)$. We will discuss the transition from Eq. (16) to the integral form of the dynamic equation of state in §2–13.

If we denote integration constants by C, C', and C'' and if these depend only on the initial conditions then, in the isothermal–isobaric case ($\dot{P} = 0$, $P = P_0$, $\dot{T} = 0$, $T = T_0$), we obtain an expression for the "volume relaxation" at constant temperature and constant pressure (cf. Eq. 2–9.20) from Eq. (16):

$$V - V_0 = C\,e^{-t/\tau_{TP}}. \qquad (2\text{–}12.17)$$

In the isothermal–isochoric case ($\dot{V} = 0$, $V = V_0$, $\dot{T} = 0$, $T = T_0$), a relation for the "pressure relaxation" at constant temperature and constant volume (cf. Eq. 2–9.21) is found:

$$P - P_0 = C'\,e^{-t/\tau_{TV}} \qquad (2\text{–}12.18)$$

---

† Such a process is naturally neither isothermal nor adiabatic.

and in the isobaric–isochoric case ($\dot{V} = 0$, $V = V_0$, $\dot{P} = 0$, $P = P_0$) an equation for the "temperature relaxation" at constant pressure and constant volume results:

$$T - T_0 = C'' \, e^{-t/\tau_{PV}}. \tag{2–12.19}$$

From these relations the physical significance of the three relaxation times is available. Also, these equations show how τ_{TP}, τ_{TV}, and τ_{PV} can be determined experimentally.

If we consider the following relations (cf. Eq. 2–9.30):

$$\frac{\tau_{TP}}{\tau_{TV}} = 1 - \frac{(v_{TP})^2}{\kappa_T V_0} \left(\frac{\partial \xi}{\partial A}\right)_{T,P}, \tag{2–12.20}$$

$$\frac{\tau_{TP}}{\tau_{PV}} = 1 - \frac{s_{TP} v_{TP}}{\beta V_0} \left(\frac{\partial \xi}{\partial A}\right)_{T,P}, \tag{2–12.21}$$

which follow from Eqs. (10), (11), (13), (14), and (15) then we derive from Eq. (16):

$$V - V_0 + \tau_{TP}\dot{V} + \left[\kappa_T V_0 - (v_{TP})^2 \left(\frac{\partial \xi}{\partial A}\right)_{T,P}\right](P - P_0 + \tau_{TV}\dot{P})$$

$$- \left[\beta V_0 - s_{TP} v_{TP} \left(\frac{\partial \xi}{\partial A}\right)_{T,P}\right](T - T_0 + \tau_{PV}\dot{T}) = 0. \tag{2–12.22}$$

In this second form of the dynamic equation of state, the relaxation times τ_{TP}, τ_{TV}, and τ_{PV} are present as factors before the time derivatives $\dot{V}$, $\dot{P}$, and $\dot{T}$ and this is found to be advantageous for many purposes. There are, however, thermodynamic functions (v_{TP}, s_{TP}, and $\partial \xi/\partial A$) which are contained in this equation and which cannot be determined from experimental data without knowledge of the relaxation mechanism. Therefore a further transformation is necessary.

With the help of Eqs. (5) to (8) the following mathematical identities can be proved:

$$\kappa_T V_0 - (v_{TP})^2 \left(\frac{\partial \xi}{\partial A}\right)_{T,P} = -\left(\frac{\partial V}{\partial P}\right)_{T,A}, \tag{2–12.23}$$

$$\beta V_0 - s_{TP} v_{TP} \left(\frac{\partial \xi}{\partial A}\right)_{T,P} = \left(\frac{\partial V}{\partial T}\right)_{P,A}, \tag{2–12.24}$$

where the condition $A = $ const prescribes the state of equilibrium $A = 0$, since all differentiations are carried out for the reference state ($A = 0$). Also, we can introduce the "isothermal compressibility at equilibrium" analogous to definitions (5) and (6):

$$\kappa_{TA} \equiv -\frac{1}{V_0}\left(\frac{\partial V}{\partial P}\right)_{T,A} \tag{2–12.25}$$

and the "thermal expansivity at equilibrium"

$$\beta_A \equiv \frac{1}{V_0} \left(\frac{\partial V}{\partial T} \right)_{P,A}. \qquad (2\text{-}12.26)$$

For very fast changes of state, κ_T and β will be measured according to Eqs. (5) and (6) (equivalent to inhibited equilibrium, $\xi = \text{const}$); κ_{TA} and β_A are determined according to Eqs. (25) and (26) for very slow state changes (equilibrium, $A = 0$).

From Eqs. (22) to (26) we derive the third form of the dynamic equation of state:

$$V - V_0 + \tau_{TP}\dot{V} + \kappa_{TA}V_0(P - P_0 + \tau_{TV}\dot{P})$$
$$- \beta_A V_0(T - T_0 + \tau_{PV}\dot{T}) = 0. \qquad (2\text{-}12.27)$$

In this relation, only quantities which can be determined from equilibrium measurements are included with the relaxation times, as in Eq. (16). Equation (27) and the following discussion are essentially due to Meixner[16].

For very rapid state changes (inhibited equilibrium, $\xi = \text{const}$), the terms with $V - V_0$, $P - P_0$, and $T - T_0$ can be neglected. Using Eqs. (5) and (6) we then find from Eq. (16):

$$\dot{V} = \left(\frac{\partial V}{\partial P} \right)_{T,\xi} \dot{P} + \left(\frac{\partial V}{\partial T} \right)_{P,\xi} \dot{T}$$

or in integrated form:

$$V - V_0 = \left(\frac{\partial V}{\partial P} \right)_{T,\xi} (P - P_0) + \left(\frac{\partial V}{\partial T} \right)_{P,\xi} (T - T_0).$$

This is the usual (static) equation of state (4) for the case $\xi = \text{const}$.

For slow state changes (equilibrium, $A = 0$), the terms with $\dot{V}$, $\dot{P}$, and $\dot{T}$ should be neglected. Using Eqs. (25) and (26) with Eq. (27) we obtain:

$$V - V_0 = \left(\frac{\partial V}{\partial P} \right)_{T,A} (P - P_0) + \left(\frac{\partial V}{\partial T} \right)_{P,A} (T - T_0).$$

That is, the static equation of state for the case $A = 0$.

The special case $v_{TP} = 0$ can be experimentally recognized by the fact that then the two compressibilities κ_T and κ_{TA} and the two expansivities β and β_A coincide. We thus have:

$$v_{TP} = 0, \quad \kappa_T = \kappa_{TA}, \quad \beta = \beta_A. \qquad (2\text{-}12.28)$$

With this and Eqs. (20) and (21) there results:

$$\tau_{TP} = \tau_{TV} = \tau_{PV} = \tau. \qquad (2\text{-}12.29)$$

Thus the three relaxation times are equal. We can now integrate Eq. (16) or (27) immediately:

$$V - V_0 = -\kappa_T V_0(P - P_0) + \beta V_0(T - T_0) + C e^{-t/\tau}, \qquad (2\text{-}12.30)$$

where C denotes an integration constant and κ_T and β can be replaced by κ_{TA} and β_A respectively. If we start from the reference state $V = V_0$, $P = P_0$, $T = T_0$, the constant C is zero and the exponential term disappears. Thus we are again left with a static equation of state in which there is no trace of an "after-effect." This is true, in general, for any experiment in which there has been some passage of time.

2–13 AFTER-EFFECT FUNCTIONS

We want to deduce the relation between the thermodynamic–phenomenological theory of relaxation and the after-effect theory (cf. §2–11) for the simple case of a single internal transformation as treated in §2–12[16].

We proceed from Eq. (2–12.16), a differential form of the dynamic equation of state, and try to obtain the integral form of this equation.

Using the abbreviations

$$x \equiv V - V_0, \quad y \equiv P - P_0, \quad z \equiv T - T_0; \left.\begin{array}{l} \\ \\ \\ \\ \end{array}\right\}$$

$$\left.\begin{array}{ll} a \equiv \dfrac{1}{\tau_{TP}}, & b \equiv \dfrac{1}{\tau_{TV}}, \quad c \equiv \dfrac{1}{\tau_{PV}}; \\[2mm] A \equiv \kappa_T V_0, & B \equiv -\beta V_0, \end{array}\right\} \qquad (2\text{–}13.1)$$

we obtain from Eq. (2–12.16):

$$\dot{x} + ax + A\dot{y} + Aby + B\dot{z} + Bcz = 0, \qquad (2\text{–}13.2)$$

a differential equation which will be further examined in the following.

If we regard the quantities y and z as experimentally available functions of time, we can replace Eq. (2) by

$$\frac{dx}{dt} + ax = f(t), \qquad (2\text{–}13.3)$$

where

$$f(t) \equiv -A\left(by + \frac{dy}{dt}\right) - B\left(cz + \frac{dz}{dt}\right). \qquad (2\text{–}13.4)$$

Equation (3) is a normal nonhomogeneous differential equation of the first order. It is well known in classical mechanics since, for example, it describes —with appropriately changed symbols—the motion of a body in a viscous medium under the influence of a time-dependent force. Integration of Eq. (3) produces:

$$x = C e^{-at} + e^{-at} \int e^{at} f(t) \, dt, \qquad (2\text{–}13.5)$$

where C is an integration constant. Following Eq. (4) we have:

$$-e^{-at} \int e^{at} f(t)\, dt = Ay(t) + Bz(t) + A(b - a)\, e^{-at} \int e^{at} y(t)\, dt$$

$$+ B(c - a)\, e^{-at} \int e^{at} z(t)\, dt. \qquad (2\text{–}13.6)$$

The value of x for an arbitrary but fixed value of the time t_0 follows from Eqs. (5) and (6) after insertion of the integration limits:†

$$\left.\begin{aligned}
x(t_0) = {}& C\, e^{-at_0} - Ay(t_0) - Bz(t_0) \\
& - A(b - a)\, e^{-at_0} \int_{-\infty}^{t_0} e^{at} y(t)\, dt \\
& - B(c - a)\, e^{-at_0} \int_{-\infty}^{t_0} e^{at} z(t)\, dt.
\end{aligned}\right\} \qquad (2\text{–}13.7)$$

With the auxiliary variable

$$u \equiv t_0 - t, \qquad (2\text{–}13.8)$$

which measures the time in reverse from a fixed point t_0, we obtain the final *integral form of the dynamic equation of state* containing an explicit description of the after-effects:

$$\left.\begin{aligned}
x(t_0) = {}& C\, e^{-at_0} - Ay(t_0) - Bz(t_0) \\
& - A \int_0^\infty \varphi_1(u) y(t_0 - u)\, du \\
& - B \int_0^\infty \varphi_2(u) z(t_0 - u)\, du
\end{aligned}\right\} \qquad (2\text{–}13.9)$$

with

$$\varphi_1(u) \equiv (b - a)\, e^{-au}, \qquad (2\text{–}13.10)$$

$$\varphi_2(u) \equiv (c - a)\, e^{-au}, \qquad (2\text{–}13.11)$$

where, in accordance with Eq. (1), $y(t_0 - u) = y(t)$ and $z(t_0 - u) = z(t)$ denote the time-dependent functions $P - P_0$ and $T - T_0$ respectively. Under normal experimental conditions ($t_0 \gg \tau_{TP}$), the first term on the right-hand side of Eq. (9) can be neglected in comparison to other terms.

We see from Eqs. (9) to (11) that the instantaneous value $x(t_0)$ of the quantity $x = V - V_0$ does not depend only on the momentary values $y(t_0)$ and $z(t_0)$ of the quantities $y = P - P_0$ and $z = T - T_0$ but also on all

---

† The lower limit is, in the most general case, $-\infty$. At the beginning of the experiment ($t = 0$) after-effects may indeed already have taken place, and this would remain unnoticed if the lower limit were 0 because then $x(0)$ would depend only on $y(0)$ and $z(0)$.

previous values of these variables, i.e. on the previous history. The earlier values of the variables y and z thus affect the instantaneous value of x (with exponentially decreasing significance). From this the functions (10) and (11) are called *after-effect functions*. They relate here to the after-effect of changes in pressure or temperature on volume.

With this we have shown, for a very simple example, the relation between the thermodynamics of relaxation processes and after-effect theory.†

2-14 VELOCITY OF SOUND IN FLUIDS

Strictly speaking, a fluid in which a sound wave spreads cannot be considered as a "homogeneous system"; but if, for the sake of simplicity, irreversible processes such as viscous flow, heat conduction, diffusion, thermal diffusion, etc., are excluded, the formalism is practically the same as that for closed homogeneous systems with chemical reactions or internal transformations. This follows explicitly from our later calculations of irreversible processes in continuous media.

We restrict our discussion to the simple case in which only one single irreversible process (that is, a single internal transformation) occurs in each volume element which then influences the propagation of sound. This relaxation process can, for example, be a dissociation or a recombination of molecules, or a transition between the two deepest vibration levels of a molecular species, or an internal transformation of molecules. The following considerations are true both for sound in the audible range of frequency and for ultrasonics.

The velocity of sound c in a fluid medium is given by the following general expression:

$$c^2 = -\tilde{V}^2\left(\frac{\partial P}{\partial \tilde{V}}\right)_{ad}. \tag{2-14.1}$$

Here $\tilde{V}$ denotes the specific volume (the reciprocal density) and P the pressure. The inferior "ad" shows that the differentiation refers to an adiabatic state change as it occurs in practice in sound waves. If we consider the speed of sound as a complex quantity then Eq. (1) still holds and we have the advantage that, in all following calculations, sound dispersion and absorption (the frequency dependence of the speed of sound and the partial transfer of sound energy to internal energy) are included, where the latter is related to the imaginary part of the complex velocity of sound.

The thermodynamic description of the fluid is formally the same as that of the homogeneous systems considered in §2-9 and §2-12. The state of

---

† A more elaborate theory that refers to linear viscoelastic behavior in general can be found in Meixner[20].

each volume element of the medium is determined by three quantities such as the temperature T, the pressure P, and the internal parameter ξ which belongs to the internal transformation in question. It is useful for the following considerations to choose the entropy S, the volume V, and the internal variable ξ as independent variables. Accordingly, we write the equation of state in the form:

$$P = P(S, V, \xi) \tag{2–14.2}$$

and, for the affinity A of the internal transformation,

$$A = A(S, V, \xi). \tag{2–14.3}$$

As in §2–12, we assume small deviations from a fixed reference state S_0, V_0, ξ_0, where $\xi_0 = \xi(S_0, V_0)$. Accordingly, the value of ξ in the reference state ξ_0 is equal to the equilibrium value ξ belonging to S_0 and V_0, so that the state S_0, V_0, ξ_0 corresponds to an equilibrium with reference to the internal transformation under consideration $(A = 0)$. This assumption excludes extremely high amplitudes in the sound waves and has the following important consequences:

1. Adiabatic and isentropic changes coincide, so that the inferior "ad" in Eq. (1) can be replaced by the condition $S = $ const (cf. p. 134).

2. The relations (2) and (3) can be written in the following form:

$$\Delta P = \left(\frac{\partial P}{\partial V}\right)_{S,\xi} \Delta V + \left(\frac{\partial P}{\partial \xi}\right)_{S,V} \Delta \xi, \tag{2–14.4}$$

$$\Delta A = A = \left(\frac{\partial A}{\partial V}\right)_{S,\xi} \Delta V + \left(\frac{\partial A}{\partial \xi}\right)_{S,V} \Delta \xi. \tag{2–14.5}$$

Here the operator Δ denotes the difference between the value of a thermodynamic quantity at any one time t and its value in the reference state.

3. A phenomenological equation (cf. Eq. 2–12.1):

$$\frac{d\xi}{dt} = aA \tag{2–14.6}$$

with constant phenomenological coefficient a is valid for the relation between the rate $d\xi/dt$ and the affinity A of the internal transformation.

If the sound wave has an angular velocity ω,† the thermodynamic quantities oscillate periodically about their equilibrium values. If we denote the amplitude of the oscillation by α and assume harmonic oscillations,‡ we find for ξ:

$$\Delta \xi = \xi - \xi_0 = \alpha\, e^{i\omega t}, \tag{2–14.7}$$

---

† $\omega = 2\pi\nu$ (ν = frequency).

‡ $e^{i\omega t} = \cos(\omega t) + i \sin(\omega t)$, where i is the imaginary unity.

from which it follows directly that

$$\frac{d\xi}{dt} = i\omega\Delta\xi. \tag{2-14.8}$$

With this there results from Eq. (6):

$$i\omega\Delta\xi = aA. \tag{2-14.9}$$

From Eqs. (4), (5), and (9) we eliminate $\Delta\xi$ and A and introduce the "relaxation time" τ ($= \tau_{SV}$ in Eq. 2-9.19):

$$\tau \equiv -\frac{1}{a}\left(\frac{\partial\xi}{\partial A}\right)_{S,V}. \tag{2-14.10}$$

With this we obtain:

$$\left(\frac{\Delta P}{\Delta V}\right)_{\Delta S=0} = \left(\frac{\partial P}{\partial V}\right)_{\text{ad}} = \left(\frac{\partial P}{\partial V}\right)_{S,\xi} - \frac{1}{1+i\omega\tau}\left(\frac{\partial P}{\partial\xi}\right)_{S,V}$$

$$\times \left(\frac{\partial A}{\partial V}\right)_{S,\xi}\left(\frac{\partial\xi}{\partial A}\right)_{S,V}. \tag{2-14.11}$$

On the basis of the identities

$$\left(\frac{\partial A}{\partial V}\right)_{S,\xi}\left(\frac{\partial\xi}{\partial A}\right)_{S,V} = -\left(\frac{\partial\xi}{\partial V}\right)_{S,A}, \tag{2-14.12}$$

$$\left(\frac{\partial P}{\partial V}\right)_{S,A} = \left(\frac{\partial P}{\partial V}\right)_{S,\xi} + \left(\frac{\partial P}{\partial\xi}\right)_{S,V}\left(\frac{\partial\xi}{\partial V}\right)_{S,A} \tag{2-14.13}$$

we find from Eq. (11):

$$\left(\frac{\partial P}{\partial V}\right)_{\text{ad}} = \left(\frac{\partial P}{\partial V}\right)_{S,\xi} + \frac{1}{1+i\omega\tau}\left[\left(\frac{\partial P}{\partial V}\right)_{S,A} - \left(\frac{\partial P}{\partial V}\right)_{S,\xi}\right]. \tag{2-14.14}$$

Here we can insert the specific volume $\tilde{V}$ in place of the total volume V and the specific entropy $\tilde{S}$ in place of the total entropy S.

The derivatives appearing on the right-hand side of Eq. (14) can be related to the "adiabatic compressibility" (2-9.35)

$$\kappa_S = -\frac{1}{V}\left(\frac{\partial V}{\partial P}\right)_{S,\xi} = -\frac{1}{\tilde{V}}\left(\frac{d\tilde{V}}{\partial P}\right)_{\tilde{S},\xi} \tag{2-14.15}$$

and to the "adiabatic compressibility for equilibrium"

$$\kappa_{SA} = -\frac{1}{V}\left(\frac{\partial V}{\partial P}\right)_{S,A} = -\frac{1}{\tilde{V}}\left(\frac{\partial\tilde{V}}{\partial P}\right)_{\tilde{S},A}. \tag{2-14.16}$$

The first quantity is valid for very rapid state changes (inhibited equilibrium, $\xi = $ const) and the second quantity holds for very slow state changes (equilibrium, $A = 0$).

According to Eqs. (1), (14), (15), and (16), the expression

$$c_\infty^2 = -\tilde{V}^2\left(\frac{\partial P}{\partial \tilde{V}}\right)_{\tilde{S},\xi} = \frac{\tilde{V}}{\kappa_S} \qquad (2\text{-}14.17)$$

denotes the square of the (complex) speed of sound for very rapid state changes (inhibited equilibrium, ξ = const, $\omega \to \infty$), and the expression

$$c_0^2 = -\tilde{V}^2\left(\frac{\partial P}{\partial \tilde{V}}\right)_{\tilde{S},A} = \frac{\tilde{V}}{\kappa_{SA}} \qquad (2\text{-}14.18)$$

denotes the square of the (complex) speed of sound for very slow state changes (equilibrium, $A = 0$, $\omega \to 0$). In the limiting case of Eq. (17), the internal transformation can no longer follow the state changes determined by the sound wave, but remains completely inhibited. In the limiting case of Eq. (18), the oscillations follow so slowly (theoretically infinitely slowly) that the internal equilibrium remains established at each instant.

We derive from Eq. (1) and Eqs. (14) to (18):

$$c^2 = \tilde{V}\left[\frac{1}{\kappa_S} + \frac{1}{1 + i\omega\tau}\left(\frac{1}{\kappa_{SA}} - \frac{1}{\kappa_S}\right)\right] = c_\infty^2 + \frac{c_0^2 - c_\infty^2}{1 + i\omega\tau}. \qquad (2\text{-}14.19)$$

This interesting relation is due to Meixner.† It gives the dependence of the speed of sound c on the angular velocity ω, and in fact is independent of any proposed model. The compressibilities κ_{SA} and κ_S and the limiting velocities c_0 and c_∞ as well as the relaxation time τ can, in principle, be determined independently of c ($\tau = \tau_{SV}$ from measurements of the adiabatic-isochoric pressure relaxation similar to τ_{TV} and following Eq. 2–12.18 in the isothermal–isochoric case). Many previous expressions calculated from kinetic models for the dispersion and the absorption of sound are contained in Eq. (19) and can be obtained by consideration of those special assumptions from Eq. (19).

REFERENCES

1. Th. de Donder, *Bull. Classe Sci., Acad. Roy. Belg.* **24**, 15 (1938).
2. I. Prigogine, P. Outer, and Cl. Herbo, *J. Phys. Colloid Chem.* **52**, 321 (1948).
3. R. H. Allen and L. D. Yats, *J. Am. Chem. Soc.* **81**, 5289 (1959).
4. I. Prigogine, *Introduction to Thermodynamics of Irreversible Processes*, Springfield, Illinois (1955), p. 59.
5. L. Onsager, *Phys. Rev.* **37**, 405 (1931).
6. R. H. Fowler, *Statistical Mechanics*, Cambridge (1929).

† Detailed literature references and calculations for complicated cases can be found[21, 22], in which, besides internal transformations, transport phenomena (viscous flow, heat conduction, diffusion, thermal diffusion, etc.) influence the propagation of sound.

7. R. C. TOLMAN, *Phys. Rev.* **23**, 699 (1924).
8. R. C. TOLMAN, *The Principles of Statistical Mechanics*, Oxford (1938).
9. R. L. BURWELL, JR. and R. G. PEARSON, *J. Phys. Chem.* **70**, 300 (1966).
10. J. MEIXNER, *Ann. Physik* (5) (a) **43**, 244 (1943); (b) **43**, 470 (1943).
11. W. JOST, *Z. Naturforsch.* **2a**, 159 (1947).
12. R. HAASE and W. JOST, *Z. Physik. Chem.* (*Leipzig*) **196**, 215 (1950).
13. F. O. KOENIG, F. H. HORNE, and D. M. MOHILNER, *J. Am. Chem. Soc.* **83**, 1029 (1961).
14. M. MANES, *J. Phys. Chem.* **67**, 651 (1963).
15. S. R. DE GROOT, *Thermodynamics of Irreversible Processes*, Amsterdam (1951), p. 174.
16. J. MEIXNER, *Kolloid-Z.* **134**, 3 (1953).
17. J. MEIXNER, *Z. Naturforsch.* **4a**, 594 (1949); (a) R. LEFEVER, G. NICOLIS, and I. PRIGOGINE, *J. Chem. Phys.* **47**, 1045 (1967).
18. J. MEIXNER, *Z. Naturforsch.* **9a**, 654 (1954).
19. J. MEIXNER, *Z. Physik*, **139**, 30 (1954).
20. J. MEIXNER, *Rheol. Acta*, **4**, 77 (1965).
21. J. MEIXNER, *Acustica*, **2**, 101 (1952).
22. J. MEIXNER and H. G. REIK, *Handbuch der Physik*, Vol. III/2, Berlin–Goettingen–Heidelberg (1959), p. 479.

CHAPTER 3

PROCESSES IN HETEROGENEOUS (DISCONTINUOUS) SYSTEMS

3-1 INTRODUCTION

As "processes in heterogeneous (discontinuous) systems", we will deal with irreversible processes which relate to the exchange of mass, electricity, and heat between two homogeneous subsystems ("phases"). Let the two phases be isotropic. Let electrification and magnetization as well as interfacial phenomena be excluded. Moreover, let the total system be closed, i.e. let no exchange of matter take place with the surroundings, so that mass exchange takes place only between the two (open) phases.[19] Furthermore, chemical reactions shall be excluded from consideration because everything important about reactions occurs either in the framework of homogeneous systems (Chapter 2) or in the framework of continuous systems (Chapter 4).

The two homogeneous subsystems are separated from each other either by a natural phase boundary (e.g. mass and heat exchange between liquid and its own vapor) or by a "valve". The valve can be a narrow opening, a capillary, a system of capillaries, a membrane, etc. The subsystems, separated by such a valve, can correspond to two different states of the same type of phase, such as two media which exhibit different pressure and temperature but are the same with reference to their state of aggregation and chemical nature. Here the discontinuous system represents a limiting case of a continuous system and certain simplifications in the thermodynamic description result (cf. §3–5).

In order that the methods of thermodynamics of irreversible processes can be applied to the present case, it must be assumed that the exchange of matter, electricity, and heat *between* the two system parts occurs slowly in comparison to the equalization of temperature, pressure, concentrations, and electric potential *inside* each phase (cf. §1–23). Then each of the two subsystems remains homogeneous, even during the course of the irreversible processes, that is, each phase remains in internal thermal and mechanical equilibrium and has uniform composition and uniform potential. Temperature, pressure, concentrations, and potential change discontinuously on crossing the phase boundary and are otherwise functions of time only. This eliminates solid phases from consideration since, in practice, the above conditions can be realized only for fluid systems.

Let the two homogeneous subsystems of our heterogeneous system (Fig. 4) be denoted as phase ' and phase ". According to the above, we can attach a definite value of the (absolute) temperature (T' or T''), of the pressure (P' or P''), of the composition (mole fractions x'_k or x''_k), and of the electric potential (φ' or φ'') to each phase at any arbitrary instant. Also, all other quantities, e.g. the amounts of substance, the internal energy, etc., will be provided with one prime (') or two primes (") as indexes, if they refer to phase ' or " respectively.

Phase '	Phase "
Temperature T' Pressure P'	Temperature T'' Pressure P''
Composition variables x'_k Electric potential φ'	Composition variables x''_k Electric potential φ''

Fig. 4. Heterogeneous (discontinuous) system which consists of two homogeneous subsystems (phase ' and phase ").

3–2 MASS BALANCE

Since our two-phase system (Fig. 4) is assumed to be closed and since chemical reactions are excluded, the "mass balance" can be written:

$$dn'_k + dn''_k = 0. \tag{3–2.1}$$

Here n_k is the amount of substance k.

As a measure of the mass transfer between the two phases, we introduce the *flow of matter* J_k of the species k from phase " to phase ':

$$J_k \equiv \frac{dn'_k}{dt} = -\frac{dn''_k}{dt}, \tag{3–2.2}$$

where t denotes the time. The flow of matter is thus the rate of increase of the amount of substance k in phase ' or the rate of decrease of the amount of substance k in phase ".

If there are charged species (electrons or ions) among the species of the system, we assign to each particle type k a charge number z_k. The quantity z_k is a positive integer for cations (e.g. $z_k = 2$ for Ca^{2+}), a negative integer for anions (e.g. $z_k = -2$ for SO_4^{2-} or $z_k = -1$ for electrons), and zero for uncharged species.

The electric charge transported with a charged species k for an infinitesimal transfer of mass from phase " to phase ' is:

$$z_k F\, dn'_k = -z_k F\, dn''_k, \tag{3–2.3}$$

where F is the Faraday constant.

Using Eqs. (2) and (3) we introduce the *partial electric current* of species k:

$$I_k \equiv z_k F J_k \tag{3–2.4}$$

and for the *total electric current* flowing from phase $''$ to phase $'$

$$I \equiv \sum_k I_k \tag{3–2.5}$$

there results immediately:

$$I = F \sum_k z_k J_k. \tag{3–2.6}$$

The summations in Eqs. (5) and (6) are over all species.

3–3 ENERGY BALANCE

Since our system consists of two open isotropic phases without electrification and magnetization and since only internal changes without dissipative effects† come into consideration, we proceed with the setting up of the "energy balance" using Eq. (1–7.4). This relation provides a definition of the concept of heat for open phases. If we split up the total internal energy of each phase into the "true internal energy" U and the "potential electrostatic energy" E_{pot} following §1–20 then we obtain for our closed total system:

$$\left.\begin{aligned}
dQ' &= dU' + dE'_{pot} + P' \, dV' - \sum_k H'_k \, dn'_k, \\
dQ'' &= dU'' + dE''_{pot} + P'' \, dV'' - \sum_k H''_k \, dn''_k.
\end{aligned}\right\} \tag{3–3.1}$$

Here dQ denotes the heat supplied to the phase in question for an infinitesimal change, P the pressure, V the volume, and H_k the partial molar enthalpy of species k. Thus *all* electric influences are taken into account in the terms dE'_{pot} and dE''_{pot} (regardless of whether we deal with an electric field applied from outside or one arising through the migration of ions).

If we denote the electric potential by φ then we find:

$$\left.\begin{aligned}
dE'_{pot} &= F \, \varphi' \sum_k z_k \, dn'_k, \\
dE''_{pot} &= F \, \varphi'' \sum_k z_k \, dn''_k.
\end{aligned}\right\} \tag{3–3.2}$$

---

† For the case in which a position-invariant electric potential is present in each phase, we need not speak of electrical work on one of the two subsystems but only of that on the total system. Thus we can derive Eq. (5), which refers to a total (closed) system (see below), from the classical formulation (1–4.8) of the first law for closed systems, provided that we allow, besides reversible volume work, electrical work due to the potential difference between the two phases.

Furthermore, according to Eq. (1–7.8), we obtain

$$dQ' = d_aQ' + d_iQ', \\ dQ'' = d_aQ'' + d_iQ'', \Big\}$$

(3–3.3)

where d_aQ or d_iQ is the heat supplied to the phase in question from the surroundings of the total system (from "outside") or from the other phase of the system (from "inside"). From Eqs. (1), (2), and (3) it follows that

$$d_aQ' + d_iQ' = dU' + P'\,dV' + \sum_k (z_kF\varphi' - H_k')\,dn_k', \\ d_aQ'' + d_iQ'' = dU'' + P''\,dV'' + \sum_k (z_kF\varphi'' - H_k'')\,dn_k''.\Bigg\}$$

(3–3.4)

The heat taken up by the closed total system from the surroundings $(d_aQ' + d_aQ'')$ is given by Eqs. (1–4.5) and (1–7.9). In our case, the work done on the total system during an infinitesimal change amounts to (cf. p. 8)

$$-P'\,dV' - P''\,dV'' \text{ (reversible volume work)}$$

and the infinitesimal increase of the energy of the system amounts to (cf. Eqs. 1–4.6 and 1–4.7)

$$dU' + dU'' + dE'_{\text{pot}} + dE''_{\text{pot}}.$$

Thus there results from Eqs. (1–4.5), (1–7.9), and (3–2.1) with Eq. (2)

$$d_aQ' + d_aQ'' = dU' + dU'' + F(\varphi' - \varphi'')\sum_k z_k\,dn_k' \\ + P'\,dV' + P''\,dV''.$$

(3–3.5)

From the relations (4) and (5) with Eq. (3–2.1) we derive

$$d_iQ' + \sum_k H_k'\,dn_k' + d_iQ'' + \sum_k H_k''\,dn_k'' = 0.$$

(3–3.6)

With this we can, by consideration of Eq. (3–2.2), describe the heat transfer between the two phases by the quantity

$$J_Q \equiv \frac{d_iQ'}{dt} = -\frac{d_iQ''}{dt} + \sum_k (H_k'' - H_k')J_k.$$

(3–3.7)

This expression denotes the *heat flow* from phase '' to phase ', i.e. the heat given per unit time to phase ' by phase ''. It is noteworthy that the relation

$$\frac{d_iQ'}{dt} + \frac{d_iQ''}{dt} = 0$$

is only valid if $H_k' = H_k''$ or $J_k = 0$. This is connected with the definition of heat which we chose for open phases (cf. §1–7).

From Eqs. (4) and (5) with Eq. (3–2.3) there follows:

$$dU' + P'\,dV' = d_aQ' + d_iQ' + \sum_k H_k'\,dn_k' - F\varphi' \sum_k z_k\,dn_k',$$

$$dU'' + P''\,dV'' = d_aQ'' + d_iQ'' + \sum_k H_k''\,dn_k'' - F\varphi'' \sum_k z_k\,dn_k''$$

$$= d_aQ'' - \left(d_iQ' + \sum_k H_k'\,dn_k'\right) + F\varphi'' \sum_k z_k\,dn_k'. \quad (3\text{–}3.8)$$

We will use these relations in §3–4.

From Eq. (8) with the definition (3–2.2) and (7) there results:

$$J_Q = -\frac{d_aQ'}{dt} + \frac{dU'}{dt} + P'\frac{dV'}{dt} + \sum_k (z_k F\varphi' - H_k')J_k$$

$$= \frac{d_aQ''}{dt} - \frac{dU''}{dt} - P''\frac{dV''}{dt} + \sum_k (z_k F\varphi'' - H_k')J_k. \quad (3\text{–}3.9)$$

From this it can be shown that the heat flow is independent of the arbitrarily fixed zero point of the internal energy. This is one of the reasons for our definition of heat (§1–7). In §3–9 (p. 189) we give an application of Eq. (9).

3–4 ENTROPY BALANCE

Under the conditions mentioned in §3–1 the Gibbs equation (1–9.5), which is valid for isotropic regions without electrification and magnetization, can be applied to each of the two phases of our system at any arbitrary instant (cf. §1–23):

$$T'\,dS' = dU' + P'\,dV' - \sum_k \mu_k'\,dn_k',$$

$$T''\,dS'' = dU'' + P''\,dV'' - \sum_k \mu_k''\,dn_k''. \quad (3\text{–}4.1)$$

Here T denotes the absolute temperature, S the entropy, and μ_k the chemical potential of species k, each referred to the phase in question.

We introduce the electrochemical potential of species k by Eq. (1–20.15)

$$\eta_k = \mu_k + z_k F\varphi, \quad (3\text{–}4.2)$$

which for all uncharged species ($z_k = 0$) is identical to the chemical potential.

For the total entropy S of the heterogeneous system using Eq. (1–24.2) we obtain

$$S = S' + S''. \quad (3\text{–}4.3)$$

On considering Eqs. (3–2.1) and (3–3.8) we find from Eqs. (1), (2), and (3) that the total entropy is given by:

$$dS = dS' + dS'' = \frac{d_a Q'}{T'} + \frac{d_a Q''}{T''} + \left(\frac{1}{T'} - \frac{1}{T''}\right)\left(d_i Q' + \sum_k H'_k \, dn'_k\right)$$

$$- \sum_k \left(\frac{\eta'_k}{T'} - \frac{\eta''_k}{T''}\right) dn'_k. \tag{3–4.4}$$

We would also obtain this equation if in Eq. (3–3.8) we omitted the terms for the electric potential and, accordingly, assumed that U denotes the total energy (total internal energy) of the phase in question. Then, indeed, the quantity η_k would replace μ_k in Eq. (1).

If, by consideration of Eqs. (3–2.2) and (3–3.7), we introduce the time t explicitly into Eq. (4) then we obtain the "entropy balance" from the general model of Eqs. (1–24.3), (1–24.4), and (1–24.5):

$$\frac{dS}{dt} = \frac{d_a S}{dt} + \frac{d_i S}{dt} \tag{3–4.5}$$

with the "entropy flow"

$$\frac{d_a S}{dt} = \frac{1}{T'} \frac{d_a Q'}{dt} + \frac{1}{T''} \frac{d_a Q''}{dt} \tag{3–4.6}$$

and the "entropy production" (cf. Eq. 1–24.6):

$$\Theta = \frac{d_i S}{dt} = \left(J_Q + \sum_k H'_k J_k\right)\left(\frac{1}{T'} - \frac{1}{T''}\right) - \sum_k J_k\left(\frac{\eta'_k}{T'} - \frac{\eta''_k}{T''}\right) \geqslant 0. \tag{3–4.7}$$

The inequality sign in (7) is true for actual (irreversible) processes in the system, the equality sign for reversible limiting cases.

The entropy flow (6) includes the heat exchanged between system and surroundings and can accordingly be positive, negative, or zero. The entropy production (7) relates to processes inside the system (matter, electricity, and heat exchange between the two phases). It is never negative, and in fact is always positive for actual courses of the processes.[20]

Apart from the trivial case in which the phase boundary is impermeable to matter and heat ($J_k = 0$, $J_Q = 0$), according to Eq. (7) the entropy production Θ generally disappears only under the conditions

$$T' = T'', \tag{3–4.8}$$

$$\eta'_k = \eta''_k. \tag{3–4.9}$$

According to §1–20, these relations represent nothing other than the general conditions for heterogeneous equilibrium. In this case,

$$J_Q = 0, \qquad J_k = 0 \tag{3–4.10}$$

are simultaneously valid. Thus, equilibrium between the two phases corresponds to the "reversible limiting case" and thus to the equality sign in Eq. (7).

For a phase boundary which is only semipermeable (e.g. a semipermeable membrane) and is impermeable to species of the type j, the mass flows J_j disappear directly, and the conditions (9) for equilibrium for the electrochemical potentials η_j are eliminated (cf. §1–20).

Should the electric current and the electric potential be allowed to appear explicitly in Eq. (7) then we take into account Eq. (3–2.6) and Eq. (2). We thus obtain for the entropy production

$$\Theta = \left(J_Q + \sum_k H'_k J_k\right)\left(\frac{1}{T'} - \frac{1}{T''}\right) - \sum_k J_k\left(\frac{\mu'_k}{T'} - \frac{\mu''_k}{T''}\right)$$

$$-I\left(\frac{\varphi'}{T'} - \frac{\varphi''}{T''}\right) \geq 0. \qquad (3\text{-}4.11)$$

3–5 DISSIPATION FUNCTION NEAR TO EQUILIBRIUM

Let Z' or Z'' be a function of state referred to phase $'$ or phase $''$ respectively (including the electric potential φ' or φ''). We then write (cf. Fig. 5)

$$Z' \equiv Z, \qquad Z'' - Z' \equiv \Delta Z \qquad (3\text{-}5.1)$$

and, according to Eq. (1–24.7), introduce the *dissipation function*

$$\Psi \equiv T\Theta. \qquad (3\text{-}5.2)$$

From this it follows with Eqs. (3–4.7) and (3–4.11) that

$$\Psi = \left(J_Q + \sum_k H_k J_k\right)\frac{\Delta T}{T + \Delta T} + \sum_k J_k T\,\Delta\!\left(\frac{\eta_k}{T}\right)$$

$$= \left(J_Q + \sum_k H_k J_k\right)\frac{\Delta T}{T + \Delta T} + \sum_k J_k T\,\Delta\!\left(\frac{\mu_k}{T}\right) + IT\,\Delta\!\left(\frac{\varphi}{T}\right) \geq 0. \quad (3\text{-}5.3)$$

This relation is quite general for the dissipation function.

Phase $'$	Phase $''$
Temperature T Pressure P	Temperature $T + \Delta T$ Pressure $P + \Delta P$
Composition variables x_k Electric potential φ	Composition variables $x_k + \Delta x_k$ Electric potential $\varphi + \Delta\varphi$

Fig. 5. The same system as in Fig. 4 (p. 158) with new notation.

We now assume:

$$|\Delta T| \ll T, \tag{3-5.4}$$

a condition which can always be satisfied in practice, since otherwise "discontinuous" systems can hardly be experimentally realized. There results from Eqs. (3) and (4), since we can arbitrarily set $\varphi = 0$ and thus eliminate the term $-I\varphi\,\Delta T/T$,

$$\Psi = \left(J_Q + \sum_k H_k J_k\right)\frac{\Delta T}{T} + \sum_k J_k T\,\Delta\!\left(\frac{\eta_k}{T}\right)$$

$$= \left(J_Q + \sum_k H_k J_k\right)\frac{\Delta T}{T} + \sum_k J_k T\,\Delta\!\left(\frac{\mu_k}{T}\right) + I\,\Delta\varphi \geqslant 0. \tag{3-5.5}$$

Because of (4), the average absolute temperature of the system can be inserted for T.

Before we can further simplify Eq. (5) we must introduce several fundamental considerations relating to the different possible types of heterogeneous systems.

We distinguish between four classes of discontinuous systems (cf. §3–1):

1. *Liquid–Vapor System.* Of the two subsystems one is liquid, the other gas. Mass transfer is evaporation or condensation. The electrochemical potentials η_k' and η_k'' refer, in this case, to two different states of aggregation, and are thus *different* functions of the independent variables, that is, of the temperature T, the pressure P, the electric potential φ, and the independent mole fractions x_j.

2. *Two-phase Liquid System.* Both subsystems are liquid but even in equilibrium they are not identical. They are distinguished at least with respect to their compositions, i.e. they are not entirely miscible. Mass transfer here denotes exchange of matter between the two liquid phases. The electrochemical potentials η_k' and η_k'' now relate to two phases of the same state of aggregation and can thus be regarded as two different values *of the same* function.

3. *Valve System of the First Kind.* Both subsystems are either gaseous or liquid (without miscibility gaps) and are separated from one another by a valve which is permeable to all species (permeable membrane, capillary, porous wall, etc.). In the equilibrium case, both subsystems become identical. Here the electrochemical potentials η_k' and η_k'' are again different values *of the same* function.

4. *Valve System of the Second Kind.* Both systems are either gaseous or liquid (without miscibility gaps) and are separated from one another by a valve permeable only to certain species (semipermeable membrane). Equilibrium can be set up only for a *rigid* membrane, which allows the establishment of a pressure difference between the two phases ("osmotic equilibrium"). In the case of equilibrium, the two subsystems are distinguished at least with regard to their pressures and compositions. But the electrochemical potentials η'_k and η''_k here are only different values *of the same* function.

For the present, we consider only the three last classes of systems and thus exclude liquid–vapor systems. Accordingly, the quantities

$$\eta'_k = \eta_k, \qquad \eta''_k = \eta_k + \Delta\eta_k$$

are to be regarded in the following as two different values of the same function $\eta_k(T, P, \varphi, x_j)$.

Moreover, we assume *small* values of ΔT, ΔP, $\Delta\varphi$, and Δx_j. Then we can set up a Taylor expansion for $\Delta(\eta_k/T)$ in Eq. (5) and truncate the series after the linear terms. We thus obtain with Eqs. (1–12.8), (1–12.12), and (3–4.2)

$$T \Delta\left(\frac{\eta_k}{T}\right) = T \Delta\left(\frac{\mu_k}{T}\right) + z_k F \Delta\varphi, \tag{3–5.6}$$

$$T \Delta\left(\frac{\mu_k}{T}\right) = -\frac{H_k}{T} \Delta T + (\Delta\mu_k)_T, \tag{3–5.7}$$

$$(\Delta\mu_k)_T = V_k \Delta P + \sum_j \left(\frac{\partial\mu_k}{\partial x_j}\right)_{T,P} \Delta x_j. \tag{3–5.8}$$

Here V_k is the partial molar volume of species k. The sum is over all independent mole fractions. The partial differentiation with respect to the mole fraction is valid for given values of the temperature T and the pressure P with all other independent mole fractions constant. Eqs. (6) and (7) only require small values of ΔT.

Since our further calculations are based on the "phenomenological equations" which assume *proximity to equilibrium* (§3–6), we have to investigate how far the condition of proximity to equilibrium corresponds to the assumption of small values of ΔT, ΔP, $\Delta\varphi$, and Δx_j. For a valve system of the first kind, both subsystems are identical at equilibrium; here the two assumptions mean the same thing. For a valve system of the second kind, on the other hand, "proximity to equilibrium" need not correspond to the requirement of small values for ΔT, ΔP, etc., since, in the equilibrium case, high pressure differences ("osmotic pressures") correspond to large concentration differences. We must thus, for valve systems of the second kind, assume small concentration differences from the beginning, so that the condition of proximity to equilibrium is again given by the relations (6) to (8). Two-phase liquid systems, for the case of proximity to equilibrium, cannot be described by Eqs. (6) to (8) at all, because, here, the following conditions must be satisfied for equilibrium:

$$\Delta\eta_k = 0, \quad \Delta T = 0, \quad \Delta P = 0, \quad \Delta x_k \neq 0.$$

These contradict Eqs. (6) to (8) if they are applied to the environment of equilibrium.

With the abbreviations

$$X_Q \equiv \frac{\Delta T}{T}, \tag{3-5.9}$$

$$X_k \equiv (\Delta\mu_k)_T + z_k F \Delta\varphi \tag{3-5.10}$$

there results from Eqs. (5), (6), and (7)

$$\Psi = J_Q X_Q + \sum_k J_k(\Delta\mu_k)_T + I \Delta\varphi$$

$$= J_Q X_Q + \sum_k J_k X_k \geqslant 0. \tag{3-5.11}$$

Here the dissipation function Ψ has the form given in Eq. (1–24.22): it is a sum of products of "generalized fluxes" and appropriate "generalized forces."

Should the quantities

$$J_Q, \quad J_k, \quad I, \tag{3-5.12}$$

be chosen as the "fluxes" then the appropriate "forces" are, according to Eq. (11),

$$X_Q, \quad (\Delta\mu_k)_T, \quad \Delta\varphi. \tag{3-5.13}$$

The last two quantities in (13) need not disappear in the equilibrium case (e.g. valve system of the second kind with a semipermeable membrane, permeable to ions).

If we choose, however,

$$J_Q, \quad J_k \tag{3-5.14}$$

as the "fluxes" (and we will do so consistently in what follows) and accordingly as "forces"

$$X_Q, \quad X_k, \tag{3-5.15}$$

following Eq. (11), then all these quantities disappear in the equilibrium case. This arises from Eqs. (3–4.2), (3–4.8), (3–4.9), (3–4.10), as well as Eqs. (1), (9), and (10).

Since the equality in (11) relates to heterogeneous equilibrium (cf. §3–4) and since we are only interested in the irreversible (actual) course of the processes, we finally write (11) in the form:

$$\Psi = J_Q X_Q + \sum_k J_k X_k > 0. \tag{3-5.16}$$

This relation is analogous to the de Donder inequality (1–24.21) and, like this, is a special case of the general inequality (1–24.23).

From the above, the expression (16) is valid for all types of heterogeneous systems, excepting liquid–vapor systems, assuming only sufficiently small values of ΔT, ΔP, $\Delta \varphi$, and Δx_k. However, if we further assume proximity to equilibrium—as in the beginning of the next section—then we need apply (16) only to valve systems. In this case, the discontinuous system represents a limiting case of a continuous system (Chapter 4). For all other kinds of heterogeneous systems, however, we must proceed from the more general relation (5) for the dissipation function (cf. the conclusion of §3–6).

3-6 PHENOMENOLOGICAL EQUATIONS AND ONSAGER'S RECIPROCITY LAW

The "generalized fluxes" J_Q (heat flow) and J_k (material flow) as well as the "generalized forces" X_Q and X_k, which appear in the dissipation function (3–5.16), all individually disappear in the equilibrium case (see §3–5). This suggests that deviations from equilibrium, i.e. the irreversible processes which occur in our system, can, as a first approximation, be described by homogeneous linear relations between all "fluxes" and "forces." These relations, which are valid for *states near equilibrium*, are called *phenomenological equations*. When the phase boundary is permeable to N species these are as follows:

$$
\begin{aligned}
J_1 &= \alpha_{11} X_1 + \alpha_{12} X_2 + \cdots + \alpha_{1N} X_N + \alpha_{1Q} X_Q, \\
J_2 &= \alpha_{21} X_1 + \alpha_{22} X_2 + \cdots + \alpha_{2N} X_N + \alpha_{2Q} X_Q, \\
&\quad \cdot \quad \cdot \quad \cdot \quad \cdot \quad \cdot \quad \cdot \quad \cdot \quad \cdot \quad \cdot \quad \cdot \quad \cdot \quad \cdot \\
J_N &= \alpha_{N1} X_1 + \alpha_{N2} X_2 + \cdots + \alpha_{NN} X_N + \alpha_{NQ} X_Q, \\
J_Q &= \alpha_{Q1} X_1 + \alpha_{Q2} X_2 + \cdots + \alpha_{QN} X_N + \alpha_{QQ} X_Q.
\end{aligned}
\tag{3-6.1}
$$

Here the quantities α_{11}, α_{12}, ..., α_{QQ} are *phenomenological coefficients* whose significance will be explained by examples given in the following sections. They are independent of the generalized fluxes and forces but they can depend in an arbitrary way on the mean values of the state variables of the system (temperature, pressure, composition, etc.) and on the type and the dimensions of the valve.

The linear equations (1) are analogous to the relations (2–4.1) for chemical reactions and, like these, represent a special case of the general equations (1–25.1).

According to the *reciprocity law of Onsager* (1–26.1) we find:

$$\alpha_{ij} = \alpha_{ji} \quad (i, j = 1, 2, \ldots, N, Q) \tag{3–6.2}$$

(cf. Eq. 2–4.2). The coefficient matrix is thus symmetric. The meaning of this statement will also become clear in the particular examples given later.

> Since the generalized flows and forces in Eq. (1) are all independent (cf. §3–5), no linear dependences between the phenomenological coefficients exist, and the validity of the Onsager reciprocity law (2) is not subject to doubt (cf. §1–27).

If we introduce the phenomenological equations (1) with the Onsager reciprocity relations (2) into the expression (3–5.16) for the dissipation function then we find that the dissipation function for irreversible processes near to equilibrium is of a positive quadratic form. From this there results that the determinant of the phenomenological coefficients with all principal cofactors is positive. This is an example of the general inequalities (1–26.5) and (1–26.6) and is analogous to the special expressions (2–4.4) and (2–4.5) for chemical reactions:

$$\alpha_{ii} > 0, \quad \alpha_{ii}\alpha_{jj} - \alpha_{ij}^2 > 0 \quad (i, j = 1, 2, \ldots, N, Q; i \neq j), \tag{3–6.3}$$

$$\begin{vmatrix} \alpha_{11} & \alpha_{12} & \cdots & \alpha_{1N} & \alpha_{1Q} \\ \alpha_{21} & \alpha_{22} & \cdots & \alpha_{2N} & \alpha_{2Q} \\ \cdot & \cdot & \cdot & \cdot & \cdot \\ \alpha_{N1} & \alpha_{N2} & \cdots & \alpha_{NN} & \alpha_{NQ} \\ \alpha_{Q1} & \alpha_{Q2} & \cdots & \alpha_{QN} & \alpha_{QQ} \end{vmatrix} > 0. \tag{3–6.4}$$

The consequences of this system of inequalities will also be more fully discussed by examples given in the following.

According to §3–5, Eqs. (1) to (4) are valid only for valve systems with small values of the temperature difference ΔT, of the pressure difference ΔP, of the electric potential difference $\Delta \varphi$, and of the independent mole fraction differences Δx_j. The explicit expressions for the "forces" X_Q and X_k are given by Eqs. (3–5.8), (3–5.9), and (3–5.10):

$$X_Q = \frac{\Delta T}{T}, \tag{3–6.5}$$

$$X_k = V_k \, \Delta P + \sum_j \left(\frac{\partial \mu_k}{\partial x_j} \right)_{T,P} \Delta x_j + z_k F \, \Delta \varphi \quad (k = 1, 2, \ldots, N). \tag{3–6.6}$$

Here T denotes the average absolute temperature of the system, F the Faraday constant, z_k the charge number, μ_k the chemical potential, and V_k the partial molar volume of species k. The summation is over all independent mole fractions.

For systems of several species, the individual material fluxes J_k are often not measured, only the volumetric throughput ("volume flow")

$$J \equiv \frac{dV}{dt} \qquad (3\text{-}6.7)$$

being available. In this expression, V is the volume of the phase $'$ and t denotes the time. Equation (7) is the rate of increase of the volume of phase $'$ (temperature T, pressure P, etc.) and simultaneously—exactly for constant volume of the total system, otherwise approximately—the rate of decrease of the volume of phase $''$ (temperature $T + \Delta T$, pressure $P + \Delta P$, etc.). If we assume that temperature and pressure are constant in time in phase $'$ then there follows with Eqs. (1-6.3), (3-2.2), and (3-5.1):

$$\frac{dV}{dt} = \sum_{k=1}^{N} V_k \frac{dn_k}{dt} = \sum_{k=1}^{N} V_k J_k.$$

From this there results with Eq. (7):

$$J = \sum_{k=1}^{N} V_k J_k, \qquad (3\text{-}6.8)$$

a relation which we will use repeatedly.†

For the special case $\Delta x_j = 0$ (one-component or multi-component system with the same compositions in both phases) we can obtain a particularly simple form of the dissipation function (3-5.16) from Eqs. (5), (6), and (8) taking account of Eq. (3-2.6):

$$\Psi = J \Delta P + I \Delta\varphi + J_Q \frac{\Delta T}{T} > 0. \qquad (3\text{-}6.9)$$

Here I is the electric current intensity. Since the special case described in Eq. (9) arises only for valve systems of the first kind and since the quantities J, I, and J_Q or ΔP, $\Delta\varphi$, and ΔT respectively disappear in the equilibrium case (cf. §3–5), we can choose these as "fluxes" or "forces" and can formulate

---

† For many experimental arrangements, the pressure in either of the two phases is held constant in time. In this case, we regard Eq. (8) as a definition of the quantity J. For liquid media, Eq. (7) is then always a good approximation, so that the name "volume flow" for J remains justified. For gaseous media, there exists, however, an actual discrepancy between statements (7) and (8) if the time-invariant pressure requirement for phase $'$ is not satisfied. We will thus not use the "volume flow" anywhere if experiments with gases are involved (cf. §3–10). [21]

new phenomenological equations which are simpler than the previous relations (1):

$$J = a_{11}\,\Delta P + a_{12}\,\Delta\varphi + a_{13}\frac{\Delta T}{T},$$

$$I = a_{21}\,\Delta P + a_{22}\,\Delta\varphi + a_{23}\frac{\Delta T}{T},$$

$$J_Q = a_{31}\,\Delta P + a_{32}\,\Delta\varphi + a_{33}\frac{\Delta T}{T}.$$

(3-6.10)

There exist three relations between the nine phenomenological coefficients $a_{11}, a_{12}, \ldots, a_{33}$ according to the Onsager reciprocity law (1-26.1):

$$a_{12} = a_{21}, \quad a_{13} = a_{31}, \quad a_{23} = a_{32}, \tag{3-6.11}$$

so that only six independent coefficients still remain. The following inequalities result from the requirement that the dissipation function is positive (9) and from (10) and (11):

$$a_{11} > 0, \quad a_{22} > 0, \quad a_{33} > 0, \tag{3-6.12}$$

$$a_{11}a_{22} - a_{12}^2 > 0, \quad a_{11}a_{33} - a_{13}^2 > 0, \quad a_{22}a_{33} - a_{23}^2 > 0, \tag{3-6.13}$$

$$\begin{vmatrix} a_{11} & a_{12} & a_{13} \\ a_{21} & a_{22} & a_{23} \\ a_{31} & a_{32} & a_{33} \end{vmatrix} > 0. \tag{3-6.14}$$

Equations (10) to (14) contain the phenomenological description of electrokinetic effects (§3-7) and of thermomechanical effects (§3-10).

Another noteworthy special case is represented by a multi-component system at disappearing pressure difference ($\Delta P = 0$). We introduce the molarity c_k with Eqs. (1-6.4), (1-6.5), and (1-6.6):

$$c_k = \frac{x_k}{\bar{V}} \quad (k = 1, 2, \ldots, N), \tag{3-6.15}$$

where $\bar{V}$ denotes the molar volume of the mixture. We then use (1-6.11),

$$\sum_{k=1}^{N} c_k V_k = 1, \tag{3-6.16}$$

and the Gibbs–Duhem equation (1-13.6) in the form

$$\sum_{k=1}^{N} c_k \left(\frac{\partial\mu_k}{\partial x_j}\right)_{T,P} = 0 \tag{3-6.17}$$

plus the condition of electric neutrality valid for each conducting medium

$$\sum_{k=1}^{N} c_k z_k = 0. \tag{3-6.18}$$

With this it follows from Eq. (6) that

$$\sum_{k=1}^{N} c_k X_k = \Delta P. \tag{3-6.19}$$

For $\Delta P = 0$ there are only $N - 1$ independent quantities among the generalized forces $X_1, X_2, \ldots, X_N$. According to §1-27, it is then expedient to eliminate the dependent "force" before setting up the phenomenological equations. If we choose $X_1, X_2, \ldots, X_{N-1}$ as independent "forces" then Eqs. (15) and (19) yield

$$X_N = -\sum_{j=1}^{N-1} \frac{x_j}{x_N} X_j \quad (\Delta P = 0).$$

With this and Eq. (3–5.16) we find for the dissipation function:

$$\Psi = J_Q X_Q + \sum_{j=1}^{N-1} \left(J_j - \frac{x_j}{x_N} J_N \right) X_j \geq 0 \quad (\Delta P = 0), \tag{3-6.20}$$

where only independent "fluxes" and "forces" are contained. Now, the fluxes

$$J_1 - \frac{x_1}{x_N} J_N, \quad J_2 - \frac{x_2}{x_N} J_N, \ldots, J_{N-1} - \frac{x_{N-1}}{x_N} J_N$$

are assigned to the forces $X_1, X_2, \ldots, X_{N-1}$. This must be noted when setting up the phenomenological equations. Meanwhile, we do not initially assume the condition $\Delta P = 0$ for our calculations, so that we can proceed as usual from the relations (1) or (10).[22]

The cases excluded thus far in this section (liquid–vapor systems, two-phase liquid systems, or valve systems of the second kind with large pressure and concentration differences) can also be treated using the methods of thermodynamics of irreversible processes if we proceed from the more general relation (3–5.5) instead of from Eq. (3–5.16). We can regard the quantities appearing in the expression (3–5.5)

$$J_1, J_2, \ldots, J_N, J_Q + \sum_{k=1}^{N} H_k J_k$$

as "fluxes" and the conjugate quantities

$$T \Delta\left(\frac{\eta_1}{T}\right), \ T \Delta\left(\frac{\eta_2}{T}\right), \ldots, T \Delta\left(\frac{\eta_N}{T}\right), \ \frac{\Delta T}{T}$$

as "forces." Since the "fluxes" and "forces" so chosen disappear at equilibrium, we can set up phenomenological relations between these quantities which are analogous to the equations (1) and which again are valid for states near equilibrium[1].

Should we want to describe, for example, the isothermal evaporation and condensation phenomena near equilibrium for a system of two non-electrolyte components then we must follow Eqs. (3–5.1) and (3–5.5) and set:

$$J_1 = \alpha_{11}(\mu_1'' - \mu_1') + \alpha_{12}(\mu_2'' - \mu_2'),$$
$$J_2 = \alpha_{21}(\mu_1'' - \mu_1') + \alpha_{22}(\mu_2'' - \mu_2'),$$

where α_{11}, α_{12}, α_{21}, and α_{22} are phenomenological coefficients for which the Onsager reciprocity law (1–26.1) states:

$$\alpha_{12} = \alpha_{21}.$$

If phase $'$ is the liquid and phase $''$ the vapor then, according to Eq. (3–2.2), J_1 and J_2 are the rates of condensation of the components 1 and 2 respectively and μ_i' and μ_i'' the chemical potentials of the components ($i = 1, 2$) in the liquid and in the vapor respectively.

All examples which follow again refer only to valve systems with small values of the temperature difference, of the pressure difference, of the electric potential difference, and of the concentration differences, so that the relations (1) to (6) are always applicable.

3–7 ELECTROKINETIC EFFECTS

(a) Empirical Description

We consider a liquid system which is split up into two subsystems by a capillary, by a porous wall, or by a permeable membrane. These two subsystems should have the same temperatures and the same compositions. However, we allow a pressure difference ΔP and an electric potential difference $\Delta\varphi$ to exist between the two subsystems. (The liquid shall contain movable electrons or ions, such as mercury or an aqueous electrolyte solution.) Then matter or electricity can flow through the valve (capillary, etc.). We introduce the volume flow J or the electric current I (cf. §3–6 and Fig. 6) as a measure of mass or charge transport.

If matter flows through the valve at a vanishingly small electric potential difference $\Delta\varphi = 0$ due to a pressure difference $\Delta P \neq 0$, we have the phenomenon known as "permeation" or "capillary flow." We will call this phenomenon *permeation*. Since the volume flow J is inversely proportional to the thickness δ of the valve (thickness of the membrane or porous wall or

Phase ′	Phase ″
Pressure P Electric potential φ	Pressure $P + \Delta P$ Electric potential $\varphi + \Delta\varphi$

$$\leftarrow J > 0 \quad \text{(volume flow)}$$
$$\leftarrow I > 0 \quad \text{(electric current)}$$

Fig. 6. Heterogeneous (discontinuous) system with transport of matter and electricity between the two subsystems (phase ′ and phase ″) as a special case of the system in Fig. 5 (p. 163).

length of the capillary) and—at least for membranes—is proportional to the effective cross section q of the valve, we set:

$$J = a\frac{q}{\delta}\Delta P \quad (\Delta\varphi = 0) \tag{3–7.1}$$

and call a the *permeability*. For values of the pressure difference ΔP which are not too high, the permeability a is found from experience to be independent of ΔP but to depend on temperature, composition, and, to a smaller extent, on the average pressure of the system. The quantity a is positive, so that, according to Fig. 6, the material flow is always from the higher to the lower pressure.

For illustration, we will express a for Poiseuille flow by quantities which can be measured in another way. For a capillary of (effective) radius r, i.e. of effective cross section $q = \pi r^2$, Poiseuille's law states:

$$a = \frac{r^2}{8\eta}, \tag{3–7.2}$$

where η denotes the viscosity of the liquid. Here a is proportional to the cross section q of the capillary, so that Eq. (1) requires J to be proportional to q^2. A porous wall can naturally be regarded as a system of capillaries.

The situation for membranes is more complicated[2–4]. Here, depending on the ratio of the molecular diameter of the permeating substance to the pore diameter of the membrane pores, Poiseuille flow (laminar flow of macroscopic parts of the liquid) or a diffusion mechanism will describe the process. The latter is the process of diffusion within the membrane, which is then to be interpreted as a binary system "membrane substance + permeating species," with a concentration drop due to the pressure difference. A combination of both mechanisms may be present.

If an electric current flows through the valve due to a potential difference ($\Delta\varphi \neq 0$) at a vanishingly small pressure difference ($\Delta P = 0$) then the normal *electric conduction* is present and Ohm's law is valid if the potential difference is not too high. If we introduce[23] the *specific electric conductivity* or *specific conductance* in place of the resistivity ρ

$$\kappa = \frac{1}{\rho},\tag{3–7.3}$$

we obtain:

$$I = \kappa \frac{q}{\delta} \Delta\varphi \quad (\Delta P = 0).\tag{3–7.4}$$

The positive quantity κ does not depend on $\Delta\varphi$, q, and δ but on the temperature, on the composition, and, to a lesser extent, on the average pressure of the system. Equation (4) is analogous to the relation (1).

In a solution which contains ions of normal size or colloids, the migrating of charge carriers in an external electric field is called "electrolysis" or "electrophoresis" respectively. If the charged particles form assemblages which are so large that a "suspension" and finally two spatially separated "phases" are present then a motion of the material in an electric field is still found and this is designated *electroosmosis*. In our case, the liquid of which the two subsystems are composed is the one phase, while the solid "immovable" substance of which the capillary, the porous wall, or the membrane consists represents the other phase. At points of contact between the two phases, electric potential differences ("electric double layers") are formed. If an external electric field is applied then the movable phase migrates, that is, the liquid moves relative to the immovable phase, i.e. the valve. Electroosmosis is noticeable in our system if an external electric potential difference ($\Delta\varphi \neq 0$) between the two subsystems at vanishingly small pressure difference ($\Delta P = 0$) gives rise to material flow through the valve.

We describe the electroosmosis quantitatively by an equation analogous to Eqs. (1) and (4):

$$J = b \frac{q}{\delta} \Delta\varphi \quad (\Delta P = 0)\tag{3–7.5}$$

and call b the *electroosmotic permeability*. Experience shows† that the electroosmotic permeability b does not depend on $\Delta\varphi$ at reasonable potential differences, but on the temperature, on the composition, and, to a lesser degree, on the average pressure of the system. The quantity b can be either positive or negative.

There is also a reciprocal effect to electroosmosis: an electric current, the so-called "streaming current," is caused by a pressure difference

---

† The electroosmosis has been known for a long time: in 1809, Reuss observed the transport of water through a porous clay wall on applying an electric voltage. A short but good survey on the modern aspects of "electrokinetic effects" is given by Adam[5].

($\Delta P \neq 0$) between two subsystems at zero potential difference ($\Delta \varphi = 0$). We call the phenomenon *streaming conduction* and describe it in a manner analogous to Eqs. (1), (4), and (5) by the following equation:

$$I = b* \frac{q}{\delta} \Delta P \quad (\Delta \varphi = 0). \tag{3-7.6}$$

Here $b*$ is the *streaming conductivity*. Again, for sufficiently small values of ΔP, $b*$ is independent of ΔP but dependent on the temperature, on the composition, and, to a lesser degree, on the average pressure of the system. The quantity $b*$ can be either positive or negative.

The most general case ($\Delta P \neq 0$, $\Delta \varphi \neq 0$) corresponds to the simultaneous appearance of permeation, electric conduction, electroosmosis, and streaming conduction which superimpose linearly according to experience.

We obtain a generalization of Eqs. (1), (4), (5), and (6):

$$J = \frac{q}{\delta} (a \, \Delta P + b \, \Delta \varphi), \tag{3-7.7}$$

$$I = \frac{q}{\delta} (b* \, \Delta P + \kappa \, \Delta \varphi). \tag{3-7.8}$$

If we eliminate $\Delta \varphi$ from one of these two equations and ΔP from the other then we find:

$$J = \frac{b}{\kappa} I + \frac{q}{\delta} \left(a - \frac{bb*}{\kappa} \right) \Delta P, \tag{3-7.9}$$

$$I = \frac{b*}{a} J + \frac{q}{\delta} \left(\kappa - \frac{bb*}{a} \right) \Delta \varphi. \tag{3-7.10}$$

Electroosmosis can be described by Eq. (9) instead of by Eq. (5) as follows:

$$J = \frac{b}{\kappa} I \quad (\Delta P = 0), \tag{3-7.11}$$

and streaming conduction by Eq. (10) instead of by Eq. (6):

$$I = \frac{b*}{a} J \quad (\Delta \varphi = 0). \tag{3-7.12}$$

The experimental determination of the electroosmotic permeability b usually follows from Eq. (11).

The electric potential difference which is set up at zero electric current ($I = 0$) with a material flow ($\Delta P \neq 0$, $J \neq 0$) is called *streaming potential*. Using Eqs. (8) and (10) we write for this quantity:

$$\Delta \varphi = -\frac{b*}{\kappa} \Delta P = -\frac{b*}{a\kappa - bb*} \frac{\delta}{q} J (I = 0). \tag{3-7.13}$$

The streaming potential is that quantity which is used most often for the experimental determination of the streaming conductivity b^* from the relation:

$$\left(\frac{\Delta\varphi}{\Delta P}\right)_{I=0} = -\frac{b^*}{\kappa}.$$

The streaming conduction described by Eqs. (6) or (12) and the streaming potential (13) are thus only different expressions for the same phenomenon.†

If we apply an electric potential difference $\Delta\varphi$ to our system under the initial condition $\Delta P = 0$, a mass flux results from electroosmosis according to Eq. (5). Thus a pressure difference ($\Delta P \neq 0$) between the two subsystems is gradually built up, only provided that a movable valve (e.g. a membrane, attached so that it gives slightly with pressure differences) is not used. The higher pressure is present in that subsystem to which matter flows by electroosmosis. According to Eq. (1), permeation takes place on account of the pressure difference. This acts against the electroosmosis due to the requirement that $a > 0$ (see above) since material transported by permeation flows from the higher to the lower pressure. Finally, at fixed values of $\Delta\varphi$ a *stationary state* (cf. §1–17) is achieved, for which the material flow through the valve disappears ($J = 0$) and due to which not only the given electric potential but also the pressure becomes constant in each subsystem. The pressure difference ΔP_{stat} belonging to the stationary state is called *electroosmotic pressure difference* and, from Eqs. (7) and (9) with $J = 0$, it is seen to be

$$\Delta P_{\text{stat}} = -\frac{b}{a}\Delta\varphi = -\frac{b}{a\kappa - bb^*}\frac{\delta}{q}I. \tag{3–7.14}$$

In the stationary state, the electric current I thus does not disappear.‡

---

† The streaming potential was first observed by Quincke in 1859 on forcing water through porous clay membranes. Thus the reciprocal effect to the electroosmosis discovered by Reuss in 1809 was found.

‡ The contradiction to the general stationary condition (3–9.11), according to which all particle flows must disappear individually as must also the electric current, is explained as follows: we have assumed that the two subsystems have the same compositions and the same temperatures at the given electric potential difference whereas in §3–9 different, but fixed, temperatures are assumed, to which all other quantities should adjust themselves automatically. Strictly speaking, the first case is impossible, for due to the variable permeability of the valve for different species the condition of equal composition of the two subsystems at constant potential difference cannot be maintained unless we consider electron conductors, e.g. mercury. The state described by Eq. (14) can thus be realized in practice, since the volume flow can still disappear before the formation of noticeable concentration differences, but in general it represents no stationary final state in the sense of §3–9.

The different effects expressed in Eqs. (5), (6), and (11) to (14) will be called *electrokinetic effects*.

(b) Thermodynamic–Phenomenological Description

According to our purely empirical presentation thus far, all transport processes possible in the systems considered, that is, permeation, electric conduction, and the different electrokinetic effects, can be described by four coefficients: a (permeability), κ (electric conductivity), b (electroosmotic permeability), and b^* (streaming conductivity). We now want to examine any additional statements which the thermodynamics of irreversible processes can contribute to this[6].

According to §3–6, Eqs. (3–6.10) to (3–6.14) are valid for the present case under the conditions $\Delta T = 0, J_Q = 0$. We thus obtain the phenomenological equations

$$\left.\begin{array}{l} J = a_{11}\,\Delta P + a_{12}\,\Delta\varphi, \\ I = a_{21}\,\Delta P + a_{22}\,\Delta\varphi, \end{array}\right\} \tag{3–7.15}$$

with the phenomenological coefficients $a_{11}, a_{12}, a_{21}, a_{22}$, for which the Onsager reciprocity law

$$a_{12} = a_{21} \tag{3–7.16}$$

and the inequality statements (cf. 1–26.7)

$$a_{11} > 0, \quad a_{22} > 0, \quad a_{11}a_{22} - a_{12}^2 > 0 \tag{3–7.17}$$

must be satisfied.

We recognize that the empirical equations (7) and (8) which contain all the effects described thus far are identical with the phenomenological equations (15). Accordingly, the condition of "near-equilibrium," assumed in (15), corresponds exactly to the region of validity for the empirical linear relations between "fluxes" (J, I) and "forces" $(\Delta P, \Delta\varphi)$. In the case of electric conduction, for example, we exceed the region of validity for the phenomenological equations only if we go to extremely high field strengths for which Ohm's law is not applicable.

We thus obtain the relations (7) and (8), first set forth as experimental facts, by the *systematic* considerations of the thermodynamics of irreversible processes.

Statements which go beyond those already known are obtained by introducing the Onsager reciprocity relations (16) and the inequalities (17) which follow from the positive character of the entropy production (cf. §3–6).

We first express the phenomenological coefficients a_{11}, a_{12}, a_{21}, and a_{22} in terms of the conventional transport coefficients a, b, b^*, and κ. We find, on comparison of Eqs. (7) and (8) with Eq. (15),

$$\left.\begin{array}{cc} a_{11} = a\dfrac{q}{\delta}, & a_{12} = b\dfrac{q}{\delta}, \\[2mm] a_{21} = b^*\dfrac{q}{\delta}, & a_{22} = \kappa\dfrac{q}{\delta}. \end{array}\right\} \tag{3–7.18}$$

With Eq. (18) the Onsager reciprocity law (16) yields

$$b = b^*. \tag{3–7.19}$$

The electroosmotic permeability b is thus equal to the streaming conductivity b^*. According to Eqs. (5) and (6) or to Eqs. (11) and (12), the "cross phenomenon," electroosmosis, and the pertinent "reciprocal effect," streaming conduction, can be described by a single coefficient. Thus the number of the independent coefficients necessary for the presentation of all irreversible processes in our system is reduced to *three*. We can thus characterize all transport processes, for example, by the quantities a, b, and κ.

Special consequences to the different electrokinetic effects result from introducing Eq. (19) into Eqs. (5) to (14). From Eqs. (13) and (19) we find for the streaming potential:

$$\Delta\varphi = -\frac{b}{\kappa}\,\Delta P = -\frac{b}{a\kappa - b^2}\frac{\delta}{q}J \quad (I = 0) \tag{3–7.20}$$

and from Eqs. (14) and (19) we obtain for the electroosmotic pressure difference

$$\Delta P_{\text{stat}} = -\frac{b}{a}\,\Delta\varphi = -\frac{b}{a\kappa - b^2}\frac{\delta}{q}I. \tag{3–7.21}$$

From (20) and (21) we derive:

$$\frac{\Delta P_{\text{stat}}}{I} = \left(\frac{\Delta P}{I}\right)_{J=0} = \left(\frac{\Delta\varphi}{J}\right)_{I=0}. \tag{3–7.22}$$

Thus Eqs. (20) to (22) contain the Onsager reciprocity law.

(c) Measurements

From Eqs. (11) and (20) it is found that

$$\left(\frac{\Delta\varphi}{\Delta P}\right)_{I=0} = -\left(\frac{J}{I}\right)_{\Delta P=0}. \tag{3–7.23}$$

This equation is denoted as "Saxén's relation": *Saxén*[7], a pupil of Wiedemann, verified the equation experimentally in 1892. In his measurements, aqueous electrolyte solutions of equal concentration formed the two liquid subsystems which were separated from one another by a porous clay plate. He used unpolarizable electrodes of

$$Zn \text{ (for } ZnSO_4), \quad Cu \text{ (for } CuSO_4), \quad \text{and} \quad Cd \text{ (for } CdSO_4).$$

The dependence of the electrode potentials on the pressure† can be neglected for the small pressure differences of Saxén's experiments, so that in the zero current state ($I = 0$) the electric potential difference measured between two electrodes of the same kind is practically equal to the streaming potential $\Delta\varphi$ in Eq. (23). The results of the measurements are found in Table 3.

In the older presentations, Eq. (23) is introduced as a consequence of molecular–kinetic considerations derived from the Helmholtz–Gouy–Smoluchowski theory for the "diffuse electric double layer" and thus also from the "electrokinetic potential" or "zeta potential" of Perrin and Freundlich. We see, however, that Eq. (23), just as does the basic statement (19), holds independently of any kinetic model. The experimental verification of Saxén's relation thus confirms Onsager's reciprocity law but not a certain molecular physical model.

From the inequalities (17) with the help of Eq. (18) the following statements about the signs result:

$$a > 0, \quad \kappa > 0, \tag{3–7.24}$$

$$b^2 < a\kappa. \tag{3–7.25}$$

According to (24), the permeability a and the electric conductivity κ are always positive. This corresponds to experience and seems obvious. According to (25), the electroosmotic permeability b can be either positive or negative, as experimental results confirm; but there is an upper limit for the value of b.

The quantities quoted in the fourth and fifth columns of Table 3 are, according to Eqs. (11) or (20), nothing other than $|b/\kappa|$. According to Eqs. (20) and (24), a positive value of $\Delta\varphi/\Delta P$ at vanishingly small electric current ($I = 0$) corresponds to a negative value of the electroosmotic permeability b. This again means, according to Eq. (5), that in the electric field ($\Delta\varphi \neq 0$, $\Delta P = 0$) the liquid moves to the positive pole, that is, it is negatively charged with respect to the valve substance. The sign on the liquid charge, relative to the surface, and the sign of the quantity b are determined both by the kind of valve substance and by the nature and composition of the liquid. If, say, pure water is positively charged with respect to the surface of the valve material, as is usually the case, an electrolyte additive of sufficient

---

† The rigorous result is obtained by taking into account both the pressure drop in one of the terminals and the pressure dependence of the electrode potentials (cf. §4–22). The correction term in $\Delta\phi/\Delta P$ can be estimated to be of the order of magnitude of 10^{-13} cm^3 Fr^{-1} at 25°C (Fr = franklin).

concentration can produce a negative charge on the liquid. That concentration, at which the charge transition takes place and at which $b = 0$, corresponds to the "isoelectric point" for colloids.

Recently, Klemm[8] has shown that electroosmosis occurs even for liquid mercury in glass capillaries which are filled with glass spheres. In addition to electroosmosis (predicted by Eq. 11), an electroosmotic pressure difference described by Eq. (21) was observed. A positive value of the electroosmotic permeability results, i.e. a positive charging of the mercury with respect to the glass.†

Table 3

Verification of the Relation (23) by Saxén for Different Aqueous Electrolyte Solutions with Clay as the Valve Substance [24]

Electrolyte	Molality, m mol kg^{-1}	Temperature °C	$|J/I|$ for $\Delta P = 0$ cm^3 Fr^{-1}	$|\Delta\varphi/\Delta P|$ for $I = 0$ cm^3 Fr^{-1}
ZnSO$_4$	0.0174	19	0.3597	0.3515
	0.0262	26	0.3817	0.3790
	0.0350	25	0.3461	0.3438
CuSO$_4$	0.0403	20	0.3850	0.3852
	0.0811	19	0.2329	0.2371
CdSO$_4$	0.0196	20	0.5823	0.5880
	0.0393	15	0.1157	0.1153

The sign of the effects is not obvious from the original. We have therefore inserted into this table the absolute magnitudes of the quantities appearing in Eq. (23).

3-8 MEMBRANE PROCESSES IN ISOTHERMAL SYSTEMS

We consider two liquid subsystems divided from one another by a permeable membrane. Apart from a pressure difference ΔP and an electric potential difference $\Delta\varphi$ let there also be arbitrary concentration differences (mole fraction differences Δx_k) between the two subsystems. The system is isothermal. We thus treat the case of Fig. 5 (p. 163) with $\Delta T = 0$. The system presented in Fig. 6 (p. 173) and discussed in §3–7 is accordingly a special case ($\Delta x_k = 0$) of the system to be considered here. For the sake of clarity, we have specified the valve from the start as a permeable membrane but the following calculations are valid also for the case of any permeable valve (a capillary, a porous wall, etc.) with appropriate changes of notation.

---

† Further examples of experiments which produce a verification of Eq. (23) or of similar relations and thus confirm Onsager's reciprocity law are given by Miller[9]. There are also some recent experiments on electrokinetic phenomena[10–12].[25]

The irreversible processes occurring in our isothermal system ("membrane processes") are described for sufficiently small values of ΔP, Δx_k, and $\Delta \varphi$ by the dissipation function (3–5.16) with $X_Q = 0$ and accordingly by the first N equations of the system of equations (3–6.1) with $X_Q = 0$. For N species, we thus have the following phenomenological equations:

$$J_i = \sum_{k=1}^{N} \alpha_{ik} X_k \quad (i = 1, 2, \ldots, N). \tag{3–8.1}$$

Here J_i is the flow of matter of the (charged or uncharged) species i, α_{ik} a phenomenological coefficient, and X_k a "generalized force," given by Eq. (3–6.6):

$$X_k = V_k \Delta P + \sum_j \left(\frac{\partial \mu_k}{\partial x_j}\right)_{T,P} \Delta x_j + z_k F \Delta \varphi, \tag{3–8.2}$$

where V_k denotes the partial molar volume, μ_k the chemical potential, x_k the mole fraction and z_k the charge number of species k, and F denotes the Faraday constant. The partial differentiation in Eq. (2) is to be understood as a derivative with respect to the mole fraction of the indicated component with the temperature T, the pressure P, and the other independent mole fractions being constant. The summation is over all independent mole fractions. For the phenomenological coefficients α_{ik} in Eq. (1), Onsager's reciprocity law (3–6.2) holds:

$$\alpha_{ik} = \alpha_{ki} \quad (i, k = 1, 2, \ldots, N). \tag{3–8.3}$$

Together with the system of equations (1) and the inequalities (3–6.3) to (3–6.4) these relations contain all statements about the irreversible processes possible in our system insofar as they occur near to equilibrium. As indicated in our discussion of §3–7, the assumption of proximity to equilibrium (small values of ΔP, Δx_k, and $\Delta \varphi$) is satisfied in most experiments. This assumption implies the validity of linear relations such as Ohm's law.

We again use, as in §3–7, the volume flow (3–6.8)

$$J = \sum_{i=1}^{N} V_i J_i \tag{3–8.4}$$

and the electric current (3–2.6)

$$I = F \sum_{i=1}^{N} z_i J_i. \tag{3–8.5}$$

We introduce the abbreviation (cf. Eq. 3–5.8)

$$(\Delta \mu_k)_{T,P} \equiv \sum_j \left(\frac{\partial \mu_k}{\partial x_j}\right)_{T,P} \Delta x_j. \tag{3–8.6}$$

With this we find from Eqs. (1) and (2):

$$J = \sum_{i=1}^{N} \sum_{k=1}^{N} \alpha_{ik} V_i [V_k \, \Delta P + (\Delta \mu_k)_{T,P} + z_k F \, \Delta \varphi], \qquad (3\text{–}8.7)$$

$$I = F \sum_{i=1}^{N} \sum_{k=1}^{N} \alpha_{ik} z_i [V_k \, \Delta P + (\Delta \mu_k)_{T,P} + z_k F \, \Delta \varphi]. \qquad (3\text{–}8.8)$$

By means of these equations we will study the different effects in our system. For the sake of simplicity, we write the double summation in Eqs. (7) and (8) from now on in the abbreviated form $\sum_{ik}$. Also, we notice that the condition

$$(\Delta \mu_k)_{T,P} = 0 \quad (k = 1, 2, \ldots, N)$$

is identical to the assumption $\Delta x_j = 0$ (for all j), meaning equal compositions in the two subsystems (Eq. 6).

We first consider *permeation*, previously considered in §3–7. According to Eq. (7):

$$J = \sum_{i,k} \alpha_{ik} V_i V_k \, \Delta P \quad (\Delta x_j = 0, \Delta \varphi = 0). \qquad (3\text{–}8.9)$$

Comparison of this expression with Eq. (3–7.1) indicates that the *permeability* a is given by the relation

$$a = \frac{\delta}{q} \sum_{i,k} \alpha_{ik} V_i V_k. \qquad (3\text{–}8.10)$$

Here δ and q denote the thickness and the (effective) cross section of the membrane.

For the *electric conduction* (cf. §3–7), we find from Eq. (8):

$$I = F^2 \sum_{i,k} \alpha_{ik} z_i z_k \, \Delta \varphi \quad (\Delta P = 0, \Delta x_j = 0). \qquad (3\text{–}8.11)$$

According to Eq. (3–7.4), the *specific electric conductivity* (specific conductance) is given by:

$$\kappa = \frac{\delta}{q} F^2 \sum_{i,k} \alpha_{ik} z_i z_k. \qquad (3\text{–}8.12)$$

The *electroosmosis* (cf. §3–7) is described according to Eq. (7) by the equation:

$$J = F \sum_{i,k} \alpha_{ik} V_i z_k \, \Delta \varphi \quad (\Delta P = 0, \Delta x_j = 0). \qquad (3\text{–}8.13)$$

Comparison with Eq. (3–7.5) yields the *electroosmotic permeability*:

$$b = \frac{\delta}{q} F \sum_{i,k} \alpha_{ik} V_i z_k. \qquad (3\text{–}8.14)$$

According to Eq. (8), there is contained in the equation

$$I = F \sum_{i,k} \alpha_{ik} z_i V_k \Delta P \quad (\Delta x_j = 0, \Delta \varphi = 0) \tag{3-8.15}$$

the description of *streaming conduction* (cf. §3–7). According to Eq. (3–7.6), we find for the *streaming conductivity*:

$$b^* = \frac{\delta}{q} F \sum_{i,k} \alpha_{ik} z_i V_k. \tag{3-8.16}$$

With the help of Onsager's reciprocity relations (3) and Eqs. (14) and (16) we write

$$b = b^*, \tag{3-8.17}$$

in agreement with Eq. (3–7.19).

If we set $J = 0$, $\Delta x_j = 0$ in Eq. (7) or $I = 0$, $\Delta x_j = 0$, in Eq. (8) then we obtain with Eqs. (10) and (14) the electroosomotic pressure difference (3–7.14) or, with Eqs. (12) and (16), the streaming potential (3–7.13).

We thus discover the phenomena dealt with in §3–7, namely, the permeation, the electric conduction, and the electrokinetic effects, for the special case $\Delta x_j = 0$ from Eqs. (7) and (8). We now turn to those phenomena which are related to the condition $\Delta x_j \neq 0$ (unequal compositions of the two subsystems).

From Eq. (7) it follows that

$$J = \sum_{i,k} \alpha_{ik} V_i (\Delta \mu_k)_{T,P} \quad (\Delta P = 0, \Delta \varphi = 0). \tag{3-8.18}$$

We call the effect described by this relation "osmosis," i.e. material transport through the membrane due to concentration differences at zero values of pressure difference and of potential difference. Often the combination of permeation, electroosmosis, and osmosis represented by Eq. (7) for the case $\Delta P \neq 0$, $\Delta x_j \neq 0$, $\Delta \varphi \neq 0$ is also denoted as "osmosis."

From Eq. (8) there results:

$$I = F \sum_{i,k} \alpha_{ik} z_i (\Delta \mu_k)_{T,P} \quad (\Delta P = 0, \Delta \varphi = 0). \tag{3-8.19}$$

This equation describes the electric flow through the membrane which accompanies osmosis.

We designate the *membrane potential* $\Delta \varphi_M$ as that value of the electric potential difference $\Delta \varphi$ which is obtained under the condition $I = 0$, $\Delta P = 0$. It corresponds to the diffusion potential for continuous systems (cf. p. 291) and represents a quantity which cannot be measured by itself, since it can only be determined experimentally in combination with other potential differences (electrode potentials, etc.) giving rise to the "electromotive force" of a galvanic cell. From Eqs. (8) and (12) with $I = 0$, $\Delta P = 0$ there follows for the membrane potential:

$$\Delta\varphi_M = -\frac{F}{\kappa}\frac{\delta}{q}\sum_{i,k}\alpha_{ik}z_i(\Delta\mu_k)_{T,P}. \qquad (3\text{–}8.20)$$

We now introduce for each (charged or uncharged) species i a *reduced transport number*

$$\tau_i \equiv F\frac{J_i}{I} \quad (\Delta P = 0, \Delta x_j = 0) \qquad (3\text{–}8.21)$$

(cf. p. 267). From Eqs. (1), (2), (11), and (12) with Eq. (21) we obtain:

$$\tau_i = \frac{F^2}{\kappa}\frac{\delta}{q}\sum_{k}\alpha_{ik}z_k, \qquad (3\text{–}8.22)$$

from which, on exchanging the inferior labeling, there follows:

$$\tau_k = \frac{F^2}{\kappa}\frac{\delta}{q}\sum_{i}\alpha_{ki}z_i. \qquad (3\text{–}8.23)$$

The summations in (22) and (23) are over all species. Because of the Onsager reciprocity law (3) we can replace Eq. (23) by

$$\tau_k = \frac{F^2}{\kappa}\frac{\delta}{q}\sum_{i}\alpha_{ik}z_i. \qquad (3\text{–}8.24)$$

Insertion of Eq. (24) in Eq. (20) results in the final expression for the membrane potential:

$$F\,\Delta\varphi_M = -\sum_{k}\tau_k(\Delta\mu_k)_{T,P}. \qquad (3\text{–}8.25)$$

This relation, which is due to Staverman[13–16], is analogous to the general expression for the diffusion potential for continuous systems (p. 293).

Other problems of this kind are considered in detail by other authors [13–17]. We cannot go further here.† [26]

3–9 PROCESSES IN NONISOTHERMAL SYSTEMS

Consider the general case of a valve system (Fig. 5, p. 163) with nonzero values of the temperature difference ΔT, of the pressure difference ΔP, of the electric potential difference $\Delta\varphi$, and of the concentration differences (mole

---

† An investigation on membrane potentials in nonisothermal systems has been presented[18].

fraction differences) Δx_k. If the above-mentioned differences are sufficiently small then the phenomenological equations (3–6.1) may be written:

$$J_i = \sum_{k=1}^{N} \alpha_{ik} X_k + \alpha_{iQ} X_Q \quad (i = 1, 2, \ldots, N), \tag{3–9.1}$$

$$J_Q = \sum_{k=1}^{N} \alpha_{Qi} X_i + \alpha_{QQ} X_Q, \tag{3–9.2}$$

in which J_i is the flux of species i and J_Q the heat flux; N denotes the number of species. For the generalized forces X_Q and X_k, we have, according to Eqs. (3–6.5) and (3–6.6),

$$X_Q = \frac{\Delta T}{T}, \tag{3–9.3}$$

$$X_k = V_k \Delta P + \sum_j \left(\frac{\partial \mu_k}{\partial x_j}\right)_{T,P} \Delta x_j + z_k F \Delta \varphi \quad (k = 1, 2, \ldots, N). \tag{3–9.4}$$

Here T is the average absolute temperature of the system. The other quantities have the significance given in Eq. (3–8.2). For the phenomenological coefficients α_{ik}, α_{iQ}, α_{Qi}, and α_{QQ}, the reciprocity relations (3–6.2) hold:

$$\alpha_{ik} = \alpha_{ki} \quad (i, k = 1, 2, \ldots, N), \tag{3–9.5}$$

$$\alpha_{iQ} = \alpha_{Qi} \, (i = 1, 2, \ldots, N) \tag{3–9.6}$$

as well as the inequalities (3–6.3) and (3–6.4):

$$\left.\begin{array}{l} \alpha_{ii} > 0, \quad \alpha_{ii}\alpha_{jj} - \alpha_{ij}^2 > 0 \\ (i, j = 1, 2, \ldots, N, Q; i \neq j), \\ \cdot \quad \cdot \quad \cdot \quad \cdot \quad \cdot \quad \cdot \quad \cdot \quad \cdot \\ |\alpha_{ik}| > 0, \\ |\alpha_{ij}| > 0, \end{array}\right\} \tag{3–9.7}$$

where $|\alpha_{ij}|$ represents the abbreviation for the determinant in (3–6.4) and $|\alpha_{ik}|$ represents the principal cofactor of N rows of this determinant which arises by elimination of the last row and column.

We define N quantities Q_k^* by the N independent equations

$$\alpha_{iQ} = \sum_{k=1}^{N} \alpha_{ik} Q_k^* \quad (i = 1, 2, \ldots, N). \tag{3–9.8}$$

These have one and only one solution $Q_1^*, Q_2^*, \ldots, Q_N^*$ because of the second from the last inequality in (7). If we insert Eq. (8) into Eq. (2), on consideration of the reciprocity relations (5) and (6), we find:

$$J_Q = \sum_{i=1}^{N} \sum_{k=1}^{N} \alpha_{ik} X_k Q_i^* + \alpha_{QQ} X_Q.$$

On comparison with Eq. (1) there results:

$$J_Q = \sum_{i=1}^{N} Q_i^* J_i + \left(\alpha_{QQ} - \sum_{i=1}^{N} \alpha_{iQ} Q_i^* \right) X_Q. \qquad (3\text{–}9.9)$$

From this and Eq. (3) we immediately recognize the physical significance of the quantities introduced formally by Eq. (8): Q_i^* is the *heat of transport*, that is, the heat transferred through the valve, at zero temperature difference ($X_Q = 0$), divided by the amount of substance of species i, as far as that heat transfer is due to species i. Eastman[19a,b] and Wagner[20a,b], who introduced the concept of the "heat of transport" in the framework of a quasi-thermodynamic theory, in principle defined the quantity Q_i^* by an equation which follows from Eq. (9), viz:

$$Q_i^* = \left(\frac{J_Q}{J_i} \right)_{J_j = 0, \, \Delta T = 0},$$

where the inferior j stands for all species apart from the particle type i in question.

From Eqs. (1) and (8) we derive:

$$J_i = \sum_{k=1}^{N} \alpha_{ik} (X_k + Q_k^* X_Q) \quad (i = 1, 2, \ldots, N). \qquad (3\text{–}9.10)$$

As we will see, the heats of transport formally introduced by Eq. (8) together with this result yield the formulas for the stationary state (Eqs. 22 and 23). It is thus noteworthy that in the derivation of Eq. (10) the reciprocity relations were not used. Meanwhile, it should be noticed that the physical interpretation of the heats of transport is only made possible by Eq. (9) in which Onsager's reciprocity law is contained.

We consider a system in which the temperatures of the two subsystems and thus also the quantities T and ΔT have fixed values, while quantities such as pressure, concentrations, etc., are variable. Let there now be a stationary state possible, due to the combined action of different irreversible processes, at which all material flows through the membrane disappear:

$$J_i = 0 \quad (i = 1, 2, \ldots, N). \qquad (3\text{–}9.11)$$

Thus, the intensive quantities (temperature, pressure, concentrations, etc.) are constant in time although the system is not at equilibrium (cf. §1–17).

Next, there follows with the help of Eqs. (3) and (9) for the stationary state characterized by (11):

$$J_Q = \lambda_\infty \frac{q}{\delta} \Delta T \quad \text{(stationary state)} \qquad (3\text{–}9.12)$$

with

$$\lambda_\infty \equiv \frac{\delta}{qT}\left(\alpha_{QQ} - \sum_{i=1}^{N} \alpha_{iQ}Q_i^*\right), \tag{3-9.13}$$

where q denotes the (effective) cross section and δ the thickness of the valve. Obviously, Eq. (12) describes the *heat conduction* through the valve for the stationary case. Accordingly, λ_∞ is the *thermal conductivity for the stationary state*. The need for the symbol λ_∞ arises from the fact that the stationary state considered in relation to a nonstationary initial state (at the time $t = 0$), will, strictly speaking, only be reached after an infinitely long time $(t \to \infty)$.

If the system is in a "uniform" nonstationary state ($\Delta P = 0$, $\Delta x_j = 0$, $\Delta\varphi = 0$) in which the two subsystems are distinguished only by their temperatures then, using Eqs. (2), (3), and (4), the heat conduction through the valve is found to be given by:

$$J_Q = \lambda_0 \frac{q}{\delta} \Delta T \quad (\Delta P = 0, \Delta x_j = 0, \Delta\varphi = 0) \tag{3-9.14}$$

with

$$\lambda_0 \equiv \frac{\delta}{q}\frac{\alpha_{QQ}}{T}. \tag{3-9.15}$$

Accordingly, we denote the quantity λ_0 as the *thermal conductivity for the uniform state*. As is obvious from comparison of this result with Eq. (13), λ_0 is different from λ_∞.

If our system is initially $(t = 0)$ homogeneous except for the temperature difference, and at the end of the experiment $(t \to \infty)$ is in a stationary state (in which pressure, concentration, and potential differences will have been set up), thus, λ_0 is to be interpreted as the "initial value" and λ_∞ as the "final value" of the thermal conductivity.

From Eqs. (8), (13), and (15) we derive a general equation for the difference of these two thermal conductivities:

$$\lambda_0 - \lambda_\infty = \frac{\delta}{qT}\sum_{i=1}^{N} \alpha_{iQ}Q_i^* = \frac{\delta}{qT}\sum_{i=1}^{N}\sum_{k=1}^{N} \alpha_{ik}Q_i^*Q_k^*. \tag{3-9.16}$$

This relation[21] relates the difference $\lambda_0 - \lambda_\infty$ to the other transport coefficients, namely, the heats of transport Q_i^* and the coefficients α_{iQ} which, according to Eq. (1), describe the mass transport due to a temperature difference ("thermoosmosis" and related phenomena, cf. §3–10 and §3–12), or the coefficients α_{ik} which are associated with the isothermal transport phenomena (electric conduction, permeation, osmosis, etc., cf. §3–8).

We now need further statements about these thermal conductivities[21]. Here we use the system of inequalities (7) which depends on the positive-only character of the entropy production or dissipation function.

First, there results from Eqs. (7) and (15):

$$\lambda_0 > 0. \tag{3–9.17}$$

Since, moreover, the last summation in Eq. (16) represents a positive quadratic form due to (7), we obtain the interesting general result:

$$\lambda_0 > \lambda_\infty. \tag{3–9.18}$$

If we solve the system of equations (8) for $Q_1^*, Q_2^*, \ldots, Q_N^*$ and add the products $\alpha_{iQ} Q_i^*$ using the rules of determinant calculations and considering Eqs. (6) and (7) we find:

$$\alpha_{QQ} - \sum_{i=1}^{N} \alpha_{iQ} Q_i^* = \frac{|\alpha_{ij}|}{|\alpha_{ik}|} > 0. \tag{3–9.19}$$

From (13) and (19) there immediately follows:

$$\lambda_\infty > 0. \tag{3–9.20}$$

Because of (20) the inequality (17) can also be regarded as a consequence of (18).

The statement that the thermal conductivity λ_0 or λ_∞ is always positive, i.e. the heat flow is always from the higher to the lower temperature, seems obvious. It must, however, be noted that, for another choice of the "heat flux," Eq. (17) is no longer guaranteed. The inequality (18) is then likewise uncertain. This is another reason for our specification of the concept of heat for open systems in §1–7. Since the heat flux is so chosen as to be independent of the arbitrary zero point of the internal energy (cf. §3–3), quantities such as the heats of transport Q_k^* and the thermal conductivities λ_0 and λ_∞ do not depend on the energy zero point. This physically reasonable invariance property is also lost if we use quantities other than J_Q as the "heat flow" (cf. §1–7). This other choice is common in the literature.

We now return to the problem of the stationary state. From Eqs. (10) and (11) there results with the second last inequality in (7) that the N bracketed expressions in (10) must individually disappear:

$$X_k + Q_k^* X_Q = 0 \quad (k = 1, 2, \ldots, N) \quad \text{(stationary state)}. \tag{3–9.21}$$

With Eqs. (3) and (4) we obtain from this:

$$V_k \, \Delta P + \sum_j \left(\frac{\partial \mu_k}{\partial x_j} \right)_{T,P} \Delta x_j + z_k F \, \Delta \varphi + Q_k^* \frac{\Delta T}{T} = 0$$

$$(k = 1, 2, \ldots, N) \text{ (stationary state)}. \tag{3–9.22}$$

We will discuss these fundamental equations, stated in principle by Wagner [20a,b], in §3–10 and §3–13 for some special cases.

Using Eqs. (3–6.15) and (3–6.19) we find the general equation for the stationary pressure difference from Eqs. (3) and (21) to be:

$$\Delta P_{\text{stat}} = -\frac{\Delta T}{T} \sum_{k=1}^{N} c_k Q_k^* = -\frac{\Delta T}{T\bar{V}} \sum_{k=1}^{N} x_k Q_k^*, \qquad (3\text{–}9.23)$$

where c_k denotes the molarity of species k and $\bar{V}$ the molar volume. Equation (23) is also due to Wagner.

The heats of transport (at least for electrically neutral species) based on Eqs. (22) and (23) are directly measurable quantities.

The *relation between the heats of transport and the heat supplied from the surroundings* follows from Eqs. (3–3.9), (3), and (9) at vanishing temperature difference ($\Delta T = T'' - T' = 0$). We obtain for the subsystem ' or ":

$$\sum_{k=1}^{N} (Q_k^* + H_k' - z_k F \varphi') J_k = -\frac{d_a Q'}{dt} + \frac{dU'}{dt} + P' \frac{dV'}{dt},$$

$$\qquad (3\text{–}9.24)$$

$$\sum_{k=1}^{N} (Q_k^* + H_k' - z_k F \varphi'') J_k = \frac{d_a Q''}{dt} - \frac{dU''}{dt} - P'' \frac{dV''}{dt}.$$

Here H_k is the partial molar enthalpy of species k, φ the electric potential, $d_a Q$ the heat received by the subsystem in question from the surroundings during a time element dt, U the internal energy, V the volume, and P the pressure. We now assume the pressures P' and P'' as well as the temperature $T(= T' = T'')$ to be constant in time:

$$\frac{dT}{dt} = 0, \quad \frac{dP'}{dt} = 0, \quad \frac{dP''}{dt} = 0, \qquad (3\text{–}9.25)$$

where the condition $P' = P''$ is not necessarily fulfilled. With the help of the relations following from Eqs. (1–5.3), (1–6.3), and (3–2.2):

$$dH' = dU' + P' \, dV' = \sum_{k=1}^{N} H_k' J_k \, dt,$$

$$dH'' = dU'' + P'' \, dV'' = -\sum_{k=1}^{N} H_k'' J_k \, dt$$

we derive from Eqs. (24) and (25)

$$\left. \begin{aligned} \sum_{k=1}^{N} (Q_k^* - z_k F \varphi') J_k &= -\frac{d_a Q'}{dt}, \\ \sum_{k=1}^{N} (Q_k^* - z_k F \varphi'') J_k &= \frac{d_a Q''}{dt} + \sum_{k=1}^{N} (H_k'' - H_k') J_k. \end{aligned} \right\} \qquad (3\text{–}9.26)$$

With Eq. (3–2.6) there results from this:

$$\sum_{k=1}^{N} Q_k^* J_k - I\varphi' = -\frac{d_a Q'}{dt},$$ (3–9.27)

where I denotes the total electric current.

For zero current we have:

$$I = F \sum_{k=1}^{N} z_k J_k = 0.$$ (3–9.28)

From this there follows from Eq. (27) on consideration of Eq. (3–2.2):

$$\sum_{k=1}^{N} Q_k^* \frac{dn_k'}{dt} = -\frac{d_a Q'}{dt}$$ (3–9.29)

$$(T = \text{const}, dP'/dt = dP''/dt = 0, I = 0).$$

The heat of transport of species k (Q_k^*) is thus equal to the heat flowing from phase ' into the surroundings of the total system divided by the amount of substance of species k, when species k migrates isothermally from phase " (pressure P'') to phase ' (pressure P') through the valve, the electric current being zero. This interpretation of the heat of transport due to Eastman and Wagner seems to be quite easy to grasp at first sight, but it does not open the way to a calculation of Q^* from other quantities. The heat of transport, as a matter of fact, may not be generally derived from classical thermodynamic functions except for a few simple cases (cf. §3–11) but must—like the phenomenological coefficients to which it is related—be measured for each case and be interpreted on the basis of a kinetic model.

For a molecular physical interpretation of the heat of transport, another special case of Eq. (24) is important: we consider a system of uniform temperature in which phase ' is at constant volume in the state without current. Then we find from Eqs. (24) and (28) that

$$\sum_{k=1}^{N} (Q_k^* + H_k') \frac{dn_k'}{dt} = -\frac{d_a Q'}{dt} + \frac{dU'}{dt} \quad (T' = T'', dV'/dt = 0, I = 0).$$

According to this, the quantity

$$U_k^* \equiv Q_k^* + H_k'$$ (3–9.30)

is the internal energy transported by species k from phase " to phase ' through the valve divided by the amount of substance of species k (i.e. the "energy of transport").

While the quasithermodynamic method uses hypotheses which are difficult to justify and actually leads only to the formulas for the stationary state, our systematic approach, based on the entropy balance, the phenomeno-logical laws, and Onsager's reciprocity relations, is general and fundamental.

The *calorimetric measurement* of the heats of transport can be under-stood by means of Eq. (29), too. Thus, for example, we have for one-component systems ($N = 1$, $Q_k^* = Q^*$, $n_k' = n'$):

$$Q^* \frac{dn'}{dt} = -\frac{d_a Q'}{dt} \quad (T = \text{const}, \, dP'/dt = 0). \tag{3-9.31}$$

If we determine the flow of matter from phase " to phase ' through the valve, dn'/dt, and the flow of heat from the surroundings into phase ', $d_a Q'/dt$, temperature and pressure being constant, we obtain experimental values of Q^*. In the experiments, we have to choose the external conditions so that the left-hand side of Eq. (31) is negative. Then the heat flow $d_a Q'/dt$ is positive and can be calculated from the voltage and current of an electric heater.

3-10 THERMOMECHANICAL EFFECTS (EMPIRICAL AND THERMODYNAMIC–PHENOMENOLOGICAL DESCRIPTION)

(a) Empirical Description

Let us consider a one-component fluid system. The valve may be a small opening, a capillary, a porous wall, or a membrane. The two subsystems will generally be at different temperatures (T and $T + \Delta T$) and different pressures (P and $P + \Delta P$) but no electric potential difference shall be present (Fig. 7). Thus transport of matter and of heat can take place through the valve.

Apart from the phenomena of permeation (§3–7) and of heat conduction (§3–9) previously considered, there is a "cross effect" still possible, namely, a mass transport due to the temperature difference, and the appropriate "reciprocal effect", i.e. a heat transport due to the pressure difference.

Phase '	Phase "
Temperature T Pressure P	Temperature $T + \Delta T$ Pressure $P + \Delta P$

$$\leftarrow J_1 > 0 \quad \text{(material flow)}$$
$$\leftarrow J_Q > 0 \quad \text{(heat flow)}$$

Fig. 7. Heterogeneous (discontinuous) system with transport of matter and heat between the two subsystems (phase ' and phase ") as a special case of the system in Fig. 5 (p. 163).

In 1873, Feddersen[22] published qualitative results from observations of a flow of air through a porous medium (platinum sponge, gypsum, etc.)

due to a difference of temperature on the two opposite sides of the porous plug, the pressure being uniform. The gas flow always took place in the direction from the lower to the higher temperature. Feddersen called the phenomenon "thermal diffusion." This expression is used today for another effect (migration of particles inside any continuous mixture in a temperature field at constant pressure and constant concentrations without a valve being present, cf. p. 355). We denote the phenomena discovered by Feddersen, and, in fact, any mass flow in a one-component fluid system due to a temperature difference through a porous wall or a membrane, as *thermal osmosis* (or "thermoosmosis").

> Feddersen was the first to detect experimentally a mass flow in gases due to temperature gradients. Therefore the concluding sentences of his work (dated the 26.12.1872) are reproduced here.
>
> "From present experiments with the most heterogeneous materials it apparently can be deduced that there is a general property of porous bodies, as soon as they are brought into the form of diaphragms, to draw the gases through themselves in the direction from the cold to the warm side. One has thus a diffusion phenomenon, which, in contrast to normal diffusion, becomes apparent if there is the same gas at the same pressure on both sides of the diaphragm. This is a special phenomenon unknown till now and one has thus the right to bestow it with the name of *thermal diffusion*."

Obviously without knowledge of the works of Feddersen and Reynolds [23] on the thermal osmosis of gases through porous media, Lippmann (1907) observed thermal osmosis in air[24a] and in water[24b] through membranes of paper, gelatin, and gold beater's skin. He thus detected thermal osmosis in liquids for the first time. These and later experiments with liquids[25] are all in all somewhat doubtful because of the possibility of osmosis and electroosmosis due to electrolytic impurities[26].

More recent experiments on thermal osmosis in gases are reported by Denbigh and Raumann[27], those on thermal osmosis in liquids (water) by Alexander and Wirtz[28] as well as from the laboratory of the author [29,30]. In these experiments, the valves were membranes of rubber (for the gases) or of cellulose preparations and gold beater's skin (for the liquids). These measurements show that thermal osmosis does not always—as was thought by the previous experimenters—occur from the cold to the warm side. The sign of the effect depends rather on the type of the fluid medium, on the nature of the membrane, and, finally, as we know now, on the mean temperature[29,30] (cf. §3–11 for more details).

Since porous media are to be regarded as systems of capillaries, this suggests that experiments also be set up with capillaries. Thus there is a continuous transition from thermal osmosis to the *Knudsen effect*. This effect was first experimentally and theoretically examined by Reynolds[23], then independently by Knudsen[31]. It consists of the following: a gas current is observed between two containers filled with gas maintained at different temperatures, which are connected to one another by a capillary, by a porous plate, or by a narrow opening in the connecting wall. The vessels are at equal pressures. Here the capillary or hole diameter is small in comparison with the mean free path of the gas molecules. If the gas is enclosed in rigid walls, the gas current leads to a pressure difference between the two subsystems at initially equal pressures. This difference in the stationary case for ideal gases can be calculated from the kinetic theory of gases.

The phenomenon analogous to thermal osmosis and to the Knudsen effect for liquid helium II (that is, for liquid helium below the lambda point), where a capillary or a powder plug represents the valve, was discovered by Allen and Jones[32]. The phenomenon is very conspicuous and is today designated as the *fountain effect*.

Since the transition from capillaries or narrow openings to porous walls or membranes is continuous and since the effects appear for all possible fluid media, it is advisable not only to introduce a general phenomenological treatment for the above-mentioned effects but also a common name. We use the designation "thermomechanical effect." (It is also called "thermomolecular pressure effect," or, in the case of gases, "thermal transpiration.")

The reciprocal phenomenon of this effect must consequently be called "mechanocaloric effect," and for thermal osmosis the "osmotic thermal effect."

The *mechanocaloric effect*, that is, the heat transport due to a pressure difference, was detected experimentally by Daunt and Mendelssohn[33] for liquid helium II, for which the thermomechanical effect is particularly large.

If we assume the temperature difference ΔT and the pressure difference ΔP to be sufficiently small, we can describe the above-mentioned effects by the following empirical equations:

$$J_1 = \frac{q}{\delta}(A\,\Delta P + B\,\Delta T), \qquad (3\text{-}10.1)$$

$$J_Q = \frac{q}{\delta}(C\,\Delta P + \lambda_0\,\Delta T). \qquad (3\text{-}10.2)$$

Here q is the (effective) cross section and δ the thickness of the valve. The flow of matter J_1 denotes the rate of increase of the amount of substance of the species in question in phase ' (cf. Fig. 7) and is related to the volume flux J according to Eq. (3-6.8) as follows:

$$J = \overline{V}J_1, \qquad (3\text{-}10.3)$$

where $\overline{V}\,(= V_1)$ represents the molar volume of the substance in question at the average temperature and at the average pressure of the system. Correspondingly, the heat flux J_Q is the heat flowing per unit time from phase " to phase $'$ through the valve (cf. Fig. 7). The meaning of the coefficients A, B, C, and λ_0 results from the following considerations.

According to §3–7, the equation

$$J_1 = A \frac{q}{\delta} \Delta P \quad (\Delta T = 0), \tag{3–10.4}$$

which results from Eq. (1), describes *permeation*. We therefore denote the quantity A as the *permeability*. As shown by a comparison with Eq. (3–7.1) and on consideration of Eq. (3), this is related to the quantity a called the "permeability" in §3–7 by:

$$a = A\overline{V}. \tag{3–10.5}$$

Both a and A are called "permeability" in the literature. Here we use only the coefficient A.

In the relation following from Eq. (2)

$$J_Q = \lambda_0 \frac{q}{\delta} \Delta T \quad (\Delta P = 0), \tag{3–10.6}$$

the *heat conduction* for the case of vanishing pressure difference is represented following §3–9. The quantity λ_0 shall be called (as in Eq. 3–9.14) the *thermal conductivity for the uniform state*.

Using the equation

$$J_1 = B \frac{q}{\delta} \Delta T \quad (\Delta P = 0), \tag{3–10.7}$$

which again results from Eq. (1), we can describe the phenomenon denoted above as the *thermomechanical effect* (thermoosmosis, Knudsen effect, fountain effect). The coefficient B is called—at least for the case of thermoosmosis—the *thermoosmotic permeability*.

Finally, the relation resulting from Eq. (2)

$$J_Q = C \frac{q}{\delta} \Delta P \quad (\Delta T = 0) \tag{3–10.8}$$

describes the *mechanocaloric effect*. In the case of a membrane, we denote this phenomenon as the *osmotic thermal effect* and the quantity C as the *osmotic thermal coefficient*.

From experience we note that the transport coefficients A, B, C, and λ_0 depend on the type of fluid, on the valve material, on the average temperature, and, to a lesser extent, on the average pressure of the system.†

---

† For capillaries, this statement is nevertheless incorrect. Thus comparison of Eq. (4) with Poiseuille's law shows that A still depends on the cross section q of the capillary (cf. p. 173).

If we eliminate ΔP from Eqs. (1) and (2), we obtain:

$$J_Q = \frac{C}{A} J_1 + \lambda_\infty \frac{q}{\delta} \Delta T \qquad (3\text{-}10.9)$$

with

$$\lambda_\infty \equiv \lambda_0 - \frac{BC}{A}. \qquad (3\text{-}10.10)$$

Since according to Eq. (3–9.11) the *stationary state* in the present special case is characterized by

$$J_1 = 0 \quad \text{(stationary state)}, \qquad (3\text{-}10.11)$$

according to §3–9, the relation

$$J_Q = \lambda_\infty \frac{q}{\delta} \Delta T \quad \text{(stationary state)} \qquad (3\text{-}10.12)$$

describes the *heat conduction* in the stationary state. The quantity λ_∞ shall therefore be called, as in Eq. (3–9.12), *thermal conductivity for the stationary state*.

From Eqs. (1) and (11) the stationary pressure difference is given by

$$\Delta P_{\text{stat}} = -\frac{B}{A} \Delta T. \qquad (3\text{-}10.13)$$

We generally denote the quantity ΔP_{stat} as the *thermomechanical pressure difference* and, in the case of thermoosmosis, as the *thermoosmotic pressure difference*. (It is also called "thermomolecular pressure difference.")

(b) Thermodynamic–Phenomenological Description

The presentation thus far has been purely empirical. We now apply the methods of the thermodynamics of irreversible processes to the phenomena under consideration[34–40].

For sufficiently small values of ΔT and ΔP, the phenomenological equations (3–9.1) and (3–9.2) hold and they are simplified on the basis of Eqs. (3–9.3) and (3–9.4) for the present special case ($N = 1$, $V_1 = \overline{V}$, $\Delta x_j = 0$, $\Delta \varphi = 0$) to yield the following expressions (see also Eq. 3–6.10):

$$J_1 = \alpha_{11} \overline{V} \Delta P + \alpha_{1Q} \frac{\Delta T}{T}, \qquad (3\text{-}10.14)$$

$$J_Q = \alpha_{Q1} \overline{V} \Delta P + \alpha_{QQ} \frac{\Delta T}{T}. \qquad (3\text{-}10.15)$$

Here the quantities α_{11}, α_{1Q}, α_{Q1}, and α_{QQ} are the phenomenological coefficients, while T can be regarded as the average absolute temperature of the system. If we set

$$\alpha_{11}\overline{V}\frac{\delta}{q} = A, \qquad \frac{\alpha_{1Q}}{T}\frac{\delta}{q} = B,$$

$$\alpha_{Q1}\overline{V}\frac{\delta}{q} = C, \qquad \frac{\alpha_{QQ}}{T}\frac{\delta}{q} = \lambda_0, \tag{3–10.16}$$

we recognize that our empirical linear equations (1) and (2) are contained in Eqs. (14) and (15) of the thermodynamic–phenomenological theory of irreversible processes (cf. §3–7).

We now apply Onsager's reciprocity relation (3–9.6)

$$\alpha_{1Q} = \alpha_{Q1} \tag{3–10.17}$$

and the inequality statements (3–9.7)

$$\alpha_{11} > 0, \quad \alpha_{QQ} > 0, \quad \alpha_{11}\alpha_{QQ} - \alpha_{1Q}^2 > 0. \tag{3–10.18}$$

We first consider the results from the reciprocity law (17).

From Eqs. (16) and (17) we have

$$C = BT\overline{V}, \tag{3–10.19}$$

which is a general relation between the thermomechanical effect and the mechanocaloric effect. This important relation is analogous to the connection (3–7.19) between electroosmosis and streaming conduction. All irreversible processes which are possible in our system are thus characterized by three coefficients (for example, A, B, and λ_0).

If, according to Eq. (3–9.8), we formally introduce the *heat of transport* Q^* (for the single substance present) then, using Eq. (16), we find that

$$Q^* \equiv \frac{\alpha_{1Q}}{\alpha_{11}} = \frac{B}{A}T\overline{V}. \tag{3–10.20}$$

Inserting this relation in Eq. (13) produces:

$$\Delta P_{\text{stat}} = -\frac{Q^*}{T\overline{V}}\,\Delta T. \tag{3–10.21}$$

The quantity Q^*, defined by Eq. (20), yields the formula for the stationary state without the Onsager reciprocity law (cf. §3–9). Equation (21) is a special case of Eq. (3–9.22) and Eq. (3–9.23).

If we now use the Onsager reciprocity relation in the form of Eq. (19) then we obtain from Eq. (20)

$$Q^* = \frac{C}{A}. \tag{3–10.22}$$

From this with Eq. (9) it follows directly that

$$J_Q = Q^* J_1 + \lambda_\infty \frac{q}{\delta} \Delta T. \tag{3-10.23}$$

On this basis we have:

$$Q^* = \left(\frac{J_Q}{J_1}\right)_{\Delta T = 0}. \tag{3-10.24}$$

The quantity Q^* formally introduced by Eq. (20) has thus the physical meaning of a "heat of transport". In order to show this, we had to use the reciprocity law of Onsager (cf. §3–9).

From Eqs. (10), (19), and (22) we derive the difference between the two thermal conductivities to be

$$\lambda_0 - \lambda_\infty = \frac{B^2}{A} T\overline{V} = BQ^*. \tag{3-10.25}$$

This relation is a special case of Eq. (3–9.16).

All transport quantities are obtained from three independent coefficients (A, B, and λ_0) by the relations (19), (20), and (25).

With the help of the inequalities (18) we find the following statements from Eqs. (16), (17), and (25):

$$A > 0, \tag{3-10.26}$$

$$\lambda_0 > 0, \tag{3-10.27}$$

$$\lambda_\infty > 0, \tag{3-10.28}$$

$$\lambda_0 > \lambda_\infty, \tag{3-10.29}$$

$$B^2 < \frac{A\lambda_0}{T\overline{V}}, \tag{3-10.30}$$

where (27) can be regarded as a consequence of (28) and (29).

Equation (26), according to which matter flows from the higher to the lower pressure, seems obvious (cf. p. 173). Equations (27) to (29) are special cases of the general inequalities (3–9.17), (3–9.18), and (3–9.20). Equation (30) is most interesting since it appears from this that B can be both positive and negative (as found experimentally) and that there is an upper limit to its absolute value.

According to Eqs. (19), (20), and (26) the heat of transport Q^* as well as the coefficients B and C always have the same sign.

3–11 THERMOMECHANICAL EFFECTS (EXPERIMENTAL EXAMPLES)

All recent experimental studies on the Knudsen effect, on the fountain effect, and on thermoosmosis involve pure gases or pure liquids, well-defined valves (capillaries or membranes of known material which can be easily reproduced), and temperature and pressure differences which can be exactly measured. In most cases, the approach to the stationary state is observed and, from the thermomechanical pressure difference, the heat of transfer is determined using Eq. (3–10.21). We thus need a more exact analysis of the processes which lead to a stationary nonequilibrium state.

Consider a single-component fluid system with fixed values of the temperatures in both subsystems, so that the temperature difference ΔT and the average temperature T are constant. Let the system be initially homogeneous up to the temperature difference $\Delta T \neq 0$ (pressure difference $\Delta P = 0$ at time $t = 0$). In this initial state, a thermomechanical effect (thermoosmosis, etc.) appears, as required by Eq. (3–10.7). If a pressure difference between the two subsystems can now be built up (e.g. a system with a rigid membrane with fixed ends, or a system with capillaries at constant volumes of the two subsystems) then at any arbitrary instant Eq. (3–10.1) demands:

$$-\frac{\Delta P}{\Delta T} = \frac{B}{A} - \frac{\delta}{q} \frac{J_1}{A \, \Delta T}, \tag{3–11.1}$$

where the coefficients A and B, the temperature difference ΔT, the thickness δ, and the cross section q of the valve are constant for a given experimental arrangement, whereas the pressure difference ΔP and the flow of matter J_1 are variable. The higher pressure exists in that subsystem into which the material flows, due to the thermomechanical effect. However, because of permeation which (§3–10) promotes a transport of matter from the higher to the lower pressure, an opposing mass transport comes into play and this leads to a permanent decrease in the amount of the quantity J_1. After a sufficiently long time ($t \rightarrow \infty$), a *stationary state* is achieved which is characterized by the condition $J_1 = 0$. Thus, the expression (1) initially ($t = 0$, $\Delta P = 0$) has a zero value, but in the final stationary state ($t \rightarrow \infty$, $\Delta P \neq 0$, $J_1 = 0$) it has the value

$$-\frac{\Delta P}{\Delta T} = \frac{B}{A} \quad \text{(stationary state)} \tag{3–11.2}$$

as required by Eq. (3–10.13). At an arbitrary instant before reaching the stationary state the thermomechanical effect outweighs permeation, so that the quantities B, $J_1/\Delta T$, and $-\Delta P/\Delta T$ have the same sign, as fixed by Eq. (3–10.1). From this with Eq. (3–10.26) and Eq. (1) it follows that in each case the quantity $|\Delta P/\Delta T|$ increases with time until it assumes the maximum value $|B/A|$ in the stationary state (Fig. 8).

For the quantity $y \equiv -\Delta P/\Delta T$ as a function of the time t, there follows from Eqs. (3–2.2), (3–5.1), (3–6.7), (3–10.3), and Eq. (1) after integration[30]:

$$y = \frac{B}{A}(1 - e^{-t/\tau}) \quad \text{with} \quad \tau \equiv \frac{q_0 \delta}{2qMgA}, \tag{3–11.2a}$$

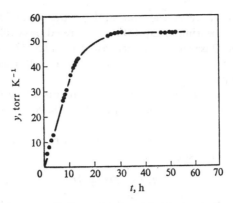

Fig. 8. The quantity $y = -\Delta P/\Delta T$ (as a function of time t) at the average temperature of 21°C and at a constant temperature difference of 0.95°C. Membrane substance: cellophane 600 g m^{-2}, wetted specimen. Permeating substance: water. From Haase and Steinert[29].

where q_0 is the cross section of the vertical capillaries used for the pressure measurement, M the molar mass of the permeating substance, and g the gravitational acceleration.

The permeability A and the coefficient B (i.e. the thermoosmotic permeability in the case of thermoosmosis) can be determined directly with the help of Eq. (3–10.4)

$$J_1 = A \frac{q}{\delta} \Delta P \quad (\Delta T = 0) \tag{3–11.3}$$

from the permeation and with the help of Eq. (3–10.7)

$$J_1 = B \frac{q}{\delta} \Delta T \quad (\Delta P = 0) \tag{3–11.4}$$

from the thermoosmosis respectively.† Comparison of the coefficients so obtained with the ratio B/A determined from the stationary state according to Eq. (2) represents a criterion of the quality of the measurements. A verification of the Onsager reciprocity law does not come into question by such a comparison, even if Eq. (2) is written in the equivalent form (3–10.21).

$$-\frac{\Delta P}{\Delta T} = \frac{Q^*}{T\overline{V}} \quad \text{(stationary state)}, \tag{3–11.5}$$

where T denotes the average absolute temperature of the system, $\overline{V}$ the molar

---

† The permeability A may also be derived from thermoosmosis experiments using Eq. (2a) or similar formulas. This has been done by Denbigh and Raumann[27], by Bearman[41], and by the author[30].

volume of the substance in question, and Q^* the heat of transport. When the quantity Q^* appearing in Eq. (5) is interpreted physically in the sense of Eq. (3–10.24), the Onsager reciprocity law is used (cf. §3–10). How this law is proved experimentally will be shown below (p. 201) using the fountain effect as an example.

If Q^* is positive or negative, according to Eq. (5), a higher pressure is established in the stationary state on the side of the lower or higher temperature respectively.

Consider as the first concrete example the *Knudsen effect* for ideal gases. For the "energy of transport"

$$U^* = Q^* + \bar{H} \qquad (3\text{–}11.6)$$

($\bar{H}$ = molar enthalpy), in the case of monatomic ideal gases, the following expression is obtained from kinetic gas theory:

$$U^* = 2RT \qquad (3\text{–}11.7)$$

(R = gas constant). Since in this case the relation

$$\bar{H} = \tfrac{5}{2}RT \qquad (3\text{–}11.8)$$

holds (using the same zero point of energy as in the previous equation), we find for the heat of transport:

$$Q^* = -\tfrac{1}{2}RT, \qquad (3\text{–}11.9)$$

that is, a negative value. This expression is independent of the arbitrarily chosen zero point of the internal energy (cf. p. 188) as required by Eq. (5). It can be shown that Eq. (9) is also valid for polyatomic ideal gases. Insertion of Eq. (9) in Eq. (5) yields (cf. Eq. 1–23.5):

$$\frac{\Delta P}{\Delta T} = \frac{R}{2\bar{V}} = \frac{P}{2T} \quad \text{(stationary Knudsen effect for ideal gases)}. \qquad (3\text{–}11.10)$$

Because of the assumption (3–5.4)

$$\left| \frac{\Delta T}{T} \right| \ll 1$$

we have:

$$1 + \frac{1}{2}\frac{\Delta T}{T} \approx \sqrt{1 + \frac{\Delta T}{T}}.$$

Thus we obtain:

$$\frac{P + \Delta P}{P} = \sqrt{\frac{T + \Delta T}{T}}$$

or with Eq. (3–5.1):

$$\frac{P''}{P'} = \sqrt{\frac{T''}{T'}} \quad \text{(stationary Knudsen effect for ideal gases).}$$

(3–11.11)

Equation (11) can be derived from kinetic arguments directly and can also be experimentally verified.

If the fundamental assumption for the validity of Eqs. (9) and (11)—the diameter of the opening being small in comparison with the mean free path of the gas molecules (cf. §3–10)—is no longer satisfied, then the expressions for the heat of transport and the thermomechanical pressure difference become more complicated even if we consider only ideal gases[42]. If, in the extreme case, the opening is so wide that the gas can pass through the valve in macroscopic flow, we find:

$$Q^* = 0$$

and thus from Eq. (5)

$$\Delta P_{\text{stat}} = 0,$$

as is obvious from the beginning. Between this extreme case and the cases to which the Knudsen effect applies, the heat of transport Q^* depends on the ratio of the opening diameter to the mean free path of the gas molecules, but as before, is negative. This can be verified[43–45] by experiments on different gases with capillaries of different diameters.

Even for the *fountain effect* of liquid helium II, the heat of transport Q^* is always negative and depends—since this deals with a given substance—only on the capillary radius and on the average temperature. Measurements show[46] that the heat of transport increases with decreasing capillary width, at given mean temperature, until the maximum value is reached at a critical capillary diameter.

When considering the fountain effect, the studies of Brewer and Edwards[47] should be noted because a particularly impressive verification of the Onsager reciprocity law results from this[9]. These authors determined Q^* both by direct calorimetry, using Eq. (3–9.31), and indirectly, using Eq. (5), from the thermomechanical pressure difference. They found agreement between the two sets of Q^* values within the limits of error of the measurements in the temperature range from 1.1°K to 1.7°K. Since the same quantity Q^* appears in Eq. (3–9.31) and in Eq. (5) as a result of the Onsager reciprocity relations, these measurements present another verification of the Onsager law.

Finally, we turn to *thermoosmosis* which has been experimentally examined recently both for gases and liquids.

Denbigh and Raumann[27] as well as Bearman[48,49] conducted very careful experiments on the thermoosmosis of gases through natural rubber membranes. The gases were H_2, N_2, CO_2, and H_2O as well as He, Ne, Ar, Kr, and Xe. The temperature differences were between 2 K and 19 K, the average temperature being approximately 300 K. The rounded values of the heats of transport Q^* found from the stationary effect according to Eq. (5) are:

H_2:	$+ 100$ cal mol^{-1}	He:	1375 cal mol^{-1}
N_2:	$- 260$ cal mol^{-1}	Ne:	1204 cal mol^{-1}
CO_2:	-1800 cal mol^{-1}	Ar:	$- 58$ cal mol^{-1}
H_2O (steam, uncertain):		Kr:	$- 275$ cal mol^{-1}
	-3400 cal mol^{-1}	Xe:	$- 589$ cal mol^{-1}

It is particularly interesting that for the same type of membrane different signs of Q^* are to be observed for different permeating gases.

In the experiments of Alexander and Wirtz[28] on the thermoosmosis of water through different types of membranes, it appears that the sign of Q^* (again determined from the thermoosmotic pressure difference) is not fixed

Table 4

Measured Values of the Heat of Transport Q^* for the System Water–Cellophane (type 600 g m^{-2}) at Different Mean Temperatures (ϑ = Celsius temperature, T = absolute or Kelvin temperature) (effective temperature differences being about 1.3 K). After Haase and de Greiff[30]

ϑ °C	T K	Q^* cal mol^{-1}
10.87	284.02	2.43
14.16	287.31	2.24
19.28	292.43	1.84
24.19	297.34	1.49
29.17	302.32	1.19
32.18	305.33	1.03
37.26	310.41	0.79
41.96	315.11	0.55
47.24	320.39	0.34
52.31	325.46	0.16
57.69	330.84	0.016
66.11	339.26	$- 0.143$
71.40	344.55	$- 0.399$
76.87	350.02	$- 0.459$

for liquids. Thus positive values of the heats of transport for water were obtained for cellulose preparations (membranes of cellophane and cuprophane) and negative values for animal albumin (membranes of gold beater's skin). Here the sign of the effect was, for the same liquid, dependent on the kind of membrane. Various filters of cellulose and nitrocellulose, unwetted cellulose preparations, and burnt clay showed no effects or changes with time.

More recent measurements[29, 30, 50, 51] have been made on the thermoosmosis† of water through membranes of different cellophane types, some of which contain copper ferrocyanide, $Cu_2Fe(CN)_6$ (cf. §3–14). The heat of transport Q^* was again determined from the thermoosmotic pressure difference using Eq. (5). It is noteworthy that we have here a change in the sign of the effect for a *given* system, the only variable being the mean temperature (see Table 4).[27]

3–12 THERMOOSMOSIS IN BINARY SYSTEMS (EMPIRICAL DESCRIPTION)

Let the system consist of a pure gas or a pure liquid (component 1) and a gaseous or liquid binary mixture (components 1 and 2). Let the two subsystems be separated from one another by a semipermeable membrane which is only permeable to component 1. Let the two components be nonelectrolytes, and let an external electric potential difference be excluded. Then in the most general case the phases can show a temperature difference ΔT, a pressure difference ΔP, and a composition difference (mole fraction difference) Δx_1. Accordingly, matter and heat exchange between the two subsystems is possible (Fig. 9). Here component 1 is designated as the "solvent," component 2 as the "solute." The mixture is called a "solution."

In the special case $x_1 = 1$, $\Delta x_1 = 0$, our binary system is reduced to the one-component system in Fig. 7 (p. 191).

As examples of the arrangement sketched in Fig. 9 we mention: a solution of polystyrene in benzene separated from pure benzene by a cellophane foil (which is permeable only to benzene) or an aqueous sugar solution separated from pure water by a cellophane foil impregnated with copper ferrocyanide (which is permeable only to water) (cf. §3–14).

In the first case, benzene represents component 1; in the second case, water represents component 1. If the concentration of component 2 (of the polystyrene or of the sugar) assumes the limiting value zero in the solution then we have the one-component system in Fig. 7 and the "semipermeable" membrane becomes a normal (permeable) membrane.

Apart from the effects known from §3–10 for one-component systems, there are two new phenomena here: at vanishingly small temperature and

† For a theoretical interpretation of thermal osmosis, see Haase[4].

Phase ' (solution)	Phase " (pure solvent)
Temperature T	Temperature $T + \Delta T$
Pressure P	Pressure $P + \Delta P$
Composition	Composition $x_1 + \Delta x_1 = 1$
(mole fraction of component 1) x_1	

$$\leftarrow J_1 > 0 \quad \text{(material flow)}$$
$$\leftarrow J_Q > 0 \quad \text{(heat flow)}$$

Fig. 9. Heterogeneous (discontinuous) system with transport of matter and heat between the two subsystems (phase ' and phase ") as a special case of the system in Fig. 5 (p. 163).

pressure differences, the concentration difference between the two subsystems leads to a mass transport known as *osmosis* (cf. §3–8) and to a heat transfer which we will call the *osmotic Dufour effect* (see also §4–26). Since permeation, heat conduction, thermal osmosis, and the osmotic thermal effect can also appear in our system, we have to deal with six different transport processes from the purely empirical point of view.

For sufficiently small values of ΔT, ΔP, and Δx_1, we describe the simultaneous occurrence of the six possible effects by the following empirical equations which represent the natural extension of the linear laws (3–10.1) and (3–10.2):

$$J_1 = \frac{q}{\delta} (A \, \Delta P + A^* \, \Delta x_1 + B \, \Delta T), \tag{3–12.1}$$

$$J_Q = \frac{q}{\delta} (C \, \Delta P + C^* \, \Delta x_1 + \lambda_0 \, \Delta T). \tag{3–12.2}$$

Here (cf. Fig. 9) J_1 is the flow of component 1, J_Q the heat flux, q the (effective) cross section of the membrane, and δ the thickness of the membrane. The six conventional transport coefficients A, A^*, B, C, C^*, and λ_0 depend on the average temperature and (to a lesser extent) on the average pressure of the system as well as on the nature of the membrane and of the solvent. The notation chosen for these quantities, for the pertinent transport processes, and for the analogous phenomena in an ordinary solution (continuous binary system) may be summarized in a table (p. 205).

If the mass flow through the membrane vanishes then at given temperatures of the two subsystems no further pressure and concentration changes can occur. The condition $J_1 = 0$ thus describes the *stationary state* in agreement with the general condition (3–9.11) and we find from Eq. (1):

$$A \, \Delta P + A^* \, \Delta x_1 + B \, \Delta T = 0 \quad \text{(stationary state)}. \tag{3–12.3}$$

Three interesting special cases are implied in this equation.

Symbol	Denotation	Effect	Analogous effect in ordinary solution
A	Permeability	Permeation	Pressure diffusion (p. 321)
A^*	Osmotic permeability	Osmosis	Diffusion (p. 271)
B	Thermoosmotic permeability	Thermal osmosis	Thermal diffusion (p. 355)
C	Osmotic thermal coefficient	Osmotic thermal effect	Pressure thermal effect (p. 356)
C^*	Osmotic Dufour coefficient	Osmotic Dufour effect	Dufour effect (p. 356)
λ_0	Thermal conductivity for uniform state	Heat conduction in the case $\Delta P = 0$, $\Delta x_1 = 0$	Heat conduction at zero gradients of pressure and concentration (p. 332)

In the first case, we deal with a nonisothermal one-component system ($\Delta x_1 = 0$). Then there results from Eq. (3) for the stationary pressure difference:

$$\Delta P_{\text{stat}} = -\frac{B}{A} \Delta T \quad (\Delta x_1 = 0). \qquad (3\text{–}12.4)$$

This is Eq. (3–10.13) for the *thermoosmotic pressure difference*. Such a steady nonequilibrium state arises from the opposing effects of permeation and thermal osmosis (cf. §3–11).

The second case concerns a nonisothermal binary system at zero pressure difference ($\Delta P = 0$). From Eq. (3) we then derive the mole fraction difference for the stationary state:

$$(\Delta x_1)_{\text{stat}} = -\frac{B}{A^*} \Delta T \quad (\Delta P = 0). \qquad (3\text{–}12.5)$$

Such a stationary nonequilibrium state arises from the opposing actions of osmosis and thermal osmosis. If we regard not ΔT but Δx_1 as the independent variable then there follows:

$$\Delta T_{\text{stat}} = -\frac{A^*}{B} \Delta x_1 \quad (\Delta P = 0). \qquad (3\text{–}12.6)$$

The steady temperature difference ΔT_{stat} is called the *osmotic temperature*[52]. It represents that temperature difference which must arise so that at a given concentration difference (i.e. at fixed concentration of the solution) the matter flow through the membrane disappears. The fundamental difference between "osmotic temperature" and "osmotic pressure" is that in the first case we deal with a stationary nonequilibrium state while in the second case with an equilibrium state. This is also expressed in the fact that the formula (6) for the osmotic temperature contains transport coefficients, while only thermodynamic quantities appear in the equation for the osmotic pressure (see below).

The third case refers to an isothermal binary system ($\Delta T = 0$) with vanishing heat flux ($J_Q = 0$). Here the stationary state ($J_1 = 0$) becomes identical with the equilibrium state and we are dealing with an *osmotic equilibrium* (cf. §1–20). The equilibrium pressure difference

$$-\Delta P \equiv \Pi \quad \text{(equilibrium)} \tag{3–12.7}$$

is the *osmotic pressure* (§1–20). From Eqs. (2) and (3) we obtain ($\Delta T = 0$, $J_Q = 0$):

$$-\Delta P = \Pi = \frac{A^*}{A} \Delta x_1 = \frac{C^*}{C} \Delta x_1 \quad \text{(equilibrium)}. \tag{3–12.8}$$

Thus the equilibrium state $J_1 = 0$, $J_Q = 0$, $\Delta T = 0$ arises from the opposing actions of permeation and osmosis or of the osmotic thermal effect and the osmotic Dufour effect.

According to classical equilibrium conditions (see Eqs. 3–4.2 and 3–4.9), the chemical potentials of component 1 in both the subsystems (μ_1 and $\mu_1 + \Delta\mu_1$) must have the same value at osmotic equilibrium. Since $\Delta T = 0$, we need only consider the expression (3–5.8)

$$(\Delta\mu_1)_T = V_1 \Delta P + \left(\frac{\partial\mu_1}{\partial x_1}\right)_{T,P} \Delta x_1. \tag{3–12.9}$$

Thus:

$$(\Delta\mu_1)_T = 0 \quad \text{(equilibrium)}. \tag{3–12.10}$$

Here V_1 is the partial molar volume of component 1 in the solution which, in the present case of dilute solutions (small values of Δx_1), is practically equal to the molar volume of pure component 1 (of the pure solvent) at the average temperature and at the average pressure of the system. The partial derivative refers to the solution.

From Eqs. (7), (9), and (10) we obtain:

$$-\Delta P = \Pi = \frac{1}{V_1}\left(\frac{\partial\mu_1}{\partial x_1}\right)_{T,P} \Delta x_1 \quad \text{(equilibrium)}, \tag{3–12.11}$$

which is the formula for the osmotic pressure of a dilute solution. By comparison of Eq. (8) and Eq. (11) we find:

$$\frac{A^*}{A} = \frac{C^*}{C} = \frac{1}{V_1} \left(\frac{\partial \mu_1}{\partial x_1}\right)_{T,P}. \tag{3-12.12}$$

Thus, by the application of *classical* thermodynamics, we reach the conclusion that of the six transport coefficients A, A^*, B, C, C^*, and λ_0 only *four* are independent (cf. §1–28). A further reduction of the number of independent coefficients occurs in §3–13 with the help of the thermodynamics of irreversible processes. In the meantime, we describe the irreversible processes in our system by the four quantities A, B, C, and λ_0.[28]

According to Eq. (12), we find the coefficients A (permeability) and B (thermoosmotic permeability) in Eqs. (4), (5), and (6) for the steady state. These coefficients are also determined directly from permeation or osmosis and from thermal osmosis using Eq. (1). Thus we may check the consistency of measurements.

From Eqs. (1), (2), (9), and (12) we derive:

$$J_1 = \frac{q}{\delta}\left[\frac{A}{V_1}(\Delta\mu_1)_T + B\,\Delta T\right], \tag{3-12.13}$$

$$J_Q = \frac{q}{\delta}\left[\frac{C}{V_1}(\Delta\mu_1)_T + \lambda_0\,\Delta T\right]. \tag{3-12.14}$$

From these relations it is obvious again that we end up with four transport coefficients.

If we eliminate $(\Delta\mu_1)_T$ from Eqs. (13) and (14) we obtain:

$$J_Q = \frac{C}{A}J_1 + \lambda_\infty\frac{q}{\delta}\Delta T \tag{3-12.15}$$

with

$$\lambda_\infty \equiv \lambda_0 - \frac{BC}{A}. \tag{3-12.16}$$

According to the above (cf. also §3–9), the quantity λ_∞ represents the *thermal conductivity for the stationary state*. For the stationary nonequilibrium states described by Eqs. (4) and (5) or (6), the heat flux is found to be

$$J_Q = \lambda_\infty\frac{q}{\delta}\Delta T \quad\text{(stationary state)}, \tag{3-12.17}$$

in agreement with Eq. (3–9.12).

3-13 THERMOOSMOSIS IN BINARY SYSTEMS (THERMODYNAMIC–PHENOMENOLOGICAL DESCRIPTION)

The thermodynamic–phenomenological description of the irreversible processes mentioned in the previous section[1] proceeds from the phenomenological equations (3–9.1) and (3–9.2). These hold for sufficiently small values of ΔT, ΔP, and Δx_1. On consideration of Eqs. (3–9.3), (3–9.4), and (3–12.9) we find:

$$J_1 = \alpha_{11}(\Delta\mu_1)_T + \alpha_{1Q}\frac{\Delta T}{T}, \tag{3-13.1}$$

$$J_Q = \alpha_{Q1}(\Delta\mu_1)_T + \alpha_{QQ}\frac{\Delta T}{T}. \tag{3-13.2}$$

Here the quantities α_{11}, α_{1Q}, α_{Q1}, and α_{QQ} are the phenomenological coefficients and T the average absolute temperature of the system. If we set:

$$\left.\begin{array}{ll} \alpha_{11}V_1\dfrac{\delta}{q} = A, & \dfrac{\alpha_{1Q}}{T}\dfrac{\delta}{q} = B, \\[2ex] \alpha_{Q1}V_1\dfrac{\delta}{q} = C, & \dfrac{\alpha_{QQ}}{T}\dfrac{\delta}{q} = \lambda_0, \end{array}\right\} \tag{3-13.3}$$

then we recognize that the empirical equations (3–12.13) and (3–12.14) (extended only by the statements of classical thermodynamics) are contained in the phenomenological equations (1) and (2) of the thermodynamics of irreversible processes. We reached a completely analogous conclusion for the treatment of the electrokinetic phenomena (§3–7) and of the thermomechanical effects in one-component systems (§3–10). We must introduce here also the Onsager reciprocity law and the inequality statements following from the positive-only character of the dissipation function, in order to obtain new results.

From the reciprocity relation (3–9.6)

$$\alpha_{1Q} = \alpha_{Q1} \tag{3-13.4}$$

there follows with Eq. (3):

$$C = BTV_1. \tag{3-13.5}$$

This is a general relation between the osmotic thermal coefficient and the thermoosmotic permeability. Thus the number of independent transport coefficients is reduced from 4 to 3.

The formal definition of the *heat of transport* Q^* (for the component 1, which, here, is the only one in question) according to Eq. (3–9.8) and Eq. (3) is:

$$Q^* \equiv \frac{\alpha_{1Q}}{\alpha_{11}} = \frac{B}{A}TV_1. \tag{3-13.6}$$

Insertion of Eq. (6) into Eqs. (3–12.4), (3–12.5), and (3–12.6), on consideration of Eq. (3–12.12), produces the following relations for the stationary states without the use of the Onsager reciprocity law (cf. §3–9 and §3–10), where the first equation, which refers to a one-component system, is identical to Eq. (3–10.21):

$$\Delta P_{\text{stat}} = -\frac{Q^*}{TV_1}\Delta T \quad (\Delta x_1 = 0), \tag{3–13.7}$$

$$(\Delta x_1)_{\text{stat}} = -\frac{Q^*}{T\left(\dfrac{\partial \mu_1}{\partial x_1}\right)_{T,P}}\Delta T \quad (\Delta P = 0), \tag{3–13.8}$$

$$\Delta T_{\text{stat}} = -\frac{T}{Q^*}\left(\frac{\partial \mu_1}{\partial x_1}\right)_{T,P}\Delta x_1 \quad (\Delta P = 0). \tag{3–13.9}$$

These relations, which represent special cases of Eqs. (3–9.22) and (3–9.23), have been known for a long time, since they were obtained partly by quasi-thermodynamic, partly by kinetic arguments, where, however, the quantity Q^* was usually interpreted in a very special way[38].

We now want to show explicitly (cf. §3–9 and §3–10) that the heat of transport Q^* defined by Eq. (6) has the physical significance corresponding to its name

$$Q^* = \left(\frac{J_Q}{J_1}\right)_{\Delta T=0}. \tag{3–13.10}$$

For this, we require the Onsager reciprocity law in the form of Eq. (5). From Eqs. (5) and (6) there results:

$$Q^* = \frac{C}{A}. \tag{3–13.11}$$

With this it follows from Eq. (3–12.15) that

$$J_Q = Q^*J_1 + \lambda_\infty \frac{q}{\delta}\Delta T, \tag{3–13.12}$$

with which Eq. (10) is proved.

From Eq. (3–12.16) and Eqs. (5) and (11) the difference between the two heat conductivities is found to be:

$$\lambda_0 - \lambda_\infty = TV_1 \frac{B^2}{A} = BQ^*. \tag{3–13.13}$$

This relation is a special case of Eq. (3–9.16).

From the inequalities (3–9.7)

$$\alpha_{11} > 0, \quad \alpha_{QQ} > 0, \quad \alpha_{11}\alpha_{QQ} - \alpha_{1Q}^2 > 0 \tag{3–13.14}$$

we finally obtain, with the help of Eqs. (3), (4), and (13), the inequality statements:

$$A > 0, \tag{3-13.15}$$

$$\lambda_0 > 0, \tag{3-13.16}$$

$$\lambda_\infty > 0, \tag{3-13.17}$$

$$\lambda_0 > \lambda_\infty, \tag{3-13.18}$$

$$B^2 < \frac{A\lambda_0}{TV_1}. \tag{3-13.19}$$

The same remarks apply, in principle, to these inequalities as to Eqs. (3–10.26) to (3–10.30) (p. 197).

From Eqs. (5), (6), and (15) it follows that B, C, and Q^* always have the same sign. With the help of Eqs. (1–22.38) and (1–22.41)

$$\left(\frac{\partial \mu_1}{\partial x_1}\right)_{T,P} > 0 \quad \text{(stable or metastable noncritical phase),} \tag{3-13.20}$$

$$\left(\frac{\partial \mu_1}{\partial x_1}\right)_{T,P} = 0 \quad \text{(critical phase)} \tag{3-13.21}$$

and with Eq. (3–12.12) using Eqs. (5), (6), (9), and (15) for the case of a non-critical solution we find that the quantities B and C^* have the same sign, while A^* is always positive; at the critical point, however, A^*, C^*, and the osmotic temperature ΔT_{stat} disappear, provided we assume that A and B are finite at this point and different from zero.†

3–14 THERMOOSMOSIS IN BINARY SYSTEMS (EXPERIMENTAL EXAMPLES)

As a concrete example of the binary system dealt with in §3–12 and §3–13, we consider an aqueous, dilute cane sugar solution, which is separated from pure water by a cellophane foil impregnated with copper ferrocyanide, $Cu_2Fe(CN)_6$. Such a membrane is—at least for short times—permeable to water (solvent, component 1), but impermeable to cane sugar (solute, component 2). It is a modification of the classical "semipermeable wall."

If the experiment[29] is restricted to very dilute solutions (cane sugar concentrations up to about 1% by weight) then *ideal dilute solutions* can be assumed. For these, the equation

$$\mu_1 = \mu_{01} + RT \ln x_1 \quad (x_2 \ll 1) \tag{3-14.1}$$

---

† This assumption is in no way self-evident, for in the analogous case of thermal diffusion the quantity corresponding to B is infinite at the critical consolute point. The diffusion coefficient, the analog of A^*, disappears at the critical point (cf. p. 279).

holds, where μ_1 and μ_{01} denote the chemical potentials of component 1 (water) in the solution at temperature T and pressure P and the chemical potential of the pure liquid component 1 respectively (of the pure water) at the same temperature and at the same pressure, R the gas constant, and x_i $(i = 1, 2)$ the mole fraction of the component i in the solution. From Eq. (1) it follows by virtue of Eq. (1–12.12) that

$$\left(\frac{\partial \mu_1}{\partial x_1}\right)_{T,P} = \frac{RT}{x_1} \approx RT \quad (x_2 \ll 1), \tag{3–14.2}$$

$$\left(\frac{\partial \mu_1}{\partial P}\right)_{T,x_1} = V_1 = \left(\frac{\partial \mu_{01}}{\partial P}\right)_T = V_{01} \quad (x_2 \ll 1), \tag{3–14.3}$$

where V_{01} is the molar volume of the pure liquid component 1. Furthermore, for the molarity c_2 of the component 2, as for any very dilute solution, we find from Eq. (3–6.15) and Fig. 9 (p. 204)

$$c_2 = \frac{x_2}{V_{01}} = \frac{\Delta x_1}{V_{01}}. \tag{3–14.4}$$

Many of the formulas in §3–12 and §3–13 are simplified with the relations (2), (3), and (4).

The condition $\Delta x_1 = 0$, $(c_2 = 0)$ in the following equations always means that we assume pure water at both sides of the given membrane. The impregnated cellophane foil then becomes a "permeable" membrane after being a "semipermeable" membrane, since it is then indeed in each case permeable to water. The conditions $\Delta P = 0$ and $\Delta P \neq 0$ will be realized by a movable and a fixed position of the foil, respectively.

The permeability A can either be determined from the permeation using Eq. (3–12.1)

$$J_1 = A \frac{q}{\delta} \Delta P \quad (\Delta x_1 = 0, \Delta T = 0) \tag{3–14.5}$$

or from the osmosis using Eqs. (3–12.1) and (3–12.12) as well as Eqs. (2), (3), and (4)

$$J_1 = A \frac{q}{\delta} RTc_2 \quad (\Delta P = 0, \Delta T = 0). \tag{3–14.6}$$

Here J_1 is the flux of component 1, q the (effective) cross section, and δ the thickness of the membrane.

The thermoosmotic permeability B can be gained from the thermal osmosis with the relation

$$J_1 = B \frac{q}{\delta} \Delta T \quad (\Delta x_1 = 0, \Delta P = 0) \tag{3–14.7}$$

which follows from Eq. (3–12.1).

The heat of transport Q^* results either from the measured values of A and B using Eq. (3-13.6) and Eq. (3):

$$Q^* = TV_{01} \frac{B}{A} \qquad (3-14.8)$$

or directly from the stationary effects, namely from the thermoosmotic pressure difference ΔP_{stat} and from the stationary composition difference $(\Delta x_1)_{stat}$ using Eqs. (3-13.7) and (3-13.8). It follows from these equations, on consideration of Eqs. (2), (3), and (4), that

$$\Delta P_{stat} = -\frac{Q^*}{TV_{01}} \Delta T \quad (\Delta x_1 = 0), \qquad (3-14.9)$$

$$(\Delta x_1)_{stat} = (x_2)_{stat} = -\frac{Q^*}{RT^2} \Delta T \quad (\Delta P = 0). \qquad (3-14.10)$$

In Eqs. (8) to (10), the average absolute temperature of the system can be inserted for T and the molar volume of pure water at the average temperature and at the average pressure of the system for V_{01}.

Equation (10) will be evaluated at given values of T and ΔT under the condition $\Delta P = 0$, the water flow J_1 will be measured as a function of the concentration x_2 of the sugar, and the value of x_2 will be determined for which J_1 disappears. This value of x_2 will be inserted in Eq. (10) as $(x_2)_{stat}$.

Table 5

Selection of the Measurements on Wetted Cellophane Foils of the Type $600 \, \mathrm{g \, m^{-2}}$ (thickness $80 \, \mu m$) Impregnated with a Deposit of Copper Ferrocyanide.   Permeating Substance: Water; Solute in the Case of the Stationary Effect: Cane Sugar.   After Haase and Steinert[29]

ϑ	average Celsius temperature	A	permeability
B	thermoosmotic permeability	Q^*	heat of transport

ϑ °C	A (from permeation) mol g^{-1} s	B (from thermal osmosis) mol K^{-1} cm^{-1} s^{-1}	Q^* cal mol^{-1} from A and B according to Eq. (8)	Q^* cal mol^{-1} from stationary effect according to Eq. (10)
22.5	1.20×10^{-15}	3.15×10^{-10}	33.4	27.4
32.5	1.80×10^{-15}	3.05×10^{-10}	22.4	21.1
42.5	2.80×10^{-15}	10.29×10^{-10}	50.3	65.1

In Table 5, a few values of A, B, and Q^* obtained from such measurements on permeation, thermal osmosis, and the stationary concentration difference are presented.

Observation of the stationary state characterized by Eq. (10) is equivalent to an experimental verification of the "osmotic temperature" (§3–12).

REFERENCES

1. R. Haase, *Z. Physik. Chem.* (*Frankfurt*) **21**, 244 (1959).
2. W. Kuhn, *Z. Elektrochem.* **55**, 207 (1951).
3. L. B. Ticknor, *J. Phys. Chem.* **62**, 1483 (1958).
4. R. Haase, *Z. Physik. Chem.* (*Frankfurt*) **51**, 315 (1966).
5. N. K. Adam, *The Physics and Chemistry of Surfaces*, 3rd ed., Oxford (1941), p. 351.
6. P. Mazur and J. Th. G. Overbeek, *Rec. Trav. Chim.* **70**, 83 (1951).
7. U. Saxén, *Wiedemanns Ann. Physik Chem.* (2) **47**, 46 (1892).
8. A. Klemm, *Z. Naturforsch.* **13a**, 1039 (1958).
9. D. G. Miller, *Chem. Rev.* **60**, 15 (1960).
10. H. F. Holmes, C. S. Shoup, Jr., and C. H. Secoy, *J. Phys. Chem.* **69**, 3148 (1965).
11. R. P. Rastogi and K. M. Jha, *Trans. Faraday Soc.* **62**, 585 (1966).
12. R. P. Rastogi and K. M. Jha, *J. Phys. Chem.* **70**, 1017 (1966).
13. A. J. Staverman, *Rec. Trav. Chim.* **70**, 344 (1951).
14. A. J. Staverman, *Trans. Faraday Soc.* **48**, 176 (1952).
15. J. L. Talen and A. J. Staverman, *Trans. Faraday Soc.* **61**, 2794, 2800 (1965).
16. A. J. Staverman, C. A. Kruissink, and D. T. F. Pals, *Trans. Faraday Soc.* **61**, 2805 (1965).
17. R. Schloegl, *Stofftransport durch Membranen, Fortschritte der Physikalischen Chemie*, Vol. 9, Darmstadt (1963).
18. M. Tasaka, S. Morita, and M. Nagasawa, *J. Phys. Chem.* **69**, 4191 (1965).
19. E. D. Eastman, *J. Am. Chem. Soc.* (a) **48**, 1482 (1926); (b) **50**, 283, 293 (1928).
20. C. Wagner, *Ann. Physik* (5) (a) **3**, 629 (1929); (b) **6**, 370 (1930).
21. R. Haase, *Z. Naturforsch.* **6a**, 420 (1951).
22. W. Feddersen, *Pogg. Ann. Physik Chem.* (5) **148**, 302 (1873).
23. O. Reynolds, *Phil. Trans. Roy. Soc. London, Ser. B* **170**, 727 (1879).
24. G. Lippmann, *Compt. Rend.* (a) **145**, 105 (1907); (b) **145**, 104 (1907).
25. M. Aubert, *Ann. Chim. Phys.* (8) **26**, 145, 551 (1912).
26. H. P. Hutchison, I. S. Nixon, and K. G. Denbigh, *Discussions Faraday Soc.* **3**, 86 (1948).
27. K. G. Denbigh and G. Raumann, *Proc. Roy. Soc.* (*London*), *Ser. A* **210**, 377, 518 (1951).

28. K. F. ALEXANDER and K. WIRTZ, *Z. Physik. Chem. (Leipzig)* **195**, 165 (1950).

29. R. HAASE and C. STEINERT, *Z. Physik. Chem. (Frankfurt)* **21**, 270 (1959).

30. R. HAASE and H. J. DE GREIFF, *Z. Physik. Chem. (Frankfurt)* **44**, 301 (1965).

31. M. KNUDSEN, *Ann. Physik* (4) **31**, 205 (1910).

32. J. F. ALLEN and H. JONES, *Nature* **141**, 243 (1938).

33. J. G. DAUNT and K. MENDELSSOHN, *Nature* **143**, 719 (1939).

34. I. PRIGOGINE, *Etude thermodynamique des Phénomènes irréversibles*, Paris–Liége (1947).

35. I. PRIGOGINE, *Introduction to Thermodynamics of Irreversible Processes*, Springfield, Illinois (1955).

36. S. R. DE GROOT, *J. Phys. Radium* (8) **8**, 188 (1947).

37. S. R. DE GROOT, *Physica* **13**, 555 (1947).

38. S. R. DE GROOT, *Thermodynamics of Irreversible Processes*, Amsterdam (1951).

39. R. HAASE, *Ergeb. Exakt. Naturw.* **26**, 56 (1952).

40. R. HAASE, *Z. Physik. Chem. (Frankfurt)* **21**, 244 (1959).

41. M. Y. BEARMAN and R. J. BEARMAN, *J. Appl. Polymer Sci.* **10**, 773 (1966).

42. E. A. MASON, R. B. EVANS III, and G. M. WATSON, *J. Chem. Phys.* **38**, 1808 (1963).

43. A. VAN ITTERBEEK and E. DE GRANDE, *Physica* **13**, 289, 422 (1947).

44. T. TAKAISHI and Y. SENSUI, *Trans. Faraday Soc.* **59**, 2503 (1963).

45. H. J. M. HANLEY and W. A. STEELE, *Trans. Faraday Soc.* **61**, 2661 (1965).

46. L. MEYER and J. H. MELLINK, *Physica* **13**, 197 (1947).

47. D. F. BREWER and D. O. EDWARDS, *Proc. Phys. Soc. (London)* **71**, 117 (1958).

48. R. J. BEARMAN, *J. Phys. Chem.* **61**, 708 (1957).

49. M. Y. BEARMAN and R. J. BEARMAN, *J. Phys. Chem.* **70**, 3010 (1966).

50. H. VOELLMY and P. LÄUGER, *Ber. Bunsenges. Physik. Chem.* **70**, 165 (1966).

51. R. P. RASTOGI and K. SINGH, *Trans. Faraday Soc.* **62**, 1754 (1966).

52. PH. KOHNSTAMM, *Proc. Koninkl. Ned. Akad. Wetenschap.* **13**, 778 (1911).

CHAPTER 4

PROCESSES IN CONTINUOUS SYSTEMS

A. Fundamental Principles

4-1 INTRODUCTION

In a continuous system (cf. §1–1 and §1–21), intensive quantities such as density, pressure, temperature, concentrations, etc., depend on the space-coordinates in a continuous manner. The above-mentioned quantities are thus, in general, functions of time and position for irreversible processes. Only in the case of a stationary state (cf. §1–17) are intensive state functions constant in time, although they still may depend on the position-coordinates.

The mathematical requirements for the thermodynamic treatment of processes in continuous systems are considerably greater than those for homogeneous and discontinuous (heterogeneous) systems. This is obvious from our derivation of the equilibrium conditions for continuous systems (§1–21). The equations which express the mass balance, the momentum balance, the energy balance, and the entropy balance must be brought to a form which refers to *local* changes in each *volume element*, where each volume element represents an *open* region.[29] The total continuous system can also be open, that is, it can exhibit matter exchange with its surroundings.

Here we discuss processes like electric conduction, diffusion, sedimentation, heat conduction, thermal diffusion, viscous flow, etc., taking into account external force fields and the macroscopic motion of system parts. Electrically charged species and chemical reactions will also be included in the presentation.

In order that the methods of the thermodynamics of irreversible processes may be applicable, in particular so that the (generalized) Gibbs equation remains valid for an arbitrary instant during the course of irreversible processes in continuous systems, it must be assumed that the processes take place "not too rapidly," as explained in more detail in §1–23. All processes of interest to us here satisfy this condition.

In Part 4A (fundamental principles), the fundamental equations for isotropic systems without electrification and magnetization will be developed with consideration of stationary external force fields. From §4–14 transverse effects for rotating systems and transport processes in strong magnetic fields will be excluded at present for simplicity.

In Part 4B (isothermal processes), we deal with processes in systems at constant temperature, while we discuss the processes determined by unequally distributed temperatures in Part 4C (nonisothermal processes). In both cases, only isotropic systems without electrification and magnetization will be discussed for time-invariant external fields. Also, transverse effects for rotating systems and transport processes in strong magnetic fields will be excluded. Viscous flow will not be explicitly considered or will be neglected.

Part 4D (complicated processes) contains, finally, problems like viscous flow, rotating systems, arbitrary electromagnetic fields, galvanomagnetic and thermomagnetic effects, as well as processes in anisotropic media.

4–2 GENERAL FORM OF A BALANCE EQUATION

Let an arbitrary extensive state function (mass, energy, entropy, etc.) be denoted by Z. If the value of Z valid for a volume element of a continuous system is divided by the volume of this space element then we obtain an intensive quantity Z_V, the *density of Z* (cf. §1–6) which varies generally in time and in position.† If we delineate a subsystem of volume V from our continuous system then the value of Z for this subsystem becomes:

$$Z = \int_V Z_V \, dV, \qquad (4\text{–}2.1)$$

where dV denotes a volume element and the integration is over the entire space which the subsystem fills. In the simplest case, Z is the mass of the subsystem and Z_V the density in a space element.

In the most general case, the change in time of the quantity Z is due to two causes: the in- or outflow of Z through the containing surfaces of the subsystem and the production or destruction of Z inside the subsystem. We must think, for example, of the amount of a definite species which can change for an open system both by matter exchange with the surroundings and by chemical reactions inside the system.

We describe the in- or outflow through the containing surfaces of the subsystem by the vector $\mathbf{J}_Z$. This vector is then defined so that its direction corresponds to the direction of flow at any one position and its length is a measure of the intensity of the flow at the position considered; the component J_{Zn} of the vector $\mathbf{J}_Z$ in the direction normal to a reference surface

---

† The exact definition of Z_V is:

$$Z_V \equiv \lim_{V \to 0} \frac{Z}{V},$$

where V denotes volume.

gives the amount of Z flowing through an element of this surface per unit time and per unit surface area. The vector $\mathbf{J}_Z$ thus describes the *current density* of Z at the point in question. If we now limit our subsystem by a closed surface area whose total surface amounts to Ω, and if we consider a surface element $d\Omega$ whose normal is directed into the outside, then $J_{Zn}\, d\Omega$ represents the contribution of the surface element $d\Omega$ to the flow of Z through the total closed surface area Ω of the subsystem from inside to outside.

The production or destruction of Z is measured by the quantity $q(Z)$. This denotes the amount of Z formed at each unit time interval and in each unit volume inside a space element and is thus called the *local production of Z* (or internal source of Z per unit time and volume). The expression $q(Z)\, dV$ represents the contribution of the volume element dV to the production of Z inside the total volume V of the subsystem.

Thus, the general form of *a balance equation* for Z (see Eq. 1) in time t is:

$$\frac{dZ}{dt} = \frac{d}{dt}\int_V Z_V\, dV = -\int_\Omega J_{Zn}\, d\Omega + \int_V q(Z)\, dV. \qquad (4\text{–}2.2)$$

Here the integrals denoted by $\int_V$ are over the total volume V of the system and the integral distinguished by $\int_\Omega$ is taken over the closed surface Ω of the system. With the help of Gauss' theorem there follows from Eq. (2):

$$\frac{dZ}{dt} = \frac{d}{dt}\int_V Z_V\, dV = -\int_V \operatorname{div} \mathbf{J}_Z\, dV + \int_V q(Z)\, dV. \qquad (4\text{–}2.3)$$

The scalar quantity $\operatorname{div} \mathbf{J}_Z$, the *divergence*† of $\mathbf{J}_Z$, represents the "source density" of Z, i.e. the amount of Z per unit time and volume flowing out of a volume element to the outside through its bounding surfaces.

Since Eq. (3) holds for a subsystem of arbitrary size, we can let the total volume shrink to a volume element and write:

$$\frac{\partial Z_V}{\partial t} = -\operatorname{div} \mathbf{J}_Z + q(Z), \qquad (4\text{–}2.4)$$

where the operator $\partial/\partial t$ denotes differentiation with respect to time at a fixed position; Z_V, $\mathbf{J}_Z$, and $q(Z)$ depend on time and position.

Equation (4) is the general expression for a *local balance* of the quantity Z; it is known as the *general equation of continuity*. The equation states that the rate of increase of the density of Z in each volume element is equal to the

---

† For rectangular coordinates (x, y, z), we have for a position-dependent vector $\mathbf{J}$ with the components J_x, J_y, J_z:

$$\operatorname{div} \mathbf{J} = \frac{\partial J_x}{\partial x} + \frac{\partial J_y}{\partial y} + \frac{\partial J_z}{\partial z}.$$

local production of Z minus the divergence of the current density of Z. If Z represents a quantity for which a law of conservation holds, such as, for instance, the total mass or the total energy, then the term $q(Z)$ disappears.

4–3 REFERENCE VELOCITIES AND DIFFUSION CURRENTS

In each volume element of a continuous system, the average velocities of the chemical species may differ from one another, so that there appears not only a macroscopic motion of the volume element ("convection") but also a macroscopically perceptible relative motion of the individual particles ("diffusion" in the widest sense).

We denote the average velocity of the particles of type k by the vector $\mathbf{v}_k$. In order to describe the diffusion (in the widest sense) quantitatively, we introduce the relative velocity

$$\mathbf{v}_k - \boldsymbol{\omega},$$

where $\boldsymbol{\omega}$ denotes a reference velocity. The quantity

$$\mathbf{J}_k \equiv c_k(\mathbf{v}_k - \boldsymbol{\omega}) \tag{4–3.1}$$

is called the *diffusion current density* of species k. Here c_k is the molarity of species k. The vector $\mathbf{J}_k$ represents the amount of substance of chemical species k which, per unit time and unit area, flows perpendicularly through a reference surface moving with velocity $\boldsymbol{\omega}$.

For fluid systems, the reference velocity $\boldsymbol{\omega}$ is so chosen that the relations

$$\boldsymbol{\omega} = \sum_k \omega_k \mathbf{v}_k, \tag{4–3.2}$$

$$\sum_k \omega_k = 1 \tag{4–3.3}$$

hold. The ω_k are the normalized weights, i.e. "weight factors" for averaging the velocities subject to the normalization (3). The summations in Eqs. (2) and (3) are over all chemical species. For solid (crystalline) media, another choice of the reference velocity is suitable (cf. below).

From Eqs. (1), (2), and (3) it follows immediately that

$$\sum_k \frac{\omega_k}{c_k} \mathbf{J}_k = 0. \tag{4–3.4}$$

For N species in a fluid medium, there are, therefore, only $N - 1$ independent diffusion currents.

The most important examples for reference velocities are:

(a) *Barycentric Velocity* $\mathbf{v}$. That is,

$$\rho \mathbf{v} = \sum_k \rho_k \mathbf{v}_k, \tag{4–3.5}$$

where ρ_k denotes the partial density (mass concentration) of species k and ρ the total density. Thus $\mathbf{v}$ is the velocity of the local center of gravity. If we introduce the mass fraction χ_k of species k then we obtain the relations:

$$\rho = \sum_k \rho_k, \tag{4-3.6}$$

$$\chi_k = \frac{\rho_k}{\rho}, \tag{4-3.7}$$

$$\sum_k \chi_k = 1, \tag{4-3.8}$$

$$\mathbf{v} = \sum_k \chi_k \mathbf{v}_k. \tag{4-3.9}$$

Comparison of Eq. (2) and Eq. (9) produces:

$$\boldsymbol{\omega} = \mathbf{v}, \qquad \omega_k = \chi_k, \tag{4-3.10}$$

so that the condition (3) is satisfied due to Eq. (8). The appropriate diffusion current density is, according to Eq. (1),

$$_v\mathbf{J}_k \equiv c_k(\mathbf{v}_k - \mathbf{v}). \tag{4-3.11}$$

From Eqs. (4), (7), and (10) and using

$$\rho_k = M_k c_k \tag{4-3.12}$$

(M_k = molar mass of species k) there follows the identity:

$$\sum_k M_k {}_v\mathbf{J}_k = 0. \tag{4-3.13}$$

By the use of the composition variables ρ_k and χ_k we find, in place of the quantity (11), the diffusion current density:

$$\mathbf{J}_k^* \equiv \rho_k(\mathbf{v}_k - \mathbf{v}) = M_k {}_v\mathbf{J}_k, \tag{4-3.14}$$

for which there is valid, according to Eq. (13),

$$\sum_k \mathbf{J}_k^* = 0. \tag{4-3.15}$$

The barycentric velocity $\mathbf{v}$ is that quantity to be preferred as the "average velocity" for establishing the fundamental equations since, with its help, the momentum balance and the energy balance can be formulated most easily. We denote a reference system based on the velocity $\mathbf{v}$ as a "barycentric system."

(b) *Mean Molar Velocity* **u**. Here:

$$c\mathbf{u} = \sum_k c_k \mathbf{v}_k, \tag{4-3.16}$$

where

$$c = \sum_k c_k \tag{4-3.17}$$

denotes the total molarity. If we introduce the mole fraction of species k

$$x_k = \frac{c_k}{c}, \tag{4-3.18}$$

we obtain:

$$\sum_k x_k = 1, \tag{4-3.19}$$

$$\mathbf{u} = \sum_k x_k \mathbf{v}_k. \tag{4-3.20}$$

A comparison of Eq. (2) and Eq. (20) yields:

$$\boldsymbol{\omega} = \mathbf{u}, \qquad \omega_k = x_k, \tag{4-3.21}$$

so that, on the basis of Eq. (19), condition (3) is satisfied. The pertinent diffusion current density is, according to Eq. (1),

$$_u\mathbf{J}_k \equiv c_k(\mathbf{v}_k - \mathbf{u}). \tag{4-3.22}$$

From Eqs. (4), (18), and (21) there results:

$$\sum_k {}_u\mathbf{J}_k = 0. \tag{4-3.23}$$

The mean molar velocity **u** plays a role in the kinetic theory of diffusion phenomena in the widest sense (diffusion, sedimentation, thermal diffusion, etc.), since it represents the average particle velocity. We therefore call a reference system based on the velocity **u** a "molecular reference system."

(c) *Mean Volume Velocity* **w**. Here:

$$\mathbf{w} = \sum_k c_k V_k \mathbf{v}_k, \tag{4-3.24}$$

where V_k denotes the partial molar volume of species k. It follows, by comparison with Eq. (2), that:

$$\boldsymbol{\omega} = \mathbf{w}, \qquad \omega_k = c_k' V_k. \tag{4-3.25}$$

Due to the identity (1–6.11)

$$\sum_k c_k V_k = 1 \tag{4-3.26}$$

condition (3) is satisfied. The pertinent diffusion current density is, according to Eq. (1),

$$_w\mathbf{J}_k \equiv c_k(\mathbf{v}_k - \mathbf{w}). \qquad (4\text{–}3.27)$$

From Eqs. (4) and (25) we find:

$$\sum_k V_k \,_w\mathbf{J}_k = 0. \qquad (4\text{–}3.28)$$

The mean volume velocity is the most important reference velocity for calculations from experimental data, since it is practically identical to the "convection velocity." Thus the condition $\mathbf{w} = 0$ corresponds to the "absence of convection" in the sense of the experimentor.† We denote the reference system based on the velocity $\mathbf{w}$ as a "Fick reference system" (cf. §4–17).

(d) *Velocity* $\mathbf{v}_1$ *of Species* 1. Here:

$$\omega_1 = 1, \qquad \omega_j = 0 \quad (j = 2, 3, \ldots), \qquad (4\text{–}3.29)$$

so that condition (3) is immediately satisfied. The pertinent diffusion current density results from Eq. (1):

$$_1\mathbf{J}_k \equiv c_k(\mathbf{v}_k - \mathbf{v}_1). \qquad (4\text{–}3.30)$$

From Eq. (4) there results, as is already obvious from definition (30), that

$$_1\mathbf{J}_1 = 0.$$

Such a choice for the reference velocity is then suitable if we want to direct attention to one component (for electrolyte solutions, say, the electrically neutral solvent) or if, in the case of "creeping motions," all dependent quantities are to be eliminated (cf. §4–12). We call a system based on the velocity $\mathbf{v}_1$ a "Hittorf reference system" (cf. §4–16).

(e) *Displacement Velocity* $\boldsymbol{\omega}_G$ *of the Lattice Points in a Crystal.* Here, according to Eq. (1), there results for the diffusion current density

$$_G\mathbf{J}_k \equiv c_k(\mathbf{v}_k - \boldsymbol{\omega}_G). \qquad (4\text{–}3.31)$$

---

† Whether the condition $\mathbf{w} = 0$ corresponds to the absence of convection for experiments actually performed depends on the degree of approximation with which the quantity

$$\sum_k V_k \,_w\mathbf{J}_k$$

(which disappears according to Eq. (28)) can be set equal to the "volume flow" measured relative to a fixed point of the apparatus. This question has been examined fundamentally by Agar[1] for a binary system with concentration and temperature gradients. It appears that, in the framework of the problems which are interesting in practice, the two quantities do not have to be distinguished.

Since $\boldsymbol{\omega}_G$ is not directly coupled in theory to the individual velocities $\mathbf{v}_k$, the relations (2), (3), and (4) cease to be valid. For solid (crystalline) media of N chemical species, we thus have, generally, N independent diffusion currents. For special mechanisms, the number of independent diffusion currents can be reduced. Also, for undeformed crystals, we usually choose as a reference species a particle type which is bound more or less rigidly to the lattice positions, and thus it is related back to the Hittorf reference system. For conciseness, we denote a reference system based on the velocity $\boldsymbol{\omega}_G$ as a "lattice reference system."

For fluid media, a conversion from a reference velocity $\boldsymbol{\omega}'$ to a second velocity $\boldsymbol{\omega}''$ is often necessary. Therefore we must consider the transition from one diffusion current density

$$\mathbf{J}'_k = c_k(\mathbf{v}_k - \boldsymbol{\omega}') \tag{4–3.32}$$

to a second diffusion current density

$$\mathbf{J}''_k = c_k(\mathbf{v}_k - \boldsymbol{\omega}''). \tag{4–3.33}$$

Here the relations

$$\boldsymbol{\omega}' = \sum_k \omega'_k \mathbf{v}_k, \qquad \boldsymbol{\omega}'' = \sum_k \omega''_k \mathbf{v}_k, \tag{4–3.34}$$

$$\sum_k \omega'_k = 1, \qquad \sum_k \omega''_k = 1, \tag{4–3.35}$$

$$\sum_k \frac{\omega'_k}{c_k} \mathbf{J}'_k = 0, \qquad \sum_k \frac{\omega''_k}{c_k} \mathbf{J}''_k = 0 \tag{4–3.36}$$

hold, as shown by Eqs. (2), (3), and (4). With this we obtain the conversion relation:

$$\mathbf{J}''_i = \mathbf{J}'_i - c_i \sum_k \frac{\omega''_k}{c_k} \mathbf{J}'_k, \tag{4–3.37}$$

where the inferior i indicates any one particular species and the summation sign indicates, again, a summation over all particle types. The "new" diffusion currents $\mathbf{J}''_k$ are thus homogeneous linear functions of the "old" diffusion currents $\mathbf{J}'_k$ (cf. §1–27).

Consider four important special cases of Eq. (37).

First, we set (cf. Eq. 14):

$$\mathbf{J}'_i = {}_v\mathbf{J}_i = \frac{1}{M_i}\mathbf{J}^*_i, \quad \mathbf{J}''_i = {}_u\mathbf{J}_i.$$

Then there results from Eqs. (18), (21), and (37):

$${}_u\mathbf{J}_i = \frac{1}{M_i}\mathbf{J}^*_i - x_i \sum_k \frac{1}{M_k}\mathbf{J}^*_k, \tag{4–3.38}$$

yielding the conversion from the barycentric system to the molecular reference system.

Secondly, let

$$\mathbf{J}_i' = {}_v\mathbf{J}_i = \frac{1}{M_i}\mathbf{J}_i^*, \qquad \mathbf{J}_i'' = {}_w\mathbf{J}_i.$$

Then there follows from Eqs. (25) and (37):

$$_w\mathbf{J}_i = \frac{1}{M_i}\mathbf{J}_i^* - c_i\sum_k \frac{V_k}{M_k}\mathbf{J}_k^*, \tag{4-3.39}$$

yielding the conversion from the barycentric system to Fick's reference system.

Thirdly, let

$$\mathbf{J}_i' = {}_v\mathbf{J}_i = \frac{1}{M_i}\mathbf{J}_i^*, \qquad \mathbf{J}_i'' = {}_1\mathbf{J}_i.$$

Then we obtain from Eqs. (29) and (37), or also directly from Eqs. (11), (14), and (30),

$$_1\mathbf{J}_i = \frac{1}{M_i}\left(\mathbf{J}_i^* - \frac{c_i}{c_1}\mathbf{J}_1^*\right), \tag{4-3.40}$$

yielding the conversion from the barycentric system to Hittorf's reference system.

Fourthly, let

$$\mathbf{J}_i' = {}_w\mathbf{J}_i, \qquad \mathbf{J}_i'' = {}_1\mathbf{J}_i.$$

Then we derive from Eqs. (29) and (37), or also directly from Eqs. (27) and (30),

$$_1\mathbf{J}_i = {}_w\mathbf{J}_i - \frac{c_i}{c_1}{}_w\mathbf{J}_1, \tag{4-3.41}$$

yielding the conversion from the Fick to the Hittorf reference system.

If there are charged particles (ions or electrons) among the chemical species of the volume element considered then all the above equations remain applicable. We can, however, assign to each charged species i a *partial electric current density*

$$\mathbf{I}_i = z_i F c_i(\mathbf{v}_i - \boldsymbol{\omega}) = z_i F \mathbf{J}_i \tag{4-3.42}$$

(cf. §3–2), where z_i is the charge number of species i (positive or negative integers for cations or anions respectively) and F is the Faraday constant.

For the *total electric current density*, we have

$$\mathbf{I} = \sum_i \mathbf{I}_i, \tag{4-3.43}$$

where the sum is over all charged species. From Eqs. (42) and (43) there follows:

$$I = F \sum_i z_i \mathbf{J}_i \tag{4-3.44}$$

or

$$I = F \sum_i z_i c_i (\mathbf{v}_i - \boldsymbol{\omega}). \tag{4-3.45}$$

For electric conductors, the condition for electric neutrality (1–21.29)

$$\sum_i z_i c_i = 0 \tag{4-3.46}$$

must be satisfied in each volume element. Thus there results from Eq. (45):

$$I = F \sum_i z_i c_i \mathbf{v}_i, \tag{4-3.47}$$

and we arrive at the interesting result that the total electric current density is independent of the reference velocity $\boldsymbol{\omega}$.

4-4 MASS BALANCE

If we apply the considerations discussed in §4–2 to the amount of substance n_k of the chemical species k then n_k appears in place of Z and the molarity c_k of species k in place of Z_V. Furthermore, the current density $\mathbf{J}_Z$ is equal to the vector $c_k \mathbf{v}_k$ ($\mathbf{v}_k$ = average velocity of the particles of type k), whose value is equal to the amount of substance of species k which flows perpendicularly through a stationary reference surface per unit time and area. The quantity $q(Z)$, the "local production of Z," is equal to the amount of species k formed in the volume element by chemical reactions per unit time and volume. If we denote the stoichiometric number of species k in the chemical reaction r by ν_{kr} (cf. §1–14) and the reaction rate of the chemical reaction r referred to unit volume by b_r then we have:†

$$q(Z) = q(n_k) = \sum_r \nu_{kr} b_r,$$

where the summation is over all reactions. Thus the general continuity equation (4–2.4) assumes the following form:

$$\frac{\partial c_k}{\partial t} = -\operatorname{div}(c_k \mathbf{v}_k) + \sum_r \nu_{kr} b_r, \tag{4-4.1}$$

---

† Cf. §2–2, in particular the footnote to p. 106.

where the operator $\partial/\partial t$ denotes the derivative with respect to time at a fixed position. This relation represents a *local mass balance*.

If we want to measure the quantity of matter not by the amount of substance but by the mass then with the help of Eq. (4–3.12) we introduce the partial density (mass concentration) ρ_k of species k and obtain in place of Eq. (1)

$$\frac{\partial \rho_k}{\partial t} = -\operatorname{div}(\rho_k \mathbf{v}_k) + M_k \sum_r \nu_{kr} b_r. \qquad (4\text{–}4.2)$$

Here M_k is the molar mass of species k.

The law of conservation of mass, applied to chemical reactions, demands that

$$\sum_k M_k \nu_{kr} = 0, \qquad (4\text{–}4.3)$$

where the summation is formed over all reacting species and the equation is valid for each single chemical reaction.

If we sum Eq. (2) over all species k then with Eqs. (4–3.5), (4–3.6), and Eq. (3) we find

$$\frac{\partial \rho}{\partial t} = -\operatorname{div}(\rho \mathbf{v}), \qquad (4\text{–}4.4)$$

where ρ is the total density and $\mathbf{v}$ the barycentric velocity. Equation (4) is the local balance for the total mass and is denoted in hydrodynamics as the "equation of continuity of matter." It has the form of Eq. (4–2.4) with $Z_V = \rho$, $\mathbf{J}_Z = \rho \mathbf{v}$, $q(Z) = 0$, where the last expression represents the mathematical formulation of the conservation of the total mass. Using the "total" or "substantial" derivative

$$\frac{d}{dt} = \frac{\partial}{\partial t} + \mathbf{v}\operatorname{grad}, \qquad (4\text{–}4.5)$$

which gives the time derivative of a quantity for an observer moving with velocity $\mathbf{v}$, and considering the relations

$$\frac{d\rho}{dt} = \frac{\partial \rho}{\partial t} + \mathbf{v}\operatorname{grad}\rho, \qquad (4\text{–}4.6)$$

$$\operatorname{div}(\rho \mathbf{v}) = \mathbf{v}\operatorname{grad}\rho + \rho\operatorname{div}\mathbf{v}, \qquad (4\text{–}4.7)$$

and Eq. (4) we obtain the total mass balance in another form:

$$\frac{d\rho}{dt} + \rho\operatorname{div}\mathbf{v} = 0. \qquad (4\text{–}4.8)$$

The continuity equation for the total mass used in hydrodynamics usually has this form.

We again return to Eq. (1). If we introduce the diffusion current density $\mathbf{J}_k$ of species k following Eq. (4–3.1) then we derive from Eq. (1):

$$\frac{\partial c_k}{\partial t} = -\operatorname{div}(c_k\boldsymbol{\omega}) - \operatorname{div}\mathbf{J}_k + \sum_r \nu_{kr}b_r. \tag{4–4.9}$$

Here $\boldsymbol{\omega}$ is any reference velocity. We see that the "mass current density" $c_k\mathbf{v}_k$ in Eq. (4–3.1) is split up into a "convective" part $c_k\boldsymbol{\omega}$ and a "nonconvective" part, the diffusion current density $\mathbf{J}_k$.

An analogous decomposition can be accomplished generally in Eq. (4–2.4), if the vector $\mathbf{J}_Z$ is split into two terms:

$$\mathbf{J}_Z = Z_V\boldsymbol{\omega} + \mathbf{J}'_Z, \tag{4–4.10}$$

where $\mathbf{J}'_Z$ is the "nonconvective" current density of Z. Equation (4–2.4) can then be written in the form

$$\frac{\partial Z_V}{\partial t} = -\operatorname{div}(Z_V\boldsymbol{\omega}) - \operatorname{div}\mathbf{J}'_Z + q(Z). \tag{4–4.11}$$

Equation (11) obviously generalizes Eq. (9).

The four most important special cases of Eq. (9) result from §4–3 if we identify $\boldsymbol{\omega}$ with $\mathbf{v}$, $\mathbf{u}$, $\mathbf{w}$, and $\mathbf{v}_1$, in that order. Using Eqs. (4–3.11), (4–3.22), (4–3.27), and (4–3.30) we find that

$$\frac{\partial c_k}{\partial t} = -\operatorname{div}(c_k\mathbf{v}) - \operatorname{div}{}_v\mathbf{J}_k + \sum_r \nu_{kr}b_r, \tag{4–4.12}$$

$$\frac{\partial c_k}{\partial t} = -\operatorname{div}(c_k\mathbf{u}) - \operatorname{div}{}_u\mathbf{J}_k + \sum_r \nu_{kr}b_r, \tag{4–4.13}$$

$$\frac{\partial c_k}{\partial t} = -\operatorname{div}(c_k\mathbf{w}) - \operatorname{div}{}_w\mathbf{J}_k + \sum_r \nu_{kr}b_r, \tag{4–4.14}$$

$$\frac{\partial c_k}{\partial t} = -\operatorname{div}(c_k\mathbf{v}_1) - \operatorname{div}{}_1\mathbf{J}_k + \sum_r \nu_{kr}b_r. \tag{4–4.15}$$

Also, the corresponding relation follows from Eqs. (4–3.14) and Eq. (2)

$$\frac{\partial \rho_k}{\partial t} = -\operatorname{div}(\rho_k\mathbf{v}) - \operatorname{div}\mathbf{J}_k^* + M_k \sum_r \nu_{kr}b_r \tag{4–4.16}$$

and is sometimes useful.

The substantial derivatives are frequently introduced in place of the local time derivative $\partial Z_V/\partial t$ (cf. Eq. 5)

$$\frac{d\tilde{Z}}{dt} = \frac{\partial \tilde{Z}}{\partial t} + \mathbf{v}\operatorname{grad}\tilde{Z}, \tag{4–4.17}$$

where $\tilde{Z}$ denotes the specific quantity

$$\tilde{Z} \equiv \frac{Z_V}{\rho}. \tag{4-4.18}$$

From Eqs. (17) and (18) and on consideration of Eq. (4) there results

$$\rho \frac{d\tilde{Z}}{dt} = \frac{\partial Z_V}{\partial t} + \text{div}\,(Z_V \mathbf{v}). \tag{4-4.19}$$

With the help of Eq. (19), Eq. (16) can be written in the following form:

$$\rho \frac{d\chi_k}{dt} = -\text{div}\,\mathbf{J}_k^* + M_k \sum_r \nu_{kr} b_r, \tag{4-4.20}$$

where χ_k is the mass fraction of species k.

In this connection, the discussion of the *stationary state* is interesting. In this case (cf. §1–17),

$$\frac{\partial c_k}{\partial t} = 0. \tag{4-4.21}$$

From this with Eq. (9) there follows:

$$\text{div}\,\mathbf{J}_k = -\text{div}\,(c_k \boldsymbol{\omega}) + \sum_r \nu_{kr} b_r \quad \text{(stationary state)}. \tag{4-4.22}$$

This relation represents the general steady-state condition.

If we assume no convection then, in practice (cf. p. 221), $\mathbf{w} = 0$ holds. With Eqs. (14) and (21) there thus results:

$$\text{div}\,_w\mathbf{J}_k = \sum_r \nu_{kr} b_r \quad \text{(convection-free stationary state)}. \tag{4-4.23}$$

If we also exclude chemical reactions ($b_r = 0$), we obtain from Eq. (23)

$$\text{div}\,_w\mathbf{J}_k = 0 \quad \begin{array}{l}\text{(convection-free stationary state without chemical} \\ \text{reactions).}\end{array} \tag{4-4.24}$$

For the experimental realization of such a stationary state, there are usually two more conditions satisfied:

a) The processes occur in a closed vessel.

b) The diffusion currents $_w\mathbf{J}_k$ of the individual species lie in *one* spatial direction because the gradients causing them have only one direction.

With assumption (b) Eq. (24) requires that the $_w\mathbf{J}_k$ are independent of position. Condition (a) requires that the $_w\mathbf{J}_k$ disappear at the vessel surfaces which surround the system perpendicular to the above-mentioned space direction. Accordingly, the diffusion current densities are zero everywhere:

$$_w\mathbf{J}_k = 0 \quad \begin{array}{l}\text{(convection-free stationary state without chemical} \\ \text{reactions in the most important case in practice).}\end{array} \tag{4-4.25}$$

From Eq. (4–3.37) it is now obvious that at the disappearance of all diffusion currents in a definite reference system the diffusion fluxes in any other arbitrary system must also be zero. We have, thus, quite generally:

$$\mathbf{J}_{k} = 0 \quad \begin{array}{l} \text{(convection-free stationary state without chemical} \\ \text{reactions in the most important case in practice).} \end{array} \quad (4\text{–}4.26)$$

4–5 MOMENTUM BALANCE

Application of Newton's law or of the momentum balance to a volume element of a continuous system requires some care, since such a space element presents variable mass and composition and the external forces can act differently on the individual chemical species. It can, however, be shown that the usual hydrodynamic equations formulated for one-component systems continue to hold if the barycentric velocity is inserted in place of the quantity usually designated in hydrodynamics merely as "velocity." We have seen this in the derivation of Eq. (4–4.4) or (4–4.8).

For the present, we consider only isotropic media without electrification or magnetization. Then, in addition to external forces acting on the medium, such as gravitational force, centrifugal force, electrostatic forces, etc., we must be concerned, in isotropic crystals,† with the forces produced by the pressure P, which is the same on all sides, and, for fluid media, also with those forces related to the viscous stresses. In order to begin with the simplest case, we will first exclude the viscous forces and thus assume "hydrodynamically ideal media."

Let ρ be the density, $\mathbf{v}$ the barycentric velocity, t the time, c_k the molarity of species k, and $\mathbf{K}_k$ the molar external force acting on species k. Then, for the case considered here,

$$\rho \frac{d\mathbf{v}}{dt} = \sum_{k} c_k \mathbf{K}_k - \text{grad } P = \mathfrak{K}_V - \text{grad } P. \quad (4\text{–}5.1)$$

Here the sum is over all species. The quantity $d\mathbf{v}/dt$ is the barycentric acceleration, i.e. the acceleration for an observer moving with the center of gravity of the volume element. This is composed of the local acceleration at a fixed position and the acceleration arising from the displacement in position of the center of gravity. $\mathfrak{K}_V = \sum_k c_k \mathbf{K}_k$ is the resultant of the "force density" of the external forces.‡

---

† If an isotropic crystal is stressed unequally, e.g. by tension in a definite direction, then it becomes a nonisotropic medium in the thermodynamic sense.

‡ The Newton law thus holds in the form "force equals mass times acceleration of center of gravity" also for a volume element of a continuous system, in which convection, diffusion, etc., are possible. In Eq. (1), the right-hand side denotes the resultant of the forces referred to unit volume. Accordingly, the density stands on the left-hand side of Eq. (1) in place of the mass.

Local mechanical equilibrium is generally distinguished by the condition

$$\frac{d\mathbf{v}}{dt} = 0. \tag{4–5.2}$$

Thus, in our case, there follows from Eq. (1):

$$\sum_k c_k \mathbf{K}_k = \text{grad } P \quad \text{(mechanical equilibrium).} \tag{4–5.3}$$

This relation is identical to condition (1–21.27) found in §1–21 for the mechanical equilibrium of a system which is in complete equilibrium. However, here we have assumed only that mechanical equilibrium is present in each volume element without the other equilibrium conditions (temperature constant, chemical equilibrium, etc.) being necessarily satisfied.

If we multiply Eq. (1) by $\mathbf{v}$ in a scalar fashion (dot product) then we obtain:

$$\rho \mathbf{v} \frac{d\mathbf{v}}{dt} = \mathbf{v} \left(\sum_k c_k \mathbf{K}_k - \text{grad } P \right). \tag{4–5.4}$$

Furthermore ($v^2 \equiv \mathbf{vv}$),

$$\rho \mathbf{v} \frac{d\mathbf{v}}{dt} = \frac{\rho}{2} \frac{dv^2}{dt} = \frac{d}{dt} \left(\frac{\rho}{2} v^2 \right) - \frac{v^2}{2} \frac{d\rho}{dt}, \tag{4–5.5}$$

from which with Eqs. (4–4.5) and (4–4.8) there follows:

$$\rho \mathbf{v} \frac{d\mathbf{v}}{dt} = \frac{\partial}{\partial t} \left(\frac{\rho}{2} v^2 \right) + \mathbf{v} \text{ grad} \left(\frac{\rho}{2} v^2 \right) + \frac{v^2}{2} \rho \text{ div } \mathbf{v}. \tag{4–5.6}$$

Due to the relation

$$\text{div} \left(\frac{\rho}{2} v^2 \mathbf{v} \right) = \frac{\rho}{2} v^2 \text{ div } \mathbf{v} + \mathbf{v} \text{ grad} \left(\frac{\rho}{2} v^2 \right) \tag{4–5.7}$$

there results from Eqs. (4) and (6):

$$\frac{\partial}{\partial t} \left(\frac{\rho}{2} v^2 \right) + \text{div} \left(\frac{\rho}{2} v^2 \mathbf{v} \right) = \mathbf{v} \left(\sum_k c_k \mathbf{K}_k - \text{grad } P \right). \tag{4–5.8}$$

With the help of this we may simplify the energy balance (§4–6).

If we deal with fluid media for which *viscous flow* has to be taken into account then we must introduce the static pressure P (which would be present without flow), and viscous pressures P_{ij} ($i, j = 1, 2, 3$), corresponding to the three additional normal pressures (P_{11}, P_{22}, P_{33}) and to the six tangential (shear) pressures ($P_{12}, P_{13}, P_{21}, P_{23}, P_{31}, P_{32}$) which act on each volume element of a streaming fluid. In place of the scalar quantity P, the *pressure tensor* appears in the form:

$$\Pi = \begin{pmatrix} P + P_{11} & P_{12} & P_{13} \\ P_{21} & P + P_{22} & P_{23} \\ P_{31} & P_{32} & P + P_{33} \end{pmatrix}. \tag{4–5.9}$$

Because of the relation $P_{ij} = P_{ji}$ following from mechanics, this tensor is symmetric. If we call the three space-coordinates in a rectangular coordinate system z_1, z_2, and z_3 then P_{12} denotes the pressure, which acts on a surface element of the volume element in question perpendicular to the z_1-axis in the z_2-axis direction. Here P_{12} is counted positive if the pertinent force (friction force) has the opposite direction to the z_2-axis.

The vector grad P in the previous equations must now be replaced by the vector Div Π, the "tensor divergence" of Π. The three cartesian components of the vector Div Π are:

$$\frac{\partial P}{\partial z_i} + \sum_{j=1}^{3} \frac{\partial P_{ji}}{\partial z_j} \quad (i = 1, 2, 3), \tag{4–5.10}$$

where the operator $\partial/\partial z_i$ denotes the derivative with respect to the position-coordinate at fixed time. The force law now states that

$$\rho \frac{d\mathbf{v}}{dt} = \sum_k c_k \mathbf{K}_k - \text{Div } \Pi. \tag{4–5.11}$$

Correspondingly, the relation

$$\frac{\partial}{\partial t} \left(\frac{\rho}{2} v^2 \right) + \text{div} \left(\frac{\rho}{2} v^2 \mathbf{v} \right) = \mathbf{v} \left(\sum_k c_k \mathbf{K}_k - \text{Div } \Pi \right) \tag{4–5.12}$$

appears in place of Eq. (8).

For many irreversible processes, we can distinguish between two stages during the course of the process. In the first stage, the macroscopic motions die away because of viscous flow. In the second stage, slow processes such as diffusion, heat conduction, slow chemical reactions, etc., are present, while all velocities and accelerations are very small. In the second phase, "local mechanical equilibrium," i.e. the disappearance of the barycentric acceleration, is approximately satisfied, corresponding to the neglection of inertial forces. Simultaneously, the viscous pressures are very small. Using Eq. (11), we thus obtain the conditions (cf. Eq. 3):

$$P_{ij} \approx 0 \quad (i, j = 1, 2, 3), \quad \sum_k c_k \mathbf{K}_k \approx \text{grad } P. \tag{4–5.13}$$

If these relations hold, we speak of "creeping motion" in the fluid.

Should there be a pressure drop forced on the system from the outside then naturally the latter considerations cease to hold (cf. §4–23).

If we introduce, with Eqs. (4–3.6) and (4–3.12), the density in the form

$$\rho = \sum_k M_k c_k, \tag{4–5.14}$$

where M_k denotes the molar mass of species k, then we find from Eq. (1–21.4):

$$\sum_k c_k \mathbf{K}_k = \rho(\mathbf{g} + \Omega^2 \mathbf{r}) + \sum_k z_k c_k F \mathfrak{E}. \tag{4–5.15}$$

Here $\mathbf{g}$ denotes the gravitational acceleration, Ω and $\mathbf{r}$ the value of the angular velocity and the distance from the axis of rotation respectively for the centrifuge, z_k the charge number of the particle type k, F the Faraday constant, and $\mathfrak{E}$ the electric field strength. In Eq. (15), the terms with the Coriolis force and the Lorentz force are neglected. If we exclude space charges (cf. p. 69) then the condition of electric neutrality (4–3.46) is valid

$$\sum_k z_k c_k = 0. \tag{4–5.16}$$

From Eqs. (13) to (16) we derive:

$$P_{ij} \approx 0, \qquad \rho(\mathbf{g} + \Omega^2 \mathbf{r}) \approx \operatorname{grad} P. \tag{4–5.17}$$

This is the explicit condition for "creeping motion" in fluid media without space charges.

Next, for isotropic crystals ($P_{ij} = 0$) Eq. (1) holds. If we also assume that, according to Eq. (2), the acceleration of the center of gravity is zero then we obtain condition (3) for mechanical equilibrium. There thus results:

$$P_{ij} = 0, \qquad \sum_k c_k \mathbf{K}_k = \operatorname{grad} P, \tag{4–5.18}$$

by analogy with Eq. (13). If, moreover, we exclude space charges as well as Coriolis and Lorentz forces then there follows from Eqs. (15), (16), and (18)

$$P_{ij} = 0, \qquad \rho(\mathbf{g} + \Omega^2 \mathbf{r}) = \operatorname{grad} P, \tag{4–5.19}$$

by analogy with Eq. (17). Nevertheless, the condition for mechanical equilibrium assumed in Eqs. (18) and (19) is often doubtful for crystalline media. We may, meanwhile, neglect the influence of a gravitational or centrifugal field for many processes which occur in isotropic crystals, as well as ignore space charges and pressure gradients. Then, according to Eqs. (15) and (16), the right-hand side of Eq. (1) disappears, so that the second equation in Eq. (18) or (19) becomes an identity. If either this situation is present or if we can assume mechanical equilibrium for isotropic crystals, we will then speak of "isotropic crystals in simple cases" for conciseness.

For further abbreviation in the mode of expression, we summarize the relations (13) and (18) or (17) and (19) in the following statement: *for fluid media with creeping motion and isotropic crystals in simple cases, we may use Eqs.* (18), *which can be reduced to the relations* (19) *in the absence of space charges and with the neglection of Coriolis and Lorentz forces.*

4–6 ENERGY BALANCE

The "energy balance" for a volume element of a continuous system arises from the concept of "heat", since such a space element represents an open region in the sense of §1–7. In §1–7, we saw that it was meaningful and useful to define the heat dQ supplied to a single open phase for an infinitesimal state change using Eq. (1–7.2) as follows:

$$dQ \equiv dE - dw - \sum_k H_k \, d_e n_k. \tag{4–6.1}$$

Here E denotes the (total) energy of the phase, dw the infinitesimal work done on the phase in question (assuming it to be closed), H_k the partial molar enthalpy of species k in the phase, and $d_e n_k$ the infinitesimal increase of amount of substance of the species k in the phase by mass exchange with the surroundings. We now examine how a meaningful transfer of convention (1) to continuous systems results.

The total energy E of an arbitrary system or system region can be split up according to Eq. (1–4.6) in the following way:

$$E = U + E_{\text{kin}} + E_{\text{pot}}. \tag{4–6.2}$$

Here U is the internal energy, E_{kin} the macroscopic kinetic energy, and E_{pot} the potential energy in external conservative force fields.

It is expedient for continuous systems to refer the macroscopic motion to the center of gravity of each volume element, since the barycentric velocity $\mathbf{v}$ assumes a preferential role in the dynamics of continuous media (cf. §4–5). We thus write for the density of the kinetic energy, i.e. for the macroscopic kinetic energy of a space element referred to unit volume,

$$E_{\text{kin}\,v} = \frac{\rho}{2}\, v^2, \tag{4–6.3}$$

where ρ is the density. Correspondingly, the change in the total energy of the volume element, the work done on the volume element when it is considered closed, and the expression analogous to the last term in Eq. (1) must be valid for an observer moving with the center of gravity. Thus the analogs to dQ are measured in the barycentric system.

If we assume isotropic systems without viscous flow, we obtain for the *heat current density* $\mathbf{J}_Q$ the following equation, which, as will be obvious immediately, represents the conversion of Eq. (1) to continuous systems:

$$-\operatorname{div} \mathbf{J}_Q \equiv \frac{\partial U_V}{\partial t} + \operatorname{div}(U_V \mathbf{v}) + \frac{\partial}{\partial t}\left(\frac{\rho}{2} v^2\right) + \operatorname{div}\left(\frac{\rho}{2} v^2 \mathbf{v}\right)$$

$$-\sum_k \mathbf{K}_k c_k \mathbf{v}_k + \operatorname{div}(P\mathbf{v}) + \operatorname{div}\left(\sum_k H_k \,_v\mathbf{J}_k\right). \qquad (4\text{–}6.4)$$

Here t denotes the time, the operator $\partial/\partial t$ denotes differentiation with respect to time at a fixed position, U_V denotes the density of the internal energy, $\mathbf{K}_k$ the molar external force for species k, c_k the molarity of substance k, $\mathbf{v}_k$ the average velocity of the particles of type k, P the pressure, and

$$_v\mathbf{J}_k \equiv c_k(\mathbf{v}_k - \mathbf{v}) \qquad (4\text{–}6.5)$$

the diffusion current density of species k in the barycentric system (cf. Eq. 4–3.11).

We now multiply Eq. (4) by the time element dt at a fixed position, integrate over the volume V of a subsystem selected from the total system to have the closed surface area Ω, we use Gauss's law and note Eq. (5). Then we find ($d\Omega$ = surface area element, dV = volume element):

$$-\int_\Omega J_{Qn}\, d\Omega\, dt = d\int_V U_V\, dV + \int_\Omega U_V v_n\, d\Omega\, dt + d\int_V \frac{\rho}{2} v^2\, dV$$

$$+ \int_\Omega \frac{\rho}{2} v^2 v_n\, d\Omega\, dt - \sum_k \int_V c_k \mathbf{K}_k \mathbf{v}_k\, dV\, dt$$

$$+ \int_\Omega P v_n\, d\Omega\, dt + \sum_k \int_\Omega H_k c_k (v_{kn} - v_n)\, d\Omega\, dt.$$

$$(4\text{–}6.6)$$

Here J_{Qn}, or v_n, or v_{kn} is the normal component of $\mathbf{J}_Q$, or $\mathbf{v}$, or $\mathbf{v}_k$ respectively at the appropriate position on the surface area of the subsystem. If we use the sign conventions of §4–2 then we recognize that for a time element dt the left-hand side denotes the heat flowing into the subsystem from the outside; the first four terms on the right-hand side represent the increase in the internal and kinetic energy of the subsystem taking account of the loss due to convection; the fifth term on the right-hand side is the increase in the potential energy of the subsystem in the external conservative force fields (valid for any reference velocity, since it depends here only on the motion of the particles relative to the force fields which are assumed to be static); the sixth term on the right-hand side is, after reversing the sign, the work done by the static pressure for a displacement of the surface area on the subsystem assumed to be closed (measured in the barycentric system); the last term represents the analog to the expression

$$-\sum_k H_k \, d_e n_k$$

for an observer moving with the center of gravity. Accordingly, Eq. (4) corresponds perfectly to Eq. (1).

With the help of the conversion relation (4–5.8) there follows from Eq. (4) and Eq. (5):

$$\frac{\partial U_V}{\partial t} = -\text{div} \, (U_V \mathbf{v}) - \text{div} \left(\mathbf{J}_Q + \sum_k H_k \, {}_v\mathbf{J}_k \right)$$

$$- P \, \text{div} \, \mathbf{v} + \sum_k \mathbf{K}_k \, {}_v\mathbf{J}_k, \tag{4–6.7}$$

where the relation

$$\text{div} \, (P\mathbf{v}) \equiv P \, \text{div} \, \mathbf{v} + \mathbf{v} \, \text{grad} \, P \tag{4–6.8}$$

was used.

For systems *with viscous flow*, the more general expression

$$\frac{\partial U_V}{\partial t} = -\text{div} \, (U_V \mathbf{v}) - \text{div} \left(\mathbf{J}_Q + \sum_k H_k \, {}_v\mathbf{J}_k \right) - P \, \text{div} \, \mathbf{v}$$

$$- \sum_{i=1}^{3} \sum_{j=1}^{3} P_{ij} \frac{\partial v_i}{\partial z_j} + \sum_k \mathbf{K}_k \, {}_v\mathbf{J}_k \tag{4–6.9}$$

holds in place of Eq. (7). Here the P_{ij} $(= P_{ji})$ are the viscous pressures $(i, j = 1, 2, 3)$, the v_i the three components of $\mathbf{v}$ in a rectangular coordinate system, the z_j the three cartesian space-coordinates, and the derivatives $\partial v_i / \partial z_j$, correspondingly, the derivatives of the components of the barycentric velocity with respect to the position-coordinates at fixed time (velocity gradients). Equation (9) arises when account is taken of the work done by viscous pressures P_{ij} as well as the work done by static pressure P in Eq. (4), and then the momentum balance is applied in the form of Eq. (4–5.12).†

---

† If we write the expression

$$\sum_{i=1}^{3} \sum_{j=1}^{3} P_{ij} \frac{\partial v_i}{\partial z_j}$$

so that it is independent of the coordinate system, we obtain:

$$(\Pi - P\delta) : \text{Grad} \, \mathbf{v}.$$

Here Π is the pressure tensor (4–5.9), δ the unity tensor, and Grad $\mathbf{v}$ the vector gradient of $\mathbf{v}$ (i.e. also a tensor). The sign : denotes the inner product of two tensors after contracting twice, and leads, accordingly, to a scalar quantity.

With the help of Eq. (8) the relation (9) can be written in the form of a local balance for the internal energy following the model of Eq. (4–4.11):

$$\frac{\partial U_V}{\partial t} = -\operatorname{div}(U_V \mathbf{v}) - \operatorname{div} \mathbf{J}_U' + q(U) \tag{4–6.10}$$

with

$$\mathbf{J}_U' \equiv \mathbf{J}_Q + P\mathbf{v} + \sum_k H_{k\ v} \mathbf{J}_k \tag{4–6.11}$$

and

$$q(U) \equiv \mathbf{v} \operatorname{grad} P - \sum_{i=1}^{3} \sum_{j=1}^{3} P_{ij} \frac{\partial v_i}{\partial z_j} + \sum_k \mathbf{K}_{k\ v} \mathbf{J}_k. \tag{4–6.12}$$

Here $\mathbf{J}_U'$ is the "nonconvective current density" and $q(U)$ the "local production" of internal energy. Let it be noted that there is a "local production" only for the internal energy but not for the total energy.

From Eq. (5) and the expression for the enthalpy density

$$H_V = U_V + P = \sum_k c_k H_k, \tag{4–6.13}$$

which follows from Eqs. (1–5.3), (1–6.7), and (1–6.10), and from Eqs. (10) and (11) we obtain:

$$\frac{\partial U_V}{\partial t} = -\operatorname{div}\left(\mathbf{J}_Q + \sum_k H_k c_k \mathbf{v}_k\right) + q(U) \tag{4–6.14}$$

or

$$\begin{aligned}
\frac{\partial H_V}{\partial t} &= \sum_k c_k \frac{\partial H_k}{\partial t} + \sum_k H_k \frac{\partial c_k}{\partial t} \\
&= \frac{\partial P}{\partial t} - \operatorname{div}\left(\mathbf{J}_Q + \sum_k H_k c_k \mathbf{v}_k\right) + q(U),
\end{aligned} \tag{4–6.15}$$

where $q(U)$ is again given by Eq. (12).

In an isotropic medium without electrification and magnetization, only the work done by the static pressure and by the viscous pressures are considered as analogs to dw in Eq. (1), because "influences from the outside"—as they were considered first in Eqs. (1–7.1) and (1–7.2)—are not of interest in continuous systems, and the friction as well as electric flows, if necessary, are already taken account of in the last two terms.

The forces $\mathbf{K}_k$ always belong here to conservative force fields (cf. §1–4). Thus, say, gravitational fields, centrifugal fields, and electrostatic fields come into question. It will be shown in §4–34 how the energy balance in the more general case of an arbitrary electromagnetic field is formed. There it will also be evident that the energy balance set up here holds for systems without electrification and magnetization in *stationary* electromagnetic fields (that is, not only in electrostatic fields).

The relations (9), (10), (14), and (15) can be regarded as general expressions for the "energy balance" in an isotropic system without electrification and magnetization and without time-variable external fields. It is no longer necessary to take account of the momentum balance since this is already contained in the above-mentioned equations.

With Eq. (4–4.19) the left-hand side and the first term of the right-hand side in Eqs. (9) and (10) can be joined together as follows:

$$\frac{\partial U_V}{\partial t} + \mathrm{div}\,(U_V \mathbf{v}) = \rho \frac{d\tilde{U}}{dt}, \tag{4–6.16}$$

where $\tilde{U}$ denotes the specific internal energy and the operator d/dt the substantial derivative (cf. Eqs. 4–4.17 and 4–4.18).

As direct results of the energy balance, we now proceed to derive two formulas which will be used later.

The first formula results with the help of the mass balance (4–4.1) and the mathematical identity

$$\mathrm{div}\,(H_k c_k \mathbf{v}_k) = H_k\,\mathrm{div}\,(c_k \mathbf{v}_k) + c_k \mathbf{v}_k\,\mathrm{grad}\,H_k$$

from Eq. (15):

$$\sum_k c_k \frac{\partial H_k}{\partial t} + \sum_k c_k \mathbf{v}_k\,\mathrm{grad}\,H_k = -\sum_r \sum_k \nu_{kr} H_k b_r + \frac{\partial P}{\partial t}$$
$$- \mathrm{div}\,\mathbf{J}_Q + q(U). \tag{4–6.17}$$

Here ν_{kr} denotes the stoichiometric number of species k in chemical reaction r, and b_r the reaction rate of reaction r. Equation (17) will be used in §4–7.

The second formula is derived from the last equation, taking account of the relations following from Eqs. (1–6.6), (1–6.16), (1–15.5), and (1–15.20)

$$\sum_k c_k \frac{\partial H_k}{\partial t} = \frac{\bar{C}_P}{\bar{V}} \frac{\partial T}{\partial t} + \left[1 - \frac{T}{\bar{V}}\left(\frac{\partial \bar{V}}{\partial T}\right)_{P,x}\right] \frac{\partial P}{\partial t}, \tag{4–6.18}$$

$$\sum_k c_k\,\mathrm{grad}\,H_k = \frac{\bar{C}_P}{\bar{V}}\,\mathrm{grad}\,T + \left[1 - \frac{T}{\bar{V}}\left(\frac{\partial \bar{V}}{\partial T}\right)_{P,x}\right]\mathrm{grad}\,P. \tag{4–6.19}$$

Here $\bar{C}_P$ denotes the molar heat capacity at constant pressure, $\bar{V}$ the molar volume, T the absolute temperature, and the inferior x $(_x)$ constant mole fractions. By combining Eqs. (5), (12), and (17) with Eqs. (18) and (19), the local rate of increase of the temperature is found to be

$$\frac{\bar{C}_P}{\bar{V}} \frac{\partial T}{\partial t} = -\mathrm{div}\,\mathbf{J}_Q - \frac{\bar{C}_P}{\bar{V}} \mathbf{v}\,\mathrm{grad}\,T + \frac{T}{\bar{V}}\left(\frac{\partial \bar{V}}{\partial T}\right)_{P,x}$$
$$\times \left(\frac{\partial P}{\partial t} + \mathbf{v}\,\mathrm{grad}\,P\right) - \sum_r \sum_k \nu_{kr} H_k b_r$$
$$- \sum_{i=1}^{3} \sum_{j=1}^{3} P_{ij} \frac{\partial v_i}{\partial z_j} + \sum_k {}_v\mathbf{J}_k(\mathbf{K}_k - \mathrm{grad}\,H_k). \tag{4–6.20}$$

This relation, which for $\mathbf{v} = 0$, $\partial P / \partial t = 0$, $b_r = 0$, $P_{ij} = 0$, $_v\mathbf{J}_k = 0$ transforms into the usual "heat conduction equation," will be used in §4–24.

4–7 INVARIANCE PROPERTIES OF THE HEAT CURRENT

For the heat current density $\mathbf{J}_Q$ defined by Eq. (4–6.9) with Eqs. (4–6.5), (4–6.12), and (4–6.17), we can write:

$$-\operatorname{div} \mathbf{J}_Q = \sum_k c_k \frac{\partial H_k}{\partial t} + \sum_k c_k \mathbf{v}_k \operatorname{grad} H_k + \sum_r \sum_k \nu_{kr} H_k b_r$$

$$- \frac{\partial P}{\partial t} - \mathbf{v} \operatorname{grad} P + \sum_{i=1}^{3} \sum_{j=1}^{3} P_{ij} \frac{\partial v_i}{\partial z_j} - \sum_k c_k \mathbf{K}_k (\mathbf{v}_k - \mathbf{v}). \quad (4\text{–}7.1)$$

All undetermined additive constants in the partial molar enthalpies H_k, which result from the arbitrary energy zero points, are eliminated in Eq. (1), since the first two terms of the right-hand side contain only time and space derivatives of H_k and the third term is the sum of the products of the reaction enthalpies and the appropriate reaction rates, where again the undetermined constants disappear. The heat current density $\mathbf{J}_Q$ is thus *invariant to a change of the energy zero points* (cf. §1–7).

The quantity $\mathbf{J}_Q$ originally denotes the heat flux density in the barycentric system. From Eq. (1) we find that, in the most general case, transferring from the center of gravity velocity $\mathbf{v}$ to another reference velocity, the right-hand side of Eq. (1) will change. It can be shown, however, that for one of the most important special cases in practice $\mathbf{J}_Q$ becomes independent of the reference velocity. This special case refers to the "creeping motion" mentioned in §4–5 and to isotropic crystals "in simple cases." For these, due to Eq. (4–5.18), we have

$$P_{ij} = 0 \quad (i, j = 1, 2, 3), \qquad \sum_k c_k \mathbf{K}_k = \operatorname{grad} P. \quad (4\text{–}7.2)$$

If we take account of Eq. (2) then there results from Eq. (1):

$$-\operatorname{div} \mathbf{J}_Q = \sum_k c_k \frac{\partial H_k}{\partial t} + \sum_k c_k \mathbf{v}_k \operatorname{grad} H_k$$

$$+ \sum_k \sum_r \nu_{kr} H_k b_r - \sum_k \mathbf{K}_k c_k \mathbf{v}_k - \frac{\partial P}{\partial t}. \quad (4\text{–}7.3)$$

This expression no longer contains the barycentric velocity $\mathbf{v}$. The heat current density $\mathbf{J}_Q$ is, accordingly, under conditions (2), *invariant to a change of the reference velocity*.

We will see later that our heat current density $\mathbf{J}_Q$ has still other simple properties which distinguish it from the quantities otherwise designated in the literature as "heat flux" or "energy flux" (cf. §1–7). These quantities are all of the general form

$$\mathbf{J}'_Q \equiv \mathbf{J}_Q + \sum_k \gamma_k c_k (\mathbf{v}_k - \mathbf{v}), \qquad (4\text{–}7.4)$$

where the γ_k represent some scalar factor. Usually, the vector

$$\mathbf{J}^*_Q \equiv \mathbf{J}_Q + \sum_k H_k c_k (\mathbf{v}_k - \mathbf{v}) = \mathbf{J}_Q + \sum_k H_k {}_v \mathbf{J}_k \qquad (4\text{–}7.5)$$

is regarded as the "heat current density" or "energy flux density" because we can abbreviate, with its help, the second term of the right-hand side of Eq. (4–6.9) and thus write the energy balance in the normal form valid for one-component systems. Also, heat conduction in reacting media (§4–28) can be formulated simply by introducing the vector $\mathbf{J}^*_Q$. As far as $\mathbf{J}^*_Q$ appears as the "heat current" in the literature, $\mathbf{J}_Q$ is called the "reduced heat current." From Eq. (5) it is immediately evident that $\mathbf{J}^*_Q$ does not have the invariance properties proved above for $\mathbf{J}_Q$. Consequently, we will consistently use the quantity $\mathbf{J}_Q$ as the heat current density.

4–8 ENTROPY BALANCE

The Gibbs equation (§1–9) holds for each volume element of an isotropic system without electrification and magnetization for sufficiently slow irreversible processes (cf. §1–23 and §4–1). We choose the form (1–9.12) for this equation which we have already used for the derivation of the equilibrium conditions (§1–21). If we introduce the partial derivatives with respect to time at fixed position in place of the differentials in (1–9.12), we obtain for the local rate of increase of the entropy density in the volume element:

$$\frac{\partial S_V}{\partial t} = \frac{1}{T}\frac{\partial U_V}{\partial t} - \frac{1}{T}\sum_k \mu_k \frac{\partial c_k}{\partial t}. \qquad (4\text{–}8.1)$$

Here T denotes the absolute temperature, U_V the density of the internal energy, and μ_k and c_k the chemical potential and the molarity of species k respectively. Equation (1) is our starting point for the derivation of the entropy balance.†

---

† The Gibbs equation is used in another form by most authors: as the derivatives of the specific quantities with respect to the time (cf. Eq. 1–9.11) for an observer moving with the center of gravity velocity $\mathbf{v}$ (operator d/dt). Moreover, there is often the comment that it is a fundamental assumption of the thermodynamics of irreversible processes that the Gibbs equation is valid in the *barycentric system*. This limitation is not necessary: the Gibbs equation can be written in the form (1) and also for an observer moving with any arbitrary reference velocity.

We determine the derivatives on the right-hand side of Eq. (1) from the mass balance (4–4.1) and the energy balance (4–6.14)—in which the momentum balance is already contained—with Eq. (4–6.12), thus:

$$\frac{\partial c_k}{\partial t} = -\operatorname{div}(c_k \mathbf{v}_k) + \sum_r \nu_{kr} b_r, \tag{4-8.2}$$

$$\frac{\partial U_v}{\partial t} = -\operatorname{div}\left(\mathbf{J}_Q + \sum_k H_k c_k \mathbf{v}_k\right) + \mathbf{v}\operatorname{grad} P$$

$$- \sum_{i=1}^{3}\sum_{j=1}^{3} P_{ij}\frac{\partial v_i}{\partial z_j} + \sum_k \mathbf{K}_k \,_v\mathbf{J}_k. \tag{4-8.3}$$

Here $\mathbf{v}_k$ is the average velocity of particles of type k, ν_{kr} the stoichiometric number of the species k in chemical reaction r, b_r the reaction rate of the reaction r, $\mathbf{J}_Q$ the heat current density, H_k the partial molar enthalpy of species k, P the static pressure, P_{ij} the viscous pressure, $\mathbf{v}$ the barycentric velocity, v_i the component of $\mathbf{v}$ in a rectangular coordinate system, z_j the cartesian space-coordinate, $\mathbf{K}_k$ the molar external force acting on species k, and $_v\mathbf{J}_k$ the diffusion current density of species k in the barycentric system. Equation (3) holds for isotropic systems without electrification and magnetization and for time-invariant external force fields.

We now take account of Eq. (1–12.6):

$$\mu_k = H_k - TS_k, \tag{4-8.4}$$

where S_k is the partial molar entropy of species k, the definition (1–14.2) of the affinity A_r of the reaction r:

$$A_r = -\sum_k \nu_{kr}\mu_k, \tag{4-8.5}$$

and the vector identities:

$$\frac{a}{T}\operatorname{div}\mathbf{J} = \operatorname{div}\left(\frac{a}{T}\mathbf{J}\right) - \mathbf{J}\operatorname{grad}\left(\frac{a}{T}\right), \tag{4-8.6}$$

$$\frac{1}{T}\operatorname{div}\mathbf{J} = \operatorname{div}\left(\frac{\mathbf{J}}{T}\right) + \frac{\mathbf{J}}{T^2}\operatorname{grad} T, \tag{4-8.7}$$

where a denotes any scalar quantity and $\mathbf{J}$ any vector. With this it follows from Eqs. (1), (2), and (3) that

$$\frac{\partial S_v}{\partial t} = -\operatorname{div}\left(\frac{\mathbf{J}_Q}{T}\right) - \frac{\mathbf{J}_Q}{T^2}\operatorname{grad} T - \operatorname{div}\left(\sum_k S_k c_k \mathbf{v}_k\right)$$

$$- \frac{1}{T^2}\sum_k H_k c_k \mathbf{v}_k \operatorname{grad} T + \frac{\mathbf{v}}{T}\operatorname{grad} P - \frac{1}{T}\sum_{i=1}^{3}\sum_{j=1}^{3} P_{ij}\frac{\partial v_i}{\partial z_j}$$

$$+ \frac{1}{T}\sum_k \mathbf{K}_k \,_v\mathbf{J}_k - \sum_k c_k \mathbf{v}_k \operatorname{grad}\left(\frac{\mu_k}{T}\right) + \frac{1}{T}\sum_r b_r A_r. \tag{4-8.8}$$

Furthermore, due to Eq. (1–12.8) we find

$$\text{grad}\left(\frac{\mu_k}{T}\right) = -\frac{H_k}{T^2}\text{grad } T + \frac{1}{T}(\text{grad } \mu_k)_T, \qquad (4\text{–}8.9)$$

where the inferior T $(_T)$ at the last gradient should show that the term containing the grad T is already subtracted and that it now deals only with the gradients of the chemical potentials which depend on pressure and composition gradients. Furthermore, we have the Gibbs–Duhem relation (1–13.6)

$$\sum_k c_k(\text{grad } \mu_k)_T = \text{grad } P. \qquad (4\text{–}8.10)$$

Finally, we note Eq. (4–3.11):

$$_v\mathbf{J}_k = c_k(\mathbf{v}_k - \mathbf{v}). \qquad (4\text{–}8.11)$$

With the help of Eqs. (9), (10), and (11) we can combine the fourth, fifth, seventh, and eighth terms of the right-hand side of Eq. (8) as follows:

$$-\frac{1}{T^2}\sum_k H_k c_k \mathbf{v}_k \text{ grad } T + \frac{\mathbf{v}}{T}\text{grad } P$$

$$+ \frac{1}{T}\sum_k \mathbf{K}_k\,{}_v\mathbf{J}_k - \sum_k c_k \mathbf{v}_k \text{ grad }\left(\frac{\mu_k}{T}\right)$$

$$= \frac{1}{T}\sum_k {}_v\mathbf{J}_k[\mathbf{K}_k - (\text{grad } \mu_k)_T]. \qquad (4\text{–}8.12)$$

From Eqs. (8) and (12) the final formula for the local rate of increase of the entropy density in a volume element of a continuous system is found to be

$$\frac{\partial S_V}{\partial t} = -\text{div } \mathbf{J}_S + \vartheta \qquad (4\text{–}8.13)$$

with the abbreviations

$$\mathbf{J}_S \equiv \frac{\mathbf{J}_Q}{T} + \sum_k c_k S_k \mathbf{v}_k \qquad (4\text{–}8.14)$$

and

$$\vartheta \equiv -\frac{1}{T^2}\mathbf{J}_Q \text{ grad } T + \frac{1}{T}\sum_k {}_v\mathbf{J}_k[\mathbf{K}_k - (\text{grad } \mu_k)_T]$$

$$+ \frac{1}{T}\sum_r b_r A_r - \frac{1}{T}\sum_{i=1}^{3}\sum_{j=1}^{3} P_{ij}\frac{\partial v_i}{\partial z_j}. \qquad (4\text{–}8.15)$$

Equation (13) takes the form of Eq. (4–2.4), and thus represents an expression for the "local balance" of the entropy or a "continuity equation" of the entropy.   Accordingly, $\mathbf{J}_S$ is the *entropy current density* and $\vartheta = q(S)$ the *local entropy production*.

The negative divergence of the entropy flux density $\mathbf{J}_S$ is that part of the local rate of increase in the entropy density which relates to the heat and mass exchange of the volume element with the surroundings.   It thus corresponds to the "entropy flux" in Eqs. (1–24.3) and (1–24.5).   The plausible form of Eq. (14) is a result of our definition of heat (§4–6) and has its analog in Eq. (1–24.13) for the entropy flux in homogeneous systems.

The local entropy production denotes that part of the local rate of increase of the entropy density which relates to processes inside the volume element and thus corresponds to the "entropy production" in Eqs. (1–24.3) and (1–24.4).   We thus have also the inequality analogous to (1–24.4):

$$\vartheta \geqslant 0, \qquad\qquad (4\text{–}8.16)$$

where the inequality sign is valid for the actual (irreversible) course of the processes inside the volume element and the equality sign holds for the reversible or limiting case.

In the inequality (16), we are dealing with the "local" formulation of the second law of thermodynamics already described in §1–8.   This is necessary for continuous systems.   While classical thermodynamics is satisfied with the statement that the total entropy of a thermally insulated system cannot decrease—such a statement is sufficient for the derivation of the equilibrium conditions in continuous systems (cf. §1–21)—the thermodynamics of irreversible processes brings about a more precise insight into the details of the procedure: it not only requires that the local entropy production ϑ be positive in each volume element or must vanish, but also produces the explicit expression (15) for ϑ which contains quantities such as the heat current density, the diffusion current density, the reaction rates, etc., and the temperature gradients, the gradients of the chemical potentials, the affinities, etc.

Equations (13) to (15) are valid, in accordance with their derivations, for isotropic systems without electric and magnetic polarization which are in stationary external fields (in the gravitational field of the earth, in a centrifugal field at constant angular velocity, or in a stationary electromagnetic field).

The most general case refers to irreversible processes in polarized anisotropic systems in both gravitational or centrifugal and electromagnetic fields of arbitrary time dependence.   As can be shown by combining the reasonings of §4–33, 4–34, and 4–37, Eq. (15) continues to hold for the local entropy production even in this complicated case provided that:

1. The molar external force $\mathbf{K}_k$ acting on species k is taken to be

$$\begin{aligned}
\mathbf{K}_k = M_k(\mathbf{g} + \Omega^2\mathbf{r} + 2\mathbf{v}_k \times \mathbf{\Omega}) \\
+ z_k F(\mathfrak{E} + \mathbf{v}_k \times \mathfrak{B}), \qquad\qquad (4\text{–}8.17)
\end{aligned}$$

where M_k denotes the molar mass of species k, $\mathbf{g}$ the gravitational acceleration, $\boldsymbol{\Omega}$ the angular velocity in a rotating system, $\mathbf{r}$ the distance from the
center of rotation, z_k the charge number of species k, F the Faraday constant,
$\mathfrak{E}$ the electric field strength, and $\mathfrak{B}$ the magnetic induction.

2. The gradient of the chemical potential of species k, taken at constant
temperature $[(\mathrm{grad}\ \mu_k)_T]$, is due to gradients of stresses, of electric or magnetic
field strength, and of concentrations.

3. Electric and magnetic relaxation effects are included in the terms
relating to chemical reactions.

4. The double sum relating to viscous flow is written in the more general
form

$$\Pi_{\mathrm{visc}} \colon \mathrm{Grad}\ \mathbf{v},$$

where Π_{visc} is the viscous pressure tensor.

4-9 ENTROPY CURRENT

If we introduce the diffusion current density valid for any reference velocity
$\boldsymbol{\omega}$ into Eq. (4-8.14) then, according to Eq. (4-3.1),

$$\mathbf{J}_k = c_k(\mathbf{v}_k - \boldsymbol{\omega}) \tag{4-9.1}$$

and for the entropy current density we obtain:

$$\mathbf{J}_S = \frac{\mathbf{J}_Q}{T} + \sum_k c_k S_k \boldsymbol{\omega} + \sum_k S_k \mathbf{J}_k. \tag{4-9.2}$$

With the help of Eq. (1-6.10)

$$S_V = \sum_k c_k S_k \tag{4-9.3}$$

we derive from this:

$$\mathbf{J}_S = S_V \boldsymbol{\omega} + \mathbf{J}_S' \tag{4-9.4}$$

with

$$\mathbf{J}_S' \equiv \frac{\mathbf{J}_Q}{T} + \sum_k S_k \mathbf{J}_k. \tag{4-9.5}$$

As comparison with Eq. (4-4.10) shows, $S_V \boldsymbol{\omega}$ is the "convective" and $\mathbf{J}_S'$ the
"nonconvective" entropy current density. Often $\mathbf{J}_S'$ is simply denoted as the
"entropy flux."

If the barycentric velocity $\mathbf{v}$ is introduced as the reference velocity into
Eqs. (2), (4), and (5) then, noting Eqs. (4-8.11) and (4-8.13), we find:

$$\frac{\partial S_V}{\partial t} + \mathrm{div}\ (S_V \mathbf{v}) = -\mathrm{div}\ {}_v\mathbf{J}_S' + \vartheta \tag{4-9.6}$$

with

$$_v\mathbf{J}'_S \equiv \frac{\mathbf{J}_Q}{T} + \sum_k S_k \,_v\mathbf{J}_k. \tag{4–9.7}$$

Here $_v\mathbf{J}'_S$ denotes the nonconvective entropy current density in the barycentric system.

With the help of Eq. (4–4.19) we can also write Eq. (6) as

$$\rho \frac{d\tilde{S}}{dt} = -\operatorname{div} \,_v\mathbf{J}'_S + \vartheta. \tag{4–9.8}$$

Here $\tilde{S}$ is the specific entropy and d/dt the operator of the total or substantial differentiation. In many presentations, the entropy balance is given in the special form (8).

For fluid media in "creeping motion" and for isotropic crystals in "simple cases", the heat flux density $\mathbf{J}_Q$, according to §4–7, is invariant to a change of the reference velocity. Accordingly, this is also valid for the total entropy current density $\mathbf{J}_S$, but not for the individual parts $S_V\boldsymbol{\omega}$ and $\mathbf{J}'_S$, as indicated by Eq. (4–8.14) as well as by Eqs. (4) and (5).

4–10 LOCAL ENTROPY PRODUCTION

According to Eq. (4–8.15), the local entropy production ϑ contains the barycentric velocity $\mathbf{v}$. Thus transferring $\mathbf{v}$ to some other reference velocity will, in general, change ϑ. But under the special conditions of (4–7.2), which are valid for fluid media in "creeping motion" and for isotropic crystals in "simple cases," ϑ is invariant to a change of the reference velocity. This follows from Eq. (4–8.13) and from the invariance of $\mathbf{J}_S$ (§4–9).

However, we can say still more. From Eq. (4–7.2) with the help of Eq. (4–8.10) we derive:

$$P_{ij} = 0 \quad (i,j = 1, 2, 3), \tag{4–10.1}$$

$$\sum_k c_k(\operatorname{grad} \mu_k)_T = \sum_k c_k\mathbf{K}_k. \tag{4–10.2}$$

On consideration of the defining equations (4–8.11) and (4–9.1) there results from this:[†]

$$\sum_k \,_v\mathbf{J}_k[\mathbf{K}_k - (\operatorname{grad} \mu)_T] = \sum_k c_k\mathbf{v}_k[\mathbf{K}_k - (\operatorname{grad} \mu_k)_T]$$

$$= \sum_k \mathbf{J}_k[\mathbf{K}_k - (\operatorname{grad} \mu_k)_T]. \tag{4–10.3}$$

---

† Equation (2) and Eq. (3), which are due to Prigogine[2], are often known as the "theorem of Prigogine."

From Eq. (4–8.15) and taking account of Eqs. (1) and (3) for the local entropy production we find

$$\vartheta = -\frac{1}{T^2} \mathbf{J}_Q \, \mathrm{grad} \, T + \frac{1}{T} \sum_k \mathbf{J}_k [\mathbf{K}_k - (\mathrm{grad} \, \mu_k)_T] + \frac{1}{T} \sum_r b_r A_r. \quad (4\text{--}10.4)$$

Under conditions (1) and (2) the heat flux $\mathbf{J}_Q$ is independent of the choice of reference velocity (§4–7). According to Eq. (3), the second term of the right-hand side of Eq. (4) also has this invariance property. Finally, the third term on the right-hand side of Eq. (4), which relates to chemical reactions inside the volume element, must also be independent of the reference system. Thus the three parts of the local entropy production in (4) do not depend on the reference velocity provided we are dealing with fluid media in "creeping motion" or with isotropic crystals in "simple cases."

Another way of writing Eq. (4) which is sometimes used results from Eq. (4–9.5). It follows from Eq. (1–27.7) that

$$\mathrm{grad} \, \mu_k = -S_k \, \mathrm{grad} \, T + (\mathrm{grad} \, \mu_k)_T. \quad (4\text{--}10.5)$$

By combination of Eq. (4–9.5) with Eqs. (4) and (5) we obtain

$$\vartheta = -\frac{1}{T} \mathbf{J}_S' \, \mathrm{grad} \, T + \frac{1}{T} \sum_k \mathbf{J}_k (\mathbf{K}_k - \mathrm{grad} \, \mu_k) + \frac{1}{T} \sum_r b_r A_r, \quad (4\text{--}10.6)$$

where $\mathbf{J}_S'$ denotes the nonconvective entropy current density.

The invariance properties of the local entropy production and of their individual constituents arise from those simplifications of the theory which result only from our choice of the heat current (cf. §4–7).

4–11 ENTROPY OF THE TOTAL SYSTEM

We arrive at the entropy change ΔS of the total system during a finite time interval by integration of the entropy balance equation over the total volume V of the continuous system and over time t. If we take account of Gauss's law and Eq. (4–8.14), we find from Eq. (4–8.13):

$$\Delta S = \Delta_a S + \Delta_i S \quad (4\text{--}11.1)$$

with

$$\Delta_a S \equiv -\int_t \int_\Omega \left(\frac{J_{Qn}}{T} + \sum_k c_k S_k v_{kn} \right) d\Omega \, dt \quad (4\text{--}11.2)$$

and

$$\Delta_i S \equiv \int_t \int_V \vartheta \, dV \, dt. \quad (4\text{--}11.3)$$

Here J_{Qn} and v_{kn} are the components of $\mathbf{J}_Q$ and $\mathbf{v}_k$ respectively normal to the surface area of the system, $d\Omega$ is a surface area element, dV a volume element, and dt a time element.

From (4–8.16) and Eq. (3) there results

$$\Delta_i S \geqslant 0, \tag{4–11.4}$$

where the inequality sign holds for the actual (irreversible) path of the processes within the system and the equality sign for the reversible limiting case.

Equations (1) to (4) correspond to Eqs. (1–10.5) to (1–10.8) derived in §1–10 for discontinuous systems. By combining (1), (2), and (4) we obtain the analog to the inequality (1–10.9):

$$\Delta S \geqslant - \int_t \int_\Omega \left(\frac{J_{Qn}}{T} + \sum_k S_k c_k v_{kn} \right) d\Omega \, dt. \tag{4–11.5}$$

The right-hand side is the entropy flowing into the total continuous system from the surroundings by heat and mass exchange during a finite state change. Thus all conclusions obtained from (1–10.9) (§1–10 and §1–11) continue to hold for continuous systems.

In particular, we derive from (5), for an *isothermal* state change (i.e. for temperature T constant in time and position) in a *closed* system (i.e. with no mass flow through the boundary surfaces of the system, $v_{kn} = 0$),

$$\Delta S \geqslant \frac{Q}{T} \tag{4–11.6}$$

with

$$Q \equiv - \int_t \int_\Omega J_{Qn} \, d\Omega \, dt. \tag{4–11.7}$$

Since the right-hand side in (7) represents the total heat absorbed by the system from the surroundings during the state change considered, the symbol Q has the same meaning as in (1–10.12) and (1–11.8).

4–12 DISSIPATION FUNCTION, GENERALIZED FLUXES, AND GENERALIZED FORCES

The parameter characterizing the course of irreversible processes in continuous systems is the local entropy production or the product

$$T\vartheta \equiv \Psi, \tag{4–12.1}$$

which we designate as the *dissipation function* (cf. §1–24). According to Eqs. (4–8.15) and (4–8.16), for all processes in isotropic media without electrification or magnetization, and excluding time-variable external force fields (cf. Eq. 1–24.22), we may write:

$$\Psi = \mathbf{J}_Q\mathbf{X}_Q + \sum_k {}_v\mathbf{J}_k\mathbf{X}_k + \sum_r b_rA_r + \sum_{i=1}^{3} \sum_{j=1}^{3} P_{ij}X_{ij} \geqslant 0 \qquad (4\text{–}12.2)$$

with

$$\mathbf{X}_Q \equiv -\frac{1}{T}\,\mathrm{grad}\,T, \qquad (4\text{–}12.3)$$

$$\mathbf{X}_k \equiv \mathbf{K}_k - (\mathrm{grad}\,\mu_k)_T, \qquad (4\text{–}12.4)$$

$$X_{ij} \equiv -\frac{1}{2}\left(\frac{\partial v_i}{\partial z_j} + \frac{\partial v_j}{\partial z_i}\right) \quad (i, j = 1, 2, 3). \qquad (4\text{–}12.5)$$

We call the quantities $\mathbf{J}_Q$ (heat current density), ${}_v\mathbf{J}_k$ (diffusion current densities), b_r (reaction rates), and $P_{ij} = P_{ji}$ (viscous pressures or "momentum flux densities") *generalized fluxes*. The expressions $\mathbf{X}_Q$ (containing the temperature gradient), $\mathbf{X}_k$ (difference between the molar external force and the gradient of the chemical potential at constant temperature), A_r (affinities), and $X_{ij} = X_{ji}$ (containing the velocity gradients) are designated the *generalized forces*. The dissipation function Ψ is thus a sum of the products of generalized fluxes and the conjugate forces (cf. §1–24).

In the case of complete equilibrium in a system at rest, the generalized fluxes vanish by definition:

$$\mathbf{J}_Q = 0, \quad {}_v\mathbf{J}_k = 0, \quad b_r = 0, \quad P_{ij} = 0 \quad \text{(equilibrium).} \qquad (4\text{–}12.6)$$

According to the equilibrium conditions (1–21.19), (1–21.25), and (1–21.26), and to Eqs. (3) and (4), we find that

$$\mathbf{X}_Q = 0, \quad \mathbf{X}_k = 0, \quad A_r = 0 \quad \text{(equilibrium).} \qquad (4\text{–}12.7)$$

For a system at rest, the condition

$$X_{ij} = 0 \quad \text{(equilibrium)} \qquad (4\text{–}12.8)$$

required by Eq. (5) is satisfied. Accordingly, for a system at equilibrium and at rest, all generalized fluxes and forces vanish. This is certainly true for the reaction rates b_r and affinities A_r for linearly independent reactions (cf. §1–19, §1–21, and §2–3) but can be proved to hold (cf. §2–7 and §2–8) for any reaction.

Consider for the sake of simplicity the chemical reactions expressed in the form of linearly independent reaction equations. Then the b_r and A_r in Eq. (2) are independent of each other. If we exclude the last term in Eq. (2), which relates to viscous flow and which will be discussed later, then in the most general case all the other generalized fluxes and forces are independent of each other, except for the diffusion flux densities ${}_v\mathbf{J}_k$ for which the identity (4–3.13) is valid.

It is appropriate for the further development of our theory to eliminate the dependent diffusion fluxes and simultaneously to introduce the quantities (4–3.14)

$$\mathbf{J}_k^* = M_k {}_v\mathbf{J}_k \qquad (4\text{–}12.9)$$

as "diffusion current densities." For these, the identity (4–3.5) is valid

$$\sum_k \mathbf{J}_k^* = 0. \qquad (4\text{–}12.10)$$

If we have N species and wish to eliminate the diffusion current density $\mathbf{J}_1^*$ then we use the relation following from (10)

$$\mathbf{J}_1^* = -\sum_{k=2}^{N} \mathbf{J}_k^*. \qquad (4\text{–}12.11)$$

By combination of Eqs. (2), (4), (9), and (11) we obtain

$$\Psi = \mathbf{J}_Q \mathbf{X}_Q + \sum_{k=2}^{N} \mathbf{J}_k^*(\mathbf{X}_k^* - \mathbf{X}_1^*) + \sum_{r=1}^{R} b_r A_r + \sum_{i=1}^{3} \sum_{j=1}^{3} P_{ij} X_{ij} \geq 0 \qquad (4\text{–}12.12)$$

with

$$\mathbf{X}_k^* \equiv \frac{1}{M_k} \mathbf{X}_k = \frac{1}{M_k} [\mathbf{K}_k - (\text{grad } \mu_k)_T] \quad (k = 1, 2, 3, \ldots, N). \qquad (4\text{–}12.13)$$

Here R is the number of independent chemical reactions. The $N - 1$ independent generalized forces $\mathbf{X}_2^* - \mathbf{X}_1^*, \ldots, \mathbf{X}_N^* - \mathbf{X}_1^*$ belong to the $N - 1$ independent diffusion flux densities $\mathbf{J}_2^*, \ldots, \mathbf{J}_N^*$.

For fluid media in "creeping motion" and for isotropic crystals in "simple cases," following Eq. (4–10.4) and Eqs. (1), (3), and (4), the dissipation function becomes

$$\Psi = \mathbf{J}_Q \mathbf{X}_Q + \sum_{k=1}^{N} \mathbf{J}_k \mathbf{X}_k + \sum_{r=1}^{R} b_r A_r \geq 0. \qquad (4\text{–}12.14)$$

This expression is obviously simplified as compared to (12). Here, according to Eq. (4–9.1), the $\mathbf{J}_k$ are diffusion current densities referred to any reference velocity

$$\mathbf{J}_k = c_k(\mathbf{v}_k - \boldsymbol{\omega}) \quad (k = 1, 2, \ldots, N). \qquad (4\text{–}12.15)$$

Since these must also be zero at equilibrium, all generalized fluxes and forces disappear from (14) at equilibrium.

In the second term on the right-hand side of Eq. (14), the $\mathbf{X}_k$ are linearly dependent; thus, according to Eq. (4–10.2) and Eq. (4), there results:

$$\sum_{k=1}^{N} c_k \mathbf{X}_k = 0. \qquad (4\text{–}12.16)$$

The diffusion current densities $\mathbf{J}_k$ are seen to be dependent for fluid media (following from 4–3.4), but in general are independent of each other for crystalline media (p. 222).

In each case, elimination of the dependent quantities is possible in Eq. (14). If, for example, using Eq. (16) we write

$$\mathbf{X}_1 = -\sum_{k=2}^{N} \frac{c_k}{c_1} \mathbf{X}_k$$

then there follows from Eq. (15) and Eq. (4–3.30) that

$$\sum_{k=1}^{N} \mathbf{J}_k \mathbf{X}_k = \sum_{k=2}^{N} \left(\mathbf{J}_k - \frac{c_k}{c_1} \mathbf{J}_1 \right) \mathbf{X}_k = \sum_{k=2}^{N} c_k (\mathbf{v}_k - \mathbf{v}_1) \mathbf{X}_k = \sum_{k=2}^{N} {}_1\mathbf{J}_k \mathbf{X}_k.$$

Here ${}_1\mathbf{J}_k$ is the diffusion current density of species k in the Hittorf reference system (reference velocity $\mathbf{v}_1$). With this we obtain the expression for the dissipation function from Eq. (14):

$$\Psi = \mathbf{J}_Q \mathbf{X}_Q + \sum_{k=2}^{N} {}_1\mathbf{J}_k \mathbf{X}_k + \sum_{r=1}^{R} b_r A_r \geqslant 0, \qquad (4\text{–}12.17)$$

in which only independent "fluxes" and "forces" are contained. Equation (17) is less general than Eq. (12) but offers, if applicable, the advantage of greater simplicity.

According to Eqs. (6) to (8), all "fluxes" and "forces" disappear at equilibrium in a system at rest. With this and Eqs. (1) and (2) or (12) the local entropy production ϑ or the dissipation function Ψ is seen to become zero. Conversely, Eqs. (6) to (8), i.e. the equilibrium conditions for a static system, do not necessarily follow from $\Psi = 0$. This can be seen from Eqs. (2) or (12) in connection with the phenomenological equations (§4–13). Actually, the condition $\Psi = 0$ still allows the possibility of "reversible changes"[3, 4a]. This is to be expected. The equality sign in (4–8.16) again denotes a "reversible limiting case," following the general formulation of the second law (§1–8), and includes, accordingly, the occurrence of "purely mechanical" processes.

In the following, we will be interested only in irreversible processes and will therefore retain only the inequality sign in (12) or (17).

4–13 PHENOMENOLOGICAL EQUATIONS

The generalized fluxes (heat current density $\mathbf{J}_Q$, diffusion current densities $\mathbf{J}_k$ or $\mathbf{J}_k^*$, reaction rates b_r, viscous pressures P_{ij}) and the generalized forces ($\mathbf{X}_Q$, $\mathbf{X}_k$ or $\mathbf{X}_k^*$, A_r, X_{ij}), which appear in the dissipation function, all individually vanish at equilibrium according to §4–12. This suggests that deviations from equilibrium, i.e. irreversible processes which occur in our system, should be described to a first approximation by homogeneous linear relations between the "fluxes" and the "forces" following the model of Eq. (1–25.1). These relations, which are valid for *states near equilibrium*, are called *phenomenological equations*. For the determination of the "fluxes" and "forces", we can, in principle, proceed from any one of the expressions for the dissipation function derived in §4–12. In order to avoid ambiguities or dependent coefficients of the phenomenological equations (cf. §1–27), we consider only those formulas for the dissipation function in which all "fluxes" and "forces" of interest are independent.

First, we deal with the most general case. Accordingly, we choose Eq. (4–12.12) as the point of departure. In this relation, all "fluxes" and "forces"—apart from those which relate to viscous flow and which are discussed later (§4–32)—are independent (cf. §4–12). Using this we obtain the following phenomenological equations (cf. Eq. 2–4.1 and 3–6.1):

$$\mathbf{J}_i^* = \sum_{k=2}^{N} \alpha_{ik}(\mathbf{X}_k^* - \mathbf{X}_1^*) + \alpha_{iQ}\mathbf{X}_Q \quad (i = 2, 3, \ldots, N), \qquad (4\text{–}13.1)$$

$$\mathbf{J}_Q = \sum_{i=2}^{N} \alpha_{Qi}(\mathbf{X}_i^* - \mathbf{X}_1^*) + \alpha_{QQ}\mathbf{X}_Q, \qquad (4\text{–}13.2)$$

$$b_r = \sum_{s=1}^{R} a_{rs}A_s \quad (r = 1, 2, \ldots, R). \qquad (4\text{–}13.3)$$

Here N denotes the number of species and R the number of independent chemical reactions. The quantities α_{ik}, α_{iQ}, α_{Qi}, α_{QQ}, and a_{rs} are the *phenomenological coefficients*. By assumption they do not depend on the "fluxes" and "forces" but can be arbitrary functions of the local state variables (temperature, pressure, and concentrations).

In the linear equations (1) to (3), we do not allow the diffusion current densities $\mathbf{J}_i^*$ and the heat current density $\mathbf{J}_Q$, which are vectors, to be dependent on the scalar affinities A_s, or the scalar reaction rates b_r to depend on the vectorial generalized forces $\mathbf{X}_k^*$ and $\mathbf{X}_Q$. Such a dependence is generally forbidden by the "symmetry principle of Curie." This theorem states that quantities of different tensorial character cannot be coupled.†

The corresponding linear equation for viscous flow will be discussed later (§4–32). It adds nothing to the previous equations apart from a relation between b_r and div **v**.

---

† More precisely, the principle of P. Curie states the following: for isotropic media, only quantities with the same transformation properties with respect to rotations in space can be linearly related to one another[5]. It will be shown in §4–32 how this statement affects the equations for viscous flow.

The vectorial "forces" in Eqs. (1) and (2) are given by Eqs. (4-12.3) and (4-12.13):

$$\mathbf{X}_k^* = \frac{1}{M_k} \mathbf{X}_k = \frac{1}{M_k} [\mathbf{K}_k - (\text{grad } \mu_k)_T] \quad (k = 1, 2, \ldots, N), \qquad (4\text{-}13.4)$$

$$\mathbf{X}_Q = -\frac{1}{T} \text{grad } T. \qquad (4\text{-}13.5)$$

Here T is the absolute temperature, M_k and μ_k the molar mass and the chemical potential respectively of species k, and $\mathbf{K}_k$ the molar external force acting on species k.

From Eqs. (4) and (5) it is obvious which phenomena are described by Eqs. (1) and (2). We are dealing with processes such as electric conduction, diffusion, sedimentation, heat conduction, thermal diffusion, etc., which will be more precisely examined later.†

Equations (3) describe chemical reactions near equilibrium (cf. Chapter 2).

For fluid media in "creeping motion" and for isotropic crystals in "simple cases", it is expedient to proceed from Eq. (4-12.17) rather than from Eq. (4-12.2). Equations (3) for chemical reactions are not changed, but the equations for transport processes will be different.

If Eq. (4-12.17) is taken as a basis then, in place of Eqs. (1) and (2), we find

$$_1\mathbf{J}_i = \sum_{k=2}^{N} a_{ik}\mathbf{X}_k + a_{iQ}\mathbf{X}_Q \quad (i = 2, 3, \ldots, N), \qquad (4\text{-}13.6)$$

$$\mathbf{J}_Q = \sum_{i=2}^{N} a_{Qi}\mathbf{X}_i + a_{QQ}\mathbf{X}_Q, \qquad (4\text{-}13.7)$$

where the quantities a_{ik}, a_{iQ}, a_{Qi}, and a_{QQ} are again phenomenological coefficients.

The connection between the phenomenological coefficients in Eqs. (1) and (2) and those in Eqs. (6) and (7) results from Eq. (4-3.40) and Eq. (4). Such a conversion need not be carried out explicitly here, but it represents an example of the transformations mentioned in §1-27.

4-14 ONSAGER'S RECIPROCITY LAW

According to §1-26, the simple equation (1-26.2) follows from the "principle of microscopic reversibility" in the absence of Coriolis and Lorentz forces for the coefficients in the phenomenological equations (§4-13). For macroscopic rotational motion and external magnetic fields, this expression must be

---

† General equations of the type (1) and (2) were first set up and discussed by Eckart and Meixner[6a, 7].

modified (cf. §1–26). However, as long as we are not interested in transverse effects in rotating systems and in transport effects in strong external magnetic fields, we can allow rotation of the system and electromagnetic fields without giving up Eqs. (1–26.2). We therefore defer discussion of the influence of rotational motion and magnetic fields to §4–35 and consider here only Eqs. (1–26.2), that is, the simplest form of the Onsager–Casimir reciprocity law.

In the case of the phenomenological equations (4–13.1), (4–13.2), and (4–13.3), we have only "Onsager coefficients." Therefore, the Onsager reciprocity relations (1–26.1) are valid:

$$\alpha_{ik} = \alpha_{ki} \quad (i, k = 2, 3, \ldots, N), \tag{4–14.1}$$

$$\alpha_{iQ} = \alpha_{Qi} \quad (i = 2, 3, \ldots, N), \tag{4–14.2}$$

$$a_{rs} = a_{sr} \quad (r, s = 1, 2, \ldots, R). \tag{4–14.3}$$

In the case of Eqs. (4–13.6) and (4–13.7), Onsager's reciprocity law states:

$$a_{ik} = a_{ki} \quad (i, k = 2, 3, \ldots, N), \tag{4–14.4}$$

$$a_{iQ} = a_{Qi} \quad (i = 2, 3, \ldots, N). \tag{4–14.5}$$

If we introduce Eqs. (4–13.1) to (4–13.3) or (4–13.6) to (4–13.7) into Eq. (4–12.12) or (4–12.27), and consider the reciprocity relations (1) to (3) or (4) and (5) then we see that the dissipation function represents a positive quadratic form for irreversible processes[†] near equilibrium. Thus, the determinant of the phenomenological coefficients with all principal cofactors is positive. There result, therefore, inequality statements of the form

$$\alpha_{ii} > 0, \quad \alpha_{ii}\alpha_{jj} - \alpha_{ij}^2 > 0 \quad (i, j = 2, 3, \ldots, N, Q; i \neq j), \tag{4–14.6}$$

$$\cdot \quad \cdot \quad \cdot \quad \cdot \quad \cdot \quad \cdot \quad \cdot \quad \cdot \quad \cdot \quad \cdot \quad \cdot$$

$$|\alpha_{ij}| = \det \alpha_{ij} > 0, \tag{4–14.7}$$

$$a_{rr} > 0, \quad a_{rr}a_{ss} - a_{rs}^2 > 0 \quad (r, s = 1, 2, \ldots, R; r \neq s), \tag{4–14.8}$$

$$\cdot \quad \cdot \quad \cdot \quad \cdot \quad \cdot \quad \cdot \quad \cdot \quad \cdot \quad \cdot \quad \cdot \quad \cdot$$

$$|a_{rs}| = \det a_{rs} > 0, \tag{4–14.9}$$

$$a_{ii} > 0, \quad a_{ii}a_{jj} - a_{ij}^2 > 0 \quad (i, j = 2, 3, \ldots, N, Q; i \neq j), \tag{4–14.10}$$

$$\cdot \quad \cdot \quad \cdot \quad \cdot \quad \cdot \quad \cdot \quad \cdot \quad \cdot \quad \cdot \quad \cdot \quad \cdot \quad \cdot$$

$$|a_{ij}| = \det a_{ij} > 0. \tag{4–14.11}$$

---

† For an actual (irreversible) path of the processes inside the volume element, the positive sign in (4–12.12) and (4–12.17) is valid according to §4–8. We thus exclude reversible motions and restricted processes. Mathematically, this is expressed in that all "fluxes" and "forces" are independent and always vanish simultaneously (namely, at equilibrium).

These inequalities correspond to the general inequality statements (1–26.5) and (1–26.6) and are analogous to the special relations (2–4.4) and (2–4.5), (3–6.3) and (3–6.4).

4–15 RANGE OF VALIDITY OF THE THEORY

The phenomenological equations (§4–13) with the Onsager reciprocity relations (§4–14) are valid according to the analysis in the previous sections using the following assumptions:

1. The system is isotropic.
2. External force fields are constant in time.
3. Electric or magnetic polarization of the material does not appear.
4. For rotational motions, the Coriolis force can be neglected (in comparison to the centrifugal force).
5. For electrical phenomena, the Lorentz force, which acts on moving charges in magnetic fields, can be neglected.
6. The irreversible processes take place near equilibrium.

In §4–33 to §4–37, we discuss the consequences of surrendering the first five assumptions on our fundamental equations. The principal form of the theory is not changed.

The sixth assumption, the assumption of "proximity to equilibrium", is fundamental and decisive, since without it the phenomenological equations could not be formulated. A few general statements, which can be derived without the phenomenological equations, are discussed in Chapter 5.

The significance of the assumption of "proximity to equilibrium" was noted in Chapter 2 with respect to chemical reactions in homogeneous systems. For reactions in continuous systems, analogous conclusions hold.

For transport processes in continuous systems, we reach conclusions similar to those found in Chapter 3 in relation to processes in discontinuous systems: processes such as electric conduction, heat conduction, diffusion, etc., almost always occur within the region of validity of the theory worked out for processes occurring near equilibrium. The phenomenological equations lead to the corresponding empirical linear relations such as the laws of Ohm, Fick, Fourier, etc. This will be shown for particular cases by application of our fundamental equations to special problems (§4–16 *et seq.*).

With the help of assumptions (3) to (5) we can bring the expression for the generalized force $\mathbf{X}_k$, which appears in the phenomenological equations (§4–13), to a simple explicit form. Under these conditions only the gravitational force, the centrifugal force, and the electrostatic force (for charged particles) enter. Then the molar external force acting on particles of type k is (cf. Eq. 4–8.17):

$$\mathbf{K}_k = M_k \mathbf{g} + M_k \Omega^2 \mathbf{r} + z_k F \mathfrak{E}. \tag{4–15.1}$$

Here M_k is the molar mass of species k, $\mathbf{g}$ the gravitational acceleration, Ω the absolute value of the angular velocity in the case of a rotating system, $\mathbf{r}$ the vector of the distance of the volume element from the rotation axis, z_k the charge number of species k, F the Faraday constant, and $\mathfrak{E}$ the electric field strength. Using Eq. (1–12.12) we obtain the gradients of the chemical potentials of species k in the form:

$$(\text{grad } \mu_k)_T = V_k \text{ grad } P + (\text{grad } \mu_k)_{T,P}. \tag{4–15.2}$$

Here V_k denotes the partial molar volume of species k, and $(\text{grad } \mu_k)_{T,P}$ the gradient of μ_k after subtraction of the terms with grad T and grad P. If we describe the composition of the volume element by the independent concentration variables ζ_j then we find:

$$(\text{grad } \mu_k)_{T,P} = \sum_j \left(\frac{\partial \mu_k}{\partial \zeta_j}\right)_{T,P} \text{grad } \zeta_j. \tag{4–15.3}$$

From Eq. (4–13.4) and Eqs. (1) and (2) we derive

$$\begin{aligned} \mathbf{X}_k &= \mathbf{K}_k - (\text{grad } \mu_k)_T \\ &= M_k\mathbf{g} + M_k\Omega^2\mathbf{r} + z_k F\mathfrak{E} - V_k \text{ grad } P - (\text{grad } \mu_k)_{T,P}. \end{aligned} \tag{4–15.4}$$

This is, under the assumptions mentioned at the beginning of the section, the most general expression for the generalized force $\mathbf{X}_k$.

As soon as we assume "creeping motions" for fluid media and "simple cases" for crystals, as well as the general lack of a space charge, there results with Eq. (4–5.19):

$$\rho(\mathbf{g} + \Omega^2\mathbf{r}) = \text{grad } P, \tag{4–15.5}$$

where ρ denotes the density. If we consider, moreover, the relation (1–21.3) between the electric field strength $\mathfrak{E}$ and the electric potential φ:

$$\mathfrak{E} = -\text{grad } \varphi \tag{4–15.6}$$

and the definition of the electrochemical potential (1–20.15)

$$\eta_k \equiv \mu_k + z_k F\varphi, \tag{4–15.7}$$

then we obtain from Eq. (4):

$$\mathbf{X}_k = (M_k - \rho V_k)(\mathbf{g} + \Omega^2\mathbf{r}) - (\text{grad } \eta_k)_{T,P} \tag{4–15.8}$$

with

$$(\text{grad } \eta_k)_{T,P} = (\text{grad } \mu_k)_{T,P} + z_k F \text{ grad } \varphi. \tag{4–15.9}$$

We can now write the phenomenological equations in the form (4–13.6) and (4–13.7), where the $\mathbf{X}_k$ are given by Eq. (8).

In the special case of an *isothermal* medium, there follows from Eq. (4–13.6):

$$_1\mathbf{J}_i = \sum_{k=2}^{N} a_{ik}\mathbf{X}_k \quad (i = 2, 3, \ldots, N), \tag{4–15.10}$$

where $_1\mathbf{J}_i$ is the diffusion current density of species i in the Hittorf reference system and N the number of chemical species. The a_{ik} are the phenomenological coefficients. The generalized forces $\mathbf{X}_k$ are determined as before by Eqs. (8) and (9).

For all problems of electric conduction and diffusion in isothermal–isobaric media (grad $T = 0$, grad $P = 0$), Eq. (10) holds with the further simplification (see Eqs. 5, 8, and 9)

$$\mathbf{X}_k = -(\text{grad } \eta_k)_{T,P} = -(\text{grad } \mu_k)_{T,P} - z_k F \text{ grad } \varphi. \tag{4–15.11}$$

For pure electric conduction, the concentration gradients drop out. Thus,

$$\mathbf{X}_k = -z_k F \text{ grad } \varphi. \tag{4–15.12}$$

If pure diffusion is present then the total electric current vanishes. We thus find with Eq. (4–3.44) that

$$\sum_i z_i \,_1\mathbf{J}_i = 0, \tag{4–15.13}$$

where the sum is over all charged species. In the dissipation function (4–12.17), all terms with grad φ drop out in the second summation. This is obvious from Eq. (11). Thus there is valid for the generalized forces in Eq. (10):

$$\mathbf{X}_k = -(\text{grad } \mu_k)_{T,P} \quad (k = 2, 3, \ldots, N), \tag{4–15.14}$$

as for nonelectrolytes ($z_k = 0$).

However, if we want to write an expression for the diffusion potential (§4–18) or for the sedimentation potential (§4–21) from Eq. (10) then we must retain the formulation of (11) or (8) for the generalized forces although Eq. (13) continues to hold.

B. Isothermal Processes

4–16 ELECTRIC CONDUCTION

(a) General

We consider electric transport in a conducting medium, e.g. in a metal or in an electrolyte solution. Gradients of temperature, of pressure, and of concentrations, as well as external force fields except for an electric field, are excluded. We thus deal with "pure electric conduction": motion of charged particles (electrons or ions) in an external electric field. Here we always assume isotropic media. (Details of anisotropic media are found in §4–38.)

In problems of electric conduction, it is useful to relate the diffusion fluxes and electric currents of all the species to the velocity v_1 of a particular species 1 ("reference species") and thus to use the Hittorf reference system. For example, in metal systems, the ions on the lattice sites are chosen as reference species,† while for solid and molten electrolytes[8, 9] a particular ion type (such as the Cl^- ion in the case of the molten salt mixture $NaCl + KCl$) and for electrolyte solutions the solvent may be chosen.

For electrolyte solutions, we must still state whether the total solvent or only the unsolvated part of the solvent is regarded as the reference species. In the first case, the average velocity of *all* solvent molecules v_1 is used as a reference velocity, and we speak, as before, of a Hittorf reference system. In the second case, in which the average speed of the *free* solvent molecules v_1^* is introduced as the reference velocity, we talk of the Washburn reference system. The latter mode of description is only mentioned for completeness. After clarification of some fundamentals we will, without exception, use the Hittorf reference system.

Let c_i be the molarity of a molecular or ionic species i and v_i the average velocity of this particle type. Then, according to Eq. (4–3.30), the diffusion current density of species i in the Hittorf reference system is:

$$_1J_i = c_i(v_i - v_1) \qquad (4\text{–}16.1)$$

and for the corresponding diffusion flux density in the Washburn reference system we have

$$_1J_i^* = c_i(v_i - v_1^*). \qquad (4\text{–}16.2)$$

With Eq. (4–3.42) the electric current density in the Hittorf or Washburn reference system becomes:

$$_1I_i = z_i F \,_1J_i = z_i F c_i(v_i - v_1) \qquad (4\text{–}16.3)$$

or

$$_1I_i^* = z_i F \,_1J_i^* = z_i F c_i(v_i - v_1^*). \qquad (4\text{–}16.4)$$

Here F denotes the Faraday constant and z_i the charge number of ionic species i (positive or negative integer for cations or anions respectively).

---

† For metals, the (Hittorf) reference system used here coincides with the lattice system according to p. 222.

For the total electric current density, according to Eqs. (4-3.43), (4-3.46), and Eqs. (1) to (4), we find:

$$
\left.
\begin{aligned}
\mathbf{I} &= \sum_i {}_1\mathbf{I}_i = \sum_i {}_1\mathbf{I}_i^* = F \sum_i z_i \, {}_1\mathbf{J}_i \\
&= F \sum_i z_i \, {}_1\mathbf{J}_i^* = F \sum_i z_i c_i (\mathbf{v}_i - \mathbf{v}_1) \\
&= F \sum_i z_i c_i (\mathbf{v}_i - \mathbf{v}_1^*) = F \sum_i z_i c_i \mathbf{v}_i,
\end{aligned}
\right\}
\qquad (4\text{--}16.5)
$$

in agreement with Eqs. (4-3.44) and (4-3.47). Here the summations are over all ionic species. As has already been proved in §4-3 (p. 224) for the general case and as is obvious also from Eq. (5), the total electric current is independent of the reference velocity.

The relative velocities or diffusion current densities in Eqs. (1) and (2) are vectors which for cations have the same, for anions, however, the opposite direction as the electric field. Accordingly, the vectors of the partial or total electric currents in Eqs. (3) and (4) or in Eq. (5) always lie in the field direction.

We designate the *solvation number* of the ionic species i by N_i, i.e. the number of the solvent molecules bound to an ion of type i, and the molarity of the solvent by c_1. Then, from the meaning of $\mathbf{v}_1$ and $\mathbf{v}_1^*$, we obtain

$$
\mathbf{v}_1 = \frac{1}{c_1} \left[\left(c_1 - \sum_i N_i c_i \right) \mathbf{v}_1^* + \sum_i N_i c_i \mathbf{v}_i \right]
$$

or

$$
\mathbf{v}_1 = \mathbf{v}_1^* + \sum_i N_i \frac{c_i}{c_1} (\mathbf{v}_i - \mathbf{v}_1^*).
\qquad (4\text{--}16.6)
$$

From Eqs. (3), (4), and (6) there follows:

$$
\frac{{}_1\mathbf{I}_i}{{}_1\mathbf{I}_i^*} = \frac{\mathbf{v}_i - \mathbf{v}_1}{\mathbf{v}_i - \mathbf{v}_1^*} = 1 - \frac{\sum_i N_i c_i (\mathbf{v}_i - \mathbf{v}_1^*)}{c_1 (\mathbf{v}_i - \mathbf{v}_1^*)}.
\qquad (4\text{--}16.7)
$$

Here the vectors may be divided by one another because they always have the same direction.

The relations formulated till now are so general that they are applicable to any transport process in electrically conducting media, i.e. they are not restricted to "pure" electric conduction.

(b) Transport Numbers

We now consider "pure" electric conduction, as it was defined in the first part of this section. The fraction of the total electric current transported by species i in such a transport process is called the *transport number*. The first experimental method for the determination of this quantity is due to Hittorf (1853) ("Hittorf method"). The direct method used more often today is the "moving boundary method" (Lodge, 1886). By an indirect method transport numbers can also be obtained by e.m.f. measurements from concentration cells with transport (§4–19) or from gravitational or centrifugal cells (§4–22). Credit is due especially to Washburn[10] and Lewis[11] as well as MacInnes and Longsworth[12] for the explanation and for the method of these transport experiments[13–17].

For metals and other electron conductors, the current flow is achieved by movement of the electrons only. The transport number of the electrons thus has the value 1 (if we use the reference system mentioned at the beginning), while the transport numbers of all other charge carriers vanish. In a similar way, for solid or molten electrolytes, we always obtain the value 0 or 1 for the transport number (cf. p. 379), provided we have only two ionic species (such as pure $NaCl$ and $AgNO_3$).†

For electrolyte solutions, on the other hand (also for solid and molten electrolytes with more than two ionic species, such as for the system $NaCl + KCl$ in the solid or liquid state), the situation is more complex. We must make a distinction here, corresponding to the two reference systems which we are using, between the *Hittorf transport number* of the ionic species i:

$$t_i \equiv \tfrac{1 \mathbf{I}_i}{\mathbf{I}} \qquad\qquad (4\text{–}16.8)$$

and the *Washburn transport number* of the ionic species i (or "true transport number"):

$$t_i^* \equiv \tfrac{1 \mathbf{I}_i^*}{\mathbf{I}} \cdot \qquad\qquad (4\text{–}16.9)$$

From Eqs. (3), (4), and (5) and Eqs. (8) and (9) we find:

$$t_i = \frac{z_i c_i (\mathbf{v}_i - \mathbf{v}_1)}{\sum_i z_i c_i (\mathbf{v}_i - \mathbf{v}_1)}, \qquad\qquad (4\text{–}16.10)$$

$$t_i^* = \frac{z_i c_i (\mathbf{v}_i - \mathbf{v}_1^*)}{\sum_i z_i c_i (\mathbf{v}_i - \mathbf{v}_1^*)} \cdot \qquad\qquad (4\text{–}16.11)$$

Equations (8) to (11) hold for "pure" electric conduction. Equations (8) and (10) can be applied to any electric conductors, while Eqs. (9) and (11) naturally have meaning only for electrolyte solutions.

----

† It can also be shown that on using another reference system for pure molten salts trivial expressions always result for the transport numbers[18]. For a fuller discussion of this point see Sundheim[19].

We obtain the relation between the two transport numbers† in any electrolyte solution from Eqs. (7), (8), (9), and (11):

$$t_i = t_i^* - \frac{z_i c_i}{c_1} \sum_i \frac{N_i}{z_i} t_i^*. \qquad (4\text{-}16.12)$$

Moreover, the identities

$$\sum_i t_i = 1, \qquad \sum_i t_i^* = 1 \qquad (4\text{-}16.13)$$

must be satisfied, because of Eqs. (5), (8), and (9).

As discussed in detail later (p. 268), the primary experimental quantities which can be determined are the "stoichiometric transport numbers." For complete dissociation, these become identical to the Hittorf transport numbers of the individual ionic species. The Washburn transport numbers, on the other hand, are never accessible from measurement. Since the solvation numbers N_i can also be obtained only from experimental data with the help of certain assumptions and correspondingly represent uncertain quantities, a hypothesis-free calculation of t_i^* according to Eq. (12) is not possible. Only in the limiting case of infinite solution ($c_i \to 0$) does t_i coincide with t_i^*. This is seen from Eq. (12).

For solutions of a single electrolyte with two ionic species (cations: inferior $+$, anions: inferior $-$), according to Eqs. (1–19.11) to (1–19.13) for complete dissociation, we find

$$\frac{c_+}{\nu_+ M_1 c_1} = \frac{c_-}{\nu_- M_1 c_1} = m. \qquad (4\text{-}16.14)$$

In this, ν_+ and ν_- denote the dissociation numbers for the cation and anion respectively (number of cations or anions which arise from the dissociation of an electrolyte molecule), M_1 the molar mass of the solvent (in practice measured in units of $kg\,mol^{-1}$), and m the molality of the electrolyte ($mol\,kg^{-1}$). From Eqs. (12) and (14) we derive:

$$\left. \begin{array}{l} t_+^* = t_+ - z_+ \nu_+ M_1 m L, \\ t_-^* = t_- - z_- \nu_- M_1 m L \end{array} \right\} \qquad (4\text{-}16.15)$$

with

$$L \equiv \frac{N_+}{z_+} t_+^* + \frac{N_-}{z_-} t_-^*.$$

The expression L is known as "electrolytic solvent transport." It has frequently been attempted to determine these quantities experimentally, but this is not

---

† The relations between the two transport numbers given previously in the literature[10] are special cases of Eq. (12).

possible without certain assumptions[20a]. With the help of the identity following from Eq. (13)

$$t_+ + t_- = 1, \qquad t_+^* + t_-^* = 1, \qquad (4\text{-}16.16)$$

we find from Eq. (15)

$$1 - t_i = (1 - t_i^*)(1 - NM_1m) + \nu_i N_i M_1 m \quad (i = +, -) \qquad (4\text{-}16.17)$$

with the abbreviation

$$N \equiv \nu_+ N_+ + \nu_- N_-. \qquad (4\text{-}16.18)$$

The quantity (18) can be designated as the "solvation number of the electrolyte." It cannot be determined without hypothesis.

In the following, we only use the Hittorf transport numbers and the Hittorf reference system (reference velocity $\mathbf{v}_1$). In this way, all uncertain solvation numbers disappear since all formulas contain only measurable quantities.

(c) Mobilities and Conductivity

We now consider the process of pure electric conduction somewhat more precisely, and, indeed, in the classical mode of presentation.

An external electric field is applied to an electric conductor in which neither temperature, nor pressure, nor concentration gradients are present. Let us call $\mathfrak{E}$ the electric field strength in any volume element of the conductor, corresponding to a local electric potential φ, connected with the field strength by the general relation (4–15.6)

$$\mathfrak{E} = - \operatorname{grad} \varphi. \qquad (4\text{-}16.19)$$

The electric transport is accomplished by the charge carriers (charged species such as electrons or ions) of the volume element.

Initially ("preliminary process"), only the electrostatic force proportional to the field strength acts on each charged species so that an accelerated motion begins. Immediately, however, retarding forces arise (friction in the medium and electrostatic interactions between the moving charged species) which are proportional to the velocity of the particles.

The proportionality between retarding force and velocity is decisive in these considerations; it can either be established by considering certain molecular kinetic models or by the general statement that this proportionality leads to Ohm's law (see below), which holds for any electric conductor except under extremely high field strengths or unsteady processes (such as in luminous arcs). By the opposing action of the electrostatic force (caused by the external field) and of the retarding force, there arises a force-free motion with constant velocity which is proportional to the field strength $\mathfrak{E}$. Since this velocity (in our present notation) represents the relative velocity

$\mathbf{v}_i - \mathbf{v}_1$ of the charged species i with reference to the particle type 1 (cations on the lattice sites for metals, solvent for electrolyte solutions, etc.), we can describe this state of affairs formally as follows:

$$|\mathbf{v}_i - \mathbf{v}_1| = u_i |\mathfrak{E}|, \tag{4–16.20}$$

where the absolute magnitudes of the vectors are inserted, so that the quantity u_i is always positive.

It follows from our presentation that u_i is independent of the field strength but can still be an arbitrary function of the state variables of the volume element (temperature, pressure, concentrations). The quantity u_i is called the *mobility* of the charged species i.

From the definition of the electric current density (cf. Eqs. 3, 5, and 20) we obtain:

$$_1\mathbf{I}_i = |z_i| F c_i u_i \mathfrak{E}, \tag{4–16.21}$$

$$\mathbf{I} = F \sum_i |z_i| c_i u_i \mathfrak{E}. \tag{4–16.22}$$

From Eqs. (21) and (22) and Eq. (8) for the Hittorf transport number of species i we find that

$$t_i = \frac{|z_i| c_i u_i}{\sum_i |z_i| c_i u_i}. \tag{4–16.23}$$

Equation (22) can be written in the form

$$\mathbf{I} = \kappa \mathfrak{E} \tag{4–16.24}$$

with

$$\kappa \equiv F \sum_i |z_i| c_i u_i. \tag{4–16.25}$$

Equation (24) is the law of Ohm (1826) in the formulation for a single volume element ("local formulation"). Accordingly, κ denotes the *specific conductance*[30] which, according to the above statements (p. 256), is independent of the reference velocity. We recognize that the underlying basis of Eq. (20) leads to Ohm's law. Equations (23) or (25) give the relation between the Hittorf transport numbers or the specific conductance and the mobilities respectively.

The specific conductance κ depends, as do the mobilities and transport numbers, on concentrations, temperature, and pressure. The last is of slight importance in condensed phases. The following values are given as examples (all valid for atmospheric pressure).

Examples for measured transport numbers and for mobilities derived from them are presented later (pp. 264–265).

Example	κ $\Omega^{-1}\,cm^{-1}$
Silver at 18°C	6.25×10^5
NaCl (solid) at 700°C	7×10^{-5}
α-AgI (solid) at 150°C	1.3
AgI (liquid) at 600°C	2.43
Water ("conductivity water") at 25°C according to Kohlrausch and Heydweiller (1894)	0.055×10^{-6}
1N aqueous KCl solution at 0°C	0.06518
1N aqueous KCl solution at 25°C	0.11134

The last two figures represent recent precise standard values for the calibration of conductivity cells.

(d) Thermodynamic–Phenomenological Theory

We now show that the phenomenological equations of the thermodynamics of irreversible processes also yield Ohm's law (24) with the relations (23) and (25). We proceed directly from the conclusion of §4–15.

If we apply the phenomenological equations (4–15.10) to pure electric conduction and relate the inferiors i and k to the charged species, noting that the contributions of the other particle types to the electric current drop out, then on consideration of Eqs. (4–15.12), (1), (3), (5), and (19) we obtain:

$$\frac{1}{z_i F}\,{}_1\mathbf{I}_i = c_i(\mathbf{v}_i - \mathbf{v}_1) = F \sum_k a_{ik} z_k \mathfrak{E}, \qquad (4\text{–}16.26)$$

$$\mathbf{I} = F^2 \sum_i \sum_k z_i z_k a_{ik} \mathfrak{E}, \qquad (4\text{–}16.27)$$

where the a_{ik} are the phenomenological coefficients. We set

$$\frac{F}{c_i}\left|\sum_k a_{ik} z_k\right| \equiv u_i, \qquad (4\text{–}16.28)$$

$$F^2 \sum_i \sum_k z_i z_k a_{ik} \equiv \kappa. \qquad (4\text{–}16.29)$$

Equation (26) is identical to Eq. (20) and Eq. (27) to Ohm's law (24).

The expression (29) is always positive due to (4–14.10) and (4–14.11). This is in agreement with experience. Thence Eq. (25) follows from Eqs. (28) and (29).

Finally, from Eqs. (26) and (27) with Eq. (8) there results

$$t_i = \frac{z_i \sum_k a_{ik} z_k}{\sum_i \sum_k a_{ik} z_i z_k} \qquad (4\text{–}16.30)$$

and with Eq. (28) and the inequality statement following from Eq. (26):

$$z_i \sum_k a_{ik} z_k > 0$$

we find (23).

Thus we have obtained all the previously derived relations using the methods of the thermodynamics of irreversible processes.

(e) Solution of a Single Electrolyte

The special case of a solution containing a single electrolyte which on dissociation produces only two kinds of ions, namely, one cation species $(+)$ and one anion species $(-)$, is important. Here the electroneutrality condition (4–3.46)

$$z_+ c_+ + z_- c_- = 0 \qquad (4\text{–}16.31)$$

is valid. If we introduce as composition variable the equivalent concentration c^* of the electrolyte then we find:

$$z_+ c_+ = -z_- c_- = \alpha c^*, \qquad (4\text{–}16.32)$$

where α is the degree of dissociation of the electrolyte. If we define as the *equivalent conductance* the quantity†

$$\Lambda \equiv \frac{\kappa}{c^*}, \qquad (4\text{–}16.33)$$

then we obtain from Eqs. (25) and (32):

$$\Lambda = \alpha F(u_+ + u_-). \qquad (4\text{–}16.34)$$

For the Hittorf transport numbers t_+ and t_-, we derive from Eqs. (23) and (32):

$$t_+ = \frac{u_+}{u_+ + u_-}, \qquad t_- = 1 - t_+ = \frac{u_-}{u_+ + u_-}. \qquad (4\text{–}16.35)$$

From Eqs. (34) and (35) it is obvious how the mobilities u_+ and u_- of the cations and anions can be determined with the knowledge of the degree of dissociation from the conductance and the transport numbers for each

---

† If c is the molarity of the electrolyte then we have

$$c^* = z_+ \nu_+ c = -z_- \nu_- c.$$

The "molar conductance" still used occasionally ($\Lambda_m \equiv \kappa/c$) is related to the equivalent conductance as follows

$$\Lambda_m = z_+ \nu_+ \Lambda = -z_- \nu_- \Lambda.$$

concentration and each temperature† (see, however, subsection (g)). If the concentration of the electrolyte (measured as equivalent concentration c^*, molarity c, or molality m) approaches zero, then the specific conductance κ vanishes‡ but not the equivalent conductance Λ. From this empirical fact and with $\alpha \to 1$ for $c \to 0$ it follows from Eq. (34) that the limiting values u_+^0 and u_-^0 of the mobilities for infinite dilution are different from zero. With the abbreviations

$$\Lambda^0 \equiv \lim_{c \to 0} \Lambda, \qquad t_i^0 \equiv \lim_{c \to 0} t_i \quad (i = +, -) \qquad (4\text{–}16.36)$$

Table 6a

Limiting Value λ_i^0 of the Ionic Conductance for Ions
in Aqueous Solution at 25°C[21a]

Ion	λ_i^0 $\Omega^{-1}\,cm^2\,mol^{-1}$
H^+	349.81
Li^+	38.68
Na^+	50.10
K^+	73.50
NH_4^+	73.55
Rb^+	77.81
Cs^+	77.26
Ag^+	61.90
Be^{2+}	45
Mg^{2+}	53.05
Ca^{2+}	59.50
Sr^{2+}	59.45
Ba^{2+}	63.63
La^{3+}	69.7
OH^-	198.3
F^-	55.4
Cl^-	76.35
Br^-	78.14
I^-	76.84
NO_3^-	71.46
ClO_4^-	67.36
SO_4^{2-}	80.02

† The pressure influence can be neglected for condensed phases.
‡ The very small electric conductivity of the solvent (e.g. of the water) is excluded here.

we obtain the relations for the limiting values following from Eqs. (34) and (35):

$$\Lambda^0 = F(u_+^0 + u_-^0), \tag{4-16.37}$$

$$t_+^0 = \frac{u_+^0}{u_+^0 + u_-^0}, \qquad t_-^0 = 1 - t_+^0 = \frac{u_-^0}{u_+^0 + u_-^0}. \tag{4-16.38}$$

As is known from experience, the limiting values u_+^0 and u_-^0 are characteristic quantities for the ion species in question, i.e. they are independent of the nature of the ions of opposite charge ("gegenions") provided that the same solvent is under consideration and that the temperature is regarded as constant. This statement is known as "Kohlrausch's law of independent ion migration" (Kohlrausch, 1893). Accordingly, the limiting value Λ^0 of the equivalent conductance is additive in ion pairs.

The quantity (see Eqs. 34 and 35)

$$\lambda_i \equiv Fu_i = \frac{1}{\alpha} \Lambda t_i \quad (i = +, -) \tag{4-16.39}$$

Table 6b

System Water–Nitric Acid at 25°C: Experimental Values[22] of the Degree of Dissociation α, of the Equivalent Conductance Λ, of the Cation Transport Number t_+, and of the Ionic Conductances λ_+ and λ_- as Functions of the Molarity c of the Acid†

c mol l^{-1}	α	Λ Ω^{-1} cm^2 mol^{-1}	t_+	λ_+ Ω^{-1} cm^2 mol^{-1}	λ_- Ω^{-1} cm^2 mol^{-1}
0	1	421.3	0.830	349.8	71.5
0.5	0.994	356.8	0.838	300.8	58.2
1	0.985	328.6	0.833	277.9	55.7
2	0.961	281.5	0.821	240.5	52.4
3	0.927	240.0	0.808	209.2	49.7
4	0.883	202.4	0.795	182.2	47.0
5	0.829	170.5	0.782	160.8	44.9
6	0.774	144.0	0.769	143.1	42.9
7	0.712	120.5	0.754	127.6	41.6
8	0.648	101.9	0.738	116.1	41.2
9	0.585	86.72	0.720	106.7	41.5
10	0.520	73.81	0.702	99.6	42.3

† Strictly speaking, the quantity measured in transference experiments is ϑ_+, not t_+ (see later). Thus the final quantities are not λ_+ and λ_-, but λ_+^*/α and λ_-^*/α (cf. below). The values for α have been derived from nuclear resonance measurements by Hood and Reilly[23].

is designated as the *ionic conductance*. Its limiting value for infinite dilution

$$\lambda_i^0 \equiv \lim_{c \to 0} \lambda_i = F u_i^0 = \Lambda^0 t_i^0$$

is usually tabulated in place of u_i^0 (see Table 6a).

The system water–nitric acid, which contains the chemical species H_2O, HNO_3, H^+, and NO_3^-, excluding the region of very high nitric acid concentrations, represents one of the few examples of electrolyte solutions for which all the experimental data for the determination of the ion mobilities, i.e. degree of dissociation, conductance, and transport numbers, are available. The data for 25°C are found in Table 6b and in Fig. 10.†

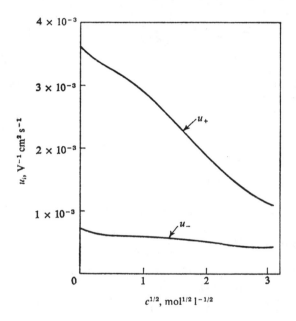

Fig. 10. Mobility u_+ or u_- of the ion H^+ or NO_3^- respectively for the system water + nitric acid as a function of the molarity c of the acid at 25°C (cf. Table 6b).

For sufficiently high dilution, the limiting law of Onsager (1927) holds which corresponds to the equation of the limiting tangent at $c = 0$ in Fig. 10:

$$\lambda_i = \lambda_i^0 - \left(z_+ |z_-| \frac{(2 + \sqrt{2})r}{1 + \sqrt{r}} \Theta \lambda_i^0 + \tfrac{1}{2}|z_i|\sigma \right)\sqrt{\Gamma} \quad (i = +, -)$$

$$(4\text{–}16.40)$$

† Recent papers on transport numbers and conductances of concentrated electrolytic solutions are the following[17, 22, 24–26].

with

$$r \equiv \frac{z_+|z_-|(\lambda_+^0 + \lambda_-^0)}{(z_+ - z_-)(z_+\lambda_-^0 - z_-\lambda_+^0)},\tag{4-16.40a}$$

$$\Theta \equiv \frac{a'}{3(1 + \tfrac{1}{2}\sqrt{2})},\tag{4-16.40b}$$

$$\sigma \equiv a'\frac{8\varepsilon_0\varepsilon RT}{3\eta},\tag{4-16.40c}$$

$$a' \equiv \frac{a}{\sqrt{\rho}},\tag{4-16.40d}$$

$$a \equiv \sqrt{2\pi L\rho}\left(\frac{e^2}{4\pi\varepsilon_0\varepsilon kT}\right)^{3/2},\tag{4-16.40e}$$

$$\Gamma \equiv \tfrac{1}{2}(z_+^2\nu_+ + z_-^2\nu_-)c.\tag{4-16.40f}$$

Here ε_0 is permittivity of vacuum in the rational system (cf. p. 423), ε the dielectric constant of the solvent, R the gas constant, T the absolute temperature, η the viscosity of the solvent, ρ the density of the solvent, L the Avogadro constant, e the elementary charge, and $k = R/L$ the Boltzmann constant. The quantity a is also present in the Debye–Hueckel limiting law for the activity coefficient of the electrolyte[27].

If we consider a solution of an electrolyte in water at 25°C (298.15 K) then there is valid:†

$$L = 6.023 \times 10^{23}\ \text{mol}^{-1}, \quad k = 1.380 \times 10^{-16}\ \text{erg K}^{-1},$$

$$e = 4.803 \times 10^{-10}\ \text{Fr}, \quad \varepsilon_0 = \frac{1}{4\pi}\frac{\text{Fr}^2}{\text{erg cm}} = \frac{10}{4\pi}\ \text{Fr}^2\ \text{erg}^{-1}\,\text{l}^{-1/3},$$

$$\varepsilon = 78.30, \quad \rho = 0.9971\ \text{g cm}^{-3}.$$

From this and Eqs. (40d) and (40e) we find

$$a = 1.1766\ \text{kg}^{1/2}\ \text{mol}^{-1/2}, \qquad a' = 1.1783\ \text{l}^{1/2}\ \text{mol}^{-1/2}.$$

Recent measurements have provided a value for the dielectric constant ε of water other than that used previously (Malmberg and Maryott, 1956). Thus the values of a and a' have, correspondingly, changed, so that the numbers given here deviate from those usually reported in the literature. The most recent values of the viscosity η of water, which are necessary for evaluation of the expression (40), are found in Table 18 (p. 412).

(f) Reduced Transport Numbers

If, in an electrolyte solution, there are present other uncharged species in addition to solvent molecules, for example, other nonelectrolyte molecules or

---

† Values of the universal constants L, k, and e according to Stille[28]. Values of ε and ρ according to Robinson and Stokes[21a].

undissociated electrolytic molecules, then their contribution to the electric current vanishes but the transport of such particle types in the electric field, measured relative to the solvent, does not necessarily vanish. This is because the ions can drag neutral molecules with them in their migration. This effect can arise indirectly, too. If, for example, an ion transports only water molecules in its motion in the electric field ("hydration") then relative motion (with reference to the water) of the other neutral molecules will take place.

We take account of this kind of transport of uncharged species in the electric field most usefully by introducing a *reduced transport number*† for any arbitrary species i:

$$\tau_i \equiv F \frac{{}_1\mathbf{J}_i}{\mathbf{I}} \tag{4-16.41}$$

(cf. p. 184). Since the vectors ${}_1\mathbf{J}_i$ (diffusion current density of species i in the Hittorf reference system) and $\mathbf{I}$ (total electric current density) have either the same or the opposite direction, τ_i represents a positive or negative number. If species i migrates from the anode to the cathode during the flow of current, τ_i is positive, otherwise it is negative. If species i is an ion ($z_i \neq 0$) then it follows from Eqs. (3), (8), and (41) that

$$t_i = z_i \tau_i, \tag{4-16.42}$$

where t_i denotes the Hittorf transport number.

The phenomenological equations, which hold for any species according to Eqs. (4-15.10), (4-15.12), (5), and (19), are:

$${}_1\mathbf{J}_i = F \sum_k a_{ik} z_k \mathfrak{E}, \tag{4-16.43}$$

$$\mathbf{I} = F^2 \sum_i \sum_k a_{ik} z_i z_k \mathfrak{E}. \tag{4-16.44}$$

Here the inferior i or k denotes, in contrast to Eqs. (26) and (27), *any one* particle type. However, while the inferior i in the expression (43) obviously includes uncharged species (apart from the particle type 1, which represents the "solvent" and thus the reference system), the neutral species ($z_i = 0$ or $z_k = 0$) again drop out in Eq. (44).

From Eqs. (41), (43), and (44) there results the important formula:

$$\tau_i = \frac{\sum_k a_{ik} z_k}{\sum_i \sum_k a_{ik} z_i z_k}, \tag{4-16.45}$$

which we will use frequently. If the inferior i refers to an ionic species then Eq. (30) follows from Eqs. (42) and (45).

---

† This quantity is called the "Washburn number" by Agar[29].

We find from Eq. (45) that

$$\sum_i z_i \tau_i = 1. \tag{4–16.46}$$

This relation corresponds to the identity (13).

The reduced transport numbers of nonelectrolytes can be experimentally determined. On the other hand, the reduced transport numbers of undissociated electrolyte parts cannot be measured.

(g) Stoichiometric Transport Numbers

For electrolytes dissociating incompletely or in several steps, the relation between the previously defined quantities, which relate to individual species, and those quantities actually measured is no longer obvious at first sight. In order to be able to formulate the problem exactly, we proceed from the concept of the "*ion constituent*," as it was introduced by Noyes and Falk[30].

The "ion constituents" of an electrolyte are the ion-forming parts of the electrolyte molecule (atoms or radicals) without consideration as to what extent these molecule parts are actually present as ions.

Thus, in an aqueous solution of sulfuric acid, the atoms or radicals H and SO_4 represent the ion constituents, where H occurs in the species H^+, HSO_4^-, and H_2SO_4 and SO_4 appears in the particle types SO_4^{2-}, HSO_4^-, and H_2SO_4. The amounts of the ion constituents can, in contrast to those of the actual species, always be determined by the methods of analytical chemistry. Thus, in the above-mentioned example, the ion constituent H can, in principle, be quantitatively determined by titration up to the second equivalence point of the sulfuric acid, while the amount of ion constituent SO_4 can be determined by precipitation as $BaSO_4$.

Another example is aqueous phosphoric acid. Here we have two ion constituents (H and PO_4), but four ionic species (H^+, $H_2PO_4^-$, HPO_4^{2-}, PO_4^{3-}). In $KHSO_4$, finally, we have three ion constituents (K, H, and SO_4) and four ionic species (K^+, H^+, HSO_4^-, and SO_4^{2-}).

Let R be an ion constituent of an electrolyte. Let the charge number of the ion which comes directly from this constituent (such as H^+ from H or SO_4^{2-} from SO_4) be z_R. We now consider the expression

$$\vartheta_R \equiv z_R F \frac{{}_1\mathbf{J}_R}{\mathbf{I}} \equiv z_R F \frac{\sum_i n_{Ri}\, {}_1\mathbf{J}_i}{\mathbf{I}}, \tag{4–16.47}$$

where ${}_1\mathbf{J}_R$ denotes the diffusion current density of the ion constituent R (in the Hittorf reference system), n_{Ri} the number of particles of R contained in one particle of i, and the summation is over all species containing R. ϑ_R is equal to the value of the equivalent amount of R, measured in the unit mol, which, on the passage of the charge of 1 faraday ($\equiv n^\dagger F$ with $n^\dagger \equiv 1$ mol), flows through an arbitrary reference plane moving with the total solvent in the solution. We designate ϑ_R as the *stoichiometric transport number* of the ion constituent R. It is this quantity which is directly measured in all transference experiments on electrolyte solutions, such as the Hittorf method[31], the moving boundary method[17], and the e.m.f. measurements on concentration, gravitational, and centrifugal cells (cf. §4–19 and §4–22).

From Eqs. (41) and (47) the relation between ϑ_R and the reduced transport numbers τ_i of the individual species follows:

$$\vartheta_R = z_R \sum_i n_{Ri}\tau_i. \qquad (4\text{--}16.48)$$

If charged particle types are designated by an inferior j and neutral molecule types by an inferior k then we obtain from Eqs. (42) and (48)

$$\vartheta_R = z_R \sum_j \frac{n_{Rj}t_j}{z_j} + z_R \sum_k n_{Rk}\tau_k, \qquad (4\text{--}16.49)$$

where t_j or z_j is the Hittorf transport number or the charge number respectively of the ion species j.

By reason of electric neutrality we write

$$z_i = \sum_R z_R n_{Ri}, \qquad (4\text{--}16.50)$$

where the summation is over all ionic constituents of the solution. From this, on combination of Eq. (46) with Eq. (48), we find

$$\sum_R \vartheta_R = 1. \qquad (4\text{--}16.51)$$

This is an identity analogous to Eq. (13).

The Hittorf transport numbers of the individual ions can be obtained from transport experiments on solutions of a single electrolyte only if we are dealing with an electrolyte containing two ionic species and, moreover, if either complete dissociation is present or if the migration of the undissociated electrolyte molecules in the electric field (relative to the total solvent) can be neglected. For the solution of an electrolyte with two kinds of ions (species $+$ and $-$) and with undissociated electrolyte molecules (species u), we see from Eq. (49) that if we choose the atoms or radicals corresponding to the cations or anions as ion constituents, then

$$\vartheta_+ = t_+ + z_+\nu_+\tau_u, \qquad \vartheta_- = t_- + z_-\nu_-\tau_u. \qquad (4\text{--}16.52)$$

Here τ_u is the reduced transport number of the undissociated electrolyte, ϑ_+ and ϑ_- the stoichiometric transport numbers of the cationic and anionic constituents of the electrolyte, t_+ or t_- the Hittorf transport number of the cations or anions respectively, and ν_+ or ν_- the dissociation number for the cations or anions. With the condition for electric neutrality (1–19.8)

$$z_+\nu_+ + z_-\nu_- = 0$$

and the identity (16)

$$t_+ + t_- = 1$$

there results from Eq. (52):

$$\vartheta_+ + \vartheta_- = 1,$$

in agreement with Eq. (51).

By combination of Eq. (39) with Eq. (52) we find that

$$u_i = \frac{\Lambda}{\alpha F}(\vartheta_i - z_i\nu_i\tau_u) \quad (i = +, -). \tag{4–16.53}$$

Accordingly, the ion mobility u_i can be obtained from experimental data only if we may set $\tau_u = 0$.

Since, in the case considered, the only quantities always accessible from experiments are Λ, ϑ_+, and ϑ_-, we may define

$$\lambda_i^* \equiv \vartheta_i\Lambda \quad (i = +, -) \tag{4–16.54}$$

and call this the *conventional ionic conductance* of the cation or anion constituent. From Eqs. (39), (51), (53), and (54) we derive

$$\lambda_+^* + \lambda_-^* = \Lambda, \tag{4–16.55}$$

$$\lambda_i^* = \alpha\lambda_i - z_i\nu_i\tau_u\Lambda \quad (i = +, -). \tag{4–16.56}$$

We now apply Eqs. (1), (20), (39), and (41) to species u (undissociated electrolyte molecules) and use Ohm's law (24) and the relations (cf. Eqs. 32 and 33)

$$c_u = (1 - \alpha)c,$$

$$\kappa = c^*\Lambda = z_+\nu_+c\Lambda = -z_-\nu_-c\Lambda.$$

We then obtain

$$z_+\nu_+\tau_u\Lambda = \pm(1 - \alpha)\lambda_u, \tag{4–16.57}$$

$$z_-\nu_-\tau_u\Lambda = \mp(1 - \alpha)\lambda_u. \tag{4–16.58}$$

Here λ_u is a quantity analogous to the ionic conductances λ_+ and λ_-. The upper or lower sign holds if the neutral electrolyte molecules migrate in the same direction as the cations or anions respectively. From Eqs. (56), (57), and (58) there follows[17]

$$\lambda_+^* = \alpha\lambda_+ \pm (1 - \alpha)\lambda_u, \tag{4–16.59}$$

$$\lambda_-^* = \alpha\lambda_- \mp (1 - \alpha)\lambda_u. \tag{4–16.60}$$

These are the relations between the measurable quantities (λ_+^* and λ_-^*) and the "molecular" quantities (λ_+, λ_-, and λ_u). By adding the last two equations, we find

$$\lambda_+^* + \lambda_-^* = \alpha(\lambda_+ + \lambda_-) = \Lambda,$$

in agreement with Eqs. (34), (39), and (55).

If, finally, we are dealing with a solution of an electrolyte in which a second nonelectrolyte, apart from the solvent, is contained then we can, using transference experiments, determine the reduced transport number of the additional nonelectrolyte since this quantity represents the amount of substance of the nonelectrolyte, measured in the unit mol, which is transposed from the anode to the cathode, relative to the total solvent, when the charge of 1 faraday is passed through the solution. The earlier attempts to determine the solvation of the ions from that kind of experiment should not be regarded as reliable since we do not know to what extent the ions carry the additional nonelectrolytes, apart from the solvent and the undissociated electrolyte, with them during migration in the electric fields[32]. For an aqueous solution of HCl containing methanol, only the following quantities are, strictly speaking, measurable: the stoichiometric transport number $\vartheta_H = 1 - \vartheta_{Cl}$ and the reduced transport number τ_{CH_3OH}.

4-17 DIFFUSION IN GASES AND NONELECTROLYTE SOLUTIONS

(a) General

Any transport of matter caused by concentration gradients is called *diffusion*. In the following, we assume fluid media without charged species and exclude temperature and pressure gradients, external force fields, and chemical reactions. We are thus dealing with processes in gases and nonelectrolyte solutions where diffusion phenomena are the only irreversible processes occurring in the medium in question.

As the reference velocity, we now choose the mean volume velocity (4-3.24) of a volume element which, in practice, corresponds to the convection velocity:

$$\mathbf{w} = \sum_k c_k V_k \mathbf{v}_k. \tag{4-17.1}$$

Here c_k denotes the molarity, V_k the partial molar volume, and $\mathbf{v}_k$ the average velocity of species k in the space element to be considered. The summation is over all chemical species. The appropriate diffusion current density of species k, according to Eq. (4-3.27), is

$$_w\mathbf{J}_k = c_k(\mathbf{v}_k - \mathbf{w}). \tag{4-17.2}$$

We call the reference system so determined the *Fick reference system*† for the sake of conciseness.

The relation between the diffusion currents is

$$\sum_k V_k \, _w\mathbf{J}_k = 0,$$ (4–17.3)

as described by Eq. (4–3.28). For N species, there are only $N - 1$ independent diffusion currents.

There also exists a general relation between the N concentration gradients grad c_k. From Eqs. (1–6.11) and (1–6.16) we derive for constant temperature and constant pressure

$$\sum_k V_k \, dc_k = 0$$ (4–17.4)

or

$$\sum_k V_k \, \text{grad } c_k = 0$$ (4–17.5)

by analogy with Eq. (3). There are, accordingly, only $N - 1$ independent concentration gradients for each N species.

(b) Fick's Laws

Experience shows that each of the independent diffusion current densities is a homogeneous linear function of all independent concentration gradients, if extremely high gradients are excluded. Thus we find for N species:

$$_w\mathbf{J}_i = - \sum_{k=2}^N D_{ik} \, \text{grad } c_k \quad (i = 2, 3, \ldots, N),$$ (4–17.6)

where $_w\mathbf{J}_1$ and grad c_1 are considered as dependent quantities. The $(N - 1)^2$ quantities D_{ik} are said to be the *diffusion coefficients*. They do not depend on the concentration gradients but can be arbitrary functions of the local state variables (temperature, pressure, and concentrations).

Inserting Eq. (6) into the relation (4–4.14) following from the local mass balance and excluding chemical reactions yields:

$$\frac{\partial c_i}{\partial t} = \text{div} \left(\sum_{k=2}^N D_{ik} \, \text{grad } c_k \right) - \text{div} \, (c_i \mathbf{w}) \quad (i = 2, 3, \ldots, N),$$ (4–17.7)

where the operator $\partial/\partial t$ denotes a differentiation with respect to time at a fixed position. Most diffusion experiments are so conducted that the last

---

† Fick discovered the basic laws of diffusion (cf. below).

term in Eq. (7) which relates to convection can be neglected. We then obtain for the local rate of increase of the concentration of species i:

$$\frac{\partial c_i}{\partial t} = \text{div} \left(\sum_{k=2}^{N} D_{ik} \text{ grad } c_k \right) \quad (i = 2, 3, \ldots, N), \quad (4-17.8)$$

which is the fundamental equation for the evaluation of the majority of all diffusion experiments[33].

Equations (6) and (8) represent a generalization of the first and second *Fick's law*. Should we wish to obtain the expression for the diffusion current or the rate of increase of the concentration of species 1, then we need only combine Eq. (3) or Eq. (4) with Eq. (6) or Eq. (8) respectively.

We now consider in somewhat more detail the case of *two species* ($N = 2$). First, from Eqs. (6) and (8) there follows with $D_{22} = D$:

$$_w\mathbf{J}_2 = -D \text{ grad } c_2, \quad (4-17.9)$$

$$\frac{\partial c_2}{\partial t} = \text{div} (D \text{ grad } c_2). \quad (4-17.10)$$

The quantity D is now merely the *diffusion coefficient*. Furthermore, from Eqs. (3), (4), and (5) there results:

$$V_1 \, _w\mathbf{J}_1 + V_2 \, _w\mathbf{J}_2 = 0, \quad (4-17.11)$$

$$V_1 \text{ grad } c_1 + V_2 \text{ grad } c_2 = 0, \quad (4-17.12)$$

$$V_1 \frac{\partial c_1}{\partial t} + V_2 \frac{\partial c_2}{\partial t} = 0. \quad (4-17.13)$$

With this we derive from Eqs. (9) and (10):†

$$_w\mathbf{J}_1 = -D \text{ grad } c_1, \quad (4-17.14)$$

$$\frac{\partial c_1}{\partial t} = \text{div} (D \text{ grad } c_1). \quad (4-17.15)$$

Comparison of Eq. (9) or (10) with Eq. (14) or (15) respectively confirms that for two species there is only one diffusion coefficient and shows that the formulation of the fundamental laws is symmetric in both components.

In the special case of the one-dimensional problem (space-coordinate z), we find from the preceding relations that

$$_w\mathbf{J}_i = -D \frac{\partial c_i}{\partial z} \quad (i = 1, 2), \quad (4-17.16)$$

$$\frac{\partial c_i}{\partial t} = \frac{\partial}{\partial z} \left(D \frac{\partial c_i}{\partial z} \right) \quad (i = 1, 2). \quad (4-17.17)$$

---

† Equation (15) results most simply from Eq. (4-4.14) and Eq. (14).

Moreover, if we assume that the diffusion coefficient D is independent of the concentration and thus of the position—an approximation disallowed in most cases—then we obtain from Eq. (17):

$$\frac{\partial c_i}{\partial t} = D \frac{\partial^2 c_i}{\partial z^2} \quad (i = 1, 2). \tag{4-17.18}$$

Equation (16) or (18) is the historic form of the first or second *Fick's law* (Adolf Fick, 1855).

The formalism above can also be applied to electrolyte solutions (§4-18) as well as to solid solutions.

The diffusion coefficient D for binary systems[†] depends on the concentration, on the temperature, and—only to a small degree in condensed phases—on the pressure. It is of the order of magnitude 1 to 10^{-1} cm^2 s^{-1} for gases, 10^{-5} cm^2 s^{-1} for liquids, and 10^{-5} to 10^{-19} cm^2 s^{-1} (smallest measurable value) for solids.

(c) Change of Reference System

For some problems (e.g. in the presence of temperature and pressure gradients), it is suitable to change to other composition variables. Should we want to retain the above definition of the diffusion coefficients then we also change the reference system. Detailed procedure will be given in the following for the special case of two species.

We introduce the diffusion current density referred to any arbitrary reference velocity $\boldsymbol{\omega}$:

$$\mathbf{J}_i = c_i(\mathbf{v}_i - \boldsymbol{\omega}) \quad (i = 1, 2). \tag{4-17.19}$$

According to Eqs. (4-3.2), (4-3.3), and (4-3.4), we find

$$\boldsymbol{\omega} = \omega_1 \mathbf{v}_1 + \omega_2 \mathbf{v}_2, \tag{4-17.20}$$

$$\omega_1 + \omega_2 = 1, \tag{4-17.21}$$

$$\frac{\omega_1}{c_1} \mathbf{J}_1 + \frac{\omega_2}{c_2} \mathbf{J}_2 = 0, \tag{4-17.22}$$

where ω_1 and ω_2 represent the "weight factors" for averaging the velocities, using the normalization (21). For $\boldsymbol{\omega} = \mathbf{w}$, there follows from Eq. (1):

$$\omega_1 = c_1 V_1, \qquad \omega_2 = c_2 V_2. \tag{4-17.23}$$

Because of (1-6.11)

$$c_1 V_1 + c_2 V_2 = 1 \tag{4-17.24}$$

[†] For the question of distinction between "species," "components," etc., in diffusion problems, see §4-18.

condition (21) is satisfied. The identity (11) also immediately results from Eqs. (2), (22), and (23).

We now define in a general way a diffusion coefficient D using the relation†

$$\mathbf{J}_i = - D \frac{1 - \omega_i}{1 - x_i} \frac{1}{\overline{V}} \operatorname{grad} x_i \quad (i = 1, 2), \qquad (4\text{-}17.25)$$

where x_i is the mole fraction of species i and $\overline{V}$ the molar volume. According to Eqs. (4-3.18) and (4-3.19), we find:

$$x_1 + x_2 = 1, \qquad (4\text{-}17.26)$$

$$x_i = c_i \overline{V} \quad (i = 1, 2). \qquad (4\text{-}17.27)$$

From the relation (T = temperature, P = pressure)

$$V_i = \overline{V} + (1 - x_i) \left(\frac{\partial \overline{V}}{\partial x_i} \right)_{T,P} \quad (i = 1, 2), \qquad (4\text{-}17.28)$$

which follows from Eqs. (1-6.9) and (1-6.13), and on consideration of (24), (26), and (27) for T = const, P = const, we obtain

$$\left. \begin{aligned} \operatorname{grad} c_1 &= \frac{V_2}{\overline{V}^2} \operatorname{grad} x_1, \\[2mm] \operatorname{grad} c_2 &= \frac{V_1}{\overline{V}^2} \operatorname{grad} x_2. \end{aligned} \right\} \qquad (4\text{-}17.29)$$

If we substitute in Eq. (25) the quantity $_w\mathbf{J}_i$ for $\mathbf{J}_i$ then, with the help of Eqs. (23), (24), (26), and (27), we find:

$$\left. \begin{aligned} _w\mathbf{J}_1 &= - D \frac{V_2}{\overline{V}^2} \operatorname{grad} x_1, \\[2mm] _w\mathbf{J}_2 &= - D \frac{V_1}{\overline{V}^2} \operatorname{grad} x_2. \end{aligned} \right\} \qquad (4\text{-}17.30)$$

Introducing Eq. (29) into these equations, we derive

$$_w\mathbf{J}_i = - D \operatorname{grad} c_i \quad (i = 1, 2), \qquad (4\text{-}17.31)$$

in agreement with Eqs. (9) and (14). We thus recognize that the quantity D defined by Eq. (25) is identical to the conventional diffusion coefficient of Fick's law. Equation (25) is, accordingly, Fick's law for any arbitrary reference system, while Eqs. (30) and (31) represent Fick's law in the Fick reference system.

---

† This elegant procedure is due to de Groot and coworkers[34].

If the Hittorf reference system (reference velocity v_1) is used, as in the phenomenological equations (conclusion of §4–15) or in the problems of electric conduction (§4–16), then with Eqs. (20), (26), (27), and (29) from Eq. (25) we obtain the relation:

$$_1\mathbf{J}_2 = c_2(\mathbf{v}_2 - \mathbf{v}_1) = -\frac{D}{x_1\bar{V}}\text{grad } x_2 = -\frac{D}{c_1 V_1}\text{grad } c_2. \quad (4\text{–}17.32)$$

If the molality m_2 of species 2 (cf. §4–16) is chosen as composition variable in Eq. (32) in place of the mole fraction x_2 or the molarity c_2 then, considering (M_1 = molar mass of species 1)

$$x_2 = \frac{M_1 m_2}{1 + M_1 m_2}$$

and on consideration of Eq. (27), we obtain the relation:

$$_1\mathbf{J}_2 = -D\frac{c_2}{m_2}\text{grad } m_2. \quad (4\text{–}17.33)$$

Equation (32) or (33) is Fick's law in the Hittorf reference system.

If the mean molar velocity $\mathbf{u}$ is used as a reference velocity, corresponding to the average particle velocity, then with the help of Eqs. (4–3.21) and (4–3.22) from Eq. (25) we find:

$$_u\mathbf{J}_i = c_i(\mathbf{v}_i - \mathbf{u}) = -\frac{D}{\bar{V}}\text{grad } x_i \quad (i = 1, 2). \quad (4\text{–}17.34)$$

This relation represents Fick's law in the molecular reference system.

Finally, if the barycentric velocity $\mathbf{v}$ is introduced, on consideration of Eqs. (4–3.6) to (4–3.15), there results from Eq. (25):

$$\mathbf{J}_i^* = M_i\,{}_v\mathbf{J}_i = \rho_i(\mathbf{v}_i - \mathbf{v}) = -D\rho\text{ grad }\chi_i \quad (i = 1, 2), \quad (4\text{–}17.35)$$

where M_i denotes the molar mass, ρ_i the partial density (mass concentration), χ_i the mass fraction of species i, and ρ the total density. Equation (35) is Fick's law in the barycentric system.

In the presence of temperature and pressure gradients, we choose the mole fraction, the mass fraction, or the molality as the composition variable, since the molarity can change alone due to compressibility and to the thermal expansion of the medium. Accordingly, the simple relations (12) and (29) cease to hold and the different formulations of Fick's law are no longer equivalent. Obviously, now, we designate that part of the matter transport which arises from gradients in the mole fraction as diffusion. The general equation (25) therefore remains intact as a description of diffusion and as the definition of the diffusion coefficient in a binary system and, indeed, in the sense that the term in $\mathbf{J}_i$ which contains grad x_i is written in the same form as the right-hand side of Eq. (25).

If the composition variables are left unaltered and the reference velocity is changed, diffusion coefficients are obtained which are different from the conventional D_{ik} in Eq. (6). The conversion calculations are complicated[35]. Also, the problem of the evaluation of the experimental data for observable volume changes for the mixing of solutions, which is connected with the transfer from the Fick reference system to the measuring cell as the reference system, is quite involved[35, 36].

(d) Thermodynamic–Phenomenological Theory

We now show that the thermodynamics of irreversible processes also leads to Eq. (6) and thus also to the same fundamental laws of diffusion as obtained by the previous purely empirical mode of consideration. Furthermore, with the help of the Onsager reciprocity law we will derive general relations between the diffusion coefficients for systems with three or more species.

First, we restrict the discussion to systems with two species. In this case, Onsager's reciprocity law produces no result, for here the diffusion processes can be described, as proved above, by a single coefficient. Thus there is only a single independent process. According to Eqs. (4–15.10) and (4–15.14), the phenomenological equation for diffusion in a system of two uncharged species states:

$$_1\mathbf{J}_2 = -a(\text{grad } \mu_2)_{T,P}. \tag{4–17.36}$$

Here $_1\mathbf{J}_2$ is the diffusion current density of species 2 in the Hittorf reference system, a $(= a_{22})$ the phenomenological coefficient, and μ_2 the chemical potential of species 2. Now we can write (cf. Eq. 4–15.3)

$$(\text{grad } \mu_2)_{T,P} = \left(\frac{\partial \mu_2}{\partial x_2}\right)_{T,P} \text{grad } x_2. \tag{4–17.37}$$

Comparison of Eq. (32) with Eq. (36) and consideration of Eq. (37) shows that the phenomenological law is identical to Fick's law. Thus we simultaneously obtain the relation between the diffusion coefficient D and the phenomenological coefficient a:

$$D = a\overline{V}x_1\left(\frac{\partial \mu_2}{\partial x_2}\right)_{T,P} = a\overline{V}\frac{x_1^2}{x_2}\left(\frac{\partial \mu_1}{\partial x_1}\right)_{T,P}, \tag{4–17.38}$$

where we have introduced the chemical potential μ_1 of species 1† with the help of Eq. (1–22.34).

---

† In connection with the kinetic theory of diffusion, we use the concept of the "mobility," which we want to denote by the symbol b for binary nonelectrolyte solutions. Then, according to definition, there is valid:

$$\mathbf{v}_2 - \boldsymbol{\omega} = -\frac{b\omega_1}{Lx_1}(\text{grad } \mu_2)_{T,P} = -\frac{b\omega_1}{Lx_1}\left(\frac{\partial \mu_2}{\partial x_2}\right)_{T,P} \text{grad } x_2.$$

[Continued on page 278

The only statement which the thermodynamics of irreversible processes produces about the quantity a follows from the positive-only character of the dissipation function Ψ for irreversible processes. According to Eqs. (4–12.17) and (4–15.14), we have

$$\Psi = a[(\operatorname{grad} \mu_2)_{T,P}]^2 > 0 \qquad (4\text{–}17.39)$$

and thus:

$$a > 0. \qquad (4\text{–}17.40)$$

Otherwise the phenomenological coefficient a—just as the diffusion coefficient D—depends in an arbitrary way on the local state variables (temperature, pressure, and concentration).

(e) Diffusion and Stability

With the help of (1–22.38) and (1–22.40) and from Eqs. (38) and (40) we obtain the following general statements:

$$D > 0 \quad \text{(stable and metastable region within the stability limit)}, \qquad (4\text{–}17.41)$$

$$D < 0 \quad \text{(unstable region)}. \qquad (4\text{–}17.42)$$

[Continued from page 277]

Here L is the Avogadro constant. Accordingly, b actually denotes the quotient of the relative velocity of a single particle of species 2 and the "driving force" $-(\operatorname{grad} \mu_2)_{T,P}/L$, which acts on this particle. (The factor ω_1/x_1 is only added on account of the arbitrariness of the reference system. In the molecular reference system, this factor is equal to unity according to Eq. 4–3.21.) Comparison of the above definition with the relation (25), on consideration of Eqs. (21), (26), and (27), leads to

$$D = \frac{b}{L} x_2 \left(\frac{\partial \mu_2}{\partial x_2}\right)_{T,P}. \qquad (4\text{–}17.38a)$$

This is the general relation between the diffusion coefficient D and the mobility b.

For ideal gas mixtures, ideal liquid mixtures, and ideal dilute solutions, we have

$$x_2 \left(\frac{\partial \mu_2}{\partial x_2}\right)_{T,P} = RT = LkT,$$

where R is the gas constant and k the Boltzmann constant. Thus we obtain from Eq. (38a) for this special case:

$$D = bkT, \qquad (4\text{–}17.38b)$$

the so-called "Fokker–Einstein Equation," which appeared for the first time in the theory of brownian motion in highly dilute solutions.

If we compare Eq. (38) to Eq. (38a) then we find the general relation between the phenomenological coefficient a and the mobility b:

$$a = \frac{bx_2}{L\bar{V}x_1}, \qquad (4\text{–}17.38c)$$

that is, a very simple relation.

The mobility b is, in contrast to the ion mobility for electrolyte solutions (§4–16), not an independently measurable quantity. Rather it can be determined by experimental means only from the diffusion coefficient D from Eq. (38a). Equally, the phenomenological coefficient a is obtained from Eq. (38) or (38c).

The inequality (41) corresponds to experience and is almost self-evident for stable regions. The inequality (42) states that in the unstable region—which cannot be realized for a fluid—the diffusion current has the same direction as the concentration gradient, i.e. it leads to a local phase separation. This statement is also reasonable. The separation which appears in the metastable region on lifting the restrictions is not contained in our description since it depends on discontinuous changes. Compared to neighboring compositions, a metastable region is stable. Thus for such a region the inequality (41) holds.

A special situation is present at a critical point or at any point on the stability limit. Here, on the basis of Eqs. (1–22.34), (1–22.39), and (1–22.41), we find:

$$\left(\frac{\partial \mu_2}{\partial x_2}\right)_{T,P} = 0, \tag{4–17.43}$$

$$\left(\frac{\partial \mu_1}{\partial x_1}\right)_{T,P} = 0. \tag{4–17.44}$$

By a "diffusion process at the critical point" we must imagine the following situation: the average composition of a volume element at the critical temperature under the given pressure corresponds exactly to the critical concentration, although a concentration gradient is still present. An analogous statement holds for any point on the stability limit.

Since the stability limit divides the stable or metastable region from the unstable region and since the critical point lies on the stability limit, the inequalities (41) and (42) require

$$D = 0 \quad \text{(stability limit including the critical state)}. \tag{4–17.45}$$

From Eqs. (38), (43), and (45) we see that, at the critical point, the phenomenological coefficient a must either remain finite or become infinite with a smaller order than the reciprocal value of $\partial \mu_2 / \partial x_2$. The actual behavior can be experimentally proved with the use of Eq. (38) if sufficiently accurate values of D and $\partial \mu_2 / \partial x_2$ are known near a critical point. It turns out that a remains finite and nonzero.†

Equation (45) has been experimentally verified for the case of the critical phase separation of binary liquid systems. This is shown in Fig. 11.‡

---

† This is concluded, say, by comparison of Eq. (38) with Eq. (4–20.28). Since with this the phenomenological coefficient a can be related back to the sedimentation coefficient s and since it is highly improbable that s should be infinite at the critical point, it thus results that a also will always remain finite. Experiments[37] show indeed that a, b, and s remain finite and nonzero.

‡ For a more recent investigation of the diffusion coefficient and related quantities in binary liquid systems near upper or lower consolute points, see Haase and Siry[37].

Finally, there follows from Eq. (45) that the very common representation of the temperature dependence of the diffusion coefficient [31]

$$D = D_0 e^{-A/RT}, \qquad (4\text{–}17.46)$$

where R is the gas constant, D_0 and A are characteristic constants at a given pressure and given composition applicable to the binary mixture in question (A = "activation energy"), cannot hold generally since, for a mixture which has exactly the critical composition at constant pressure, D must disappear at the critical temperature (see Fig. 11). This is not expressed in Eq. (46).

The statements (41), (42), and (45) are due in principle to Dehlinger [39, 40] and Becker[41]. The more rigorous derivations given here were suggested by Onsager[42] and given in detail by the author[43a].

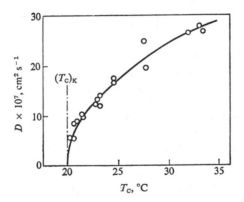

Fig. 11. Diffusion coefficient D as a function of the Celsius temperature T_C for an equimolar liquid mixture of n-hexane and nitrobenzene. According to measurements of Claesson and Sundelof[38]. The critical mixing temperature $(T_C)_K$ is $19.9 \pm 0.1°C$; the critical composition corresponds approximately to the equimolar mixture.

(f) Relations between Diffusion Coefficients of Multi-component Systems

On consideration of Eqs. (4–15.3) and (4–15.14) for diffusion in a system of N species, the phenomenological equations (4–15.10) produce

$$
{}_1\mathbf{J}_i = -\sum_{j=2}^{N} a_{ij}(\text{grad } \mu_j)_{T,P}
$$

$$
= -\sum_{j=2}^{N}\sum_{l=2}^{N} a_{ij}\, \mu_{jl}\, \text{grad } c_l \quad (i = 2, 3, \ldots, N) \qquad (4\text{–}17.47)
$$

with

$$\mu_{jl} \equiv \left(\frac{\partial \mu_j}{\partial c_l}\right)_{T,P,c_m} \quad (j, l, m = 2, 3, \ldots, N; m \neq l). \quad (4\text{-}17.48)$$

Here $_1\mathbf{J}_i$ is the diffusion current density of species i in the Hittorf reference system, μ_j the chemical potential of species j, and the a_{ij} are the phenomenological coefficients.

The generalized form of Fick's law (6) states:

$$_w\mathbf{J}_k = -\sum_{l=2}^{N} D_{kl} \operatorname{grad} c_l \quad (k = 2, 3, \ldots, N). \quad (4\text{-}17.49)$$

Here $_w\mathbf{J}_k$ is the diffusion current density of species k in the Fick reference system and the D_{kl} are the diffusion coefficients numbering to $(N-1)^2$.

For the conversion from the Fick to the Hittorf reference system, following Eq. (4–3.41) and Eq. (3) we obtain

$$_1\mathbf{J}_i = \sum_{k=2}^{N} \left(\delta_{ik} + \frac{c_i V_k}{c_1 V_1}\right)_w\mathbf{J}_k \quad (i = 2, 3, \ldots, N). \quad (4\text{-}17.50)$$

Here δ_{ik} denotes the Kronecker delta ($\delta_{ik} = 1$ for $i = k$, $\delta_{ik} = 0$ for $i \neq k$).

With the abbreviation

$$\varepsilon_{ik} \equiv \delta_{ik} + \frac{c_i V_k}{c_1 V_1} \quad (4\text{-}17.51)$$

there follows from Eqs. (49) and (50):

$$_1\mathbf{J}_i = -\sum_{k=2}^{N} \sum_{l=2}^{N} \varepsilon_{ik} D_{kl} \operatorname{grad} c_l \quad (i = 2, 3, \ldots, N). \quad (4\text{-}17.52)$$

By comparison of Eq. (47) with Eq. (52) it is obvious that the phenomenological equations lead to the generalized form of Fick's law. Thus, on comparison of the coefficients in (47) and (52), the relation between the phenomenological coefficients a_{ij} and the diffusion coefficients D_{kl} is obtained:

$$\sum_{j=2}^{N} a_{ij}\mu_{jl} = \sum_{k=2}^{N} \varepsilon_{ik} D_{kl} \quad (i, l = 2, 3, \ldots, N). \quad (4\text{-}17.53)$$

If we rewrite this relation in the form

$$\sum_{j=2}^{N} a_{ij}\mu_{jm} = \sum_{k=2}^{N} \varepsilon_{ik} D_{km} \quad (i, m = 2, 3, \ldots, N) \quad -17.54)$$

and multiply Eq. (53) by μ_{im}, Eq. (54) by μ_{il}, and sum over the index i then we find:

$$\sum_{i=2}^{N}\sum_{j=2}^{N} a_{ij}\mu_{jl}\mu_{im} = \sum_{i=2}^{N}\sum_{k=2}^{N} \varepsilon_{ik}D_{kl}\mu_{im}, \tag{4–17.55}$$

$$\sum_{i=2}^{N}\sum_{j=2}^{N} a_{ij}\mu_{jm}\mu_{il} = \sum_{i=2}^{N}\sum_{k=2}^{N} \varepsilon_{ik}D_{km}\mu_{il}, \tag{4–17.56}$$

and, after exchanging the summation index on the left-hand side of Eq. (56),

$$\sum_{i=2}^{N}\sum_{j=2}^{N} a_{ji}\mu_{im}\mu_{jl} = \sum_{i=2}^{N}\sum_{k=2}^{N} \varepsilon_{ik}D_{km}\mu_{il}. \tag{4–17.57}$$

At this point we require Onsager's reciprocity law (4–14.4):

$$a_{ij} = a_{ji} \quad (i, j = 2, 3, \ldots, N). \tag{4–17.58}$$

From Eqs. (57) and (58) it follows immediately that

$$\sum_{i=2}^{N}\sum_{j=2}^{N} a_{ij}\mu_{im}\mu_{jl} = \sum_{i=2}^{N}\sum_{k=2}^{N} \varepsilon_{ik}D_{km}\mu_{il}. \tag{4–17.59}$$

If we compare Eq. (55) with Eq. (59), we then gain an expression in measurable quantities:

$$\sum_{i=2}^{N}\sum_{k=2}^{N} \varepsilon_{ik}\mu_{im}D_{kl} = \sum_{i=2}^{N}\sum_{k=2}^{N} \varepsilon_{ik}\mu_{il}D_{km} \quad (l, m = 2, 3, \ldots, N). \tag{4–17.60}$$

This is the precise form[44] of the relation between the diffusion coefficients of a multi-component system derived by Onsager[42]. A formula given previously by Meixner[45a] holds for ideal gas mixtures.

According to Eq. (60), there exist between the $(N - 1)^2$ diffusion coefficients $\frac{1}{2}(N - 1)(N - 2)$ independent linear relations corresponding to the same number of reciprocity relations of the form (58). With this the number of the independent diffusion coefficients is reduced to $\frac{1}{2}N(N - 1)$. For 3 species, there are thus only 3 independent diffusion coefficients. In the next section, we will show that, for a suitable definition of diffusion coefficients, Eq. (60) is valid for electrolyte solutions, too, when N denotes the number of independently moving substances.

Taking Eq. (60) for the case of $N = 3$ we have

$$(\varepsilon_{22}\mu_{2m} + \varepsilon_{32}\mu_{3m})D_{2l} + (\varepsilon_{23}\mu_{2m} + \varepsilon_{33}\mu_{3m})D_{3l}$$
$$= (\varepsilon_{22}\mu_{2l} + \varepsilon_{32}\mu_{3l})D_{2m} + (\varepsilon_{23}\mu_{2l} + \varepsilon_{33}\mu_{3l})D_{3m} \quad (l, m = 2, 3).$$

For $l = m$, this equation is an identity. In the cases $l = 2$, $m = 3$ and $l = 3$, $m = 2$, it produces the same result. Thus we are left with a single relation:

$$(\varepsilon_{22}\mu_{23} + \varepsilon_{32}\mu_{33})D_{22} + (\varepsilon_{23}\mu_{23} + \varepsilon_{33}\mu_{33})D_{32}$$
$$= (\varepsilon_{22}\mu_{22} + \varepsilon_{32}\mu_{32})D_{23} + (\varepsilon_{23}\mu_{22} + \varepsilon_{33}\mu_{32})D_{33}.$$

If we consider Eq. (51) then we obtain the explicit form of this formula:†

$$\alpha_{22}D_{22} + \alpha_{32}D_{32} = \alpha_{23}D_{23} + \alpha_{33}D_{33} \qquad (4\text{-}17.61)$$

with

$$\alpha_{22} \equiv \left(1 + \frac{c_2 V_2}{c_1 V_1}\right)\mu_{23} + \frac{c_3 V_2}{c_1 V_1}\mu_{33}, \qquad (4\text{-}17.62)$$

$$\alpha_{32} \equiv \frac{c_2 V_3}{c_1 V_1}\mu_{23} + \left(1 + \frac{c_3 V_3}{c_1 V_1}\right)\mu_{33}, \qquad (4\text{-}17.63)$$

$$\alpha_{23} \equiv \left(1 + \frac{c_2 V_2}{c_1 V_1}\right)\mu_{22} + \frac{c_3 V_2}{c_1 V_1}\mu_{32}, \qquad (4\text{-}17.64)$$

$$\alpha_{33} \equiv \frac{c_2 V_3}{c_1 V_1}\mu_{22} + \left(1 + \frac{c_3 V_3}{c_1 V_1}\right)\mu_{32}. \qquad (4\text{-}17.65)$$

Gosting and others[46–53] as well as Miller[54–55] have verified Eq. (61) using experimental data on diffusion in ternary liquid systems and thus have verified Onsager's reciprocity law for this case. Since almost all systems contain electrolytes, we defer the more detailed treatment to §4–18, where we will present numerical values (Table 7, p. 292). Here we need only mention that the diffusion coefficients D_{23} and D_{32} are usually small compared to D_{22} and D_{33} and always vanish for $c_2 \to 0$ or $c_3 \to 0$.‡ Also, D_{23} and D_{32} can—in contrast to D_{22} and D_{33}—assume negative values. (For other fundamental questions of diffusion in ternary systems, see Refs. [56–58].)

For ternary ideal gas mixtures:

$$V_1 = V_2 = V_3 = \frac{RT}{P}, \qquad (4\text{-}17.66)$$

$$\mu_{23} = \mu_{32} = 0, \qquad c_2\mu_{22} = c_3\mu_{33} = RT. \qquad (4\text{-}17.67)$$

From this with Eqs. (61) to (65) there follows:

$$D_{22} + \left(1 + \frac{c_1}{c_3}\right)D_{32} = \left(1 + \frac{c_1}{c_2}\right)D_{23} + D_{33} \qquad (4\text{-}17.68)$$

or with Eqs. (4–3.18) and (4–3.19):

$$x_2 x_3(D_{22} - D_{33}) = x_3(1 - x_3)D_{23} - x_2(1 - x_2)D_{32}, \qquad (4\text{-}17.69)$$

---

† The additional relation $\alpha_{22} = \alpha_{33}$ can be derived from classical thermodynamics.
‡ From Eq. (2) it follows that: $_w\mathbf{J}_i \to 0$ for $c_i \to 0$, that is, with Eq. (6): $D_{ik} \to 0$ for $c_i \to 0$ ($i \neq k$, $c_k \neq 0$).

where x_i is the mole fraction of component i. This very simple relation[5a] does not appear to have been experimentally verified yet. We recognize from Eqs. (68) and (69) that the assumption $D_{23} = D_{32} = 0$ implies $D_{22} = D_{33}$ and thus cannot be generally correct.

As seen from Eq. (61) with $\alpha_{22} = \alpha_{33}$, the assumption

$$D_{23} = D_{32} = 0$$

leads *generally* to the statement $D_{22} = D_{33} = D$, so that the diffusion processes in a ternary system can be described by a single diffusion coefficient D. Such simple conditions are found only for types of molecules which are very similar to one another, i.e. for ideal ternary mixtures. Since it is known that in binary ideal liquid systems the diffusion coefficient is a linear function of the mole fraction,† we would also expect a linear dependence of the diffusion coefficient D on the mole fractions x_1, x_2, x_3 in ternary ideal liquid mixtures. In fact, measurements[60–62] show that, in the nearly ideal liquid system toluene (1)–chlorobenzene (2)–bromobenzene (3), all expected simplifications hold within the accuracy of the measurements. Thus, for the diffusion coefficient at 29.6°C:

$$\frac{D}{10^{-5} \text{ cm}^2 \text{ s}^{-1}} = 2.36x_1 + 1.84x_2 + 1.30x_3.$$

4-18 DIFFUSION IN ELECTROLYTE SOLUTIONS

(a) General

Consider an electrolyte solution at fixed values of temperature and pressure in which diffusion processes excluding external force fields may occur. Since the dissociation equilibrium is set up very rapidly in comparison with equalization of concentrations by diffusion, we can everywhere assure local chemical equilibrium with respect to dissociation of the various electrolytes. Moreover, electric neutrality must be satisfied at each position. Thus as long as we are not dealing with the problem of the "diffusion potential" (see below) the migrating units will not be chosen to be the ionic and molecular species present but the "independently moving substances" (electrolytes and nonelectrolytes). We shall see later that these are not necessarily identical to the "components" in the sense of the phase rule. If we characterize one substance present in excess or otherwise distinguishable (normally the "solvent") by an inferior 1 and the other independently moving substances by inferiors 2, 3, . . . , N, then the reference velocity, the diffusion currents, and the concentration gradients are defined as in Eqs. (4–17.1) to (4–17.5), and Fick's

---

† Other authors, however, maintain that the quantity $D\eta/T$ (η = viscosity) is a linear function of the mole fraction, in a binary ideal liquid system such as water–glycol between 25°C and 70°C[59].

laws (4–17.6) to (4–17.18) are valid. The number of independent diffusion current densities or concentration gradients is then, according to §4–17, $N - 1$, and the number of diffusion coefficients is $(N - 1)^2$.

In order to transform the description of the diffusion processes using chemical species (§4–17) into a representation in terms of "independently moving substances," we first consider the simplest case: the solution of a single electrolyte (component 2) with two ionic species in a single neutral solvent (component 1). This corresponds to a system of four chemical species. There are, thus: solvent molecules, undissociated electrolyte molecules (indicated by inferior u), cations (indicated by inferior $+$), and anions (indicated by inferior $-$). The condition of electric neutrality (4–16.31) states:

$$z_+ c_+ + z_- c_- = 0, \tag{4–18.1}$$

where z_i or c_i denotes the charge number or molarity of species i respectively. To describe the composition of the solution we require one concentration variable. We choose the molarity c_2 of the electrolyte:

$$c_2 = \frac{c_u}{1 - \alpha} = \frac{c_+}{\nu_+ \alpha} = \frac{c_-}{\nu_- \alpha}. \tag{4–18.2}$$

Here α denotes the degree of dissociation of the electrolyte and ν_i the dissociation number of ionic species i (cf. Eq. 1–19.13). Since, finally, no electric current flows, according to Eq. (4–3.47) we find

$$z_+ c_+ \mathbf{v}_+ + z_- c_- \mathbf{v}_- = 0.$$

Here $\mathbf{v}_i$ is the average velocity of species i. With Eq. (1) there results:

$$\mathbf{v}_+ = \mathbf{v}_- \equiv \mathbf{v}_2.$$

The quantity $\mathbf{v}_2$ denotes the common velocity of the cations and anions. The undissociated electrolyte molecules move with the same velocity, since the assumed condition of local dissociation equilibrium covers the fact that at each position (i.e. at each composition) the equilibrium concentration of the undissociated electrolyte belonging to the appropriate ion concentrations is present.† We thus obtain:

$$\mathbf{v}_+ = \mathbf{v}_- = \mathbf{v}_u = \mathbf{v}_2. \tag{4–18.3}$$

---

† If the neutral molecules moved ahead during the diffusion, ions would immediately be formed at the position in question corresponding to the local dissociation equilibrium. If the neutral molecules lagged behind the ions then there would again simultaneously arise neutral molecules in the ion front which is moving ahead. The diffusion velocity $\mathbf{v}_u$ of the neutral electrolyte must therefore be equal everywhere to the migration velocity $\mathbf{v}_2$ of the ions.

Accordingly, we can characterize the migration of the electrolyte by a single velocity v_2. We therefore introduce the quantity

$$\mathbf{J}_2 \equiv c_2(\mathbf{v}_2 - \boldsymbol{\omega})$$

as the "diffusion current density of the electrolyte" wherein $\boldsymbol{\omega}$ denotes an arbitrary reference velocity. According to Eq. (4–3.1), the diffusion current density of the individual species is found to be

$$\mathbf{J}_+ = c_+(\mathbf{v}_+ - \boldsymbol{\omega}),$$
$$\mathbf{J}_- = c_-(\mathbf{v}_- - \boldsymbol{\omega}),$$
$$\mathbf{J}_u = c_u(\mathbf{v}_u - \boldsymbol{\omega}).$$

From this with Eqs. (2) and (3) there follows ($\nu \equiv \nu_+ + \nu_-$):

$$\mathbf{J}_+ = \nu_+ \alpha \mathbf{J}_2, \quad \mathbf{J}_- = \nu_- \alpha \mathbf{J}_2, \quad \mathbf{J}_u = (1 - \alpha)\mathbf{J}_2, \\ \nu \mathbf{J}_u + \mathbf{J}_+ + \mathbf{J}_- = \nu \mathbf{J}_2. \Bigg\} \qquad (4\text{–}18.4)$$

This relation holds, as is obvious from its derivation, for quite arbitrary transport processes in electrolyte solutions with two ion types so long as no electric current flows. It can thus be applied to sedimentation (§4–20) and to thermal diffusion (§4–27).

If we set $\boldsymbol{\omega} = \mathbf{w}$ (mean volume velocity) and accordingly $\mathbf{J}_2 = {}_w\mathbf{J}_2$ (diffusion current density of the electrolyte in the Fick reference system) then, for the diffusion, we can write Fick's law in the original form (4–17.9):

$$_w\mathbf{J}_2 = -D \text{ grad } c_2. \qquad (4\text{–}18.5)$$

Here D is the diffusion coefficient. We here have one independently migrating electrolyte, two independently moving substances (the electrolyte and the solvent), and, accordingly, only one independent diffusion current, or concentration gradient, or diffusion coefficient. The number of the independently moving substances in this case corresponds to the number of components in the sense of the phase rule.

The conditions for more than two ion species are more complicated. The number of independently moving substances (N) need no longer be identical to the number of components (N') in the sense of the phase rule. We now consider a few examples for systems with three or more ionic species.

The simplest example is a system which consists of one neutral solvent and one electrolyte with three ion species, e.g. water–sulfuric acid. Here there are five chemical species: H_2O, H^+, HSO_4^-, SO_4^{2-}, and H_2SO_4. The assumption of local dissociation equilibrium for the reactions

$$H_2SO_4 \rightleftharpoons H^+ + HSO_4^-$$
$$HSO_4^- \rightleftharpoons H^+ + SO_4^{2-}$$

as well as the condition of electric neutrality (4-3.46) or the condition for the disappearance of the total electric current (4-3.47) represent three independent relations between the concentrations or velocities of the individual species. Thus there remain two independently moving substances ($N = 2$) which can be chosen to be H_2O and H_2SO_4. Also, in the sense of the phase rule, there are two components ($N' = 2$).

The next example relates to a system of one neutral solvent and one electrolyte with four kinds of ions, e.g. water–potassium bisulfate. Here we have six species (H_2O, K^+, H^+, HSO_4^-, SO_4^{2-}, $KHSO_4$), two local equilibrium conditions and one electric neutrality condition or condition for the disappearance of the total current. Thus we obtain three independently moving substances ($N = 3$), e.g. H_2O, $KHSO_4$, and K_2SO_4. Now, however, the number of components, in the sense of the phase rule, is $N' = 2$. For equilibrium problems (i.e. homogeneous solutions), the further condition that $c_{H^+} = c_{SO_4^{2-}}$ is added to the above-mentioned relations and this results from the nature of the system (solution of potassium bisulfate in water).

For the systems

$$H_2O + NaCl + KCl \tag{A}$$

and

$$H_2O + NaCl + KBr \tag{B}$$

the number of components must obviously be $N' = 3$ in the sense of the phase rule. For diffusion problems, however, system (B) behaves differently from system (A). The electric neutrality condition or the condition of vanishing total current produces two independent ion concentrations or ion velocities in case (A) but three in case (B). Accordingly, we can characterize system (A) by three independently moving substances (H_2O, NaCl, and KCl) ($N = 3$) while for system (B) we must introduce four independently moving substances ($N = 4$) (e.g. H_2O, NaCl, KCl, and KBr). The additional condition which reduces the number of components for the equilibrium problem in case (B) to three, but which is invalid for diffusion problems, results from the manner of making up a homogeneous system (B). Indeed, the condition

$$c_{Na^+} + c_{NaCl} + c_{NaBr} = c_{Cl^-} + c_{NaCl} + c_{KCl}$$

must be valid, in which c_{NaCl}, c_{NaBr}, and c_{KCl} denote the molarities of the undissociated electrolytes. This relation expresses the fact that the concentrations of ion constituents Na and Cl in the case of a homogeneous solution are equal. The equality of the concentrations of the other ion constituents K and Br follows automatically from electric neutrality.

In the most general case, a fluid mixture contains nonelectrolyte molecules, undissociated electrolyte molecules, and ions. Chemical reactions, in which ions take part, always occur very rapidly in comparison with diffusion. Local chemical equilibrium can therefore be assumed. For nonelectrolyte molecules, the concentration changes by chemical reactions in any one volume element can either be comparable to or substantially larger or smaller than the concentration changes caused by diffusion. In the first case, no simple general statement can be made. In the second case, we must either assume local chemical equilibrium or completely inhibited reactions. Only these extreme cases shall be considered here.

If there are K' molecular species from nonelectrolytes in the mixture and if there are R independent (rapid) reactions possible between them then we have $K' - R = K$ independently migrating nonelectrolytes. Furthermore, let there be n ion species and n' types of undissociated electrolyte molecules. The requirement of electric neutrality produces first a reduction in the number of independent concentrations or velocities by one. Furthermore, we have n' local equilibrium conditions for dissociation reactions in which undissociated electrolyte molecules participate. Finally, there can still be a number B of equations of condition which describe the local equilibrium for ion reactions of the type $HSO_4^- \rightleftharpoons H^+ + SO_4^{2-}$. With this the number of independent concentrations or velocities is decreased by $n' + B$. Fast reactions between electrolytes and nonelectrolytes (solvation, etc.) need not be considered, since to each newly formed species (solvate complex, etc.) there belongs a new equation corresponding to the local chemical equilibrium condition. We thus find, for the number N of the independently moving substances,[32]

$$N = K + n - B - 1. \tag{4–18.6}$$

Since N is, from the above, equal to the number of independent velocities, the number of independent diffusion fluxes or concentration gradients is $N - 1$, according to Eq. (4–17.3) or (4–17.5) respectively. We will thus introduce $(N - 1)^2$ diffusion coefficients using Eq. (4–17.6) in this general case.

A mixture of one solvent which produces a single independently moving substance (e.g. H_2O), with any number of electrolytes (n ion species) corresponds to the "normal case" ($K = 1$). Here, according to Eq. (6), $N = n - B$.

For molten salts, $K = 0$ and we obtain from Eq. (6): $N = n - B - 1$. Thus, as an example, in the case of the liquid mixture (molten salt) $NaCl + KCl$ there are three ionic species but because $B = 0$ there are only two independently moving substances ($N = 2$); in the mixture $NaCl + KCl + KBr$ we have four ionic species and thus three independently moving substances ($N = 3$); in the mixture $NaCl + KBr$, however, there are four ionic species and three independently moving substances ($N = 3$), say, $NaCl$, KCl, and KBr.

The above considerations are valid not only for diffusion but also for other transport processes insofar as they occur without electric current, that is, for sedimentation (§4–20) or for thermal diffusion (§4–27).

(b) Thermodynamic–Phenomenological Theory

The formula (4–12.17) for the dissipation function is simplified in the present case to (cf. the conclusion of §4–15):

$$\Psi = \sum_{i=2}^{Z} {}_1\mathbf{J}_i\mathbf{X}_i. \tag{4–18.7}$$

Here Z denotes the total number of species, ${}_1\mathbf{J}_i$ the diffusion current density of species i in the Hittorf reference system, and $\mathbf{X}_i$ the appropriate "generalized force," which is given by Eq. (4–15.11),

$$\mathbf{X}_i = -\operatorname{grad}\mu_i - z_iF\operatorname{grad}\varphi, \tag{4–18.8}$$

where μ_i is the chemical potential of species i, z_i the charge number of the particle type i, F the Faraday constant, and φ the electric potential. Accordingly, the phenomenological equations are in agreement with Eq. (4–15.10) and state:

$$_1\mathbf{J}_i = -\sum_{k=2}^{Z} a_{ik}(\operatorname{grad}\mu_k + z_kF\operatorname{grad}\varphi) \quad (i = 2, 3, \ldots, Z). \tag{4–18.9}$$

The a_{ik} are the phenomenological coefficients for which the Onsager reciprocity law (4–14.4)

$$a_{ik} = a_{ki} \quad (i, k = 2, 3, \ldots, Z) \tag{4–18.10}$$

holds. Equation (4–16.43) is a special case of Eq. (9) for $\operatorname{grad}\mu_k = 0$.

If we wish to deal with the problems of the diffusion potential or with the relations between the diffusion coefficients and ion mobilities, we proceed from Eqs. (9) and (10), since these formulas also describe electric conduction (§4–16). If, however, we are dealing with diffusion only then it is expedient to choose another method of presentation, one used in §4–15 and which we now present in more detail.

First, the condition for the disappearance of the total electric current is obtained, according to Eq. (4–15.3),

$$\sum_{i=2}^{Z} z_i\,{}_1\mathbf{J}_i = 0. \tag{4–18.11}$$

By insertion of Eq. (8) and Eq. (11) in Eq. (7) there follows immediately:

$$\Psi = -\sum_{i=2}^{Z} {}_1\mathbf{J}_i\operatorname{grad}\mu_i. \tag{4–18.12}$$

On further application of Eq. (11) we can eliminate one more term in Eq. (12), e.g. the term with $_1\mathbf{J}_Z$, where $z_Z \neq 0$; the species Z should thus be an ion. We then obtain:

$$\Psi = -\sum_{i=2}^{Z-1} {}_1\mathbf{J}_i\left(\operatorname{grad} \mu_i - \frac{z_i}{z_Z} \operatorname{grad} \mu_Z\right).\qquad(4\text{–}18.13)$$

We can now introduce the idea of "independently moving substances" defined previously. The number of these amounts to N. In this way, Eq. (13) assumes the short form

$$\Psi = -\sum_{k=2}^{N} {}_1\mathbf{J}_k \operatorname{grad} \mu_k,\qquad(4\text{–}18.14)$$

where the inferior k denotes either a nonelectrolytic component or an electrolyte as a whole. If only a single neutral solvent is present ("normal case") then the summation in Eq. (14) contains only terms relating to independently migrating electrolytes, since the solvent is always chosen as substance 1.

The direct proof of the equivalence of Eqs. (13) and (14) is complicated[63]. We will therefore restrict the discussion to the simplest case. However, the general validity of Eq. (14) is apparent without needing a calculated proof if we consider that the independently moving substances can be introduced in place of the individual chemical species. Thus, instead of Eq. (7), Eq. (14) results immediately. This corresponds to the process known from classical thermodynamics in which the components of the region in question are introduced in place of the individual species in the Gibbs equation, provided that electrical processes and chemical reactions are not under consideration.

We now take an illustration of the direct transfer from Eq. (13) to Eq. (14). The solution of an electrolyte (component 2) consisting of cations (indicated by an inferior $+$), anions (indicated by an inferior $-$), and undissociated electrolyte molecules (indicated by an inferior u) in a neutral solvent (component 1 or species 1) is considered. If we choose the anions as species Z then we obtain from Eq. (13):

$$\Psi = -{}_1\mathbf{J}_u \operatorname{grad} \mu_u - {}_1\mathbf{J}_+\left(\operatorname{grad} \mu_+ - \frac{z_+}{z_-} \operatorname{grad} \mu_-\right).\qquad(4\text{–}18.15)$$

We have according to Eq. (4):

$$_1\mathbf{J}_u = (1 - \alpha)\,{}_1\mathbf{J}_2, \qquad {}_1\mathbf{J}_+ = \alpha\nu_+\,{}_1\mathbf{J}_2,\qquad(4\text{–}18.16)$$

where $_1\mathbf{J}_2$ is the diffusion current density of the electrolyte in the Hittorf reference system. The condition of local dissociation equilibrium yields the relation (cf. Eqs. 1–19.19, 1–19.20, and 1–21.16):

$$\nu_+\mu_+ + \nu_-\mu_- = \mu_u = \mu_2.\qquad(4\text{–}18.17)$$

With the help of the condition for electric neutrality (1–19.8)

$$z_+ \nu_+ + z_- \nu_- = 0 \qquad (4\text{–}18.18)$$

we obtain from Eqs. (15) to (17)

$$\Psi = - {}_1\mathbf{J}_2 \, \text{grad} \, \mu_2. \qquad (4\text{–}18.19)$$

The dissipation function thus has the form of (14).

On the basis of Eq. (19) we can set up a phenomenological relation

$$_1\mathbf{J}_2 = -a \, \text{grad} \, \mu_2 \qquad (4\text{–}18.20)$$

which corresponds completely to Eq. (4–17.36) and thus also leads to a relation analogous to Eq. (4–17.38) between the phenomenological coefficient a and the diffusion coefficient D defined by Eq. (5).

In general, by proceeding from Eq. (14) we can write the phenomenological equations in the form (4–17.47). Since the diffusion coefficients were generally defined by Eq. (4–17.49), all the relations of subsection (f) in §4–17 can be applied to liquid mixtures with electrolytes, provided that the number of independently moving substances is designated as N. Accordingly, the important relation (4–17.60) which follows from the Onsager reciprocity law and which for $N = 3$ is reduced to Eq. (4–17.61) holds for the diffusion coefficients.

As an example of diffusion among three independently moving substances, we consider the system

$$\text{H}_2\text{O} \ (1) + \text{Na}_2\text{SO}_4 \ (2) + \text{H}_2\text{SO}_4 \ (3)$$

Here the four diffusion coefficients D_{22}, D_{23}, D_{32}, and D_{33} for four compositions at 25°C have been experimentally determined (Table 7) and the relation (4–17.61) has been verified within the accuracy of the measurements[51]. In the present system,† the values of D_{23} and D_{32} are exceptionally large.‡

(c) Diffusion Potential

During the diffusion of electrolytes, an electric potential arises within the solution due to the different mobilities of the individual ion species.

---

† For the sake of clarity, let the system of equations (4–17.49) be again written down explicitly for $N = 3$:

$$_w\mathbf{J}_2 = -D_{22} \, \text{grad} \, c_2 - D_{23} \, \text{grad} \, c_3,$$

$$_w\mathbf{J}_3 = -D_{32} \, \text{grad} \, c_2 - D_{33} \, \text{grad} \, c_3.$$

Here c_i or $_w\mathbf{J}_i$ denotes the molarity or the diffusion current density (in the Fick reference system) of the independently moving substance i.

‡ Many other papers referring to ternary systems containing electrolytes have been quoted in §4–17 (p. 283).

This will be called the *diffusion potential*. This phenomena is exemplified most simply in a solution of a single electrolyte with two ionic species in a neutral solvent. The two ion species (cations and anions) move in the presence of a concentration gradient with velocities in the same direction. The internal

Table 7

Diffusion Coefficients D_{22}, D_{33}, D_{23}, and D_{32} in the System $H_2O(1)$–$Na_2SO_4(2)$–$H_2SO_4(3)$ at $25°C$ for four Compositions. After Wendt[51]

Molarities mol l^{-1}	D_{22}	D_{33}	D_{23}	D_{32}
	10^{-5} cm^2 s^{-1}			
$c_2 = 0.5$; $c_3 = 0.5$	1.09_{32}	2.73_{17}	-0.55_{01}	-0.62_{01}
$c_2 = 1$; $c_3 = 0.5$	0.84_{84}	2.53_{84}	-0.60_{45}	-0.31_{23}
$c_2 = 0.5$; $c_3 = 1$	1.04_{20}	2.50_{78}	-0.33_{38}	-0.50_{87}
$c_2 = 1$; $c_3 = 1$	0.90_{55}	2.61_{22}	-0.51_{19}	-0.39_{69}

electric field arising therefrom retards the more rapid ions and accelerates the slower ions until cations and anions migrate at equal rates due to the condition of electric neutrality (see Eq. 3). The electrostatic field present is the microscopic cause of the diffusion potential. For more than two ionic species, a diffusion potential must also arise and the number of independent ion velocities will be diminished by one by the electric neutrality condition (see above). The different ionic species need not move with the same velocity in this general case.

The general expression for the diffusion potential arises from the phenomenological equations (9) and condition (11) for the disappearance of the total electric current. By combination of the two equations we obtain

$$\text{grad } \varphi = -\frac{\sum_i \sum_k a_{ik} z_i \text{ grad } \mu_k}{F \sum_i \sum_k a_{ik} z_i z_k}, \tag{4–18.21}$$

where the summations are over all (charged and uncharged) species with the exception of species 1. According to Eq. (4–16.45), the reduced transport number of particle type k is found to be

$$\tau_k = \frac{\sum_i a_{ki} z_i}{\sum_i \sum_k a_{ik} z_i z_k}. \tag{4–18.22}$$

On the basis of Onsager's reciprocity law (10)

$$a_{ik} = a_{ki}$$

we recover the expression (22) in Eq. (21). It thus follows that if we set $\varphi = \varphi_D$:

$$F \text{ grad } \varphi_D = -\sum_k \tau_k \text{ grad } \mu_k. \qquad (4\text{--}18.23)$$

This general formula for the diffusion potential φ_D corresponds to Eq. (3–8.25) for the membrane potential (p. 184).

The diffusion potential itself cannot be measured. In §4–19 we will show that a combination of diffusion potentials and electrode potentials yields the formulas for the measurable e.m.f. of concentration cells with transference. In place of the reduced transport numbers (which cannot be measured for all species), the stoichiometric transport numbers (which can always be determined) appear.

The best-known special case of Eq. (23) relates to solutions of completely dissociated electrolytes in a neutral solvent. The solvent is always chosen as species 1. In the summation expression on the right-hand side of Eq. (23), only the terms referring to ions appear, so that on application of Eq. (4–16.42) we find

$$F \text{ grad } \varphi_D = -\sum_k \frac{t_k}{z_k} \text{ grad } \mu_k. \qquad (4\text{--}18.24)$$

This well-known formula relates the diffusion potential to the Hittorf transport numbers t_k of the individual ionic species.

(d) Diffusion Coefficient for a Solution of a Single Electrolyte

Consider the solution of a single electrolyte which is completely dissociated in a neutral solvent and which gives two kinds of ions upon dissociation. We shall discuss the connection between the diffusion coefficient of such a solution and other measurable quantities. These include the mobilities of the ions present.

According to Eq. (5), we find for the diffusion coefficient D:

$$_w\mathbf{J}_2 = -D \text{ grad } c, \qquad (4\text{--}18.25)$$

where we set $c \equiv c_2$. On the other hand, the phenomenological equations (9), which in the present special case are written only for the cations (indicated by inferior $+$) and the anions (indicated by inferior $-$) of the electrolyte, state:

$$_1\mathbf{J}_+ = -a_{++}(\text{grad } \mu_+ + z_+ F \text{ grad } \varphi) \Big\} \\ \qquad - a_{+-}(\text{grad } \mu_- + z_- F \text{ grad } \varphi), \qquad (4\text{--}18.26)$$

$$_1\mathbf{J}_- = -a_{-+}(\text{grad } \mu_+ + z_+ F \text{ grad } \varphi) \Big\} \\ \qquad - a_{--}(\text{grad } \mu_- + z_- F \text{ grad } \varphi). \qquad (4\text{--}18.27)$$

The Onsager reciprocity law is, according to Eq. (10),

$$a_{+-} = a_{-+}. \qquad (4\text{-}18.28)$$

For diffusion processes, φ is the diffusion potential. It is obtained from Eqs. (21) and (28):

$$-F(a_{++}z_+^2 + 2a_{+-}z_+z_- + a_{--}z_-^2)\,\mathrm{grad}\,\varphi$$
$$= (a_{++}z_+ + a_{+-}z_-)\,\mathrm{grad}\,\mu_+ + (a_{+-}z_+ + a_{--}z_-)\,\mathrm{grad}\,\mu_-. \qquad (4\text{-}18.29)$$

According to Eq. (4), we have for complete dissociation:

$$\nu_1 \mathbf{J}_2 = {_1}\mathbf{J}_+ + {_1}\mathbf{J}_-. \qquad (4\text{-}18.30)$$

We eliminate $\mathrm{grad}\,\varphi$ from Eqs. (26), (27), and (29) with the help of Eqs. (28) and (30).

With the abbreviation

$$A = \frac{a_{++}a_{--} - a_{+-}^2}{z_+^2 a_{++} + 2z_+z_- a_{+-} + z_-^2 a_{--}} \qquad (4\text{-}18.31)$$

we find

$$\nu_1 \mathbf{J}_2 = (z_- - z_+)A(z_+\,\mathrm{grad}\,\mu_- - z_-\,\mathrm{grad}\,\mu_+). \qquad (4\text{-}18.32)$$

Moreover, according to Eqs. (17) and (18) we have

$$\frac{z_+ - z_-}{\nu} = \frac{z_+}{\nu_-} = -\frac{z_-}{\nu_+} \equiv q, \qquad (4\text{-}18.33)$$

$$z_+\,\mathrm{grad}\,\mu_- - z_-\,\mathrm{grad}\,\mu_+ = q\,\mathrm{grad}\,\mu_2 = q\left(\frac{\partial \mu_2}{\partial c}\right)_{T,P}\mathrm{grad}\,c, \qquad (4\text{-}18.34)$$

where q is a positive rational number,† T the temperature, and P the pressure. Finally, with the help of Eqs. (4-17.24) and (4-17.50) we derive

$$_w\mathbf{J}_2 = (1 - cV_2)\,_1\mathbf{J}_2, \qquad (4\text{-}18.35)$$

where V_2 is the partial molar volume of the electrolyte. If we combine Eqs. (32) to (35) and compare with Eq. (25), we obtain:

$$D = Aq^2(1 - cV_2)\left(\frac{\partial \mu_2}{\partial c}\right)_{T,P}. \qquad (4\text{-}18.36)$$

If with Eqs. (1-19.12) and (1-19.36) we introduce the molality m and the conventional activity coefficient γ of the electrolyte then we find (cf. Eqs. 4-17.32 and 4-17.33):

$$(1 - cV_2)\left(\frac{\partial \mu_2}{\partial c}\right)_{T,P} = \frac{m}{c}\left(\frac{\partial \mu_2}{\partial m}\right)_{T,P} = \frac{\nu RT}{c}\left[1 + m\left(\frac{\partial \ln \gamma/\gamma^\dagger}{\partial m}\right)_{T,P}\right]. \qquad (4\text{-}18.37)$$

---

† For electrolytes like HCl, K_2SO_4, $LaCl_3$, etc., we have: $q = 1$, for $CaSO_4$; $q = 2$, for $LaPO_4$; $q = 3$; etc.

Here R is the gas constant and $\gamma^{\dagger} \equiv 1 \text{ kg mol}^{-1}$. With this it follows from Eq. (36) that

$$D = Aq^2\nu RT \frac{1}{c}\left[1 + m\left(\frac{\partial \ln \gamma/\gamma^{\dagger}}{\partial m}\right)_{T,P}\right]. \tag{4-18.38}$$

This formula[64–67]† gives the general connection between the diffusion coefficient D and the phenomenological coefficients a_{++}, a_{+-}, a_{--}.‡ The latter quantities also appear for electric conduction (§4–16). If we have a molecular kinetic theory of electric conduction in electrolyte solutions then we can calculate a_{++}, a_{+-}, and a_{--} and with the help of Eqs. (31) and (38) determine the diffusion coefficient D from a knowledge of the activity coefficient γ.[33]

The phenomenological coefficients in Eq. (38) can only be expressed in terms of measurable quantities for the case‖

$$a_{+-} = 0. \tag{4-18.39a}$$

Then we find from Eq. (4–16.28):

$$u_+ = F\frac{z_+ a_{++}}{c_+}, \quad u_- = -F\frac{z_- a_{--}}{c_-}, \tag{4-18.39b}$$

---

† If a diffusion coefficient D' is defined by the equation

$$_1\mathbf{J}_2 = -D' \text{ grad } c$$

then it follows from Eqs. (25) and (35) that

$$D = (1 - cV_2)D',$$

and we obtain the relation derived earlier[43b], which is equivalent to Eq. (38).
‡ An analogous treatment of a more complicated case (system H_2O–CdI_2 at 25°C) has been given[68].
‖ We may, however, solve the equations (4–16.29), (4–16.30), and (38) to give the phenomenological coefficients in terms of the measurable quantities t_+ (t_-), κ, and D[69]:

$$a_{++} = \frac{t_+^2 \kappa}{z_+^2 F^2} + \frac{\nu_+^2 D}{B},$$

$$a_{--} = \frac{t_-^2 \kappa}{z_-^2 F^2} + \frac{\nu_-^2 D}{B},$$

$$a_{+-} = \frac{t_+ t_- \kappa}{z_+ z_- F^2} + \frac{\nu_+ \nu_- D}{B},$$

where

$$B \equiv \frac{\nu RT}{c}\left[1 + m\left(\frac{\partial \ln \gamma/\gamma^{\dagger}}{\partial m}\right)_{T,P}\right].$$

A more general treatment involving incomplete dissociation is given by Haase and Richter[69a].

where u_i denotes the mobility of ionic species i. Equations (39), as the kinetic theory shows[70], are valid for very dilute (but not necessarily ideal dilute) solutions in which an "electrophoretic effect" can be neglected. For such solutions, u_i becomes the limiting value u_i^0 for infinite dilution without the activity coefficient having to assume a value of unity. According to Eqs. (1), (2), (31), (33), (38), and (39), for a highly dilute solution of a completely dissociated electrolyte we may write[70, 71]:

$$D = \frac{RT}{F} \frac{z_- - z_+}{z_+ z_-} \frac{u_+^0 u_-^0}{u_+^0 + u_-^0} \left[1 + m\left(\frac{\partial \ln \gamma/\gamma^\dagger}{\partial m}\right)_{T,P}\right]. \qquad (4\text{–}18.40)$$

Equation (40), simplified in comparison with the rigorous formula (38) for very dilute electrolyte solutions, makes the major contribution† to the concentration dependence of the diffusion coefficient (see Fig. 12).

For infinite dilution ($c \to 0$, $m \to 0$), we find a limiting value for the diffusion coefficient:

$$D^0 \equiv \lim_{m \to 0} D. \qquad (4\text{–}18.41)$$

Furthermore[43c],

$$\lim_{m \to 0} m\left(\frac{\partial \ln \gamma/\gamma^\dagger}{\partial m}\right)_{T,P} = 0. \qquad (4\text{–}18.42)$$

From Eqs. (40) to (42) we obtain the limiting law of Noyes (1908):

$$D^0 = \frac{RT}{F} \frac{z_- - z_+}{z_+ z_-} \frac{u_+^0 u_-^0}{u_+^0 + u_-^0}. \qquad (4\text{–}18.43)$$

For uni-univalent electrolytes ($z_+ = -z_- = 1$), there follows from this the formula of Nernst (1888):

$$D^0 = \frac{2RT}{F} \frac{u_+^0 u_-^0}{u_+^0 + u_-^0}. \qquad (4\text{–}18.44)$$

4–19 CONCENTRATION CELLS WITH TRANSFERENCE

(a) General

A *concentration cell with transference* is an isothermal–isobaric galvanic cell with two chemically similar electrodes which are immersed in two electrolyte solutions, whose compositions are different but which have at least one common ion species, that being the one for which the electrodes are reversible ("potential-determining ion"). Otherwise the solutions can consist of arbitrary substances (electrolytes and nonelectrolytes). The "terminals" should be chemically identical. Furthermore, let the two solutions which

---

† A critical analysis of the remaining difference is given by Guggenheim[72] and also Lorenz[67].

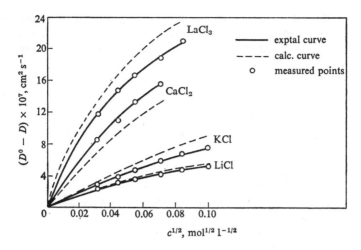

Fig. 12. Diffusion coefficient D as a function of the molarity c of the electrolyte for aqueous solutions of LiCl, KCl, $CaCl_2$, and $LaCl_3$ at 25°C. After data by Robinson and Stokes[21b]. The broken curves are according to Eq. (40); the D^0-values are from Eq. (43).

surround the electrodes ("electrode solutions") be connected by a series of mixtures the composition of which changes continuously from that of the first electrode solution to that of the second electrode solution ("bridge solutions"). Finally, let there be electrochemical equilibrium with respect to two directly adjoining phases (electrode surface/solution and terminal/electrode). We then obtain the following general phase diagram for a concentration cell with transference:

Terminal	Electrode (Phase III)	Solution I (Phase I)	Bridge solutions	Solution II (Phase II)	Electrode (Phase IV)	Terminal
Potential:	φ_{III}	φ_I		φ_{II}	φ_{IV}	(A)

There is an electric potential difference between the two terminals, and this, when it is measured without flow of current, is called the "electromotive force" (e.m.f.) of the galvanic cell.† Since the potential difference between the

---

† We use the following convention: The electromotive force (e.m.f.) of any galvanic cell is the difference between the electric potential in the terminal on the right and the electric potential in the terminal on the left, the cell being open and the local heterogeneous equilibria being established for all phase boundaries. It should be added that homogeneous chemical equilibria (such as $Cu^+ + \ominus \rightleftharpoons Cu$ in metallic copper or $H^+ + Cl^- \rightleftharpoons HCl$ in aqueous hydrochloric acid) are assumed to be established, too.

[Continued on page 298]

terminal on the left and the adjoining electrode (phase III) is exactly compensated by the potential difference between the terminal on the right and the adjoining electrode (phase IV), the measurable e.m.f. Φ of the cell amounts to:

$$\Phi = \varphi_{IV} - \varphi_{III} = (\varphi_{IV} - \varphi_{II}) + (\varphi_I - \varphi_{III}) + (\varphi_{II} - \varphi_I), \quad (4\text{-}19.1)$$

which is independent of the nature of the terminals.

As examples of concentration cells with transference, i.e. as special cases of the cell (A), let us mention the following galvanic cells:

	Cu	Ag	Solution I (with Ag$^+$)	Bridge solutions	Solution II (with Ag$^+$)	Ag	Cu	(B)
Potential:		φ_{III}	φ_I		φ_{II}	φ_{IV}		

	Cu	Ag	AgCl	Solution I (with Cl$^-$)	Bridge solutions	Solution II (with Cl$^-$)	AgCl	Ag	Cu	(C)
Potential:		φ_{III}		φ_I		φ_{II}	φ_{IV}			

[*Continued from page 297*]

There is, however, *no equilibrium* with respect to a chemical heterogeneous reaction corresponding to the *whole* cell. An example for such a chemical reaction is

$$\tfrac{1}{2} H_2 + AgCl \to Ag + HCl \quad (I)$$

This reaction accompanies the flow of the charge 1 faraday ($= n^\dagger F$, $F =$ Faraday constant, $n^\dagger = 1$ mol) from left to right through the isothermal–isobaric cell

$$Cu \mid H_2(Pt) \mid HCl(aq) \mid AgCl \mid Ag \mid Cu \quad (II)$$

Let a galvanic cell be open (no flow of electric current) and let all the local equilibria be established, as explained above. Then, if irreversible processes, such as diffusion, sedimentation, heat conduction, etc., can occur inside the cell, it is called an "irreversible cell." If such processes cannot take place within the cell, it is called a "reversible cell." Thus the galvanic cell (II) is a reversible cell. Reversible cells must be isothermal, but they are not necessarily isobaric. Thus gravitational and centrifugal cells (§4–22), in the case of sedimentation equilibrium, represent nonisobaric reversible cells.

We obtain the formula for the e.m.f. Φ of an *isobaric reversible cell* from classical thermodynamics. In fact, using Eqs. (1–11.14) and (1–14.9) and extending the definition (1–14.2) to heterogeneous reactions, we obtain:

$$F\Phi = A. \quad (III)$$

Here A is the affinity of the chemical reaction that accompanies the flow of 1 faraday from left to right. For the cell (II) the corresponding reaction is (I). We have for this example:

$$F\Phi = A = \tfrac{1}{2}\mu_{H_2} + \mu_{AgCl} - \mu_{Ag} - \mu_{HCl}. \quad (IV)$$

After rearranging formulas of the type (IV) we find Eq. (4–19.24).

Examples for irreversible galvanic cells are: concentration cells with transference (as discussed here), gravitational or centrifugal cells in the general case (§4–22), thermocouples (§4–25), and thermocells (§4–29). To obtain the equations for the e.m.f. of such cells we must apply the methods of nonequilibrium thermodynamics.

In both cells, the terminals are copper wires. For the cell (B), we have electrodes of the first kind (silver electrodes); for the cell (C), electrodes of the second kind (silver–silver chloride electrodes). The potential differences at the phase boundaries Ag/AgCl in (C) exactly compensate. Equation (1) accordingly remains applicable in each case.

The first two expressions in parentheses in Eq. (1) denote potential differences between electrodes and adjoining electrolyte solutions ("electrode potentials"). They can be calculated using the assumptions made initially for equilibrium and thus derived from classical thermodynamics. If we designate the potential-determining ionic species—Ag^+ in the cell (B) or Cl^- in the cell (C)—by the inferior i ($_i$), we obtain from the condition (1–20.13) for electrochemical equilibrium, applied to the phase boundaries III/I and II/IV in the cell (A),

$$\left.\begin{aligned}
\varphi_I - \varphi_{III} &= \frac{1}{z_i F} [(\mu_i)_{III} - (\mu_i)_I], \\[1em]
\varphi_{IV} - \varphi_{II} &= \frac{1}{z_i F} [(\mu_i)_{II} - (\mu_i)_{IV}],
\end{aligned}\right\} \tag{4-19.2}$$

where F denotes the Faraday constant and z_i or μ_i the charge number or the chemical potential of species i respectively. Since the two electrodes (phases III and IV) consist of the same material, we find:

$$(\mu_i)_{III} = (\mu_i)_{IV}, \tag{4-19.3}$$

while the chemical potential of the ion species i in solution I, namely $(\mu_i)_I$ must be different from that in solution II, namely $(\mu_i)_{II}$.

From Eqs. (2) and (3) we derive

$$(\varphi_{IV} - \varphi_{II}) + (\varphi_I - \varphi_{III}) = \frac{1}{z_i F} [(\mu_i)_{II} - (\mu_i)_I] = \frac{1}{z_i F} \int_I^{II} d\mu_i, \tag{4-19.4}$$

where the integration is from the composition of solution I to that of solution II.

The last expression in parentheses in Eq. (1), that is,

$$\varphi_{II} - \varphi_I \equiv \Delta\varphi_D, \tag{4-19.5}$$

is the potential difference between solution II and solution I. This quantity results from integration of the differential equation (4–18.23) which was derived using the thermodynamics of irreversible processes. For simplicity, we assume that the gradient of the diffusion potential grad φ_D has the same direction as the concentration gradients (and thus as the grad μ_k) at each point in the bridge solutions. We then find from Eq. (4–18.23):

$$\Delta\varphi_D = \varphi_{II} - \varphi_I = -\frac{1}{F} \int_I^{II} \sum_k \tau_k \, d\mu_k. \tag{4-19.6}$$

Here τ_k or μ_k denotes the reduced transport number or the chemical potential of any (charged or uncharged) species k. The summation is over all species, with the exception of the particle type 1 which can normally be regarded as the "solvent." The quantity $\Delta\varphi_D$ is often called simply the "diffusion potential." It depends—apart from simple special cases (see below)—on the type and structure of the bridge solutions. The integral in Eq. (6) in general is indefinite.

By combining Eqs. (1), (4), and (6) we obtain:

$$F\Phi = \frac{1}{z_i} \int_{\text{I}}^{\text{II}} d\mu_i - \int_{\text{I}}^{\text{II}} \sum_k \tau_k \, d\mu_k. \qquad (4\text{-}19.7)$$

This is the general formula for the measurable e.m.f. of a concentration cell with transference.

If the solutions contain only completely dissociated electrolytes in one single neutral solvent, the inferior k ($_k$) in Eq. (7) refers only to ions. According to Eqs. (4-16.42) and (4-16.46), we have

$$t_k = z_k \tau_k, \qquad (4\text{-}19.8)$$

$$\sum_k t_k = 1, \qquad (4\text{-}19.9)$$

where t_k denotes the Hittorf transport number of the ionic species k and the summation is over all ion types of the solutions. We find from Eqs. (7), (8), and (9):

$$F\Phi = \int_{\text{I}}^{\text{II}} \sum_k t_k \left(\frac{1}{z_i} \, d\mu_i - \frac{1}{z_k} \, d\mu_k \right). \qquad (4\text{-}19.10)$$

This well-known formula[73] replaces Eq. (7) under the above assumptions.

(b) Special Case of a Single Electrolyte with two Ion Species

As a most useful application of Eq. (7), let us consider a concentration cell containing solutions of a single electrolyte with two kinds of ions in a neutral solvent. Since, in this case, there are only two independently moving substances and thus only one single independent concentration variable (cf. p. 285), all the quantities τ_k and μ_k appearing in the second integral of Eq. (7) depend only on this variable. Accordingly, the integral is uniquely determined by the two values of the electrolyte concentration in the two electrode solutions and thus is independent of the nature and of the origin of the bridge solutions. Thus also the e.m.f. of the cell is uniquely specified.

Let the electrolyte contain per molecule ν_+ cations of charge number z_+ and ν_- anions of charge number z_-. Apart from the cations (indicated by inferior $+$) and anions (indicated by inferior $-$), let the undissociated part of the electrolyte be present in solution (indicated by inferior u). The electrodes shall be reversible either to the cations or to the anions.†

As examples, we mention the following concentration cells:

$$Ag \left| \begin{array}{c} AgNO_3 \ (aq) \\ m_I \end{array} \right\| \begin{array}{c} AgNO_3 \ (aq) \\ m_{II} \end{array} \right| Ag$$

$$Ag \left| AgCl \left| \begin{array}{c} KCl \ (aq) \\ m_I \end{array} \right\| \begin{array}{c} KCl \ (aq) \\ m_{II} \end{array} \right| AgCl \right| Ag$$

Here m_I or m_{II} denotes the molality of the electrolyte (dissolved in water) in the first or the second electrode solution respectively. The bridge solutions are indicated by the symbol ∥.‡

For this case, z_i and μ_i should be replaced in Eq. (7) either by z_+ and μ_+ respectively, or by z_- and μ_- respectively. Equation (1) is then, on the basis of Eqs. (8) and (9),

$$F\Phi = \int_I^{II} \left(\frac{1}{z_+} d\mu_+ - \frac{1-t_-}{z_+} d\mu_+ - \frac{t_-}{z_-} d\mu_- - \tau_u \, du_u \right)$$

$$= \int_I^{II} \left[t_- \left(\frac{d\mu_+}{z_+} - \frac{d\mu_-}{z_-} \right) - \tau_u \, d\mu_u \right] \tag{4–19.11}$$

(electrodes reversible for cations),

$$F\Phi = \int_I^{II} \left(\frac{1}{z_-} d\mu_- - \frac{t_+}{z_+} d\mu_\mathrm{I} - \frac{1-t_+}{z_-} d\mu_- - \tau_u \, d\mu_u \right)$$

$$= \int_I^{II} \left[t_+ \left(\frac{d\mu_-}{z_-} - \frac{d\mu_+}{z_+} \right) - \tau_u \, d\mu_u \right] \tag{4–19.12}$$

(electrodes reversible for anions),

where t_+ or t_- is the Hittorf transport number of the cations or anions respectively and τ_u the reduced transport number of the undissociated electrolyte.

---

† The more general case, for which an electrolyte produces more than two ion species (such as H_3PO_4) and for which the electrodes can be reversible also to complex ions (e.g. HPO_4^{2-}), has been dealt with by Spiro[74].

‡ In the cell with the silver–silver chloride electrodes, the solubility of AgCl has been neglected. If platinum–hydrogen electrodes are used, as in the cell

$$H_2(Pt) \left| \begin{array}{c} HCl \\ m_I \end{array} \right\| \begin{array}{c} HCl \\ m_{II} \end{array} \right| H_2(Pt)$$

then we assume implicitly that:

(a) The solubility of H_2 in the electrolyte solution can be neglected.

(b) The partial pressure of H_2 over the electrode, and thus the composition of the solid phase Pt–H_2, which serves as an electrode, have been fixed by the experimental conditions.

The condition for local dissociation equilibrium, which may always be assumed here, according to Eqs. (4–18.17) and (4–18.18), states:

$$\frac{d\mu_+}{z_+} - \frac{d\mu_-}{z_-} = \frac{d\mu_2}{z_+\nu_+} = -\frac{d\mu_2}{z_-\nu_-}, \tag{4-19.13}$$

$$d\mu_u = d\mu_2. \tag{4-19.14}$$

Here μ_2 denotes the chemical potential of the electrolyte. Accordingly, the integrand in Eq. (11) or (12) assumes the form

$$\left(\frac{t_-}{z_+\nu_+} - \tau_u\right) d\mu_2 = -\left(\frac{t_-}{z_-\nu_-} + \tau_u\right) d\mu_2 \tag{4-19.15}$$

or

$$-\left(\frac{t_+}{z_+\nu_+} + \tau_u\right) d\mu_2 = \left(\frac{t_+}{z_-\nu_-} - \tau_u\right) d\mu_2. \tag{4-19.16}$$

Obviously, we choose as "ion constituents" of the electrolyte (cf. p. 268) the atoms or radicals corresponding to the cations or anions and denote these for the sake of simplicity by inferior $+$ or $-$ respectively. Then, for the stoichiometric transport numbers, according to Eq. (4–16.52) we find:

$$\vartheta_+ = z_+\left(\frac{t_+}{z_+} + \nu_+\tau_u\right), \tag{4-19.17}$$

$$\vartheta_- = 1 - \vartheta_+ = z_-\left(\frac{t_-}{z_-} + \nu_-\tau_u\right). \tag{4-19.18}$$

The integrand (15) or (16) can thus be written in the form

$$-\frac{\vartheta_-}{z_-\nu_-} d\mu_2 = \frac{1-\vartheta_+}{z_+\nu_+} d\mu_2$$

or

$$-\frac{\vartheta_+}{z_+\nu_+} d\mu_2 = \frac{1-\vartheta_-}{z_-\nu_-} d\mu_2.$$

Accordingly, we obtain from Eqs. (11) and (12) in a single formula an expression for the e.m.f. of the concentration cell:

$$F\Phi = \frac{1}{z_i\nu_i} \int_{\mathrm{I}}^{\mathrm{II}} (1 - \vartheta_i) \, d\mu_2. \tag{4-19.19}$$

The differential $d\mu_2$, which refers to the isothermal–isobaric change of the chemical potential of the electrolyte with the concentration, can be replaced by an expression which contains the measurable activity coefficient of the electrolyte (cf. below).

For complete dissociation or when neglecting the migration of the undissociated molecules in the electric field ($\tau_u = 0$) from Eqs. (17), (18), and (19) there results:

$$F\Phi = \frac{1}{z_i \nu_i} \int_{\mathrm{I}}^{\mathrm{II}} (1 - t_i)\, d\mu_2. \qquad (4\text{-}19.20)$$

This equation is particularly applicable to dilute solutions of strong electrolytes. It was derived previously[75].

From Eq. (1-19.26) it follows that

$$d\mu_2 = \nu RT\, d\ln(m\gamma), \qquad (4\text{-}19.21)$$

where ν ($= \nu_+ + \nu_-$) denotes the sum of the dissociation numbers, R the gas constant, T the absolute temperature, m the molality of the electrolyte, and γ the conventional activity coefficient. The identities:

$$\frac{\nu}{z_+ \nu_+} = \frac{1}{z_+} - \frac{1}{z_-}, \qquad \frac{\nu}{z_- \nu_-} = \frac{1}{z_-} - \frac{1}{z_+} \qquad (4\text{-}19.22)$$

hold. We derive from Eqs. (19), (21), and (22):

$$\Phi = \pm\left(\frac{1}{z_+} - \frac{1}{z_-}\right) \frac{RT}{F} \int_{\mathrm{I}}^{\mathrm{II}} (1 - \vartheta_i)\, d\ln(m\gamma), \qquad (4\text{-}19.23)$$

where the upper or lower sign is valid if the ion i (for which the electrodes are reversible) represents the cation or the anion of the electrolyte respectively. In the first case, we have $1 - \vartheta_i = 1 - \vartheta_+ = \vartheta_-$; in the second case, we have $1 - \vartheta_i = 1 - \vartheta_- = \vartheta_+$.

With the knowledge of the activity coefficient γ (e.g. from vapor pressure determinations or from e.m.f. measurements on reversible cells), following Eq. (23), the transport number ϑ_i can be derived from e.m.f. measurements on concentration cells. Examples for transport numbers so obtained are found in Table 9 (p. 321). According to Eq. (20), the quantity ϑ_i coincides with the Hittorf transport number t_i of species i only if complete dissociation is assumed or if migration of the undissociated electrolyte molecules in the electric field is neglected.[34]

If, conversely, the transport number is known from other sources then the activity coefficient can be determined from Eq. (23). This procedure is particularly advantageous for low electrolyte concentrations. (The transport number is then best determined by the moving boundary method.)

The dependence of the e.m.f. on electrolyte concentration for a reversible isobaric cell has a certain similarity to Eq. (23) (cf. footnote to p. 297). This relation can be obtained from classical thermodynamics and naturally does not contain the transport number. Let there be present in the solution only a single electrolyte with two ion species, as assumed in Eq. (23). Then one electrode of the cell must be reversible to the cations of the electrolyte and the other to the anions. For the e.m.f. Φ of such a cell, there results[43d]:

$$\Phi = \Phi^{\ominus} \pm \left(\frac{1}{z_+} - \frac{1}{z_-}\right) \frac{RT}{F} \ln (\nu_{\pm} \, m\gamma) \qquad (4\text{-}19.24)$$

with

$$\nu_{\pm}^{\nu} \equiv \nu_+^{\nu^+} \nu_-^{\nu^-}.$$

Here $\Phi^{\ominus}$ is the standard value of the e.m.f. (dependent only on temperature and pressure). The upper or lower sign applies if the electrode reversible to the cations is written on the right or on the left respectively of the diagram representing the cell. According to Eq. (24), the activity coefficient can be obtained from e.m.f. measurements on reversible galvanic cells.

4-20 DIFFUSION AND SEDIMENTATION IN ARBITRARY FLUID SYSTEMS

(a) Definitions

To describe *diffusion* in an arbitrary fluid, following §4–17 and §4–18, we first define a diffusion current density $_wJ_i$ (in the Fick reference system) by Eq. (4–17.2), valid for each independently moving substance i. We then write the $N - 1$ independent diffusion current densities $_wJ_2, _wJ_3, \ldots, _wJ_N$ as homogeneous linear functions of the $N - 1$ independent concentration gradients grad c_2, grad $c_3, \ldots$, grad c_N (c_i = molarity of substance i), N being the number of independently moving substances. We thus obtain the generalized form of Fick's law (4–17.6):

$$_wJ_i = \sum_{k=2}^{N} D_{ik} \, \text{grad} \, c_k \quad (i = 2, 3, \ldots, N). \qquad (4\text{-}20.1)$$

The D_{ik} are the *diffusion coefficients*.

While diffusion denotes migration of substances due to concentration gradients, *sedimentation* means the transport of matter due to an external force field (gravitational or centrifugal field). Experience indicates that the relative velocity of any independently moving substance ($\mathbf{v}_i - \mathbf{w} = _wJ_i/c_i$, $\mathbf{w}$ being the mean volume velocity) is proportional to the acceleration $\mathbf{g}$ due to gravity or centrifugal force ($\mathbf{g}$ is either gravitational acceleration or is equal to $\Omega^2\mathbf{r}$, where Ω is the absolute magnitude of the angular velocity of the centrifuge and $\mathbf{r}$ is the distance from the rotation axis). We thus find the following empirical equation for sedimentation:†

$$_wJ_i = c_i s_i \mathbf{g} \quad (i = 2, 3, \ldots, N). \qquad (4\text{-}20.2)$$

---

† For two independently moving substances ($N = 2$, $s_2 \equiv s$), there results from Eq. (2):

$$\mathbf{v}_2 - \mathbf{w} = \frac{1}{c_2} \, _wJ_2 = s\mathbf{g}. \qquad (4\text{-}20.2a)$$

Here the s_i are the *sedimentation coefficients*. These quantities are independent of $\mathbf{g}$ but can still be functions of the local state variables (temperature, pressure, concentrations).

Equation (1) is valid if $\mathbf{g} = 0$, and Eq. (2) with the assumption that grad $c_k = 0$ $(k = 2, 3, \ldots, N)$. Let diffusion and sedimentation occur simultaneously in the volume element considered (as is always the case for initially homogeneous systems in a gravitational or centrifugal field, since then concentration gradients are built up). Then, for the two types of irreversible processes, we have a linear superposition, as experience shows. We thus find for isothermal transport of matter in any fluid in a gravitational or centrifugal field:

$$_w\mathbf{J}_i = c_i s_i \mathbf{g} - \sum_{k=2}^{N} D_{ik} \text{ grad } c_k \quad (i = 2, 3, \ldots, N). \quad (4\text{-}20.3)$$

Here grad c_k denotes the concentration gradient after subtraction of the term containing the pressure gradient (cf. below).

In the definition of the "independently moving substances" (§4–18), we assumed that there is local chemical equilibrium for all species which can react. We also make this assumption here, since chemical reactions (particularly ion reactions) tend to occur rapidly in comparison with diffusion and sedimentation processes. Under these circumstances we can further assume that the condition for local mechanical equilibrium is satisfied. From Eq. (1–21.30) and Eq. (4–15.5) this condition is identical to the

---

Now we can find a way of writing this expression independently of the reference system as we did for the diffusion (see Eq. 4–17.25). If ω is an arbitrary reference velocity and if ω_1 and ω_2 represent the weight factors defined by Eqs. (4–17.20) and (4–17.21) for averaging the velocities, then we obtain with Eqs. (4–17.19), (4–17.23), and (4–17.24):

$$\mathbf{v}_2 - \boldsymbol{\omega} = \frac{1}{c_2} \mathbf{J}_2 = \frac{\omega_1}{c_1 V_1} s\mathbf{g}, \quad (4\text{-}20.2b)$$

where V_1 denotes the partial molar volume of substance 1. This expression is reduced to Eq. (2a) on using the Fick reference system ($\omega = \mathbf{w}$, $\omega_1 = c_1 V_1$). We can regard Eq. (2b) as the most general definition of the sedimentation coefficient s for two independently moving substances. We recognize that this definition is independent of the reference velocity, on consideration of Eqs. (4–17.20) and (4–17.21). There follows from this:

$$\mathbf{v}_2 - \boldsymbol{\omega} = \mathbf{v}_2 - (\omega_1 \mathbf{v}_1 + \omega_2 \mathbf{v}_2) = -\omega_1 \mathbf{v}_1 + (1 - \omega_2) \mathbf{v}_2 = \omega_1 (\mathbf{v}_2 - \mathbf{v}_1).$$

With this we derive from Eq. (2b):

$$\mathbf{v}_2 - \mathbf{v}_1 = \frac{s}{c_1 V_1} \mathbf{g}, \quad (4\text{-}20.2c)$$

in which the weight factor ω_1 no longer appears. Equation (2c) simultaneously represents the equation (2b) in the Hittorf reference system ($\omega = \mathbf{v}_1$, $\omega_1 = 1$).

assumption of "creeping motion" in a fluid medium that exhibits no space charges. Thus we obtain:

$$\rho \mathbf{g} = \operatorname{grad} P. \tag{4-20.4}$$

Here ρ is the density and P the pressure.

We now recognize that the term with grad P in the grad c_k of Eq. (3) must be eliminated. This term appears in the most general case because of the pressure dependence of the volume. It has nothing to do with diffusion. According to Eq. (1–6.5) and Eq. (1–22.12), the following quantities must be inserted in place of the grad c_k in Eq. (3):

$$(\operatorname{grad} c_k)_P = \operatorname{grad} c_k - c_k \kappa_T \operatorname{grad} P. \tag{4-20.5}$$

Here κ_T denotes the (isothermal) compressibility. For two independently moving substances ($N = 2$, $s_2 \equiv s$, $D_{22} \equiv D$), there follows from Eq. (3):

$$_w\mathbf{J}_2 = c_2 s \mathbf{g} - D(\operatorname{grad} c_2)_P. \tag{4-20.6}$$

If, on the other hand, we define the diffusion coefficient D by Eq. (4–17.25) using Eqs. (4–17.23) to (4–17.27), we find, in agreement with Eq. (4–17.30),

$$_w\mathbf{J}_2 = c_2 s \mathbf{g} - D \frac{V_1}{\overline{V}^2} \operatorname{grad} x_2, \tag{4-20.7}$$

where $\overline{V}$ is the molar volume, V_1 the partial molar volume of substance 1, and x_2 the mole fraction† of substance 2. According to Eq. (4–17.29), we have:

$$(\operatorname{grad} c_2)_P = \frac{V_1}{\overline{V}^2} \operatorname{grad} x_2. \tag{4-20.8}$$

Thus Eqs. (6) and (7) are identical.

---

† The most appropriate variable for a general description of the composition of a binary mixture is without doubt the mole fraction. For electrolyte solutions, however, the molality offers certain advantages, particularly when we do not wish or are not able to measure the properties in question up to the pure liquid electrolyte—which often corresponds only to a hypothetical state. Finally, the molarity is almost without exception used in the treatment of conductivities, transport numbers, diffusion and sedimentation coefficients, as is clearly seen from our discussion hitherto. This is due to certain theoretical aspects (integration of the differential equations, molecular interpretation of the properties of dilute electrolyte solutions). Thus, in particular for electrolyte solutions, we have three composition variables: x_2 (mole fraction of component 2, for electrolyte solutions: stoichiometric mole fraction of electrolyte), m (molality of substance 2, in particular that of the electrolyte), and $c = c_2$ (molarity of component 2, in particular that of the electrolyte). Let n_i or M_i be the amount of substance or molar mass respectively of component i and V or $\overline{V}$ the volume or molar volume

(b) Sedimentation Equilibrium

Sedimentation equilibrium has been attained if, in each volume element of the isothermal fluid, diffusion and sedimentation have compensated, i.e. when the vector $_w\mathbf{J}_i$ vanishes everywhere for each independently moving substance. Accordingly, by means of Eq. (3) we obtain the conditions for sedimentation equilibrium:

$$c_i s_i \mathbf{g} = \sum_{k=2}^{N} D_{ik} \operatorname{grad} c_k \quad (i = 2, 3, \ldots, N). \tag{4-20.9}$$

This system of $N - 1$ equations represents the equilibrium conditions from the standpoint of irreversible processes.

On the other hand, classical thermodynamics also produces definite equilibrium conditions due to Eq. (1–21.31) and these, in our present notation, state that:

$$(M_j - V_j \rho)\mathbf{g} = (\operatorname{grad} \mu_j)_{T,P} + z_j \mathbf{F} \operatorname{grad} \varphi. \tag{4-20.10}$$

Here M_j and V_j are the molar mass and the partial molar volume of species j, μ_j and z_j the chemical potential and charge number of species j, $\mathbf{F}$ the Faraday constant, and φ the equilibrium value of the sedimentation potential. The inferiors T,P at grad μ_j signify that the terms involving grad T (which here disappear) and grad P ($= \rho\mathbf{g}$) have already been subtracted.

From the system of equations (10), which encompasses as many equations as there are chemical species, we must obtain a system of equations for the independently moving substances so that we can compare it with Eq. (9).

Let an electrically neutral molecule of the independently moving substance i, which may arise by combining certain species (cf. §4–18), contain ν_{ji} molecules or ions of type j. Then, on the basis of electric neutrality, from

---

respectively of the mixture. Then, according to definition, there is valid (cf. Eqs. 1–6.4, 1–6.5, and 1–19.12):

$$x_2 = \frac{n_2}{n_1 + n_2}, \quad m = \frac{n_2}{M_1 n_1}, \quad c = \frac{n_2}{V}.$$

If we introduce the density ρ of the mixture ($x_1 = 1 - x_2$)

$$\rho = \frac{M_1 n_1 + M_2 n_2}{V} = \frac{M_1 x_1 + M_2 x_2}{\overline{V}}$$

then we obtain the following relations:

$$m = \frac{x_2}{M_1(1 - x_2)}, \quad c = \frac{x_2}{\overline{V}}, \quad c = \frac{\rho m}{1 + M_2 m}.$$

We may add that for solutions of polymers there are further composition variables, such as the mass concentration (partial density) or the "volume fraction." These essentially trivial complications often frighten the new student.

the conservation of mass, and from the condition of local chemical equilibrium (cf. Eq. 1–12.12 and Eq. 1–21.16), we find:

$$\sum_j \nu_{ji} z_j = 0, \tag{4-20.11}$$

$$\sum_j \nu_{ji} M_j = M_i, \tag{4-20.12}$$

$$\sum_j \nu_{ji} \mu_j = \mu_i, \tag{4-20.13}$$

$$\sum_j \nu_{ji} V_j = V_i, \tag{4-20.14}$$

where M_i denotes the molar mass, μ_i the chemical potential, and V_i the partial molar volume of the independently moving substance i.

Combination of Eqs. (11) to (14) with Eq. (10) produces:

$$(M_i - V_i \rho)\mathbf{g} = (\text{grad } \mu_i)_{T,P} \quad (i = 1, 2, \ldots, N). \tag{4-20.15}$$

From Eqs. (4–3.6), (4–3.12), (4–3.26), (4–8.10), (4–15.2), and (4–15.3) there follows:

$$\sum_{i=1}^{N} c_i M_i = \rho, \tag{4-20.16}$$

$$\sum_{i=1}^{N} c_i V_i = 1, \tag{4-20.17}$$

$$\sum_{i=1}^{N} c_i (\text{grad } \mu_i)_{T,P} = 0, \tag{4-20.18}$$

$$(\text{grad } \mu_i)_{T,P} = \sum_{k=2}^{N} \mu_{ik} \text{ grad } c_k \tag{4-20.19}$$

with

$$\mu_{ik} \equiv \left(\frac{\partial \mu_i}{\partial c_k} \right)_{T,P,c_l}. \tag{4-20.20}$$

Here c_l stands for all independent concentrations except c_k. The grad c_k in Eq. (19) should, as in Eq. (9), be replaced by $(\text{grad } c_k)_P$ from Eq. (5). According to Eqs. (16) to (18), of the N equations in Eq. (15) only $N - 1$ are independent. It is therefore sufficient if we write Eq. (15) for substances $2, 3, \ldots, N$. With the abbreviation

$$\psi_i \equiv M_i - V_i \rho \tag{4-20.21}$$

we find from Eqs. (15) and (19):

$$\psi_i \mathbf{g} = \sum_{k=2}^{N} \mu_{ik} \, \text{grad } c_k \quad (i = 2, 3, \ldots, N). \qquad (4\text{-}20.22)$$

This system of $N - 1$ equations is suitable for comparison with the system of equations (9). Equation (22) represents the general expression for the equilibrium distribution of the different substances in the gravitational or centrifugal field.

(c) Relation between Diffusion and Sedimentation Coefficients

We define the determinants

$$\Gamma \equiv \begin{vmatrix} \mu_{22} & \mu_{23} & \cdots & \mu_{2N} \\ \mu_{32} & \mu_{33} & \cdots & \mu_{3N} \\ \cdot & \cdot & \cdots & \cdot \\ \mu_{N2} & \mu_{N3} & \cdots & \mu_{NN} \end{vmatrix} \qquad (4\text{-}20.23)$$

and

$$\Gamma_k \equiv \begin{vmatrix} \mu_{22} & \cdots & \mu_{2,k-1} & \psi_2 & \mu_{2,k+1} & \cdots & \mu_{2N} \\ \mu_{32} & \cdots & \mu_{3,k-1} & \psi_3 & \mu_{3,k+1} & \cdots & \mu_{3N} \\ \cdot & \cdot & \cdot & \cdot & \cdot & \cdot & \cdot \\ \mu_{N2} & \cdots & \mu_{N,k-1} & \psi_N & \mu_{N,k+1} & \cdots & \mu_{NN} \end{vmatrix} \qquad (4\text{-}20.24)$$

and solve the system of equations (22) for $\text{grad } c_k$. We insert the result into Eq. (9) and find:

$$s_i = \frac{1}{c_i \Gamma} \sum_{k=2}^{N} \Gamma_k D_{ik} \quad (i = 2, 3, \ldots, N). \qquad (4\text{-}20.25)$$

This general relation between the measurable sedimentation coefficients s_i and the measurable diffusion coefficients D_{ik} contains only measurable quantities. It should be noted that all quantities refer to the same pressure. Thus, the sedimentation coefficients measured at the high pressures in an ultracentrifuge must be converted to atmospheric pressure, since diffusion coefficients and the other quantities in Eq. (25) are usually determined at atmospheric pressure.

It is noteworthy that Eq. (25) can be derived from classical conditions for sedimentation equilibrium. The derivation[76–78] represents an example of the case mentioned in §1–28, namely, of the deduction of statements for irreversible processes from the equilibrium conditions of classical thermodynamics. Previous derivations which refer only to certain special cases of Eq. (25) depend on kinetic or quasithermodynamic considerations or on the methods of the thermodynamics of irreversible processes[79]. The last method which proceeds from Eqs. (4–15.10) is generally valid but much more troublesome than the methods described here.

For two independently moving substances ($N = 2$), from Eqs. (20), (21), (23), (24), and (25) we obtain:[†]

$$\frac{D}{s} = \frac{c_2}{M_2 - V_2\rho}\left(\frac{\partial \mu_2}{\partial c_2}\right)_{T,P}. \qquad (4\text{-}20.26)$$

This formula[‡] which is identical to Eq. (1–28.8) with the assumption of an ideal dilute solution is transformed into the well-known equation of Svedberg[81]. We find the exact formulation of this relation (as the limiting law for infinite dilution) if we combine Eqs. (4–18.37) and (4–18.42) with Eq. (26), and consider that V_2/M_2 represents the partial specific volume $\tilde{V}_2$ of substance 2 (the "solute"). Furthermore, we set

$$\lim_{c_2 \to 0} \tilde{V}_2 \equiv \tilde{V}_2^0, \quad \lim_{c_2 \to 0} \rho = \rho_{01}, \quad \lim_{c_2 \to 0} D \equiv D^0, \quad \lim_{c_2 \to 0} s \equiv s^0,$$

where ρ_{01} denotes the density of the pure substance 1 (of the pure "solvent"). Then we obtain the following limiting law:

$$\frac{D^0}{s^0} = \frac{\nu R T}{M_2(1 - \tilde{V}_2^0 \rho_{01})}. \qquad (4\text{-}20.27)$$

Here R is the gas constant and ν a number which is unity for a dissolved nonelectrolyte but which is the sum of the dissociation numbers for a dissolved electrolyte. According to Eq. (27), the molecular weight of the solute can be determined from diffusion and sedimentation experiments.

A more detailed discussion of Eqs. (25) to (27) is found in the quoted papers. Here we need only point out that, with the help of Eq. (26), we can calculate the sedimentation coefficient s of a high-polymer solution—the experimental determination of which requires the use of an ultracentrifuge. s can be interpreted more easily than the diffusion coefficient D. For this, the following quantities must be known:

---

† An equation equivalent to Eq. (26), but more suitable for electrolyte solutions[80], can be derived by combining Eq. (26) with Eq. (4–18.37):

$$\frac{D}{s} = \frac{(1 + M_1 m)\overline{V}}{V_1(M_2 - V_2\rho)} \nu R T \left[1 + m\left(\frac{\partial \ln \gamma/\gamma^\dagger}{\partial m}\right)_{T,P}\right], \qquad (4\text{-}20.26a)$$

where the identity

$$\frac{1}{1 - c_2 V_2} = (1 + M_1 m)\frac{\overline{V}}{V_1} \qquad (4\text{-}20.26b)$$

has been used (see footnote to p. 306).

‡ Equation (26) was first obtained—by quasithermodynamic arguments[82–84]— for binary nonelectrolyte solutions. Later[85a, b], it was shown that it follows from the thermodynamics of irreversible processes. Finally[76–78], it was shown that the formula can be derived in the way mentioned above and that it holds for arbitrary systems with two independently moving substances (e.g. also for aqueous sulfuric acid).

1. Partial molar volume V_2 of the polymer and density ρ of the solution (from density measurements).

2. Molecular weight M_2 of the polymer (e.g. from measurements of the osmotic pressure).

3. $(\partial\mu_2/\partial c_2)_{T,P}$ for the polymer (e.g. from measurements of the osmotic pressure or of the vapor pressure).

4. Diffusion coefficient D (from diffusion measurements).

Sedimentation coefficients calculated in this way are compared in Table 8 to sedimentation coefficients measured directly in the ultracentrifuge.†

Table 8

Sedimentation Coefficients s (calc), Calculated from Eq. (26) or (27), and their Limiting Values s^0 (calc) Compared with the Directly Measured Values s (meas) and s^0 (meas) for the System Toluene (1)–Polystyrene (2) at 20°C for Several Mass Concentrations $\rho_2 = M_2c_2$ and Molecular Weights (molar masses M_2) According to Rehage[88] and Ernst[89]. (The M_2-values noted in the third and fifth columns belong to the s-values of the second and fourth columns respectively)

ρ_2 g l^{-1}	s (calc) $\times 10^{13}$ s	$M_2 \times 10^{-5}$ g mol^{-1}	s (meas) $\times 10^{13}$ s	$M_2 \times 10^{-5}$ g mol^{-1}
0 ($s = s^0$)	4.3	1.8	4.45	1.24
			5.9	2.5
			7.00	3.13
			8.75	5.36
2.2	3.8	1.8	5.83	3.8
4.4	2.7	1.8	4.27	3.8

With the help of a few conversion formulas (see Eqs. 4–17.27, 4–17.29, 4–23.13, and 4–26.6), Eq. (26) can be transformed to:

$$\frac{D}{s} = \frac{x_2\bar{V}^2}{M_1^2 M_2 x_1 \tilde{V}_1(\bar{V}_1 - \bar{V}_2)}\left(\frac{\partial\mu_2}{\partial x_2}\right)_{T,P} \qquad (4\text{–}20.28)$$

---

† Creeth[86] verified Eq. (26) for the system $H_2O + Tl_2SO_4$ at 25°C for the concentration region from 8 to 50 g l^{-1}. He used measurements of the sedimentation coefficient s, of the diffusion coefficient D, of the density (for the determination of $V_2\rho$), and of the e.m.f. of the isothermal–isobaric cell

$$Tl(Hg) \mid Tl_2SO_4(aq) \mid Hg_2SO_4 \mid Hg$$

In this case, the e.m.f. measurements served for the determination of the quantity $(\partial\mu_2/\partial c_2)_{T,P}$ with the help of Eqs. (4–18.37) and (4–19.24). La Bar and Baldwin[87] checked Eq. (26) for aqueous cane sugar solutions at 25°C (concentration range: 36 to 56 g l^{-1}).

(x_i = mole fraction, $\tilde{V}_i$ = partial specific volume of substance i). Since for a stable mixture

$$D > 0, \qquad \left(\frac{\partial \mu_2}{\partial x_2}\right)_{T,P} > 0$$

and since we may also assume $\tilde{V}_1 > 0$, we conclude from Eq. (28) that s is positive if $\tilde{V}_1$ is greater than $\tilde{V}_2$. According to Eq. (6) or (7), at vanishing concentration gradients that component which has the smaller partial specific volume migrates downwards in the gravitational field or outwards in the centrifugal field. For solutions of high polymers, this component is almost always the polymer.

For an ideal binary mixture (of arbitrary concentration),

$$x_2\left(\frac{\partial \mu_2}{\partial x_2}\right)_{T,P} = RT, \qquad \overline{V} = x_1 V_{01} + x_2 V_{02}, \left.\begin{array}{c} \\ \\ \end{array}\right\} \qquad (4\text{–}20.29)$$
$$V_1 = V_{01}, \qquad V_2 = V_{02}.$$

Here V_{0i} denotes the molar volume of the pure liquid component i (at the same values of T and P as the liquid mixture). Inserting Eq. (29) into Eq. (28) produces[85b]

$$\frac{D}{s} = \frac{RT(x_1 V_{01} + x_2 V_{02})^2}{x_1 V_{01}(M_2 V_{01} - M_1 V_{02})}. \qquad (4\text{–}20.30)$$

Ideal mixtures only arise if there are very similar species, i.e. they never occur for high-polymer solutions or for electrolyte solutions.

By comparison of Eq. (4–18.40) with Eq. (26) and on consideration of Eqs. (4–18.33) and (4–18.37) we find for the sedimentation coefficient of a highly dilute solution ($c_2 V_2 \ll 1$) of a completely dissociated electrolyte:

$$s = \frac{M_2^*}{F}(1 - \tilde{V}_2\rho)\frac{u_+^0 u_-^0}{u_+^0 + u_-^0}, \qquad (4\text{–}20.31)$$

where

$$M_2^* \equiv \frac{M_2}{z_+ \nu_+} = -\frac{M_2}{z_- \nu_-} \qquad (4\text{–}20.32)$$

denotes the equivalent mass of the electrolyte and u_i^0 the limiting value of the mobility of ionic species i for infinite dilution. Since $\tilde{V}_2\rho$ is not strongly concentration dependent at highly dilute solutions, s varies considerably less with the composition of the solution than does the diffusion coefficient D in the same region of concentration (note Eq. 4–18.40).

From Eq. (31) there follows immediately that:

$$s^0 = \frac{M_2^*}{F}(1 - \tilde{V}_2^0\rho_{01})\frac{u_+^0 u_-^0}{u_-^0 + u_-^0}. \qquad (4\text{–}20.33)$$

This equation can scarcely be distinguished from Eq. (31) but it provides a direct relation between the limiting value of the sedimentation coefficient s^0, the limiting values u_-^0 and u_+^0 of the ion mobilities, and the equivalent mass M_2^* of the electrolyte. This is also obtained using Eqs. (4–18.33) and (32) and the two limiting laws (4–18.43) and (27).

The theory behind the evaluation of experiments in the ultracentrifuge is discussed in the monograph of Fujita[90], which deals with the integration of the system of equations following from Eq. (3) with the mass balance under the existing boundary conditions.

4–21 SEDIMENTATION POTENTIAL

From the motion of charged particles in a gravitational or centrifugal field there arises an internal electric field which gives rise to a *sedimentation potential*. This is analogous to the diffusion potential (p. 291). If the system is initially homogeneous, i.e. if it shows neither temperature nor concentration gradients (a pressure gradient is formed practically instantaneously), then there is first a "sedimentation potential in the narrower sense." In the course of time, however, concentration gradients are built up by sedimentation, and they result in diffusion and in a diffusion potential. Thus, in isothermal media, there is, under arbitrary conditions, a "sedimentation potential in the wider sense" which is composed of the sedimentation potential in the narrower sense and the diffusion potential. The general equation for the above-mentioned electric potential follows from the equations of the thermodynamics of irreversible processes in the same way as do the formulas for the diffusion potential (see p. 292).

We combine the phenomenological equations (4–15.10) with the condition (4–15.13) for the disappearance of the total electric current and we note Eqs. (4–15.8) and (4–15.9). Then, for the gradient of the sedimentation potential φ_S in the wider sense, we obtain:

$$\text{grad } \varphi_S = \frac{\sum_i \sum_k a_{ik} z_i [(M_k - V_k \rho)\mathbf{g} - (\text{grad } \mu_k)_{T,P}]}{F \sum_i \sum_k a_{ik} z_i z_k} \qquad (4\text{–}21.1)$$

with the same meaning of the symbols as in §4–15 except that the vector $\mathbf{g}$ stands for gravitational and centrifugal acceleration. The summations are over all (charged and uncharged) species—with the exception of particle type 1 which normally corresponds to the "solvent." According to Eq. (4–16.45), for the reduced transport number τ_k of species k, there follows:

$$\tau_k = \frac{\sum_i a_{ki} z_i}{\sum_i \sum_k a_{ik} z_i z_k} . \qquad (4\text{–}21.2)$$

With the Onsager reciprocity law (4–14.4)

$$a_{ik} = a_{ki} \qquad (4\text{–}21.3)$$

we recover the expression (2) in Eq. (1). Therefore:

$$\boldsymbol{F} \operatorname{grad} \varphi_S = \sum_k \tau_k [(M_k - V_k \rho)\mathbf{g} - (\operatorname{grad} \mu_k)_{T,P}]. \tag{4–21.4}$$

On comparison with Eq. (4–18.23) we see that when $\mathbf{g} = 0$ ($\operatorname{grad} P = 0$) we obtain the formula for the diffusion potential. At vanishing concentration gradients $(\operatorname{grad} \mu_k)_{T,P} = 0$ we obtain the expression for the sedimentation potential in the narrower sense. This is valid for the initial state of a homogeneous medium with respect to the concentrations:

$$\boldsymbol{F} \operatorname{grad} \varphi_S = \sum_k \tau_k (M_k - V_k \rho)\mathbf{g} \quad \text{(initial state).} \tag{4–21.5}$$

After a sufficiently long time, by the counterbalance of diffusion and sedimentation, a stationary state appears which corresponds to the sedimentation equilibrium (§4–20). With the help of the classical equilibrium condition (4–20.10) we find:

$$z_k F \operatorname{grad} \varphi_S = (M_k - V_k \rho)\mathbf{g} - (\operatorname{grad} \mu_k)_{T,P} \quad \text{(final state),} \tag{4–21.6}$$

where the inferior k refers to any charged species and φ_S denotes the equilibrium value of the sedimentation potential.

With the condition (4–20.4)

$$\rho\mathbf{g} = \operatorname{grad} P \tag{4–21.7}$$

we write Eqs. (4), (5), and (6) in the following form:

$$\boldsymbol{F} \operatorname{grad} \varphi_S = \sum_k \tau_k \left[\left(\frac{M_k}{\rho} - V_k \right) \operatorname{grad} P - (\operatorname{grad} \mu_k)_{T,P} \right], \tag{4–21.8}$$

$$\boldsymbol{F} \operatorname{grad} \varphi_S = \sum_k \tau_k \left(\frac{M_k}{\rho} - V_k \right) \operatorname{grad} P \quad \text{(initial state),} \tag{4–21.9}$$

$$z_k F \operatorname{grad} \varphi_S = \left(\frac{M_k}{\rho} - V_k \right) \operatorname{grad} P - (\operatorname{grad} \mu_k)_{T,P} \quad \text{(final state).} \tag{4–21.10}$$

The sedimentation potential cannot be measured by itself, but can be determined experimentally only when combined with other potential differences (electrode potentials, etc.) (§4–22). In this respect, the sedimentation potential is analogous to the membrane potential (p. 183) and to the diffusion potential (p. 293).

The validity of Eqs. (4–15.10), our point of departure in the derivation of the above formulas, is, according to §4–15, restricted to fluid media in creeping motion and to isotropic crystals in simple cases (cf. also §4–5). The assumption of mechanical equilibrium, i.e. the assumption that the center of gravity acceleration vanishes and thus Eq. (7) holds, is not necessarily valid for the solid phases of gravitational and centrifugal cells. Actually, we must proceed from Eq. (4–13.1) where the generalized forces are given by (4–15.4). Such a process would unnecessarily complicate the calculations, for, in the last analysis, the contribution of the homogeneous effects caused by pressure drop and external fields in the solid phases, as compared to the total e.m.f. of the cell, are quite insignificant because the mass of the electrons is very small compared to the masses of the other species. We thus apply Eq. (8) to electronic conductors of a gravitational or centrifugal cell and this, due to the absence of concentration gradients, is reduced to Eq. (9). For electron conductors, the Hittorf reference system becomes identical to the lattice reference system (cf. p. 222). For the Hittorf transport number $t_\ominus$ of the electrons (the inferior $\ominus$ always relates to electrons), we have $t_\ominus = 1$. Thus, it follows from Eq. (4–16.42) with $z_\ominus = -1$ that $\tau_\ominus = -1$. We obtain:

$$F \operatorname{grad} \varphi_S = \left(V_\ominus - \frac{M_\ominus}{\rho} \right) \operatorname{grad} P \quad \text{(electronic conductors)}. \quad (4\text{–}21.11)$$

4-22 GRAVITATIONAL AND CENTRIFUGAL CELLS

A *gravitational cell* or a *centrifugal cell* is an isothermal galvanic cell with two chemically similar electrodes which are at various heights in the gravitational field or at different distances from the axis of rotation in a rotating system. The electrolyte solution in which the electrodes are immersed thus develops pressure and concentration gradients. If we start with an electrolyte solution which is homogeneous with reference to the concentrations then the e.m.f. of the cell measured at the beginning of the experiment will be called the "initial value of the e.m.f." After a sufficiently long (theoretically infinitely long) time a stationary state is set up, which corresponds to a sedimentation equilibrium (cf. §4–20 and §4–21). The e.m.f. of the cell measured for this case is called the "final value of the e.m.f." In the following, we will derive an expression for the e.m.f. valid for an arbitrary instant and thus for arbitrary pressure and concentration drops.

The cells described here are, in the most general case, irreversible galvanic cells since, when no current is flowing, irreversible processes (diffusion and sedimentation) occur in the solution. Only in the special case of sedimentation equilibrium are we dealing with reversible galvanic cells. Accordingly, the methods of the thermodynamics of irreversible processes must be used (for the calculation of the sedimentation potential, cf. §4–21). The formulas of classical thermodynamics are only sufficient for the determination of the final value of the e.m.f.

We write the phase diagram of a gravitational or centrifugal cell in the following form:

	a	b	c	d	
	Terminal (A)	Electrode (B)	Solution	Electrode (B)	Terminal (A)
Pressure:	P		P	$P + \Delta P$ $P + \Delta P$	P
Potential:	φ_I	φ_{II}	φ_{III} φ_{IV}	φ_V	φ_{VI} φ_{VII}

Here the electrodes and terminals are electronic conductors. In the most general case (see above), there still are concentration gradients in the solution besides the pressure drop. The pressure drop in the right-hand terminal arises from the fact that the metallic leads at the measuring apparatus must be under the same pressure (normal pressure). (The normal pressure corresponds to the electrode at the height of the measuring instrument or to the electrode near the rotation axis, for gravitational or centrifugal cells respectively.) The electric potential differences at the phase boundaries a, b, c, and d are determined under the assumption of local electrochemical equilibrium.

The e.m.f. of the cell, which we denote by Φ, is the difference between the electric potential of the right-hand terminal and that of the left-hand terminal, measured for zero current and for local heterogeneous equilibrium at the phase boundaries (cf. p. 297). We thus have:

$$\Phi = \varphi_{VII} - \varphi_I = (\varphi_{VII} - \varphi_{VI}) + (\varphi_{VI} - \varphi_V) + (\varphi_V - \varphi_{IV})$$
$$+ (\varphi_{IV} - \varphi_{III}) + (\varphi_{III} - \varphi_{II}) + (\varphi_{II} - \varphi_I). \tag{4-22.1}$$

The potential differences $\varphi_{VII} - \varphi_{VI}$ and $\varphi_{IV} - \varphi_{III}$ depend on sedimentation potentials (§4–21). All other electric tensions are assumed to be equilibrium potential differences which can be calculated with the help of classical thermodynamics.

From Eq. (4–21.11) we find, for the sedimentation potential† in the terminal on the right,

$$F(\varphi_{VII} - \varphi_{VI}) = \int_P^{P+\Delta P} \left[\frac{M_\ominus}{\rho_A} - (V_\ominus)_A \right] dP, \tag{4-22.2}$$

while we derive from Eq. (4–21.8), for the sedimentation potential in the solution,

$$F(\varphi_{IV} - \varphi_{III}) = \int_P^{P+\Delta P} \sum_k \tau_k \left[\left(\frac{M_k}{\rho} - V_k \right) dP - (d\mu_k)_{T,P} \right]. \tag{4-22.3}$$

Here F denotes the Faraday constant, ρ the density, M_k and V_k the molar mass and partial molar volume of species k respectively, and τ_k and μ_k the

---

† For the sake of brevity, we refer to "sedimentation potential" instead of "sedimentation potential difference."

reduced transport number and the chemical potential of species k. The inferior $\ominus$ ($_\ominus$) relates to electrons, the inferior A ($_A$) to the terminal substance. The summation in Eq. (3) is over all species of the solution except the solvent. The inferior T,P ($_{T,P}$) labeling of ($d\mu_k$) in Eq. (3) indicates that we are dealing with the change in the chemical potential at constant temperature T and at constant pressure P. That is, the second part of the integral in (3) depends on the concentration gradients in the solution (diffusion potential) and contains the concentrations as implicit integration variables.

The phase boundaries a and d are reversible to the electrons (indicated by inferior $\ominus$); the phase boundaries b and c are reversible to a definite ionic species of the solution (indicated by inferior i). We denote quantities which relate to the electrode substance and to the terminal substance by the inferior B and A respectively, while quantities without inferior labeling characterize the solution. From the classical condition (1–20.13) for electrochemical equilibrium we obtain for the potential jumps at the four phase boundaries:

$$F(\varphi_{VI} - \varphi_V) = (\mu_\ominus)_A - (\mu_\ominus)_B \quad \text{(pressure } P + \Delta P), \qquad (4\text{–}22.4)$$

$$F(\varphi_{II} - \varphi_I) = (\mu_\ominus)_B - (\mu_\ominus)_A \quad \text{(pressure } P), \qquad (4\text{–}22.5)$$

$$F(\varphi_V - \varphi_{IV}) = \frac{1}{z_i} [\mu_i - (\mu_i)_B] \quad \text{(pressure } P + \Delta P), \qquad (4\text{–}22.6)$$

$$F(\varphi_{III} - \varphi_{II}) = \frac{1}{z_i} [(\mu_i)_B - \mu_i] \quad \text{(pressure P)}, \qquad (4\text{–}22.7)$$

where z_i denotes the charge number of ionic species i. According to this, the values of μ_i in Eqs. (6) and (7) refer not only to different pressures but also to different concentrations.

From Eq. (4–15.2) there follows:

$$(d\mu_k)_T = V_k \, dP + (d\mu_k)_{T,P}. \qquad (4\text{–}22.8)$$

If we note this relation then we derive the general formula[91] for the e.m.f. of a gravitational or centrifugal cell from Eqs. (1) to (7) to be:

$$F\Phi = \int_P^{P+\Delta P} \left[\frac{M_\ominus}{\rho_A} - (V_\ominus)_B - \frac{(V_i)_B}{z_i} + \frac{V_i}{z_i} \right] dP$$

$$+ \int_P^{P+\Delta P} \sum_k \tau_k \left(\frac{M_k}{\rho} - V_k \right) dP$$

$$+ \frac{1}{z_i} \int_b^c (d\mu_i)_{T,P} - \int_b^c \sum_k \tau_k (d\mu_k)_{T,P}. \qquad (4\text{–}22.9)$$

In this, the last two integrals are taken over the concentrations of the solution as independent variables from the phase boundary b to the phase boundary c. As we recognize on comparison with Eq. (4–19.7), these two integrals

represent the expression for the e.m.f. of a concentration cell with transference. This result could be expected from the beginning, since for $\Delta P = 0$
the cell considered here becomes a concentration cell. If the last two integrals
in Eq. (9) are omitted then we have the expression for the initial value of the
e.m.f. The pressure difference ΔP can be calculated from the difference in
height or from the distances from the rotation axis.

In the electrode B, there is a homogeneous chemical equilibrium between
species a (the neutral atomic species belonging to the ion type i), the ion
species i, and the electrons:

$$a \rightleftharpoons i + z_i \ominus \quad \text{(phase B)}.$$

From this there follows with Eqs. (1–12.12) and (1–21.16):

$$\mu_a = (\mu_i)_B + z_i(\mu_\ominus)_B, \tag{4–22.10}$$

$$V_a = (V_i)_B + z_i(V_\ominus)_B. \tag{4–22.11}$$

Here μ_a and V_a denote the chemical potential and the partial molar volume
of the neutral atomic species in question in the electrode substance. If
electrodes of the second kind are present then Eqs. (10) and (11) remain
valid, provided that by μ_a and V_a we understand a certain linear combination
of chemical potentials and partial molar volumes. It can be shown, for
example[92], that the following relations hold:

$\mu_a = \mu_{Ag}$ for Ag electrodes (reversible for Ag^+),

$\mu_a = \mu_{Cd}$ for Cd amalgam electrodes (reversible for Cd^{2+}),

$\mu_a = \frac{1}{2}(\mu_{H_2})_{Gas}$ for H_2 electrodes (reversible for H^+),

$\mu_a = \mu_{AgCl} - \mu_{Ag}$ for Ag/AgCl electrodes (reversible for Cl^-).

With Eq. (11) we can make the following substitution in Eq. (9)

$$-(V_\ominus)_B - \frac{(V_i)_B}{z_i} + \frac{V_i}{z_i} = \frac{1}{z_i}(V_i - V_a). \tag{4–22.12}$$

As an example, consider the solution of a single electrolyte (component
2) in a neutral solvent (component 1). Let the electrolyte contain one cation
type (species $+$), one anion type (species $-$), and undissociated electrolyte
molecules (species u). Let the electrodes be reversible either for the cations or
for the anions of the electrolyte. Furthermore, we note (4–19.8) and (4–19.9):

$$t_k = z_k \tau_k \quad (k = +, -), \tag{4–22.13}$$

$$t_+ + t_- = 1, \tag{4–22.14}$$

where t_k denotes the Hittorf transport number of the ionic species k. Since
the last two integrals in Eq. (9) have already been worked out for the special
case considered in §4–19, we need now take account only of the first two

integrals in (9), which relate to the initial value of the e.m.f. Φ_0. We obtain from Eqs. (9), (12), (13), and (14):

$$F\Phi_0 = \int_P^{P+\Delta P} \left[\frac{M_\ominus}{\rho_A} - \frac{V_a}{z_+} + \frac{M_+}{z_+\rho} - \frac{t_-}{\rho}\left(\frac{M_+}{z_+} - \frac{M_-}{z_-}\right) \right.$$
$$\left. + \frac{\tau_u}{\rho} M_u + t_-\left(\frac{V_+}{z_+} - \frac{V_-}{z_-}\right) - \tau_u V_u \right] dP \qquad (4\text{--}22.15)$$

(electrodes reversible for cations),

$$F\Phi_0 = \int_P^{P+\Delta P} \left[\frac{M_\ominus}{\rho_A} - \frac{V_a}{z_-} + \frac{M_-}{z_-\rho} + \frac{t_+}{\rho}\left(\frac{M_+}{z_+} - \frac{M_-}{z_-}\right) \right.$$
$$\left. + \frac{\tau_u}{\rho} M_u - t_+\left(\frac{V_+}{z_+} - \frac{V_-}{z_-}\right) - \tau_u V_u \right] dP \qquad (4\text{--}22.16)$$

(electrodes reversible for anions).

Now there is valid (cf. Eqs. 4–19.13 and 4–19.14):

$$M_u = M_2, \qquad \frac{M_+}{z_+} - \frac{M_-}{z_-} = \frac{M_2}{z_+\nu_+} = -\frac{M_2}{z_-\nu_-}, \qquad (4\text{--}22.17)$$

$$V_u = V_2, \qquad \frac{V_+}{z_+} - \frac{V_-}{z_-} = \frac{V_2}{z_+\nu_+} = -\frac{V_2}{z_-\nu_-}. \qquad (4\text{--}22.18)$$

Here ν_+ and ν_- denote the dissociation numbers for the cations and anions respectively, and M_2 and V_2 the molar mass and the partial molar volume of the electrolyte. Furthermore, for the stoichiometric transport numbers ϑ_+ and ϑ_-, Eqs. (4–19.17) and (4–19.18) yield:

$$\frac{\vartheta_+}{z_+\nu_+} = \frac{1-\vartheta_-}{z_+\nu_+} = \frac{t_+}{z_+\nu_+} + \tau_u, \qquad (4\text{--}22.19)$$

$$\frac{\vartheta_-}{z_-\nu_-} = -\frac{1-\vartheta_+}{z_+\nu_+} = \frac{t_-}{z_-\nu_-} + \tau_u. \qquad (4\text{--}22.20)$$

Combination of formulas (15) to (20) produces:

$$F\Phi_0 = \int_P^{P+\Delta P} \left[\frac{M_\ominus}{\rho_A} + \frac{M_i}{z_i\rho} - \frac{V_a}{z_i} - \frac{1-\vartheta_i}{z_i\nu_i}\left(\frac{M_2}{\rho} - V_2\right) \right] dP. \quad (4\text{--}22.21)$$

Here the inferior i ($_i$) again refers to that ionic species for which the electrodes are reversible. As is obvious, the stoichiometric transport numbers for an electrolyte can be obtained from measurements of initial values of the e.m.f. of gravitational or centrifugal cells, provided that certain data on the density of the terminal, of the electrode substances, and of the electrolyte solution are available. A knowledge of activity coefficients, as with concentration cells with transference (§4–19), is not necessary. The term $M_\ominus/\rho_A$ containing the electron mass can usually be neglected. This is shown by a numerical comparison with the other terms of the integrand in Eq. (21). For the determination of the Hittorf transport numbers (t_+ and t_-), the considerations applying to concentration cells (p. 303) remain valid.

By adding Eq. (21) to Eq. (4–19.19) we derive an expression for the e.m.f. at an arbitrary instant (cf. above):

$$F\Phi = F\Phi_0 + \frac{1}{z_i \nu_i} \int_b^c (1 - \vartheta_i)(d\mu_2)_{T,P}. \qquad (4\text{--}22.22)$$

Here μ_2 is the chemical potential of the electrolyte.

There follows from Eqs. (4–20.4), (4–20.19), and (4–20.22) for the sedimentation equilibrium:

$$\left(\frac{M_2}{\rho} - V_2\right) dP = (d\mu_2)_{T,P}. \qquad (4\text{--}22.23)$$

With this Eqs. (21) and (22) produce for the final value of the e.m.f. Φ_∞:

$$F\Phi_\infty = \int_P^{P+\Delta P} \left(\frac{M_\ominus}{\rho_A} + \frac{M_i}{z_i \rho} - \frac{V_a}{z_i}\right) dP. \qquad (4\text{--}22.24)$$

Naturally, the transport numbers are no longer present in this expression.

From Eq. (24) with the help of Eq. (21) we derive immediately:

$$F(\Phi_\infty - \Phi_0) = \frac{1}{z_i \nu_i} \int_P^{P+\Delta P} (1 - \vartheta_i)\left(\frac{M_2}{\rho} - V_2\right) dP. \qquad (4\text{--}22.25)$$

The determination of transport numbers according to this equation is obviously simpler than the corresponding operation using Eq. (21). In practice, however, Φ_∞ is difficult to obtain from measurements, while Φ_0 has been experimentally determined many times.

Des Coudres[93] and Grinnell and Koenig[94] have published measurements of Φ_0 for gravitational cells. Corresponding experiments on centrifugal cells stem from des Coudres[95], Tolman[96, 97], and MacInnes, Ray and others[98–100]. However, Miller[101] was the first to develop the rigorous theory. Then Schoenert[91] independently developed the theory in the sense described here.

If we reassemble all procedures for the determination of transport numbers (cf. §4–16 and §4–19), we obtain the following summary:

a) Hittorf method (abbreviated "Hittorf").
b) Moving boundary method ("mov. boundary").
c) Evaluation from e.m.f. measurements on concentration cells according to Eq. (4–19.23) ("conc. cells").
d) Evaluation from e.m.f. measurements on gravitational or centrifugal cells according to Eq. (21) ("grav. cells" or "centr. cells").

In order to show the agreement attained using the present methods of measurement and the different experimental values from the above-mentioned methods, we present a few examples of measured transport numbers in Table 9. We tabulate the stoichiometric transport number of the cation constituent of the electrolyte which, for dilute solutions of a strong electrolyte, coincides with the Hittorf transport number of the cation.

Table 9

Cation Transport Numbers for Aqueous Electrolyte Solutions at 25°C According to Various Methods[94, 102, 103]

Electrolyte	Molarity, c mol l^{-1}	Hittorf	Mov. boundary	Conc. cell	Grav. cell	Centr. cell
HCl	0.005		0.8239	0.824		
	0.01		0.8251	0.825		
	0.02		0.8266	0.827		
	0.05		0.8292	0.830		
	0.1		0.8314	0.830		
LiCl	0.01	0.329	0.3289			
	0.02	0.327	0.3261			
	0.05	0.323	0.3211			
	0.1	0.319	0.3168			
KI	0.1941		0.4887			0.4873
KI (20°C) with smaller addition of I$_2$	~1				0.490	0.486

4–23 PRESSURE DIFFUSION

We consider an isothermal fluid mixture in which a pressure drop is maintained. The pressure gradient shall not, as with sedimentation, be tied to the gravitational field or to a centrifugal field. If the pressure gradient arises indirectly from gravity, it need not lie in the direction of the earth's field. Accordingly, the conditions for "creeping motion" or "mechanical equilibrium" are no longer valid. The thermodynamic–phenomenological description of the transport processes must result from Eqs. (4–13.1).

The phenomena of interest to us here are *diffusion*, i.e. the particle migration due to the concentration gradients, as already described, and *pressure diffusion*, i.e. particle transport due to pressure gradients. For the sake of brevity, we restrict the discussion to a mixture of two uncharged species.

We denote the diffusion current density of species i, defined by Eq. (4–3.1), by $\mathbf{J}_i$ ($i = 1, 2$), the weight factors introduced in Eq. (4–3.3) for averaging the velocities by ω_1 and ω_2, the molar volume of the mixture by

$\overline{V}$, the mole fraction of species i by x_i, and the pressure by P. We then describe the simultaneous occurrence of diffusion and pressure diffusion in an isothermal fluid with two uncharged species by an extension of Eq. (4–17.25):

$$\mathbf{J}_i = -\frac{1 - \omega_i}{1 - x_i}\frac{1}{\overline{V}}\left(D \text{ grad } x_i \mp D_P \frac{\text{grad } P}{P}\right) \quad (i = 1, 2), \quad (4\text{–}23.1)$$

where the upper or lower sign holds for $i = 1$ or $i = 2$ respectively.† The *diffusion coefficient* D (§4–17) and the *pressure diffusion coefficient* D_P are introduced by Eq. (1). Equation (1) is in accord with results from the kinetic theory of gases. Experimental examinations on the pressure diffusion do not seem to be available yet.‡

If the Fick reference system is taken as a basis, there follows from Eq. (1) using Eqs. (4–17.23), (4–17.24), (4–17.26), and (4–17.27):

$$_w\mathbf{J}_1 = -\frac{V_2}{\overline{V}^2}\left(D \text{ grad } x_1 - D_P \frac{\text{grad } P}{P}\right), \quad (4\text{–}23.2\text{a})$$

$$_w\mathbf{J}_2 = -\frac{V_1}{\overline{V}^2}\left(D \text{ grad } x_2 + D_P \frac{\text{grad } P}{P}\right). \quad (4\text{–}23.2\text{b})$$

Here V_i denotes the partial molar volume of species i. From Eq. (2) with grad $P = 0$ we obtain Fick's law in the form (4–17.30).

If we proceed from the molecular reference system, we find from Eq. (1) with the help of Eq. (4–3.21):

$$_u\mathbf{J}_i = -\frac{1}{\overline{V}}\left(D \text{ grad } x_i \mp D_P \frac{\text{grad } P}{P}\right) \quad (i = 1, 2). \quad (4\text{–}23.3)$$

When grad $P = 0$ this relation becomes Fick's law in the form of (4–17.34).

---

† The identity (4–17.22)

$$\frac{\omega_1}{c_1}\mathbf{J}_1 + \frac{\omega_2}{c_2}\mathbf{J}_2 = 0$$

must be satisfied.

‡ If we stipulate $\rho\mathbf{g} = $ grad P (mechanical equilibrium), we obtain the special case of sedimentation. For grad $x_2 = 0$, Eq. (1) then reads (see Eq. 4–20.2b):

$$\mathbf{J}_2 = \frac{c_2\omega_1}{c_1 V_1}s\mathbf{g} = -\frac{\omega_1\rho}{x_1 P\overline{V}}D_P\mathbf{g} \quad (\text{grad } x_2 = 0),$$

where c_i denotes the molarity of species i, V_1 the partial molar volume of species 1, s the sedimentation coefficient, $\mathbf{g}$ the gravitational or centrifugal acceleration, and ρ the density. From this and Eqs. (1–6.4) and (1–6.5) we derive:

$$s = -\frac{\rho V_1}{P\overline{V}x_2}D_P. \quad (4\text{–}23.1\text{a})$$

This is the general relation between the two transport coefficients s and D_P.

Finally, if we use the barycentric system then we derive from Eq. (1) with Eqs. (4–3.6), (4–3.7), (4–3.8), (4–3.10), (4–3.12), (4–3.14), and (4–3.18):

$$M_i \,_v\!J_i = J_i^* = - \frac{M_1 M_2}{(M_1 x_1 + M_2 x_2)^2} \, \rho \left(D \, \text{grad} \, x_i \mp D_P \frac{\text{grad} \, P}{P} \right)$$

$$(i = 1, 2). \qquad (4\text{–}23.4)$$

Here ρ is the density of the mixture and M_i the molar mass of species i. If we note the relation following from Eqs. (4–3.6), (4–3.7), (4–3.8), (4–3.12), and (4–3.18)

$$\text{grad} \, x_i = \frac{(M_1 x_1 + M_2 x_2)^2}{M_1 M_2} \, \text{grad} \, \chi_i \quad (i = 1, 2), \qquad (4\text{–}23.5)$$

where χ_i is the mass fraction of species i, then from Eq. (4) we find:

$$J_i^* = - D\rho \, \text{grad} \, \chi_i \pm \frac{M_1 M_2 \rho}{(M_1 x_1 + M_2 x_2)^2} \, D_P \frac{\text{grad} \, P}{P} \quad (i = 1, 2). \qquad (4\text{–}23.6)$$

For $\text{grad} \, P = 0$, Eq. (6) is reduced to Fick's law in the form of (4–17.35).

From the phenomenological equations (4–13.1) with Eq. (4–13.4) for our case ($N = 2$, $\text{grad} \, T = 0$, $\mathbf{K}_k = 0$) we find the following relation:†

$$J_2^* = a \left[\frac{1}{M_1} (\text{grad} \, \mu_1)_T - \frac{1}{M_2} (\text{grad} \, \mu_2)_T \right]. \qquad (4\text{–}23.7)$$

Here μ_i is the chemical potential of substance i and a is the phenomenological coefficient. The inferior T $(_T)$ indicates that the gradient is formed at constant temperature T.

With the help of Eqs. (4–3.18), (4–20.18), and (4–22.8) there follows from Eq. (7)

$$J_2^* = -a \left[\left(\frac{V_2}{M_2} - \frac{V_1}{M_1} \right) \text{grad} \, P + \frac{M_1 x_1 + M_2 x_2}{M_1 M_2 x_1} \left(\frac{\partial \mu_2}{\partial x_2} \right)_{T,P} \text{grad} \, x_2 \right]. \qquad (4\text{–}23.8)$$

On elimination of a from Eqs. (4) and (8) there results:

$$\frac{D_P}{D} = \frac{M_1 M_2 P x_1}{M_1 x_1 + M_2 x_2} \left(\frac{V_2}{M_2} - \frac{V_1}{M_1} \right) \frac{1}{(\partial \mu_2 / \partial x_2)_{T,P}}. \qquad (4\text{–}23.9)$$

This is the general relation[104] between the pressure diffusion coefficient D_P and the diffusion coefficient D.‡

---

† Equation (7) also remains valid if we allow a gravitational or centrifugal field, for, according to Eqs. (4–13.1), (4–13.4), and (4–15.4), the term relating to this drops out in the expression for J_2^*.

‡ From Eqs. (1a) and (9) we obtain the general relation between the diffusion coefficient D and the sedimentation coefficient s:

$$\frac{D}{s} = \frac{x_2 V}{V_1 (M_2 - V_2 \rho)} \left(\frac{\partial \mu_2}{\partial x_2} \right)_{T,P}. \qquad (4\text{–}23.9a)$$

This formula is equivalent to Eq. (4–20.28) derived previously by another method.

According to Eq. (9) and the inequalities (1–22.38) and (4–17.41), D_P has the same sign as the expression

$$\frac{V_2}{M_2} - \frac{V_1}{M_1} = \tilde{V}_2 - \tilde{V}_1,$$

where $\tilde{V}_i$ denotes the partial specific volume of species i. According to Eqs. (1) to (4), that substance whose partial specific volume is greater migrates, at vanishing concentration gradient, from the region of higher pressure to the region of lower pressure.

For *ideal gas mixtures*:

$$V_1 = V_2 = \frac{RT}{P}, \qquad \left(\frac{\partial\mu_2}{\partial x_2}\right)_{T,P} = \frac{RT}{x_2}. \tag{4-23.10}$$

Here R is the gas constant. From Eqs. (9) and (10) there follows:†

$$\frac{D_P}{D} = \frac{(M_1 - M_2)x_1 x_2}{M_1 x_1 + M_2 x_2}. \tag{4-23.11}$$

For pressure diffusion in binary ideal gas mixtures, accordingly, that component whose molecular weight is smaller migrates from the higher to the lower pressure.

If at a given pressure gradient there is no concentration gradient, a concentration displacement appears due to the pressure diffusion. This results in a diffusion which again seeks to equalize concentrations. We thus expect that a *stationary state* arises after a sufficiently long time due to the opposing effects of pressure diffusion and diffusion and that this is characterized by Eq. (4–4.26) for the local disappearance of the diffusion flux ($\mathbf{J}_i = 0$). From Eqs. (1) and (9) with $\mathbf{J}_i = 0$ we obtain the stationary concentration distribution in an isothermal fluid binary mixture:

$$\left(\frac{\partial\mu_2}{\partial x_2}\right)_{T,P} \text{grad } x_2 = \frac{M_1 M_2 x_1}{M_1 x_1 + M_2 x_2}\left(\frac{V_1}{M_1} - \frac{V_2}{M_2}\right)\text{grad } P. \tag{4-23.12}$$

With the help of Eqs. (4–3.6), (4–3.12), (4–3.17), (4–3.18), and (4–3.26) we derive

$$\frac{M_1 M_2 x_1}{M_1 x_1 + M_2 x_2}\left(\frac{V_1}{M_2} - \frac{V_2}{M_2}\right) = \frac{M_2}{\rho} - V_2. \tag{4-23.13}$$

Equation (12) is, accordingly, formally identical to condition (4–22.23) for a sedimentation equilibrium. This fact is not surprising. If the pressure drop for a pressure diffusion experiment is produced by a gravitational or centrifugal field and we then wait for the stationary state, we must obtain the concentration distribution characteristic of sedimentation equilibrium.

---

† The kinetic theory of gases[105] also yields Eq. (11).

For many purposes, it is practical to introduce the dimensionless quantity

$$\alpha_P \equiv \frac{D_P}{Dx_1x_2} \qquad (4\text{-}23.14)$$

which we call the *pressure diffusion factor*. For gases, we can write Eqs. (9) and (11) in the following form:

$$\alpha_P = \frac{P(M_1V_2 - M_2V_1)}{(M_1x_1 + M_2x_2)x_2(\partial\mu_2/\partial x_2)_{T,P}}, \qquad (4\text{-}23.15)$$

$$\alpha_P^0 = \lim_{P \to 0} \alpha_P = \frac{M_1 - M_2}{M_1x_1 + M_2x_2}. \qquad (4\text{-}23.16)$$

In Eq. (16), we have used the statement that any gas mixture in the limit $P \to 0$ becomes an ideal gas mixture. We have valid (cf. Eq. 10)

$$RT = PV_{0i} \quad (i = 1, 2), \qquad (4\text{-}23.17)$$

where V_{0i} denotes the molar volume of the pure gas i at the temperature T for $P \to 0$ (i.e. in the ideal gas state). We obtain by combination of Eqs. (15), (16), and (17):

$$\alpha_P = \frac{\alpha_P^0(M_1x_1 + M_2x_2)RT + M_1P(V_2 - V_{02}) - M_2P(V_1 - V_{01})}{(M_1x_1 + M_2x_2)x_2(\partial\mu_2/\partial x_2)_{T,P}}. \qquad (4\text{-}23.18)$$

According to this relation, the pressure diffusion factor α_P for an arbitrary pressure P can be determined from the thermodynamic properties of the (real) gas mixture, i.e. from the equation of state, provided the pressure diffusion factor α_P^0 at vanishing pressure ($P \to 0$) is known for a given composition and temperature. Equation (18) is significant as the basis for an analogous equation for the thermal diffusion factor (p. 365).

C. Nonisothermal Processes

4–24 GENERAL

(a) Fundamental Equations

If arbitrary irreversible processes occur in a nonisothermal system, we must consider transport phenomena other than those which we already know (electric conduction, diffusion, sedimentation, pressure diffusion). Thus we shall deal with heat conduction, i.e. the heat transport by temperature gradients, as well as various cross phenomena and their reciprocal effects (thermoelectric effects, thermal diffusion, Dufour effect, etc.). Chemical reactions can also take place in our system.

All these processes are, under the assumptions mentioned at the beginning of §4–15, described by the phenomenological equations in §4–13 with the Onsager reciprocity relations in §4–14. We now discuss the individual equations as they occur.

From Eq. (4–13.1) we obtain the relation for the diffusion current density $\mathbf{J}_i^*$ of species i (in the barycentric system) in a mixture of N particle types

$$\mathbf{J}_i^* = \sum_{k=2}^{N} \alpha_{ik}(\mathbf{X}_k^* - \mathbf{X}_1^*) + \alpha_{iQ}\mathbf{X}_Q \quad (i = 2, 3, \ldots, N). \qquad (4\text{–}24.1)$$

Here the α_{ik} and α_{iQ} are the phenomenological coefficients. For the generalized forces $\mathbf{X}_i^*$ and $\mathbf{X}_Q$, according to Eqs. (4–13.4), (4–13.5), (4–15.4), and (4–15.6), we find:

$$\mathbf{X}_i^* = \mathbf{g} - \frac{z_i}{M_i} F \operatorname{grad} \varphi - \frac{V_i}{M_i} \operatorname{grad} P - \frac{1}{M_i} (\operatorname{grad} \mu_i)_{T,P}$$

$$(i = 1, 2, \ldots, N), \qquad (4\text{–}24.2)$$

$$\mathbf{X}_Q = -\frac{1}{T} \operatorname{grad} T. \qquad (4\text{–}24.3)$$

Here $\mathbf{g}$ denotes the gravitational or centrifugal acceleration, F the Faraday constant, φ the electric potential, P the pressure, T the absolute temperature, z_i the charge number, M_i the molar mass, V_i the partial molar volume, and μ_i the chemical potential of species i. The particle type 1 plays the role of the "solvent" or, more generally, that of the "reference species."

According to Eq. (4–13.2), we write for the heat current density:

$$\mathbf{J}_Q = \sum_{i=2}^{N} \alpha_{Qi}(\mathbf{X}_i^* - \mathbf{X}_1^*) + \alpha_{QQ}\mathbf{X}_Q \qquad (4\text{–}24.4)$$

with the phenomenological coefficients α_{Qi} and α_{QQ}.

The Onsager reciprocity law (4–14.1) and (4–14.2), states, for the coefficients in Eqs. (1) and (4), that

$$\alpha_{ik} = \alpha_{ki} \quad (i, k = 2, 3, \ldots, N), \qquad (4\text{–}24.5)$$

$$\alpha_{iQ} = \alpha_{Qi} \quad (i = 2, 3, \ldots, N). \qquad (4\text{–}24.6)$$

Finally, according to Eq. (4–13.3) for states near to equilibrium, we can set up the reaction rate b_r of each chemical reaction r as a homogeneous linear function of the affinities $A_1, A_2, \ldots, A_R$:

$$b_r = \sum_{s=1}^{R} a_{rs}A_s \quad (r = 1, 2, \ldots, R), \qquad (4\text{–}24.7)$$

where R is the number of the independent reactions. Onsager's reciprocity law (4–14.3) gives for the phenomenological coefficients a_{rs}:

$$a_{rs} = a_{sr} \quad (r, s = 1, 2, \ldots, R). \tag{4-24.8}$$

In most cases, we can assume "creeping motion" for fluids and "simple cases" for isotropic crystals as well as the absence of space charges. Then, in place of Eqs. (1) and (4), Eqs. (4–13.6) and (4–13.7) hold:

$$_1\mathbf{J}_i = \sum_{k=2}^{N} a_{ik}\mathbf{X}_k + a_{iQ}\mathbf{X}_Q \quad (i = 2, 3, \ldots, N), \tag{4-24.9}$$

$$\mathbf{J}_Q = \sum_{i=2}^{N} a_{Qi}\mathbf{X}_i + a_{QQ}\mathbf{X}_Q. \tag{4-24.10}$$

Here $_1\mathbf{J}_i$ is the diffusion current density of species i in the Hittorf reference system and $\mathbf{X}_i$ the appropriate generalized force, given by Eqs. (4–15.8) and (4–15.9):

$$\mathbf{X}_i = (M_i - \rho V_i)\mathbf{g} - z_i F \operatorname{grad} \varphi - (\operatorname{grad} \mu_i)_{T,P} \quad (i = 2, 3, \ldots, N), \tag{4-24.11}$$

where ρ denotes the density. The a_{ik}, a_{iQ}, a_{Qi}, and a_{QQ} are the phenomenological coefficients connected to one another by the Onsager reciprocity relations (4–14.4) and (4–14.5):

$$a_{ik} = a_{ki} \quad (i, k = 2, 3, \ldots, N), \tag{4-24.12}$$

$$a_{iQ} = a_{Qi} \quad (i = 2, 3, \ldots, N). \tag{4-24.13}$$

We also require the inequality statements which follow from (4–14.10) and (4–14.11). We write these in the form:

$$a_{ii} > 0, \quad a_{ii}a_{jj} - a_{ij}^2 > 0, \quad (i, j = 2, 3, \ldots, N, Q; i \neq j), \tag{4-24.14}$$

. . . .

$$|a_{ik}| > 0, \qquad |a_{ij}| > 0. \tag{4-24.15}$$

Here

$$|a_{ij}| \equiv \begin{vmatrix} a_{22} & a_{23} & \cdots & a_{2N} & a_{2Q} \\ a_{32} & a_{33} & \cdots & a_{3N} & a_{3Q} \\ \cdot & \cdot & \cdot & \cdot & \cdot \\ a_{N2} & a_{N3} & \cdots & a_{NN} & a_{NQ} \\ a_{Q2} & a_{Q3} & \cdots & a_{QN} & a_{QQ} \end{vmatrix} \tag{4-24.16}$$

and $|a_{ik}|$ is the principal cofactor arising by the omission of the last row and column of the determinant (16).

For further considerations in this section, we will proceed from the relations (9) to (16).

(b) Heats of Transport

We define $N - 1$ quantities $*Q_k$ by the following equations (cf. Eq. 3–9.8):

$$a_{iQ} = \sum_{k=2}^{N} a_{ik} *Q_k \quad (i = 2, 3, \ldots, N).\qquad (4\text{–}24.17)$$

This system of $N - 1$ independent equations, on the basis of the first inequality in (15), has one and only one solution $*Q_2, *Q_3, \ldots, *Q_N$. On noting the reciprocity relations (12) and (13) (cf. Eq. 3–9.9), Eqs. (9), (10), and (17) yield:

$$\mathbf{J}_Q = \sum_{i=2}^{N} *Q_{i\,1}\mathbf{J}_i + \left(a_{QQ} - \sum_{i=2}^{N} a_{iQ} *Q_i\right)\mathbf{X}_Q.\qquad (4\text{–}24.18)$$

From this the physical significance of the quantities formally defined by Eq. (17) results: $*Q_i$ is the heat transported through a solvent-fixed reference plane at vanishing temperature gradients ($\mathbf{X}_Q = 0$) divided by the amount of substance i. Thus $*Q_i$ is called the *heat of transport* of species i (in the Hittorf reference system).

For brevity, we denote here and in the following the particle type 1, i.e. the reference species of the Hittorf reference system, as the "solvent." This nomenclature is to be taken literally only in the case of electrolyte solutions or other liquid mixtures with a predominant concentration of one species. For solid electron conductors (cf. below), the ions in the crystal lattice will be chosen as the reference particle type; for liquid electron conductors (e.g. mercury), the positive ions; for solid and molten electrolytes (cf. p. 257), certain ions. For a gas mixture, the choice of the reference species is quite arbitrary.

In the literature, there is an abundance of quantities known as "heat of transport" or "heat of transfer" (symbol Q_i^* or similar). Determinations of the concept "heat of transport" deviating from our definition arise from three different possibilities:

1. Another choice of the heat current density (§4–7).
2. Another choice of diffusion current density and thus of the reference system (§4–3).
3. Not noting the dependencies among the "forces" and "fluxes," so that linearly dependent phenomenological coefficients and thus dependent quantities appear as heats of transport.

Our definition is the most practical for applications and corresponds to the original determination of the concept "heat of transfer" by Eastman and Wagner (cf. §3–9). Also, our quantity $*Q_i$ is independent of the arbitrary energy zero points (§4–7).

From Eqs. (9) and (17) there follows directly (cf. Eq. 3–9.10):

$$_1\mathbf{J}_i = \sum_{k=2}^{N} a_{ik}(\mathbf{X}_k + {}^*Q_k\mathbf{X}_Q) \quad (i = 2, 3, \ldots, N). \qquad (4\text{–}24.19)$$

From these equations, which do not contain the Onsager reciprocity law, the ${}^*Q_k$ enter the formulas for the stationary state (see below). Meanwhile, it should be noted that the physical interpretation of ${}^*Q_k$ as "heat of transport" was first made possible by Eq. (18) and that in the derivation of this equation the reciprocity relations were used.

(c) Transported Entropies

We define, moreover,

$$ {}^*S_i \equiv \frac{{}^*Q_i}{T} + S_i \quad (i = 2, 3, \ldots, N), \qquad (4\text{–}24.20) $$

where S_i is the partial molar entropy of species i. On considering Eqs. (4–3.30) and (4–9.5), insertion of Eq. (20) into Eq. (18) produces

$$ _1\mathbf{J}'_S = \frac{\mathbf{J}_Q}{T} + \sum_{i=2}^{N} S_i \,_1\mathbf{J}_i $$

$$ = \sum_{i=2}^{N} {}^*S_i \,_1\mathbf{J}_i + \frac{1}{T}\left(a_{QQ} - \sum_{i=2}^{N} a_{iQ} \,{}^*Q_i\right)\mathbf{X}_Q. \qquad (4\text{–}24.21) $$

Here $_1\mathbf{J}'_S$ denotes the nonconvective part of the entropy current density in the Hittorf reference system. On the basis of Eq. (21) we give the following physical interpretation of the quantity ${}^*S_i$ formally introduced by Eq. (20): ${}^*S_i$ is the entropy transported through a solvent-fixed reference plane at vanishing temperature gradients divided by the amount of substance i. Thus we denote ${}^*S_i$ concisely as the *transported entropy* of species i (in the Hittorf reference system).

　　　Also, in this case, there are many other definitions in the literature. In particular, it should be noted that Eastman's "entropy of transfer" or "entropy of transport" S^* represents the quantity ${}^*Q_i/T$. Thus we denote our quantity ${}^*S_i$ as the ' transported entropy," following Agar.

The transported entropy ${}^*S_i$ plays a central role in nonisothermal transport processes in electron conductors (§4–25) and electrolytes (§4–31), because it can be experimentally determined for each individual species, provided we assign definite values to certain arbitrary constants, as will be explained later.

(d) Stationary State

If the temperature distribution at the boundaries of a closed continuous system is fixed then after a sufficiently long time a *stationary state* will be set up by the combined effects of the different irreversible processes inside the system, i.e. a nonequilibrium state, in which all intensive quantities (temperature, pressure, concentrations, etc.) are constant in time although varying from place to place (cf. §1–17). Here it is assumed that any external force fields do not change in time. If chemical reactions and convection are excluded then we have according to Eq. (4–4.26):

$$_1\mathbf{J}_i = 0 \quad (i = 2, 3, \ldots, N). \tag{4–24.22}$$

From this with Eq. (19) there results from the first inequality (15):

$$\mathbf{X}_i + {}^*Q_i\mathbf{X}_Q = 0 \quad (i = 2, 3, \ldots, N) \tag{4–24.23}$$

or with Eqs. (3) and (11):

$$(\rho V_i - M_i)\mathbf{g} + z_i F \operatorname{grad} \varphi + (\operatorname{grad} \mu_i)_{T,P} + \frac{{}^*Q_i}{T} \operatorname{grad} T = 0$$

$$(i = 2, 3, \ldots, N). \tag{4–24.24}$$

In this formula, which holds for a stationary state in a nonisothermal system, the heat of transport ${}^*Q_i$ appears (cf. Eq. 3–9.22). With $\operatorname{grad} T = 0$ the classical condition (4–20.10) for the sedimentation equilibrium results.

The specialization of Eq. (24) for $\operatorname{grad} T \neq 0$ which is most important in practice relates to the case $\mathbf{g} = 0$ ($\operatorname{grad} P = 0$). With the formula thus simplified the heats of transport of a nonelectrolyte or of an electrolyte (but not that of a single ion species) can be determined experimentally from the stationary concentration distribution in a temperature field. Two simple examples should illustrate this.

In the first case, let a mixture of two uncharged species be present. We then obtain from Eq. (24) with $\mathbf{g} = 0$, $z_i = 0$, $N = 2$:

$$(\operatorname{grad} \mu_2)_{T,P} + \frac{{}^*Q_2}{T} \operatorname{grad} T = 0. \tag{4–24.25}$$

The heat of transport ${}^*Q_2$ of species 2 can thus be measured.

As a second example, we consider the solution of an electrolyte (component 2) of two ion species in a neutral solvent (component 1 = species 1). We characterize the cations, anions, and undissociated molecules by the inferior $+$, $-$, and u respectively. Then there follows from Eq. (24) with $\mathbf{g} = 0$†

$$z_+ F \operatorname{grad} \varphi + (\operatorname{grad} \mu_+)_{T,P} + \frac{*Q_+}{T} \operatorname{grad} T = 0, \qquad (4\text{–}24.26a)$$

$$z_- F \operatorname{grad} \varphi + (\operatorname{grad} \mu_-)_{T,P} + \frac{*Q_-}{T} \operatorname{grad} T = 0, \qquad (4\text{–}24.26b)$$

$$(\operatorname{grad} \mu_u)_{T,P} + \frac{*Q_u}{T} \operatorname{grad} T = 0. \qquad (4\text{–}24.27)$$

According to Eqs. (4–18.17) and (4–18.18), we have:

$$\nu_+ z_+ + \nu_- z_- = 0, \qquad (4\text{–}24.28)$$

$$\nu_+ \mu_+ + \nu_- \mu_- = \mu_u = \mu_2. \qquad (4\text{–}24.29)$$

Here ν_+ and ν_- denote the dissociation numbers for the cations and anions respectively and μ_2 the chemical potential of the electrolyte. With

$$*Q_2 \equiv \nu_+ \, *Q_+ + \nu_- \, *Q_- \qquad (4\text{–}24.30)$$

we obtain from Eqs. (26) to (29):

$$(\operatorname{grad} \mu_2)_{T,P} = -\frac{*Q_2}{T} \operatorname{grad} T = -\frac{*Q_u}{T} \operatorname{grad} T. \qquad (4\text{–}24.31)$$

The quantity $*Q_2 = *Q_u$, the "heat of transport of the electrolyte," can be determined experimentally from Eq. (31).

   If we combine Eq. (4–18.4) with Eq. (18) and take account of Eq. (3) and Eq. (30), we then find for the present case:

$$\begin{aligned} \mathbf{J}_Q &= [\alpha(\nu_+ \, *Q_+ + \nu_- \, *Q_- - *Q_u) + *Q_u] \, _1\mathbf{J}_2 \\ &= [\alpha(*Q_2 - *Q_u) + *Q_u] \, _1\mathbf{J}_2 \quad (\operatorname{grad} T = 0), \end{aligned}$$

where $_1\mathbf{J}_2$ is the diffusion current density of the electrolyte in the Hittorf reference system and α the degree of dissociation of the electrolyte. With the relation $*Q_u = *Q_2$ obtained above (which in spite of its derivation from the stationary state conditions must be generally valid, since the heats of transport depend only on the local state variables) there results from this:

$$\mathbf{J}_Q = *Q_2 \, _1\mathbf{J}_2 \quad (\operatorname{grad} T = 0).$$

Thus it becomes clear that $*Q_2$ may be denoted as the "heat of transport of the electrolyte."

---

† In Eqs. (22) and (24), chemical reactions and thus, in particular, dissociation reactions are indeed excluded. It can, however, be shown (§4–27) that at local dissociation equilibrium the relations (22) and (24) are valid also for an incompletely dissociated electrolyte.

(e) Thermal Conductivity

By *heat conduction* we mean the heat flow caused by temperature gradients. For systems in which no chemical reactions occur, it is expedient to denote the coefficient of $-\operatorname{grad} T$ in the general equation for $\mathbf{J}_Q$ as the *thermal conductivity* λ. The quantity λ, however, is unambiguously defined only if we state the particular conditions under which the heat transport occurs. If the system is not under the influence of external force fields and if all intensive quantities apart from the temperature (pressure, electric potential, concentrations) are constant in position† then, for the sake of brevity, we speak of a "uniform state" and call the value of λ the "thermal conductivity for the uniform state λ_0." If, by the superposition of different transport processes (diffusion, thermal diffusion, etc.), a stationary state is set up then the value of λ is denoted as the "thermal conductivity for the stationary state λ_∞."

Under other conditions λ has another value. We here restrict the discussion to the quantities λ_0 and λ_∞ and assume fluids in creeping motion or isotropic crystals in simple cases, as well as the lack of space charges and the absence of chemical reactions. Under these conditions we can proceed from Eqs. (10), (11), (18), and (22). The heat conduction in chemically reacting media will be discussed in §4–28.

If a system without external force fields is initially (at the time $t = 0$) in a uniform state then the thermal conductivity λ_0 is measured. In the course of time, concentration gradients arise by phenomena such as thermal diffusion, etc. (also potential gradients if possible), and we find some intermediate value of λ. After a sufficiently long (theoretically infinitely long) time ($t \to \infty$) a stationary state is set up in which the diffusion currents disappear due to the opposing effects of diffusion and thermal diffusion. Now the thermal conductivity λ_∞ is measured. In each case, λ is independent of the temperature gradient and of the heat current density (excluding extremely high gradients) but can still be an arbitrary function of the local state variables (temperature, pressure, and concentrations).

The uniform state is, according to Eq. (11), described by the condition $\mathbf{X}_i = 0$. We thus derive from Eqs. (3) and (10):

$$\mathbf{J}_Q = -\lambda_0 \operatorname{grad} T = -\frac{a_{QQ}}{T} \operatorname{grad} T \quad \text{(uniform state)}. \quad (4\text{–}24.32)$$

From this with (14) there follows:

$$\lambda_0 = \frac{a_{QQ}}{T} > 0. \quad (4\text{–}24.33)$$

---

† As function of temperature, the density naturally varies from place to place.

Equation (32) is Fourier's law which was originally set up as an empirical equation.†

The stationary state according to Eq. (22) is described by the condition $_1J_i = 0$. Thus from Eqs. (3) and (18) we find:

$$\mathbf{J}_Q = -\lambda_\infty \text{ grad } T = -\frac{1}{T}\left(a_{QQ} - \sum_{i=2}^{N} a_{iQ} {}^*Q_i\right) \text{ grad } T \quad \text{(stationary state)},$$

(4–24.34)

hence:

$$\lambda_\infty = \frac{1}{T}\left(a_{QQ} - \sum_{i=2}^{N} a_{iQ} {}^*Q_i\right).$$

(4–24.35)

Equation (34) again takes the form of Fourier's law.

Equations (18) and (21) can, with Eqs. (3) and (35), be written in the following form:

$$\mathbf{J}_Q = \sum_{i=2}^{N} {}^*Q_i \, _1\mathbf{J}_i - \lambda_\infty \text{ grad } T,$$

(4–24.36a)

$$_1\mathbf{J}'_S = \sum_{i=2}^{N} {}^*S_i \, _1\mathbf{J}_i - \frac{\lambda_\infty}{T} \text{ grad } T.$$

(4–24.36b)

Combination of Eq. (33) with Eq. (35) produces the relation between the two thermal conductivities:

$$\lambda_0 - \lambda_\infty = \frac{1}{T}\sum_{i=2}^{N} a_{iQ} {}^*Q_i.$$

(4–24.37)

From Eq. (17) and the first inequality in (15) there follows:

$$\sum_{i=2}^{N} a_{iQ} {}^*Q_i = \sum_{i=2}^{N} \sum_{k=2}^{N} a_{ik} {}^*Q_i {}^*Q_k > 0.$$

With Eq. (37) we derive:

$$\lambda_0 - \lambda_\infty = \frac{1}{T}\sum_{i=2}^{N} \sum_{k=2}^{N} a_{ik} {}^*Q_i {}^*Q_k > 0.$$

(4–24.38)

The thermal conductivity for the uniform state λ_0 is thus greater than that for the stationary state λ_∞.

To prove that λ_∞ is always positive we must eliminate ${}^*Q_i$ from Eq. (35). For this, we solve the system of equations (17) with respect to ${}^*Q_2, {}^*Q_3, \ldots$,

---

† The law $\mathbf{J}_Q = -\lambda \text{ grad } T$ was first proposed by Biot (1804) and then more precisely examined by Fourier (1822). It is thus older than Ohm's law (1826) and Fick's law (1855).

$^*Q_N$ and add the products $a_{iQ} {}^*Q_i$. According to the rules of determinant calculations and from Eqs. (13) and (15), we find:

$$a_{QQ} - \sum_{i=2}^{N} a_{iQ} {}^*Q_i = \frac{|a_{ij}|}{|a_{ik}|} > 0. \qquad (4\text{–}24.39)$$

From Eqs. (35) and (39) there results:

$$\lambda_\infty = \frac{1}{T} \frac{|a_{ij}|}{|a_{ik}|} > 0. \qquad (4\text{–}24.40)$$

Thus all general statements on λ_0 and λ_∞ have been derived[106].

Equations (33), (37), (38), and (40) correspond to (3–9.16), (3–9.17), (3–9.18), and (3–9.20). The thermal conductivities λ_0 and λ_∞ are independent of the arbitrary energy zero points and of the reference velocity (cf. p. 237). Also, the quantity λ_0 or λ_∞ appears in the term div (λ grad T) of the "heat conduction equation" which is important for the evaluation of data and which presents a relation between temperature, position, and time, as will be shown in the following section.

In one-component systems, no concentration gradients are possible. But in electrically conducting media, e.g. electron conductors or solid and liquid salts, we have potential gradients which arise instantaneously in a temperature field. For pure insulators, the quantities λ_0 and λ_∞ coincide; however, for electrically conducting one-component systems, we must distinguish between λ_0 (thermal conductivity for zero field) and λ_∞ (thermal conductivity for zero current). Since, however, only the latter can be measured (cf. p. 345), there is in this case only *one* experimentally accessible value of the thermal conductivity.

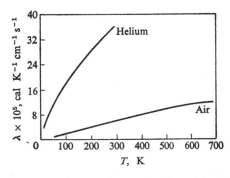

Fig. 13. Thermal conductivity λ of gases as a function of the absolute temperature T. After Zemansky[107a].

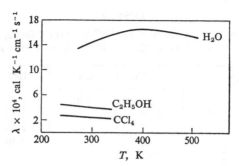

Fig. 14. Thermal conductivity λ of liquids as a function of the absolute temperature T. After Zemansky[107a] and Tyrrell[108].

Typical examples† of measured values of the thermal conductivity λ as a function of the temperature T for pure substances (or phases of constant composition) are shown in Fig. 13 (gases), Fig. 14 (liquids), and Fig. 15 (crystalline phases). All curves relate to atmospheric pressure. The curves in Fig. 15 represent low-temperature values. At higher temperatures λ can increase with increasing temperature (e.g. for quartz) or remain almost constant. Thus for silver a λ-value of about 1.0 cal K^{-1} cm^{-1} s^{-1} is found in the temperature interval from $-160°C$ to $+100°C$ and for nickel a constant value of about 0.14 cal K^{-1} cm^{-1} s^{-1} is valid from $-160°C$ to $+700°C$.

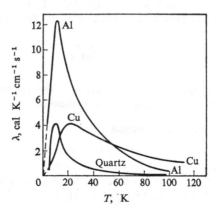

Fig. 15. Thermal conductivity λ of crystals as a function of the absolute temperature T. After Zemansky[107a].

† Since we here assume isotropic media, the crystalline phases must represent either cubic crystals or a polycrystalline substance (without preferential directions of the crystallites). The heat conduction in anisotropic crystals will be treated in §4–39.

Table 10

Measured Values of $\lambda/\kappa T$ in the unit 10^{-8} V^2 K^{-2} for Metals, for Comparison with the Theoretical Value 2.44×10^{-8} V^2 K^{-2}. After Zemansky[107a]

Metal	$-170°C$	$-100°C$	$0°C$	$18°C$	$100°C$
Copper	1.85	2.17	2.30	2.32	2.32
Silver	2.04	2.29	2.33	2.33	2.37
Zinc	2.20	2.39	2.45	2.43	2.33
Cadmium	2.39	2.43	2.40	2.39	2.44
Tin	2.48	2.51	2.49	2.47	2.49
Lead	2.55	2.54	2.53	2.51	2.51

In Table 10, some experimental values of the quantity $\lambda/(\kappa T)$ for metals are shown, where κ denotes the specific conductance (cf. p. 260). According to the Wiedemann–Franz–Sommerfeld law for all metals,[35]

$$\frac{\lambda}{\kappa T} = \frac{\pi^2}{3}\left(\frac{R}{F}\right)^2 \approx 2.44 \times 10^{-8} \text{ V}^2 \text{ K}^{-2},$$

where R is the gas constant. We see that this regularity is approximately satisfied.

The above law was derived by Sommerfeld (1928) from the electron theory of metals. Wiedemann and Franz had already found the empirical relation $\lambda/\kappa = \text{const}$ (for $T = \text{const}$) in 1853. L. Lorenz in 1872 gave the more exact formulation $\lambda/(\kappa T) = \text{const}$. Finally, Drude (1900) obtained the equation

$$\frac{\lambda}{\kappa} = 3\left(\frac{R}{F}\right)^2 T,$$

which was then improved by Sommerfeld. The thermodynamics of irreversible processes does not lead to any relation between λ and κ.

(f) General Heat Conduction Equation

We assume fluid media with creeping motions or isotropic crystals in simple cases and no space charges. Furthermore, we suppose that apart from an electrostatic field there are no external force fields present. Then there follows from Eqs. (4–5.19), (4–15.1), (4–15.5), and (4–15.6):

$$P_{ij} = 0, \quad \mathbf{K}_k = -z_k F \text{ grad } \varphi, \quad \text{grad } P = 0.$$

From this the last term in Eq. (4–6.20) is reduced with the help of Eqs. (4–3.11) and (4–3.45) to:

$$-\sum_k {}_v\mathbf{J}_k \text{ grad } H_k - \mathbf{I} \text{ grad } \varphi,$$

where $_v\mathbf{J}_k = \mathbf{J}_k^*/M_k$, while H_k is the partial molar enthalpy of species k, and $\mathbf{I}$ the total electric current density. The general heat conduction equation (4–6.20) thus assumes the following form:

$$\frac{\bar{C}_P}{\bar{V}} \frac{\partial T}{\partial t} = -\operatorname{div} \mathbf{J}_Q - \frac{\bar{C}_P}{\bar{V}} \mathbf{v} \operatorname{grad} T - \sum_{k=1}^{N} {}_v\mathbf{J}_k \operatorname{grad} H_k$$

$$- \mathbf{I} \operatorname{grad} \varphi + \frac{T}{\bar{V}}\left(\frac{\partial \bar{V}}{\partial T}\right)_{P,x} \frac{\partial P}{\partial t} - \sum_{r=1}^{R} \sum_{k=1}^{N} \nu_{kr} H_k b_r. \quad (4\text{–}24.41)$$

Here t is the time, $\bar{V}$ the molar volume, $\bar{C}_P$ the molar heat capacity for constant pressure, $\mathbf{v}$ the barycentric velocity, and ν_{kr} the stoichiometric coefficient of species k in reaction r. The inferior x ($_x$) in the differentiation indicates constant mole fractions, the operator $(\partial/\partial t)$ is the local derivative with respect to the time (at a fixed position).

According to Eq. (4–3.1), the diffusion current density of species k in an arbitrary reference system is:

$$\mathbf{J}_k = c_k(\mathbf{v}_k - \boldsymbol{\omega}) \quad (k = 1, 2, \ldots, N). \quad (4\text{–}24.42)$$

Here c_k and $\mathbf{v}_k$ denote the molarity and the average velocity of the particle type k, while $\boldsymbol{\omega}$ is the reference velocity. If we set $\boldsymbol{\omega} = \mathbf{v}$ or $\boldsymbol{\omega} = \mathbf{v}_1$ then we obtain the diffusion current density $_v\mathbf{J}_k$ or $_1\mathbf{J}_k$ in the barycentric system or in the Hittorf reference system. With $\boldsymbol{\omega} = \mathbf{w}$ (mean volume velocity) the diffusion current density $_w\mathbf{J}_k$ in the Fick reference system results from Eq. (42).

Furthermore, with $\operatorname{grad} P = 0$ there results from Eqs. (1–6.3), (1–6.6), (1–6.7), (1–6.10), (1–6.16), (1–15.5), and (4–4.18):

$$\operatorname{grad} H_k = C_{Pk} \operatorname{grad} T + (\operatorname{grad} H_k)_{T,P}, \quad (4\text{–}24.43)$$

$$\sum_{k=1}^{N} c_k C_{Pk} = \frac{\bar{C}_P}{\bar{V}} = \rho c_P, \quad (4\text{–}24.44)$$

$$\sum_{k=1}^{N} c_k(\operatorname{grad} H_k)_{T,P} = 0, \quad (4\text{–}24.45)$$

where C_{Pk} is the partial molar heat capacity (for constant pressure) of species k and c_P the specific heat capacity (for constant pressure).

From Eqs. (42) to (45) we derive:

$$\sum_{k=1}^{N} {}_v\mathbf{J}_k \operatorname{grad} H_k + \frac{\bar{C}_P}{\bar{V}} \mathbf{v} \operatorname{grad} T$$

$$= \sum_{k=1}^{N} c_k \mathbf{v}_k (\operatorname{grad} H_k)_{T,P} + \sum_{k=1}^{N} c_k C_{Pk} \mathbf{v}_k \operatorname{grad} T$$

$$= \sum_{k=1}^{N} \mathbf{J}_k [C_{Pk} \operatorname{grad} T + (\operatorname{grad} H_k)_{T,P}] + \frac{\bar{C}_P}{\bar{V}} \boldsymbol{\omega} \operatorname{grad} T. \quad (4\text{–}24.46)$$

Combining Eqs. (41), (44), and (46) leads to the following form of the general equation for heat conduction:

$$\rho c_P\left(\frac{\partial T}{\partial t} + \omega \text{ grad } T\right) = \frac{\overline{C}_P}{\overline{V}}\left(\frac{\partial T}{\partial t} + \omega \text{ grad } T\right)$$

$$= -\text{div } \mathbf{J}_Q - \mathbf{I} \text{ grad } \varphi + \frac{T}{\overline{V}}\left(\frac{\partial \overline{V}}{\partial T}\right)_{P,x}\frac{\partial P}{\partial t}$$

$$- \sum_{k=1}^{N} \mathbf{J}_k[C_{Pk} \text{ grad } T + (\text{grad } H_k)_{T,P}]$$

$$- \sum_{r=1}^{R}\sum_{k=1}^{N} b_r \nu_{kr} H_k. \qquad (4\text{–}24.47)$$

For practical purposes, Eq. (47) is easier to use than Eq. (41). For temperature measurements in liquids and gases, the experiment can often be arranged to satisfy the following conditions: $\mathbf{w} = 0$ (no convection), $\mathbf{I} = 0$ (no electric current), $\partial P/\partial t = 0$ (no pressure change).

If we set $\omega = \mathbf{w}$ in Eq. (47) and $\mathbf{J}_k = {}_w\mathbf{J}_k$, we thus obtain:

$$\rho c_P \frac{\partial T}{\partial t} = \frac{\overline{C}_P}{\overline{V}}\frac{\partial T}{\partial t}$$

$$= -\text{div } \mathbf{J}_Q - \sum_{k=1}^{N} {}_w\mathbf{J}_k[C_{Pk} \text{ grad } T + (\text{grad } H_k)_{T,P}]$$

$$- \sum_{r=1}^{R}\sum_{k=1}^{N} b_r \nu_{kr} H_k. \qquad (4\text{–}24.48)$$

We will require Eq. (48) in §4–26.

Let us consider the case of "pure" heat conduction in gases or liquids without chemical reactions. We can then drop the double summation in Eq. (48) and neglect the first summation term which, compared to the term $-\text{div } \mathbf{J}_Q$, is of minor importance. We thus find the well-known simple relation:

$$\rho c_P \frac{\partial T}{\partial t} = -\text{div } \mathbf{J}_Q. \qquad (4\text{–}24.49)$$

For the uniform state or for the stationary state ($\partial T/\partial t = 0$), there follows, according to Eq. (32) or (34), from Eq. (49):

$$\rho c_P \frac{\partial T}{\partial t} = \text{div }(\lambda_0 \text{ grad } T) \quad \text{(uniform state)}, \qquad (4\text{–}24.50a)$$

$$\rho c_P \frac{\partial T}{\partial t} = \text{div }(\lambda_\infty \text{ grad } T) = 0 \quad \text{(stationary state)}. \qquad (4\text{–}24.50b)$$

Equation (50a) or (50b) represents the usual form of the heat conduction equation with λ_0 or λ_∞ as the thermal conductivity.

(g) Transport Processes in Electron Conductors

For isobaric and isotropic electron conductors with no external force fields apart from an electrostatic field, the assumption of a "simple case" in the sense of §4–5 (p. 231) is satisfied. We may thus apply both the phenomenological equations (9) and (10) and the general heat conduction equation (47). Since migrations of other species relative to one another are slow in comparison with electron motions, we write:

$$\mathbf{v}_1 = \mathbf{v}_3 = \cdots = \mathbf{v}_N \equiv \mathbf{v}_G, \qquad \mathbf{v}_2 \equiv \mathbf{v}_\ominus, \qquad (4\text{–}24.51)$$

where the inferior G or $\ominus$ denotes the lattice or the electrons respectively. Here, according to Eqs. (4–3.30) and (4–3.31), the Hittorf reference system (particle type 1 as the reference species) and the lattice reference system coincide. We can thus, according to Eqs. (4–3.30) and (4–3.45), set up:

$$\boldsymbol{\omega} = \mathbf{v}_1, \qquad {}_1\mathbf{J}_i = c_i(\mathbf{v}_i - \mathbf{v}_1) = 0 \quad (i = 1, 3, \ldots, N),$$

$$
{}_1\mathbf{J}_2 = {}_1\mathbf{J}_\ominus = c_\ominus(\mathbf{v}_\ominus - \mathbf{v}_1) = -\frac{1}{F}\mathbf{I}. \qquad (4\text{–}24.52)
$$

Moreover,

$$\operatorname{grad} P = 0, \quad \partial P/\partial t = 0, \quad b_r = 0. \qquad (4\text{–}24.53)$$

From Eqs. (42), (47), (51), (52), and (53) we derive:

$$\rho c_P \frac{\partial T}{\partial t} = \frac{\bar{C}_P}{\bar{V}} \frac{\partial T}{\partial t}$$

$$= -\operatorname{div} \mathbf{J}_Q - \mathbf{I} \operatorname{grad} \varphi + \frac{1}{F} \mathbf{I}[C_{P\ominus} \operatorname{grad} T + (\operatorname{grad} H_\ominus)_T]. \qquad (4\text{–}24.54)$$

We will use this special form of the heat conduction equation in the next section.

Under the same conditions for which Eq. (54) holds, the phenomenological equations (9) and (10) assume the following simple form on the basis of Eqs. (3), (11), (52), and (53), since only the expression $\mathbf{J}_Q \mathbf{X}_Q + {}_1\mathbf{J}_\ominus \mathbf{X}_\ominus$ remains in the dissipation function (4–12.17):

$$\mathbf{I} = -\alpha F[F \operatorname{grad} \varphi - (\operatorname{grad} \mu_\ominus)_T] + \beta \frac{F}{T} \operatorname{grad} T,$$

$$\mathbf{J}_Q = \beta'[F \operatorname{grad} \varphi - (\operatorname{grad} \mu_\ominus)_T] - \frac{\gamma}{T} \operatorname{grad} T. \qquad (4\text{–}24.55)$$

Here we have set $\alpha_{22} \equiv \alpha$, $\alpha_{2Q} \equiv \beta$, $\alpha_{Q2} \equiv \beta'$. The reciprocity relation (13) produces the statement

$$\beta = \beta'. \tag{4–24.56}$$

Equations (55) and (56) will also be used in the following section.

Equations (54) to (56) are valid for electron conductors, in which concentration gradients, temperature gradients, and an electric field may exist.

(h) Electric Potentials inside Nonisothermal Conductors

From the migration of charged species due to concentration gradients, external fields, and temperature gradients, electric fields arise inside conducting media. That is, potential gradients which are already known to us in special cases may exist. Examples include the "diffusion potential" (§4–18) and the "sedimentation potential" (§4–21). The general expression for that kind of electric potential gradient will now be derived.

With condition (4–15.33), which formulates the disappearance of the total electric current, there follows from Eqs. (3), (11), and (19) that:

$$F \operatorname{grad} \varphi = \frac{\sum_{i,k} z_i a_{ik}}{\sum_{i,k} z_i z_k a_{ik}} \left[-(\operatorname{grad} \mu_k)_{T,P} + (M_k - \rho V_k)\mathbf{g} - \frac{{}^*Q_k}{T} \operatorname{grad} T \right]$$

$$(\mathbf{I} = 0), \qquad (4–24.57)$$

where the double summation is from 2 to N. The basic phenomenological equations (9) are now reduced for pure electric conduction ($\mathbf{g} = 0$, $(\operatorname{grad} \mu_1)_{T,P} = 0$, $\operatorname{grad} T = 0$) to Eqs. (4–16.43) and (4–16.44). Also, Eq. (4–16.45) holds. Finally, if the Onsager reciprocity laws (12) are noted then there results from Eq. (57):

$$F \operatorname{grad} \varphi = \sum_{k=2}^{N} \tau_k \left[-(\operatorname{grad} \mu_k)_{T,P} + (M_k - \rho V_k)\mathbf{g} - \frac{{}^*Q_k}{T} \operatorname{grad} T \right]$$

$$(\mathbf{I} = 0). \qquad (4–24.58)$$

Here τ_k is the reduced transport number of species k. The formulas (4–18.23) and (4–21.4) are special cases of this expression.

The first term in the square brackets in Eq. (58) relates to the diffusion potential, the second to the sedimentation potential (in the narrower sense), while the last term containing the heats of transport ${}^*Q_k$ is to be assigned to the "thermal diffusion potential" (§4–27).

According to its derivation, Eq. (58) is valid for isotropic crystals in "simple cases" or fluid media in "creeping motion" in the absence of space charges. It is applicable at any arbitrary instant. If after a sufficiently long time ($t \to \infty$) a stationary state is set up, according to Eq. (24) for the stationary electric potential φ_∞ we obtain:

$$z_i F \operatorname{grad} \varphi_\infty = -(\operatorname{grad} \mu_i)_{T,P} + (M_i - \rho V_i)\mathbf{g} - \frac{{}^*Q_i}{T} \operatorname{grad} T$$

$$(i = 2, 3, \ldots, N). \quad (4\text{-}24.59)$$

For electronic conductors, the stationary state coincides with the state without current. Accordingly, Eq. (58) and Eq. (59) must lead to the same result. Indeed, with the assumption (cf. p. 315)

$$\tau_k = 0 \quad (k = 3, 4, \ldots, N), \qquad \tau_\ominus = -1$$

and with $\mathbf{g} = 0$ we find from Eq. (58):

$$F \operatorname{grad} \varphi = (\operatorname{grad} \mu_\ominus)_{T,P} + \frac{{}^*Q_\ominus}{T} \operatorname{grad} T. \quad (4\text{-}24.60)$$

This can also be derived directly from Eq. (59) with $z_i = z_\ominus = -1$. The quantity φ in Eq. (60) represents the "thermoelectric potential" (§4–25).

4–25 THERMOELECTRIC EFFECTS

(a) Fundamental Equations

Consider a metal or another electron conductor in which both a temperature gradient and an electric field can exist. Let the medium be isotropic and show no concentration and pressure gradients. Also, apart from the electric field, there should be no external force fields present. Then the following irreversible processes can occur in each volume element of the conductor:

1. Electric conduction (electric current due to an electric potential gradient).
2. Heat conduction (heat flow due to a temperature gradient).
3. A cross effect (electric current due to a temperature gradient).
4. The appropriate reciprocal effect (heat flow due to an electric potential gradient).

The first two phenomena are known to us. The last two phenomena, however, belong to a new class of irreversible processes which comprise the *thermoelectric effects*. These effects are partly measurable only if two different electronic conductors are combined, each one of which individually contributes its "homogeneous effect" into a "thermocouple" (see below). Then "heterogeneous effects" become effective.

We begin with the thermodynamic–phenomenological treatment of the homogeneous effects.†

With the condition $(\text{grad } \mu_\ominus)_T = 0$ (no concentration gradients) we obtain from Eq. (4–24.55) for electron conductors under the initial assumptions:

$$I = -\alpha F^2 \, \text{grad } \varphi + \frac{\beta F}{T} \, \text{grad } T, \qquad (4\text{–}25.1)$$

$$J_Q = \beta' F \, \text{grad } \varphi - \frac{\gamma}{T} \text{grad } T. \qquad (4\text{–}25.2)$$

Here F denotes the Faraday constant, I the electric current density, J_Q the heat current density, φ the electric potential, and T the absolute temperature. The quantities α, β, β', and γ are the phenomenological coefficients, for which Onsager's reciprocity law (4–24.56) and the system of inequalities (4–24.14) produce the following statements:

$$\beta = \beta', \qquad (4\text{–}25.3)$$

$$\alpha > 0, \quad \gamma > 0, \quad \alpha\gamma - \beta^2 > 0. \qquad (4\text{–}25.4)$$

The thermodynamic–phenomenological description of the transport phenomena in electron conductors is contained completely in Eqs. (1) to (4); according to the initial statements, α is assigned to electric conduction and γ to heat conduction, while β characterizes the cross effect and β' the appropriate reciprocal effect.

Comparison of Eq. (1) with Eq. (4–16.24) shows that for grad $T = 0$ Ohm's law follows from Eq. (1):

$$I = -\kappa \, \text{grad } \varphi \quad (\text{grad } T = 0), \qquad (4\text{–}25.5)$$

where

$$\kappa \equiv \alpha F^2 \qquad (4\text{–}25.6)$$

represents the *specific conductance*, which is a quantity measured in a known manner.

From Eq. (2) with grad $\varphi = 0$ Fourier's law results for the field-free state:

$$J_Q = -\lambda_0 \, \text{grad } T \quad (\text{grad } \varphi = 0) \qquad (4\text{–}25.7)$$

with

$$\lambda_0 \equiv \frac{\gamma}{T}. \qquad (4\text{–}25.8)$$

---

† After the quasithermodynamic theory (cf. p. 190) of the thermoelectric effects, elaborated by William Thomson (Lord Kelvin), Eastman, and Wagner, it was the development of the electron theory of metals which first enabled an explanation of the phenomena[109]. Onsager[110a, b] showed the way to the model-free method of the thermodynamic–phenomenological theory of thermoelectricity[2, 111, 112].

λ_0 is the *thermal conductivity for zero field.* According to Eq. (4–24.32), it corresponds to the "thermal conductivity for the uniform state" in §4–24. (The measurability of λ_0 will be discussed later.)

If we eliminate grad φ from Eqs. (1) and (2), take account of the reciprocity relation (3), and introduce Eq. (4–24.52) for the electron flux density $_1 J_\Theta$ (vector of the electron current density in the Hittorf reference system), we find:

$$\mathbf{J}_Q = -\frac{{}^*Q_\Theta}{F}\,\mathbf{I} - \lambda_\infty\,\text{grad}\,T = {}^*Q_\Theta\,{}_1\mathbf{J}_\Theta - \lambda_\infty\,\text{grad}\,T \qquad (4\text{–}25.9)$$

with

$$ {}^*Q_\Theta \equiv \frac{\beta'}{\alpha} = \frac{\beta}{\alpha}, \qquad\qquad\qquad (4\text{–}25.10)$$

$$ \lambda_\infty \equiv \frac{1}{T}\left(\gamma - \frac{\beta^2}{\alpha}\right). \qquad\qquad (4\text{–}25.11)$$

As a comparison with Eq. (4–24.36a) indicates, ${}^*Q_\Theta$ is the *heat of transport of the electrons* and λ_∞ the *thermal conductivity for zero current.* The last quantity corresponds to the "thermal conductivity for the stationary state." (The question of the measurability of λ_∞ and ${}^*Q_\Theta$ will be discussed below.) From Eq. (9) with $\mathbf{I} = 0$ there follows that

$$\mathbf{J}_Q = -\lambda_\infty\,\text{grad}\,T \quad (\mathbf{I} = 0), \qquad\qquad (4\text{–}25.12)$$

that is, Fourier's law for states without current flow.

Because of the Onsager relation (3) only three of the four phenomenological coefficients are independent. Accordingly, there must still be a relation between the four transport quantities κ, λ_0, λ_∞, and ${}^*Q_\Theta$. This is derived from Eqs. (6), (8), (10), and (11)[113]:

$$\lambda_0 - \lambda_\infty = \frac{\kappa({}^*Q_\Theta)^2}{F^2 T}. \qquad\qquad (4\text{–}25.13)$$

The difference between the thermal conductivity for zero field λ_0 and that for zero current λ_∞ is thus connected to the heat of transport of the electrons ${}^*Q_\Theta$.

From (4), (6), (8), (10), (11), and (13) the following inequalities result[113]:

$$\kappa > 0, \quad \lambda_\infty > 0, \quad \lambda_0 > \lambda_\infty, \qquad\qquad (4\text{–}25.14)$$

$$ {}^*Q_\Theta{}^2 < F^2 T \frac{\lambda_0}{\kappa}. \qquad\qquad (4\text{–}25.15)$$

The last two relations in (14) are special cases of the general statements (4–24.38) and (4–24.40). The inequality (15) fixes an upper limit for the amount of the heat of transport of the electrons. The sign of ${}^*Q_\Theta$ is left open.

Equations (1) and (2) with Eqs. (6), (8), (9), and (10) can be written in the form:

$$\mathbf{I} = -\kappa \, \text{grad} \, \varphi + \frac{\kappa \, {}^{*}Q_{\ominus}}{FT} \, \text{grad} \, T, \qquad (4\text{-}25.16)$$

$$\mathbf{J}_Q = \frac{\kappa \, {}^{*}Q_{\ominus}}{F} \, \text{grad} \, \varphi - \lambda_0 \, \text{grad} \, T = -\frac{{}^{*}Q_{\ominus}}{F} \mathbf{I} - \lambda_\infty \, \text{grad} \, T. \qquad (4\text{-}25.17)$$

These equations will be very useful for the association of effects discussed thus far with measurable quantities.

Furthermore, according to Eq. (4-24.20), it is expedient to introduce the *transported entropy of the electrons* by the relation

$$^{*}S_{\ominus} \equiv \frac{{}^{*}Q_{\ominus}}{T} + S_{\ominus}. \qquad (4\text{-}25.18)$$

Here $S_{\ominus}$ is the partial molar entropy of the electrons.

For the state without current flow ($\mathbf{I} = 0$), from Eqs. (16) and (18) we obtain the following expression as a special case of Eq. (4-24.60):

$$\text{grad} \, \varphi = \frac{{}^{*}Q_{\ominus}}{FT} \, \text{grad} \, T = \frac{1}{F} ({}^{*}S_{\ominus} - S_{\ominus}) \, \text{grad} \, T \quad (\mathbf{I} = 0). \quad (4\text{-}25.19)$$

Thus, inside an electron conductor, an electric potential difference is produced by a temperature gradient which is related to the heat of transport of the electrons and, due to this, to the cross phenomenon (flow of electric current due to temperature gradients) or to the corresponding reciprocal effect (heat flow due to potential gradients). This potential difference can be regarded as a special case of a "thermal diffusion potential" (§4-27). Let it be called the *thermoelectric potential*. Neither this potential difference nor the heat of transport of the electrons can be measured, although the transported entropy can be experimentally determined, at least as a relative value (cf. below).

It is obvious from Eqs. (16) to (19) that, due to Onsager's reciprocity law, both the cross effect and its reciprocal can be described by the heat of transport or transported entropy of the electrons. All measurable thermoelectric effects in isotropic electron conductors therefore, as long as they are not determined by equilibrium phenomena at phase boundaries (cf. below) or from electric and heat conduction, must be related to one single transport coefficient. We will verify this statement in the following.

(b) Heat Conduction and Thomson Effect

The heat conduction equation (4-24.54) with the condition (grad $H_{\ominus})_T = 0$ (no concentration gradient) produces the following relation for the local rate of increase of the temperature in a volume element of the conductor:

$$\frac{\bar{C}_P}{\bar{V}} \frac{\partial T}{\partial t} = -\text{div } \mathbf{J}_Q - \mathbf{I}\left(\text{grad } \varphi - \frac{C_{P\ominus}}{F} \text{ grad } T\right). \qquad (4\text{–}25.20)$$

Here $\bar{C}_P$ is the molar heat capacity for constant pressure, $\bar{V}$ the molar volume, t the time, and $C_{P\ominus}$ the partial molar heat capacity of the electrons for constant pressure. (The operator $\partial/\partial t$ denotes the derivative with respect to the time at a fixed position.)

First, for the state without current flow ($\mathbf{I} = 0$), we recognize from Eqs. (12) and (20) that a normal heat conduction equation can be obtained of the form

$$\frac{\bar{C}_P}{\bar{V}} \frac{\partial T}{\partial t} = \text{div } (\lambda_\infty \text{ grad } T) \quad (\mathbf{I} = 0) \qquad (4\text{–}25.21)$$

with λ_∞ as the thermal conductivity, while for the field-free state (grad $\varphi = 0$) the relation

$$\frac{\bar{C}_P}{\bar{V}} \frac{\partial T}{\partial t} = \text{div } (\lambda_0 \text{ grad } T) + \frac{C_{P\ominus}}{F} \mathbf{I} \text{ grad } T \quad (\text{grad } \varphi = 0) \qquad (4\text{–}25.22)$$

results from Eqs. (7) and (20). Since $C_{P\ominus}$ represents no measurable quantity, λ_0 cannot be determined directly from Eq. (22). Moreover, it is not possible to establish a field-free state in an electron conductor with temperature gradients without knowledge of the (also unmeasurable) thermoelectric potential. Accordingly, only the thermal conductivity for zero current λ_∞ can be experimentally determined, while the thermal conductivity for zero field λ_0 cannot be measured. λ_0 cannot be determined from Eq. (13) since the heat of transport $*Q_\ominus$ of the electrons, which is connected to the thermoelectric potential by Eq. (19), is also experimentally inaccessible. Equation (13) thus has significance only in principle but not in practice.

A discussion of Eq. (20) is of interest for the general case. If we eliminate $\mathbf{J}_Q$ and grad φ with the help of Eqs. (16) and (17) we obtain, with the condition div $\mathbf{I} = 0$, valid for any electric conductor (cf. p. 428),

$$\frac{\bar{C}_P}{\bar{V}} \frac{\partial T}{\partial t} = \text{div } (\lambda_\infty \text{ grad } T) + \frac{1}{\kappa} I^2$$

$$+ \frac{1}{F} \mathbf{I}\left(T \text{ grad } \frac{*Q_\ominus}{T} + C_{P\ominus} \text{ grad } T\right), \qquad (4\text{–}25.23)$$

where the gradient of $*Q_\ominus$ refers to the temperature dependence of this quantity. Furthermore, according to Eq. (1–15.5) and Eq. (18), we have

$$T\left(\frac{\partial S_\ominus}{\partial T}\right)_P = C_{P\ominus}, \qquad (4\text{–}25.24)$$

$$\frac{*Q_\ominus}{T} = *S_\ominus - S_\ominus, \qquad (4\text{–}25.25)$$

where P denotes the pressure. In addition, we define

$$\frac{T}{F}\left(\frac{\partial {}^{*}S_{\ominus}}{\partial T}\right)_{P} \equiv -\tau. \tag{4–25.26}$$

With this there follows:

$$T \text{ grad } \frac{{}^{*}Q_{\ominus}}{T} + C_{P\ominus} \text{ grad } T = -F\tau \text{ grad } T. \tag{4–25.27}$$

Insertion of Eq. (27) in Eq. (23) finally produces:

$$\frac{\overline{C}_{P}}{\overline{V}} \frac{\partial T}{\partial t} = \text{div} (\lambda_{\infty} \text{ grad } T) + \frac{1}{\kappa} I^{2} - \tau I \text{ grad } T. \tag{4–25.28}$$

As is obvious from Eq. (28), a temperature change appears in each element of a nonisothermal conductor with current flowing through it, and this is determined by three effects:

1. Heat conduction, described by the thermal conductivity λ_{∞}.
2. Normal electric conduction, described by the electric conductivity κ.
3. A further effect, the *Thomson effect*, described by the new coefficient τ.

In the same sense in which the term I^{2}/κ was designated as the "Joule heat" (per unit volume and time) (cf. p. 86), the expression $-\tau I \text{ grad } T$ is said to be the "Thomson heat" (per unit volume and time). The quantity τ is called the "Thomson coefficient" and can be measured for each conductor using Eq. (28); the specific conductance κ and the thermal conductivity λ_{∞} can be independently measured. According to Eq. (26), the temperature coefficients of the transported entropy of the electrons ${}^{*}S_{\ominus}$ can also be thus obtained from experimental data. Since at given pressure ${}^{*}S_{\ominus}$ depends only on the temperature and on the nature of the electron conductor (Eqs. 10 and 18), τ must also be a "material constant" due to Eq. (26). This is confirmed by experience.

The Thomson effect is, in our presentation, the first thermoelectric phenomenon which is directly experimentally accessible and represents the only measurable *homogeneous* thermoelectric effect. The phenomenon was discovered in 1854 by William Thomson (Lord Kelvin).

(c) Seebeck Effect

A *thermocouple* is a system of two different electronic conductors (usually metals) A and B, whose two phase boundaries (junctions) are held at two different (absolute) temperatures T_{1} and T_{2}. We can set up such an arrangement (Fig. 16) as we did for a galvanic cell:

	C	B	A	B	C	
Temperature:	T_0	T_0	T_1	T_2	T_0	T_0
Potential:	φ_I	φ_{II}	φ_{III} φ_{IV}	φ_V φ_{VI} φ_{VII}	φ_{VIII}	

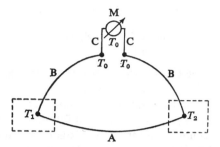

Fig. 16. Thermocouple of the electronic conductors (metals) A and B with the circuit broken for the measurement of the thermoelectric power by the measuring instrument M, to which the wires of the metal C lead. The (absolute) room temperature is T_0, while T_1 and T_2 are the (absolute) temperatures of the first and second junctions respectively.

The metal C is the terminal, i.e. the conductor to the measuring instrument; T_0 is the (absolute) room temperature, at which the terminals and the measuring instrument are fixed.

The fact that a thermocouple without current flow shows an electric potential difference was discovered in 1823 by Seebeck and is thus called the *Seebeck effect*.

The measurable potential difference of a thermocouple is called the "thermoelectric tension." It is fixed by agreement as the difference between the electric potential in the right-hand terminal and that in the left-hand terminal, measured with no current flowing, and is designated with the symbol Φ. We have thus:

$$\Phi = \varphi_{\text{VIII}} - \varphi_{\text{I}}. \tag{4–25.29}$$

The expression

$$\varepsilon_{\text{AB}} \equiv \lim_{\Delta T \to 0} \frac{\Phi}{\Delta T} \tag{4–25.30}$$

with

$$\Delta T \equiv T_2 - T_1 \tag{4–25.31}$$

is by definition the *thermoelectric power* of the thermocouple.† Accordingly, for the thermoelectric tension we write:

$$\Phi = \varepsilon_{\text{AB}} \Delta T + \text{higher terms in } \Delta T \tag{4–25.32}$$

---

† According to agreement, the "thermocouple A–B" is assigned to the diagram described here. If in the diagram we exchange the conductors A and B and keep the other symbols unchanged then we have the "thermocouple B–A." It is obvious from the above diagram that the inferior labeling and sign of ε determine the sign of the thermoelectric power and thus the direction of the appropriate current ("thermal current"). Thus a positive value of ε_{AB} denotes that the thermal current flows, at the colder junction, from A to B and, at the warmer junction, from B to A.

or

$$\Phi = \int_{T_1}^{T_2} \varepsilon_{AB}(T)\, dT. \tag{4-25.33}$$

From experience the thermoelectric power ε_{AB} depends only on the kind of conductors A and B and on the average temperature of the thermocouple, provided the pressure (whose influence is small anyway) is regarded as constant. From this and the last equations it follows that the thermoelectric tension of a thermocouple of given materials is only a function of the two temperatures T_1 and T_2 of the junctions, i.e. it is independent of the temperature distribution in the conductors ("the law of Magnus"). Furthermore, from what has been said we can assign an arbitrary value ε_X of the "thermoelectric power" at a given temperature to any one conductor X and thus numerically fix the thermoelectric power ε_Y of any other conductor Y by the relation

$$\varepsilon_{XY} = \varepsilon_X - \varepsilon_Y$$

for the given temperature.† Moreover, the thermoelectric powers are additive in material contributions ("thermoelectric voltage series"). From the above we have

$$\varepsilon_{AB} = \varepsilon_A - \varepsilon_B = (\varepsilon_A - \varepsilon_X) - (\varepsilon_B - \varepsilon_X) = \varepsilon_{AX} - \varepsilon_{BX}. \tag{4-25.34}$$

These statements will be justified by theoretical arguments in the following. We maintain that we may always write

$$\varepsilon_{AB} = \varepsilon_A - \varepsilon_B = -(\varepsilon_B - \varepsilon_A) = -\varepsilon_{BA}, \tag{4-25.35}$$

the material constants ε_A and ε_B, which are arbitrary in their absolute values, depending on temperature and denoted as "thermoelectric powers" of the conductors A and B.

From Eq. (29) and our diagram of the thermocouple there follows for the thermoelectric tension:

$$\begin{aligned}
\Phi = {} & (\varphi_{VIII} - \varphi_{VII}) + (\varphi_{VII} - \varphi_{VI}) \\
& + (\varphi_{VI} - \varphi_V) + (\varphi_V - \varphi_{IV}) + (\varphi_{IV} - \varphi_{III}) \\
& + (\varphi_{III} - \varphi_{II}) + (\varphi_{II} - \varphi_I).
\end{aligned} \tag{4-25.36}$$

The second, fourth, and sixth terms of the right-hand side of Eq. (36) refer to the thermoelectric potentials inside the conductor, i.e. to homogeneous effects. The other terms, however, correspond to heterogeneous effects, namely, potential differences at phase boundaries ("contact potentials").

---

† The normalization for the determination of "absolute values" of the thermoelectric powers of individual conductors will be reported below.

For the summation $\Delta\varphi_{\text{hom}}$ of the thermoelectric potentials in our diagram there results from Eq. (19):

$$
\begin{aligned}
F\Delta\varphi_{\text{hom}} &= F[(\varphi_{\text{VII}} - \varphi_{\text{VI}}) + (\varphi_{\text{V}} - \varphi_{\text{IV}}) + (\varphi_{\text{III}} - \varphi_{\text{II}})] \\
&= \int_{T_2}^{T_0} [(*S_\ominus)_{\text{B}} - (S_\ominus)_{\text{B}}]\, dT + \int_{T_1}^{T_2} [(*S_\ominus)_{\text{A}} - (S_\ominus)_{\text{A}}]\, dT \\
&\quad + \int_{T_0}^{T_1} [(*S_\ominus)_{\text{B}} - (S_\ominus)_{\text{B}}]\, dT \\
&= \int_{T_1}^{T_2} [(S_\ominus)_{\text{B}} - (S_\ominus)_{\text{A}} + (*S_\ominus)_{\text{A}} - (*S_\ominus)_{\text{B}}]\, dT
\end{aligned}
\tag{4–25.37}
$$

or with Eq. (31):

$$
\lim_{\Delta T \to 0} \frac{\Delta\varphi_{\text{hom}}}{\Delta T} = \frac{1}{F}[(S_\ominus)_{\text{B}} - (S_\ominus)_{\text{A}} + (*S_\ominus)_{\text{A}} - (*S_\ominus)_{\text{B}}].
\tag{4–25.38}
$$

For the calculation of the contact potentials, we assume local electrochemical equilibrium at the phase boundaries. Then we may use the classical equilibrium condition (1–20.13). If we denote the chemical potential of the electrons by $\mu_\ominus$ and note that $\varphi_{\text{VIII}} - \varphi_{\text{VII}}$ and $\varphi_{\text{II}} - \varphi_{\text{I}}$ cancel out, we obtain for the contact potentials

$$
F(\varphi_{\text{VI}} - \varphi_{\text{V}}) = (\mu_\ominus)_{\text{B}} - (\mu_\ominus)_{\text{A}} \quad \text{(temperature } T_2\text{)},
\tag{4–25.39}
$$

$$
F(\varphi_{\text{IV}} - \varphi_{\text{III}}) = (\mu_\ominus)_{\text{A}} - (\mu_\ominus)_{\text{B}} \quad \text{(temperature } T_1\text{)},
\tag{4–25.40}
$$

$$
\begin{aligned}
F\Delta\varphi_{\text{het}} &= F(\varphi_{\text{VI}} - \varphi_{\text{V}} + \varphi_{\text{IV}} - \varphi_{\text{III}}) \\
&= [(\mu_\ominus)_{\text{B}} - (\mu_\ominus)_{\text{A}}]_{T_2} - [(\mu_\ominus)_{\text{B}} - (\mu_\ominus)_{\text{A}}]_{T_1},
\end{aligned}
\tag{4–25.41}
$$

where $\Delta\varphi_{\text{het}}$ denotes the sum of the contact potentials. With the relation (1–12.7)

$$
\left(\frac{\partial\mu_\ominus}{\partial T}\right)_P = -S_\ominus
\tag{4–25.42}
$$

it follows from Eqs. (31) and (41) that

$$
\lim_{\Delta T \to 0} \frac{\Delta\varphi_{\text{het}}}{\Delta T} = \frac{1}{F}[(S_\ominus)_{\text{A}} - (S_\ominus)_{\text{B}}].
\tag{4–25.43}
$$

According to Eqs. (36), (37), and (41), we have

$$
\Phi = \Delta\varphi_{\text{hom}} + \Delta\varphi_{\text{het}},
\tag{4–25.44}
$$

that is, from Eq. (30):

$$
\varepsilon_{\text{AB}} = \lim_{\Delta T \to 0} \frac{\Delta\varphi_{\text{hom}}}{\Delta T} + \lim_{\Delta T \to 0} \frac{\Delta\varphi_{\text{het}}}{\Delta T}.
\tag{4–25.45}
$$

We derive from Eqs. (38), (43), and (45):

$$\varepsilon_{AB} = \frac{1}{F}[(^*S_\Theta)_A - (^*S_\Theta)_B].\tag{4-25.46}$$

The thermoelectric power of the thermocouple is thus related to the difference between the transported entropies of the electrons in the two conductors A and B. Since these represent temperature-dependent material constants (cf. above), by Eq. (46) we have verified Eqs. (34) and (35) and the law of Magnus.

Comparison of Eq. (35) with Eq. (46) produces the relation between the transported entropies and the thermoelectric powers of the individual conductors:

$$F\varepsilon_A = (^*S_\Theta)_A, \qquad F\varepsilon_B = (^*S_\Theta)_B.\tag{4-25.47}$$

From Eqs. (26) and (47) there follows for the Thomson coefficients of the two conductors:

$$\tau_A = -T\left(\frac{\partial \varepsilon_A}{\partial T}\right)_P, \qquad \tau_B = -T\left(\frac{\partial \varepsilon_B}{\partial T}\right)_P.\tag{4-25.48}$$

We will return to these equations.

(d) Peltier Effect

In 1834, Peltier discovered that when current flows through the isothermal junctions of a thermocouple a heating up or cooling down is noticed. If the junctions are held at constant temperatures then heat—positive or negative—("Peltier heat") is taken from the surroundings.

The correct quantitative treatment of this "Peltier effect" requires somewhat more effort than the preceding phenomenon, since now a heat conduction equation (i.e. a temperature–position–time relation) must be set up, and this must take account of a discontinuity, namely the phase boundary between the two electron conductors. We solve the problem by first assuming a "continuous distribution of inhomogeneities," i.e. a mixture of the components A and B with continuous concentration gradients, and then passing to the limiting case of a concentration jump at the phase boundaries between A and B.

We recognize that the Peltier effect is covered in principle by Eq. (17) which, when applied to two different conductors A and B for grad $T = 0$, produces the following expressions:

$$(\mathbf{J}_Q)_A = -\frac{1}{F}(^*Q_\Theta)_A\mathbf{I}, \qquad (\mathbf{J}_Q)_B = -\frac{1}{F}(^*Q_\Theta)_B\mathbf{I}.$$

Here the continuity of the electric current $\mathbf{I}$, which passes through the phase boundary between A and B, is taken into account. Since, due to the Seebeck effect, the transported entropies for the two conductors must be different, the heats of transport $(*Q_\Theta)_A$ and $(*Q_\Theta)_B$ must, according to Eq. (18), also be different from each other. Thus there results a "jump" of the heat flow at the isothermal junction which must be related to the phenomenon of the "Peltier heat."

For the exact treatment of the problem, we replace the phenomenological equations (1) and (2) and the heat conduction equation (20) by the more general relations (4–24.55) and (4–24.54) in which concentration gradients are still allowed. If we note definitions (6), (8), and (10), we obtain

$$\mathbf{I} = -\kappa \left[\text{grad } \varphi - \frac{1}{F} (\text{grad } \mu_\Theta)_T \right] + \frac{\kappa *Q_\Theta}{FT} \text{grad } T, \qquad (4\text{–}25.49)$$

$$\mathbf{J}_Q = \frac{\kappa *Q_\Theta}{F} \left[\text{grad } \varphi - \frac{1}{F} (\text{grad } \mu_\Theta)_T \right] - \lambda_0 \text{grad } T, \qquad (4\text{–}25.50)$$

$$\frac{\bar{C}_P}{\bar{V}} \frac{\partial T}{\partial t} = -\text{div } \mathbf{J}_Q - \mathbf{I} \text{ grad } \varphi + \frac{1}{F} \mathbf{I}[C_{P\Theta} \text{ grad } T + (\text{grad } H_\Theta)_T]. \quad (4\text{–}25.51)$$

Here (cf. Eq. 1–12.6)

$$H_\Theta = \mu_\Theta + TS_\Theta \qquad (4\text{–}25.52)$$

is the partial molar enthalpy of the electrons. After elimination of the expression in square brackets in Eqs. (49) and (50) and on taking account of Eq. (13) we find

$$\mathbf{J}_Q = -\frac{*Q_\Theta}{F} \mathbf{I} - \lambda_\infty \text{ grad } T. \qquad (4\text{–}25.53)$$

This relation agrees with Eq. (9). From Eq. (49) there results immediately (cf. Eq. 4–24.60):

$$\text{grad } \varphi = \frac{1}{F} (\text{grad } \mu_\Theta)_T + \frac{*Q_\Theta}{FT} \text{grad } T - \frac{1}{\kappa} \mathbf{I}. \qquad (4\text{–}25.54)$$

The result from this formula for $\mathbf{I} = 0$ is different from Eq. (19): in addition to the thermoelectric potential (second term on the right-hand side) there still exists a diffusion potential (first term on the right-hand side).

If we insert Eqs. (53) and (54) into Eq. (51) and note Eq. (52) as well as the condition div $\mathbf{I} = 0$, we then derive:

$$\frac{\bar{C}_P}{\bar{V}} \frac{\partial T}{\partial t} = \text{div } (\lambda_\infty \text{ grad } T) + \frac{1}{\kappa} I^2$$

$$+ \frac{1}{F} \mathbf{I} \left[\text{grad } *Q_\Theta + T(\text{grad } S_\Theta)_T - \frac{*Q_\Theta}{T} \text{grad } T + C_{P\Theta} \text{ grad } T \right].$$

$$(4\text{–}25.55)$$

With the help of Eqs. (24) to (26) we bring this equation to the form:

$$\frac{\overline{C}_P}{\overline{V}} \frac{\partial T}{\partial t} = \text{div}\,(\lambda_\infty\,\text{grad}\,T) + \frac{1}{\kappa}\,I^2 - \tau\mathbf{I}\,\text{grad}\,T$$

$$+ \frac{T}{F}\,\mathbf{I}(\text{grad}\,{}^*S_\ominus)_T. \tag{4–25.56}$$

A comparison with Eq. (28) shows that in the heat conduction equation a new term now appears which takes account of an isothermal change in the quantity

$$\Pi \equiv \frac{T}{F}\,{}^*S_\ominus \tag{4–25.57}$$

with concentration. This term also appears when grad $T = 0$ and should thus be added to the "Joule heat." Obviously, we are dealing with a "Peltier heat" in the case of a "continuous Peltier effect."

At an isothermal phase boundary between two conductors A and B, the last term in Eq. (56) after multiplication by the distance element in the direction of current flow becomes (cf. Eq. 57):

$$\frac{T}{F}\,[({}^*S_\ominus)_A - ({}^*S_\ominus)_B]I = (\Pi_A - \Pi_B)I, \tag{4–25.58}$$

where I denotes the absolute value of the current density from B to A. Thus the actual (discontinuous) Peltier effect is verified.

The measurable quantity

$$\Pi_{AB} \equiv -\Pi_{BA} \equiv \Pi_A - \Pi_B = \frac{T}{F}\,[({}^*S_\ominus)_A - ({}^*S_\ominus)_B] \tag{4–25.59}$$

is called the "Peltier coefficient" for the conductor system A + B. The quantities Π_A and Π_B—which are only determined to within the arbitrary additive constants—are the "Peltier coefficients" of the individual conductors A and B.

From the above the definition of the sign of the Peltier coefficient is also derived. If, for example, Π_{AB} is positive then for a flow of current from A to B a cooling of the junction appears. In the literature, there is also the opposite sign convention which is then coupled with a corresponding sign reversal for the definition of ε_{AB}, so that Eq. (60) is upheld.

As is obvious, the Peltier effect, just like the Thomson or Seebeck effects, can be related back to the transported entropy of the electrons. There must accordingly be two independent relations between the three measurable effects; we will discuss these in the next subsection.

(e) Thomson Relations

From Eqs. (46) and (59) we obtain the relation between the thermoelectric power ε_{AB} of a thermocouple of the phase diagram described above and the Peltier coefficient Π_{AB}:

$$\Pi_{AB} = T\varepsilon_{AB}. \qquad (4\text{-}25.60)$$

Furthermore, from Eqs. (46) and (48) there follows the relation between the temperature coefficient of the thermoelectric power and the difference between the Thomson coefficients τ_A and τ_B:

$$\tau_B - \tau_A = T\left(\frac{\partial \varepsilon_{AB}}{\partial T}\right)_P. \qquad (4\text{-}25.61)$$

Table 11

Experimental Verification of the Thomson Relation $\Pi_{AB} = T\varepsilon_{AB}$.
After Zemansky[107c]†

Thermocouple	T K	Π_{AB}/T μV K^{-1}	ε_{AB} μV K^{-1}
Cu–Ni	273	18.6	20.39
	287	20.2	21.0
	295	20.5	21.4
	302	22.3	21.7
	373	24.4	24.9
Fe–Hg	292	16.7	16.66
	330	16.2	16.14
	373	15.6	15.42
	405	14.9	14.81
	456	13.9	13.74

† The first mentioned metal is conductor A.

Table 12

Experimental Verification of the Thomson Relation
$-(\tau_A - \tau_B) = \tau_B - \tau_A = T(\partial \varepsilon_{AB}/\partial T)$. After Zemansky[107c]‡

Thermocouple	T K	$1/T(\tau_A - \tau_D)$ μV K^{-2}	$-\partial \varepsilon_{AB}/\partial T$ μV K^{-2}
Cu–Pt	273	0.039	0.036
	293	0.037	0.036
	373	0.030	0.036
Cu–Fe	273	0.020	0.028
	373	0.039	0.033

‡ The first mentioned metal is conductor A.

Equations (60) and (61) were derived in 1854 by William Thomson (Lord Kelvin) and are known as the "Thomson equations of thermoelectricity."

In Tables 11 and 12, we have listed the experimental data for the verification of the Thomson equations (60) and (61).

(f) Conventional Absolute Value of the Transported Entropies

From Eq. (26) the transported entropy of the electrons $*S_\ominus$ of a single electronic conductor at the temperature T (at constant pressure) is found to be:

$$*S_\ominus = *S_\ominus^0 - F \int_0^T \frac{\tau}{T} dT, \qquad (4\text{-}25.62)$$

where the superior zero indicates the limiting value for $T \to 0$ and the integral can be experimentally evaluated from the temperature dependence of the Thomson coefficient τ. Because of the empirical equation

$$\lim_{T \to 0} \tau = 0 \qquad (4\text{-}25.63)$$

the integral converges, so that the transported entropy and its limiting value at the absolute zero of temperature can be assigned a finite value.

With the help of Eq. (46) there results from Eq. (62) for the measurable thermoelectric power ε_{AB} of a thermocouple composed of the conductors A and B:

$$\varepsilon_{AB} = \frac{1}{F} [(*S_\ominus)_A - (*S_\ominus)_B]$$

$$= \frac{1}{F} [(*S_\ominus^0)_A - (*S_\ominus^0)_B] - \int_0^T \frac{\tau_A - \tau_B}{T} dT. \qquad (4\text{-}25.64)$$

The statement, postulated by Nernst and confirmed by all recent experimental results

$$\lim_{T \to 0} \varepsilon_{AB} = 0, \qquad (4\text{-}25.65)$$

with Eqs. (63) and (64) leads to the equation

$$(*S_\ominus^0)_A = (*S_\ominus^0)_B. \qquad (4\text{-}25.66)$$

The zero-point value of the transported entropy of electrons is thus independent of the nature of the conductor.

We now take account of Eq. (66) most simply by setting

$$*S_\ominus^0 = 0 \qquad (4\text{-}25.67)$$

for each conductor. With this normalization we obtain from Eq. (62) the *conventional absolute value* of the transported entropy of the electrons of an arbitrary conductor at the temperature T:

$$*S_\ominus = -F \int_0^T \frac{\tau}{T} dT. \qquad (4\text{-}25.68)$$

Thus if we have evaluated the integral of Eq. (68) for a single conductor then, according to Eq. (64), we can determine the conventional absolute values of all other electron conductors at any arbitrary temperature from thermoelectric powers of thermocouples.

With the help of (cf. Eqs. 47 and 57)

$$\Pi = T\varepsilon = \frac{T^*S_\ominus}{F} \qquad (4\text{-}25.69)$$

we find conventional absolute values of the thermoelectric powers ε and of the Peltier coefficients Π of individual conductors. The formula $\Pi = T\varepsilon$ is sometimes called the "Thomson equation" because it corresponds to Eq. (60).

In Table 13, conventional absolute values of the transported entropies for electrons are listed for some metals at 25°C.

Table 13

Conventional Absolute Transported Entropies of Electrons
$^*S_\ominus$ for Metals at 25°C[114,115]

Metal	$^*S_\ominus$ cal K^{-1} mol^{-1}
Cu	−0.045
Ag	−0.050
Pt	0.104
Bi	1.7

4-26 THERMAL DIFFUSION IN GASES AND NONELECTROLYTE SOLUTIONS

(a) Empirical Description

In a nonisothermal multi-component system, there is a transport of matter by temperature gradients which, for a mixture initially homogeneous with respect to concentrations, brings about a "separation," i.e. a nonuniform concentration distribution. This transport phenomenon is denoted as *thermal diffusion*, or, particularly for condensed systems, as the "Ludwig-Soret effect," or also more concisely as the "Soret effect."

Thermal diffusion in liquid systems was discovered in 1856 by Ludwig and examined more precisely 1879–1881 by Soret. Chapman and Dootson found the effect for gases in 1917, after Chapman (1911–1917) and Enskog (1912–1917) had previously treated the phenomenon kinetically. Thermal diffusion in solids (solid solutions) was unambiguously established by the experiments of Darken and Oriani (1954).

Thermal diffusion is a cross phenomenon. The appropriate reciprocal effect, i.e. the transport of heat by concentration gradients also exists. It was discovered in 1872 by Dufour for gases and was more precisely examined 1942–1949 by Clusius and Waldmann. The phenomenon is called the *Dufour effect*.

If, finally, a pressure gradient is maintained in a multi-component system, we then expect pressure diffusion (§4–23) and a *pressure thermal effect*, i.e. a transport of heat by pressure gradients. This phenomenon does not seem to have been experimentally studied as yet; its existence is a necessary result of the thermodynamic–phenomenological theory (see below).

In a continuous multi-component system in which gradients of concentrations, pressure, and temperature are present, we must consider the following transport processes:

1. Diffusion (mass transport by concentration gradients).
2. Pressure diffusion (mass transport by pressure gradients).
3. Thermal diffusion (mass transport by temperature gradients).
4. Heat conduction (heat transport by temperature gradients).
5. Pressure thermal effect (heat transport by pressure gradients).
6. Dufour effect (heat transport by concentration gradients).

For simplicity, we restrict the discussion to fluid systems with two uncharged species, i.e. to a binary gas mixture or a binary nonelectrolyte solution. Since the new effects (thermal diffusion, Dufour effect, etc.) are described under normal conditions, just as are diffusion, heat conduction, etc., by linear relations between the fluxes (diffusion current density, heat current density) and the gradients, we can extend the known equations for diffusion, pressure diffusion, and heat conduction by including new terms and thus incorporate the new phenomena into the normal scheme.

Simultaneous diffusion, pressure diffusion, and thermal diffusion in a fluid with two uncharged species is described by an extension of Eq. (4–23.1):

$$\mathbf{J}_i = -\frac{1-\omega_i}{1-x_i}\frac{1}{\overline{V}}\left(D\,\mathrm{grad}\,x_i \mp D_P\frac{\mathrm{grad}\,P}{P} \pm D_T\frac{\mathrm{grad}\,T}{T}\right)\quad (i=1,2),$$

$$(4\text{–}26.1)$$

where the upper or lower sign holds if $i = 1$ or $i = 2$.† Here $\mathbf{J}_i$ denotes the diffusion current density of species i in any reference system (defined by Eq. 4–3.1); ω_i is the weight factor introduced in Eqs. (4–3.2) and (4–3.3) for the averaging of velocities of the two particle types, x_i the mole fraction of the species i, $\overline{V}$ the molar volume, P the pressure, and T the absolute temperature.

---

† The identity (4–3.4) is satisfied, as follows from Eq. (1) with Eqs. (4–3.3), (4–3.18), and (4–3.19).

Furthermore, D is the diffusion coefficient and D_P the pressure diffusion coefficient. Accordingly, the first two terms on the right-hand side of Eq. (1) are assigned to diffusion and pressure diffusion; the third term relates to thermal diffusion. We call the quantity D_T the *thermal diffusion coefficient*. This coefficient is defined following Eq. (1) by the relation

$$J_i = \mp \frac{1 - \omega_i}{1 - x_i} \frac{1}{\bar{V}} D_T \frac{\text{grad } T}{T} \quad (\text{grad } x_i = 0, \text{grad } P = 0). \qquad (4\text{–}26.2)$$

(There is, in the literature, the converse sign convention.) For many purposes, it is convenient to introduce a *thermal diffusion factor* α_T by the equation

$$\alpha_T \equiv \frac{D_T}{D x_1 x_2}, \qquad (4\text{–}26.3)$$

For liquid phases, we often use the *Soret coefficient*

$$\sigma \equiv \frac{\alpha_T}{T} = \frac{D_T}{D T x_1 x_2} \qquad (4\text{–}26.4)$$

instead of D_T or α_T. For grad $P = 0$, grad $T = 0$, Eq. (1) becomes Fick's law (4–17.25).

In the literature, two more quantities are used for the description of thermal diffusion in gases or liquids, namely, the "thermal diffusion ratio"

$$k_T \equiv \frac{D_T}{D} = \alpha_T x_1 x_2$$

or the quantity

$$D' \equiv \frac{D_T}{T x_1 x_2} = D\sigma,$$

which is denoted as the "thermal diffusion coefficient" too. Here we do not use k_T and D'.

We now make Eq. (1) specific to definite reference systems. For this we require the following relations which result from Eqs. (4–3.6), (4–3.8), (4–3.12), (4–17.24), (4–17.26), (4–17.27), and (4–23.5):

$$x_1 + x_2 = 1, \quad \chi_1 + \chi_2 = 1, \quad c_1 V_1 + c_2 V_2 = 1, \qquad (4\text{–}26.5)$$

$$x_i = c_i \bar{V}, \quad \rho_i = M_i c_i, \quad \chi_i = \frac{\rho_i}{\rho}, \quad \rho \bar{V} = M_1 x_1 + M_2 x_2, \quad (4\text{–}26.6)$$

$$\text{grad } x_i = \frac{(M_1 x_1 + M_2 x_2)^2}{M_1 M_2} \text{grad } \chi_i, \qquad (4\text{–}26.7)$$

where χ_i is the mass fraction, c_i the molarity, V_i the partial molar volume, ρ_i the mass concentration or partial density, M_i the molar mass of species i, and $\bar{V}$ and ρ the molar volume and the density of the mixture respectively.

If we take the Hittorf reference system as a basis then from Eq. (1) with Eqs. (6), (4–3.29), and (4–3.30) (cf. Eq. 4–17.32) we find:

$$_1\mathbf{J}_2 = c_2(\mathbf{v}_2 - \mathbf{v}_1) = -\frac{1}{x_1\overline{V}}\left(D \text{ grad } x_2 + D_P \frac{\text{grad } P}{P} - D_T \frac{\text{grad } T}{T}\right)$$

(4–26.8)

or

$$\mathbf{v}_2 - \mathbf{v}_1 = -\frac{1}{x_1 x_2}\left(D \text{ grad } x_2 + D_P \frac{\text{grad } P}{P} - D_T \frac{\text{grad } T}{T}\right).$$ (4–26.9)

Here $\mathbf{v}_i$ denotes the mean velocity of species i. Our equation in the form (9) appears in the kinetic theory of gases.

In the Fick reference system, on the basis of Eqs. (5), (6), (4–3.25), and (4–3.27), Eq. (1) assumes the following form (cf. Eq. 4–23.2):

$$_w\mathbf{J}_1 = c_1(\mathbf{v}_1 - \mathbf{w})$$

$$= -\frac{V_2}{\overline{V}^2}\left(D \text{ grad } x_1 - D_P \frac{\text{grad } P}{P} + D_T \frac{\text{grad } T}{T}\right), \quad (4\text{–}26.10a)$$

$$_w\mathbf{J}_2 = c_2(\mathbf{v}_2 - \mathbf{w})$$

$$= -\frac{V_1}{\overline{V}^2}\left(D \text{ grad } x_2 + D_P \frac{\text{grad } P}{P} - D_T \frac{\text{grad } T}{T}\right). \quad (4\text{–}26.10b)$$

Here $\mathbf{w}$ is the mean volume velocity.

If we proceed from the molecular reference system, we obtain from Eq. (1) with the help of Eqs. (6), (4–3.21), and (4–3.22) (cf. Eq. 4–23.3):

$$_u\mathbf{J}_i = c_i(\mathbf{v}_1 - \mathbf{u})$$

$$= -\frac{1}{\overline{V}}\left(D \text{ grad } x_i \mp D_P \frac{\text{grad } P}{P} \pm D_T \frac{\text{grad } T}{T}\right) \quad (i = 1, 2), \quad (4\text{–}26.11)$$

where $\mathbf{u}$ represents the mean molar velocity (average particle velocity).

Finally, if we use the barycentric system then we derive from Eq. (1) with Eqs. (5), (6), (4–3.10), and (4–3.14) (cf. Eq. 4–23.4):

$$M_i {}_v\mathbf{J}_i = \mathbf{J}_i^* = \rho_i(\mathbf{v}_i - \mathbf{v})$$

$$= -\frac{M_1 M_2 \rho}{(M_1 x_1 + M_2 x_2)^2}\left(D \text{ grad } x_i \mp D_P \frac{\text{grad } P}{P} \pm D_T \frac{\text{grad } T}{T}\right)$$

$$(i = 1, 2). \quad (4\text{–}26.12)$$

Here $\mathbf{v}$ is the barycentric velocity.

If, according to Eq. (4–23.14) or Eq. (3), we introduce the dimensionless pressure diffusion factor ($\alpha_P = D_P/Dx_1 x_2$) or the dimensionless thermal diffusion factor ($\alpha_T = D_T/Dx_1x_2$) then there results from Eqs. (6), (7), (11), and (12):

$$_u\mathbf{J}_i = c_i(\mathbf{v}_i - \mathbf{u})$$

$$= -D(c_1 + c_2)\left(\text{grad } x_i \mp \alpha_P x_1 x_2 \frac{\text{grad } P}{P} \pm \alpha_T x_1 x_2 \frac{\text{grad } T}{T}\right)$$

$$(i = 1, 2), \quad (4\text{–}26.13\text{a})$$

$$\mathbf{J}_i^* = \rho_i(\mathbf{v}_i - \mathbf{v})$$

$$= -D(\rho_1 + \rho_2)\left(\text{grad } \chi_i \mp \alpha_P \chi_1 \chi_2 \frac{\text{grad } P}{P} \pm \alpha_T \chi_1 \chi_2 \frac{\text{grad } T}{T}\right)$$

$$(i = 1, 2). \quad (4\text{–}26.13\text{b})$$

These two equations are completely analogous to each other.

We have defined the coefficients D_T, α_T, and σ so that a positive value of these quantities has the following significance: the transport of the species 1 by thermal diffusion occurs from the higher to the lower temperature, while substance 2 migrates in the direction of the temperature gradient. We now decide on the convention that species 1 shall be that one with the greater molecular weight. Then a positive value of D_T, α_T, or σ states: the "lighter component" is enriched at the "warmer position."

We must still deal with the other transport processes, namely, heat conduction, pressure thermal effect, and Dufour effect. We describe the simultaneous appearance of these phenomena by a meaningful extension of the Fourier law (4–24.32):

$$\mathbf{J}_Q = -\lambda_0 \text{ grad } T - \zeta \text{ grad } P - \beta_T \text{ grad } x_1. \quad (4\text{–}26.14)$$

Here $\mathbf{J}_Q$ denotes the heat current density and λ_0 the thermal conductivity for uniform state. The first term on the right-hand side thus relates to the heat conduction under the conditions grad $P = 0$, grad $x_1 = 0$. The second term characterizes the pressure thermal effect, the third term the Dufour effect. Accordingly, ζ is called the *pressure thermal coefficient* and β_T the *Dufour coefficient*.

The above scheme does not exclude external force fields such as gravitational or centrifugal fields. Indeed, their influence would not appear explicitly in the formulas, as will be shown below, but would only lead to pressure gradients. However, we do not assume that there is local mechanical equilibrium—as in the consideration of sedimentation (§4–20); due to this the conditions for "creeping motion" drop out. We can thus allow, as in the treatment of pressure diffusion (§4–23), a "pressure gradient forced on the system by the outside." Conversely, if we consider grad $P = 0$ in the following, we must exclude external force fields as well as viscous flow, so that a fluid is present to which we may apply the formulas of subsections (b) to (f) in §4–24.

The above formulas remain valid if chemical reactions are allowed. Due to the assumption of two uncharged species, only a single reaction can occur: a transformation or dissociation reaction or something similar (cf. §4–28).

For grad $P = 0$ and no convection, according to the last statement, the general heat conduction equation (4–24.48) holds with $N = 2$, $R = 1$. Moreover, on the basis of Eqs. (4–3.28) and (4–24.25) we have:

$$V_1 \, _w\mathbf{J}_1 + V_2 \, _w\mathbf{J}_2 = 0,$$

$$c_1(\text{grad } H_1)_{T,P} + c_2(\text{grad } H_2)_{T,P} = 0,$$

where H_i denotes the partial molar enthalpy of species i. From this we derive with the help of Eq. (5):

$$\sum_{k=1}^{2} {}_w\mathbf{J}_k[C_{Pk} \text{ grad } T + (\text{grad } H_k)_{T,P}]$$

$$= {}_w\mathbf{J}_1\left(C_{P1} - \frac{V_1}{V_2}C_{P2}\right)\text{grad } T + {}_w\mathbf{J}_1 \frac{1}{c_2 V_2}(\text{grad } H_1)_{T,P}. \qquad (4\text{–}26.15)$$

Here C_{Pi} is the partial molar heat capacity (for constant pressure) of substance i. If we insert Eqs. (10a) and (14)—each under the condition grad $P = 0$ —as well as Eq. (15) into Eq. (4–24.48) and omit the terms with grad $x_1 \times$ grad T, and (grad T)², which experience indicates to be very small, as well as the term referring to chemical reaction then on consideration of Eq. (6) we obtain:

$$\frac{\bar{C}_P}{\bar{V}}\frac{\partial T}{\partial t} = \text{div}\,(\lambda_0 \text{ grad } T) + \text{div}\,(\beta_T \text{ grad } x_1)$$

$$+ \frac{1}{x_2 \bar{V}}\left(\frac{\partial H_1}{\partial x_1}\right)_{T,P} D(\text{grad } x_1)^2. \qquad (4\text{–}26.16)$$

Here $\bar{C}_P$ denotes the molar heat capacity (for constant pressure) of the mixture and t the time. For the uniform state (grad $x_1 = 0$), Eq. (16) becomes the usual heat conduction equation (4–24.50a) with which we can determine the thermal conductivity λ_0. Since the diffusion coefficient D can be independently measured, Eq. (16) is the basis of the experimental determination of the Dufour coefficient β_T.† Such measurements for gases‡ have been conducted by Waldmann[116a, b, 117].‖

---

† For the history of its discovery see Clusius[118]; for its evaluation see also Haase[119].

‡ The last term in Eq. (16) vanishes for ideal gas mixtures, because here $\partial H_1/\partial x_1 = 0$ (heat of mixing vanishes). Thus this term, which only plays a role for real gas mixtures, is called the "heat of mixing term."

‖ Recently, Rastogi and Madan[120] claim to have found the Dufour effect in the liquid system benzene + chlorobenzene at 35°C.

For grad $P = 0$, we find from Eqs. (5), (8), and (14):

$$\mathbf{J_Q} = {}^*Q_{2\,1}\mathbf{J_2} - \lambda_\infty \text{ grad } T \quad (\text{grad } P = 0) \qquad (4\text{-}26.17)$$

with

$$^*Q_2 \equiv -\frac{\beta_T}{D} x_1 \overline{V}, \qquad (4\text{-}26.18)$$

$$\lambda_\infty \equiv \lambda_0 - \frac{\beta_T D_T}{DT}. \qquad (4\text{-}26.19)$$

Since Eq. (17) for grad $T = 0$ is transformed into

$$\mathbf{J_Q} = {}^*Q_{2\,1}\mathbf{J_2}(\text{grad } P = 0, \text{ grad } T = 0),$$

$^*Q_2$ represents the heat of transport of species 2 (Hittorf reference system).

We now consider the stationary state. We exclude convection and chemical reactions. Then, according to Eq. (4–4.26),

$$\mathbf{J}_i = 0 \quad (i = 1, 2). \qquad (4\text{-}26.20)$$

From Eqs. (1), (5), and (20) with grad $P = 0$ there follows:

$$\text{grad } x_1 = -\text{grad } x_2 = -\frac{D_T}{DT} \text{grad } T \quad (\text{grad } P = 0, \text{ stationary state})$$

$$(4\text{-}26.21)$$

or with Eq. (4):

$$\frac{1}{x_1 x_2} \text{grad } x_2 = \frac{\alpha_T}{T} \text{grad } T = \sigma \text{ grad } T \quad (\text{grad } P = 0, \text{ stationary state}).$$

$$(4\text{-}26.22)$$

Such a stationary state arises through the opposing effects of diffusion and thermal diffusion. Its experimental verification is the most direct procedure for the measurement of the thermal diffusion factor α_T or of the Soret coefficient σ. The thermal diffusion coefficient D_T can also be determined from this provided the diffusion coefficient D is known. The method is applied both to gases and to liquids[108, 121].

From Eqs.(17) and (20) there results:

$$\mathbf{J_Q} = -\lambda_\infty \text{ grad } T \quad (\text{grad } P = 0, \text{ stationary state}).$$

This is Fourier's law in the form (4–24.34). Thus λ_∞ denotes the thermal conductivity for the stationary state.

(b) Thermodynamic–Phenomenological Description

Since we consider the possibility of a pressure gradient generated from the outside (cf. above), we must, in the thermodynamic–phenomenological description of the effects mentioned, proceed from Eqs. (4–24.1) and (4–24.4). For mixtures of two uncharged species ($N = 2$, $z_i = 0$), there results from

Eqs. (4–24.1), (4–24.3), and (4–24.4) with $\alpha_{22} \equiv A$, $\alpha_{2Q} \equiv B$, $\alpha_{Q2} \equiv B'$, $\alpha_{QQ} \equiv C$:

$$\mathbf{J}_2^* = A(\mathbf{X}_2^* - \mathbf{X}_1^*) - B \frac{\text{grad } T}{T}, \tag{4–26.23}$$

$$\mathbf{J}_Q = B'(\mathbf{X}_2^* - \mathbf{X}_1^*) - C \frac{\text{grad } T}{T}. \tag{4–26.24}$$

Here A, B, B', and C are the phenomenological coefficients. On the basis of Eqs. (6), (4–20.18), (4–24.2), and (4–24.6), there is valid:

$$\mathbf{X}_2^* - \mathbf{X}_1^* = \left(\frac{V_1}{M_1} - \frac{V_2}{M_2}\right) \text{grad } P - \frac{M_1 x_1 + M_2 x_2}{M_1 M_2 x_1} \left(\frac{\partial \mu_2}{\partial x_2}\right)_{T,P} \text{grad } x_2, \tag{4–26.25}$$

$$B = B', \tag{4–26.26}$$

where μ_2 denotes the chemical potential of species 2. In this, a gravitational or centrifugal field is allowed since, according to Eq. (4–24.2), the pertinent term in Eq. (25) drops out. Equation (26) is an expression of the Onsager reciprocity law.†

The thermodynamics of irreversible processes thus states: all transport processes considered are described by three independent coefficients (e.g. A, B, and C). In the empirical treatment of the effects, we introduced six transport quantities (D, D_P, D_T, λ_0, ζ, β_T), so there must then be three independent relations between these quantities.

If we compare the coefficients in Eqs. (12) and (23) or in Eqs. (14) and (24), on taking account of Eqs. (5), (6), and (25), we find:

$$D = A \frac{\rho^2 \overline{V}^3}{M_1^2 M_2^2 x_1} \left(\frac{\partial \mu_2}{\partial x_2}\right)_{T,P}, \tag{4–26.27}$$

$$D_P = A \frac{\rho \overline{V}^2 P}{M_1 M_2} \left(\frac{V_2}{M_2} - \frac{V_1}{M_1}\right), \tag{4–26.28}$$

$$D_T^\bullet = -B \frac{\rho \overline{V}^2}{M_1 M_2}, \tag{4–26.29}$$

$$\lambda_0 \overset{\bullet}{=} \frac{C}{T}, \tag{4–26.30}$$

$$\zeta = B' \left(\frac{V_2}{M_2} - \frac{V_1}{M_1}\right), \tag{4–26.31}$$

$$\beta_T = -B' \frac{\rho \overline{V}}{M_1 M_2 x_1} \left(\frac{\partial \mu_2}{\partial x_2}\right)_{T,P}. \tag{4–26.32}$$

---

† The thermodynamic–phenomenological treatment of thermal diffusion in the above sense is, in principle, due to Meixner[122].

Here it should be noted that both the phenomenological coefficients (A, B, B', C) and also the measurable transport quantities (D, D_P, D_T, λ_0, ζ, β_T) can be arbitrary functions of the local state variables (temperature, pressure, concentration).

From Eqs. (27) and (28) the relation between the pressure diffusion coefficient D_P and the diffusion coefficient D follows directly, in agreement with Eq. (4–23.9). Noting Eq. (4–23.13) and Eq. (6), we obtain:

$$D_P = DP\left(V_2 - \frac{M_2}{\rho}\right)\frac{1}{(\partial\mu_2/\partial x_2)_{T,P}}. \qquad (4\text{–}26.33)$$

Combination of Eq. (31) with Eq. (32) produces the relation between the pressure thermal coefficient ζ and the diffusion thermal coefficient β_T and on the basis of Eqs. (4–23.13), (6), and (33) this can be written in the following form[104]:

$$\zeta = -\frac{\beta_T D_P}{DP}. \qquad (4\text{–}26.34)$$

Equations (33) and (34) were evolved *without* the Onsager reciprocity law.

From Eqs. (3), (29), and (32), *with* consideration of the Onsager relation (26), there results the relation between the Dufour coefficient β_T and the thermal diffusion coefficient D_T or the thermal diffusion factor α_T:

$$\beta_T = D_T\frac{1}{x_1\overline{V}}\left(\frac{\partial\mu_2}{\partial x_2}\right)_{T,P} = D\alpha_T\frac{x_2}{\overline{V}}\left(\frac{\partial\mu_2}{\partial x_2}\right)_{T,P}. \qquad (4\text{–}26.35)$$

This general equation[104, 123] when applied to ideal gas mixtures is reduced with Eq. (4–23.10) to the expression

$$\beta_T = \frac{D_T P}{x_1 x_2} = D\alpha_T P, \qquad (4\text{–}26.36)$$

which was derived by Meixner[45b] after the kinetic theory of gases had previously given the same relation, although only for monatomic ideal gases. Equation (36) has been confirmed by the experiments of Waldmann (see above). With the help of Eq. (35) the thermal diffusion coefficient D_T or the thermal diffusion factor α_T can generally be determined from measurements of the Dufour effect.

Equations (33), (34), and (35) represent the three independent relations sought between the six measurable transport coefficients. We also recognize from Eqs. (33) to (35) that, assuming the existence of diffusion and thermal diffusion, the pressure diffusion, the pressure thermal effect, and the Dufour effect *must* appear.

From Eqs. (4), (18), and (35) there follows for the heat of transport *Q_2:

$$-\frac{*Q_2}{(\partial\mu_2/\partial x_2)_{T,P}} = \frac{D_T}{D} = \alpha_T x_1 x_2 = \sigma T x_1 x_2. \qquad (4\text{–}26.37)$$

Insertion of Eq. (37) in Eq. (21) or (22) produces the relation:

$$\left(\frac{\partial \mu_2}{\partial x_2}\right)_{T,P} \text{grad } x_2 = (\text{grad } \mu_2)_{T,P}$$

$$= -\frac{{}^*Q_2}{T} \text{grad } T \quad (\text{grad } P = 0, \text{stationary state}) \quad (4\text{–}26.38)$$

in agreement with Eq. (4–24.25).

From Eqs. (19), (35), and (37) we obtain the relation between the two thermal conductivities[104, 123]

$$\lambda_0 - \lambda_\infty = \frac{D_T^2}{D} \frac{1}{T\bar{V}x_1} \left(\frac{\partial \mu_2}{\partial x_2}\right)_{T,P} = \frac{D\alpha_T^2 x_1 x_2^2}{T\bar{V}} \left(\frac{\partial \mu_2}{\partial x_2}\right)_{T,P}$$

$$= \frac{D\sigma^2 T x_1 x_2^2}{\bar{V}} \left(\frac{\partial \mu_2}{\partial x_2}\right)_{T,P} = \frac{D {}^*Q_2^2}{T\bar{V}x_1(\partial \mu_2/\partial x_2)_{T,P}}. \quad (4\text{–}26.39)$$

Following (4–14.6) we see that the positive-only character of the local entropy production leads to the inequalities:

$$A > 0, \quad C > 0, \quad AC - B^2 > 0. \quad (4\text{–}26.40)$$

If stable or metastable phases (excluding critical points) are assumed then, according to Eqs. (1–22.34) and (1–22.38),

$$\left(\frac{\partial \mu_2}{\partial x_2}\right)_{T,P} > 0. \quad (4\text{–}26.41)$$

Thus the statements[104, 123]:

$$D > 0, \quad \lambda_\infty > 0, \quad \lambda_0 > \lambda_\infty, \quad (4\text{–}26.42)$$

$$D_T^2 < \frac{D\lambda_0 T\bar{V}x_1}{(\partial \mu_2/\partial x_2)_{T,P}} \quad (4\text{–}26.43)$$

result on noting Eqs. (27), (29), (30), and (39). The relations (42) are already known to us. The inequality (43) leaves the sign of D_T open, but fixes an upper limit to the absolute value of this quantity. It represents the generalization of an inequality given by Meixner[45b].†

(c) Experimental Values of Thermal Diffusion Factor for Gases

To characterize thermal diffusion in binary gas mixtures, the thermal diffusion factor α_T is used almost without exception. This quantity is dimensionless and less concentration dependent than the thermal diffusion coefficient D_T.

We find, empirically, that α_T is a function of the temperature, of the pressure, and of the concentration and—even in the same system—that it can

---

† More details on simplification of the formulas (35), (39), and (43) for the application to ideal liquid mixtures, ideal dilute solutions, and ideal gas mixtures are found in Haase[20b].

be both positive and negative. (We recall that the statement $\alpha_T > 0$ denotes enrichment of the lighter component 2 in the warmer region.)

The kinetic theory of gases, which was developed for thermal diffusion by Chapman and Enskog, in the case of ideal gas mixtures—and only under certain assumptions—permits a calculation of α_T from molecular data. Therefore the older measurements refer mainly to lower pressures.

In recent times, interest has developed in real gas mixtures and in the critical evaporation region. In connection with this, the thermal diffusion in ideal gas mixtures is characterized by the limiting value

$$\lim_{P \to 0} \alpha_T \equiv \alpha_T^0, \qquad (4\text{-}26.44)$$

which only depends on the temperature and on the composition.

In §4–23, we showed that the general relation between pressure diffusion and diffusion yields a formula (Eq. 4–23.18) which allows the pressure diffusion factor α_P at any arbitrary pressure to be calculated from the limiting value α_P^0 (valid for $P = 0$). This is true provided the equation of state of the gas mixture is known. The thermodynamics of irreversible processes produces no analogous relation for the thermal diffusion factor because a general connection between thermal diffusion and diffusion does not exist; we can, however, try to draw conclusions about a relation between α_T and α_T^0 by analogy.

Consideration of Eq. (4–23.18) makes such an analogy plain. First, let the dimensionless quantity α_P or α_P^0 be replaced by the dimensionless quantity α_T or α_T^0 respectively. Then, remembering that in classical thermodynamics the partial molar enthalpy H_i plays a role in the temperature dependence of an effect which is similar to that of the partial molar volume V_i for the pressure dependence, let the expression $H_i - H_{0i}$ replace $P(V_i - V_{0i})$, where

$$H_{0i} \equiv \lim_{P \to 0} H_i \qquad (4\text{-}26.45)$$

is the limiting value of the partial molar enthalpy of the species i ($i = 1, 2$) for vanishing pressure and thus at the same time the molar enthalpy of the pure ideal gas i at the given temperature. We thus guess the formula[119, 123]:†

$$\alpha_T = \frac{\alpha_T^0(M_1 x_1 + M_2 x_2)RT + M_1(H_2 - H_{02}) - M_2(H_1 - H_{01})}{(M_1 x_1 + M_2 x_2)x_2(\partial \mu_2/\partial x_2)_{T,P}}.$$

$$(4\text{-}26.46)$$

Here R denotes the gas constant.

---

† The argument becomes particularly plausible if we write the second and third terms in the bracket in Eq. (13a) with Eqs. (1–12.8) and (1–12.12) in the following way

$$\mp T x_1 x_2 \left[\frac{\alpha_P}{PV_i} \left(\operatorname{grad} \frac{\mu_i}{T} \right)_{T,x_i} + \frac{\alpha_T}{H_i} \left(\operatorname{grad} \frac{\mu_i}{T} \right)_{P,x_i} \right].$$

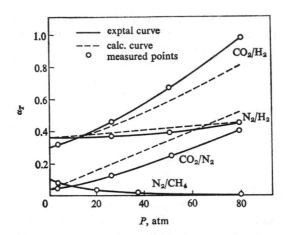

Fig. 17. Thermal diffusion factor α_T as a function of the pressure P for approximately equimolar gas mixtures at about 360 K. After Haase[119].

If the limiting value α_T^0 is known as a function of temperature and composition (from kinetic theory or from extrapolation of measured data), Eq. (46) produces the thermal diffusion factor α_T for any arbitrary pressure provided we know the equation of state of the gas mixture to evaluate $H_i - H_{0i}$ and $(\partial\mu_2/\partial x_2)_{T,P}$. We can thus obtain the function $\alpha_T(x_2, T, P)$ from the function $\alpha_T^0(x_2, T)$ without adjustable parameters.

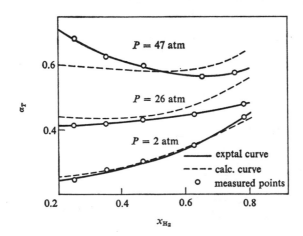

Fig. 18. Thermal diffusion factor α_T as a function of the mole fraction x_{H_2} of the hydrogen for the gas mixture $CO_2 + H_2$ at about 360 K. After Haase[119].

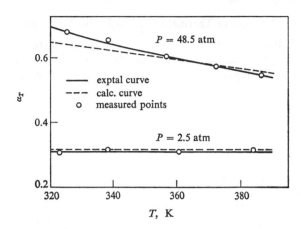

Fig. 19. Thermal diffusion factor α_T as a function of the absolute temperature T for the approximately equimolar gas mixture $CO_2 + H_2$. According to Haase[119].

Figures 17, 18, and 19 show that Eq. (46) correctly reproduces all essential features of the pressure dependence of the thermal diffusion factor for binary gas mixtures. In many cases, a quantitative agreement was almost obtained. The data points come from Becker[124, 125]. The curves were calculated using the abbreviated Beattie–Bridgeman equation of state (with three parameters).

Rutherford and Roof[126] experimentally investigated thermal diffusion in the system n-butane + methane. They approached the critical vaporization line from the liquid-state region. They showed that the thermal diffusion factor α_T became infinity at critical points. This result, too, follows from Eq. (46) since we have for any binary system, according to Eq. (1–22.41),

$$\left(\frac{\partial \mu_2}{\partial x_2}\right)_{T,P} = 0 \quad \text{(critical point)}. \qquad (4\text{–}26.47)$$

For the system in question, α_T^0 may be obtained from a kinetic formula due to Chapman and Cowling and the thermodynamic functions may be calculated from the equation of state given by Benedict, Webb, and Rubin. Thus Eq. (46) may be checked. This has been done by Rutherford[127]. The agreement between calculated and experimental values of α_T is good.

In the meantime, it was discovered in the author's laboratory[128] that Eq. (46) also describes the thermal diffusion in the gas system propane + methane in the critical vaporization region.[36]

These results are remarkable since they show that Eq. (46) holds even for the liquid state in the vicinity of critical points.

(d) Experimental Values of Thermal Diffusion Factor for Liquids

The thermal diffusion factor α_T of binary nonelectrolyte solutions also depends on temperature, pressure, and composition. But if we exclude the critical vaporization region we may now ignore the dependence on pressure.

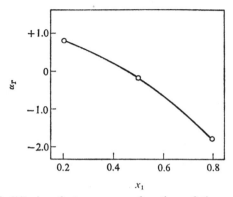

Fig. 20. Thermal diffusion factor α_T as a function of the mole fraction x_1 of benzene for the liquid mixture benzene + methanol at 40°C. After Tichacek, Kmak, and Drickamer[136].

Two examples for measured values of α_T are shown in Figs. 20 and 21. In both cases, α_T, considered as a function of composition at given temperature, changes sign. (The case $\alpha_T > 0$ again means increase of concentration of the lighter component 2 in the hotter region.) Many other examples may be found in the literature[108, 129–133]. All measurements refer to atmospheric pressure.

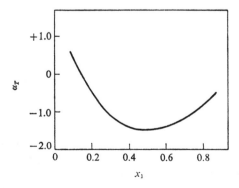

Fig. 21. Thermal diffusion factor α_T as a function of the mole fraction x_1 of ethanol for the liquid mixture ethanol + water at 25°C. After Tichacek, Kmak, and Drickamer[136].

An interesting problem is thermal diffusion in the immediate neighborhood of a consolute point in liquid–liquid equilibria. While it is certain that the diffusion coefficient D vanishes at critical points, the behavior of the thermal diffusion coefficient D_T, of the thermal diffusion factor α_T, and of the heat of transport $*Q_2$ is somewhat uncertain. According to Eqs. (4–17.38a), (3), and (37), the following relations hold:

$$D_T = D\alpha_T x_1 x_2, \qquad *Q_2 = -\frac{D_T L}{b x_2}. \qquad (4\text{–}26.48)$$

Here L is the Avogadro constant and b the "mobility" as defined on p. 277. Experience tells us (see p. 279) that b is nonzero and finite at critical points. If we accept the result

$$D_T \to \pm\infty \quad \text{(critical point)}, \qquad (4\text{–}26.49)$$

which probably but not certainly follows from the relevant experiments [134, 135], we thus conclude from Eqs. (48):

$$\alpha_T \to \pm\infty \quad \text{(critical point)}, \qquad (4\text{–}26.50)$$

$$*Q_2 \to \mp\infty \quad \text{(critical point)}. \qquad (4\text{–}26.51)$$

The two liquid systems for which the relations (49), (50), and (51) probably hold are nitrobenzene + n-hexane (which has an upper consolute point) and triethylamine + water (which has a lower consolute point). Figures 22 and 23 refer to the first system. They seem to verify (49) and (50).

Let us finally check the inequality (43) for noncritical conditions. The relation may be written

$$\alpha_T^2 < \frac{\lambda_0 T \bar{V}}{D x_1 x_2^2 \mu_{22}}, \qquad (4\text{–}26.52)$$

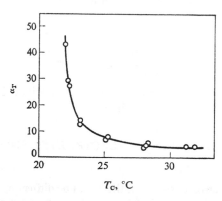

Fig. 22. Thermal diffusion factor α_T as a function of the Celsius temperature T_C for the equimolar liquid mixture nitrobenzene + n-hexane (critical point at about 20°C, approximately equimolar). After Thomaes[134].

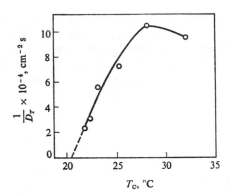

Fig. 23. Thermal diffusion coefficient D_T as a function of the Celsius temperature T_C for the equimolar liquid mixture nitrobenzene + n-hexane. Computed from the data in Figs. 11 (p. 280) and 22.

where

$$\mu_{22} \equiv \left(\frac{\partial \mu_2}{\partial x_2}\right)_{T,P}.$$

We use the measured values[135] for the system triethylamine + water (lower critical point at 18°C and $x_2 = 0.9$) at 10°C and $x_2 = 0.99$:

$$\overline{V} = 18.8 \text{ cm}^3 \text{ mol}^{-1}, \qquad \mu_{22} = 570 \text{ cal mol}^{-1},$$

$$D = 3.9 \times 10^{-6} \text{ cm}^2 \text{ s}^{-1}, \qquad \alpha_T = 37.$$

The thermal conductivity λ_0 may be taken to be that of pure water at 10°C, approximately:

$$\lambda_0 = 1.2 \times 10^{-3} \text{ cal K}^{-1} \text{ cm}^{-1} \text{ s}^{-1}.$$

Thus we obtain:

$$\alpha_T^2 \approx 1370,$$

$$\frac{\lambda_0 T \overline{V}}{D x_1 x_2^2 \mu_{22}} \approx 4 \times 10^5.$$

We conclude that the right-hand side of the inequality (52) exceeds the left-hand side by a factor of 300, approximately.

4–27 THERMAL DIFFUSION IN ELECTROLYTE SOLUTIONS

(a) General

Just as in a nonelectrolyte solution, there will be diffusion, thermal diffusion, heat flow, and the Dufour effect in an electrolyte solution, provided there are concentration gradients and a temperature gradient. As a matter of fact, thermal diffusion was discovered in an electrolyte solution (Ludwig, 1856).

We ignore pressure gradients and external fields. We further restrict the discussion to a single neutral solvent (component 1) and a single electrolyte (component 2) with two ionic species (cations: species $+$, anions: species $-$, undissociated electrolyte molecules: species u).

Owing to electroneutrality and local dissocation equilibrium (cf. §4–18), we have two independently moving substances and thus one independent concentration gradient and one independent diffusion flow. Therefore we may formally apply the scheme described in §4–26 (subsection a). Only the conditions for the stationary state require special treatment, since now we have to take account of the dissociation reaction.

Let us denote the degree of dissociation of the electrolyte by α, the number of cations and anions contained in one molecule of electrolyte by ν_+ and ν_-, and the sum $\nu_+ + \nu_-$ by ν. Then, according to Eq. (4–18.4), we have for the diffusion flow densities $\mathbf{J}_+$, $\mathbf{J}_-$, and $\mathbf{J}_u$ of the electrolyte species:

$$\mathbf{J}_+ = \nu_+\alpha\mathbf{J}_2, \quad \mathbf{J}_- = \nu_-\alpha\mathbf{J}_2, \quad \mathbf{J}_u = (1 - \alpha)\mathbf{J}_2, \qquad (4\text{–}27.1)$$

$$\nu\mathbf{J}_u + \mathbf{J}_+ + \mathbf{J}_- = \nu\mathbf{J}_2. \qquad (4\text{–}27.2)$$

Here $\mathbf{J}_2$ is the diffusion current density of the electrolyte, as defined in §4–18.

Thermal diffusion in electrolyte solutions is usually described by the Soret coefficient σ. Here we only need the Hittorf and the Fick reference systems, and we need not consider the formulas for the solvent. Thus, to describe diffusion and thermal diffusion, we just apply Eqs. (4–26.4), (4–26.8), and (4–26.10b) for constant pressure:

$$_1\mathbf{J}_2 = -\frac{D}{x_1\overline{V}}\,(\operatorname{grad} x_2 - \sigma x_1 x_2 \operatorname{grad} T), \qquad (4\text{–}27.3)$$

$$_w\mathbf{J}_2 = -D\,\frac{V_1}{\overline{V}^2}\,(\operatorname{grad} x_2 - \sigma x_1 x_2 \operatorname{grad} T), \qquad (4\text{–}27.4)$$

where $_1\mathbf{J}_2$ and $_w\mathbf{J}_2$ are the diffusion current densities of the electrolyte in the Hittorf and Fick reference systems respectively, D is the diffusion coefficient, $\overline{V}$ the molar volume, V_1 the partial molar volume of the solvent, T the temperature, and x_1 and x_2 the stoichiometric mole fractions of solvent and electrolyte respectively.

The composition variables mainly used for electrolyte solutions are the molality m and the molarity c of the electrolyte. From Eqs. (1–19.12), (1–22.20), (4–17.26) to (4–17.29) we derive:

$$\frac{1}{x_1 x_2}\operatorname{grad} x_2 = \frac{V_1}{c(1 - cV_2)\overline{V}^2}\operatorname{grad} x_2 = \frac{1}{m}\operatorname{grad} m, \qquad (4\text{–}27.5)$$

$$\operatorname{grad} x_2 = \frac{\overline{V}^2}{V_1}\,(\operatorname{grad} c + c\beta \operatorname{grad} T). \qquad (4\text{–}27.6)$$

Here V_2 is the partial molar volume of the electrolyte and β the thermal expansivity of the solution.

From Eqs. (4) to (6) we obtain, on consideration of Eq. (4–17.27),

$$_w\mathbf{J}_2 = -Dc(1 - cV_2)\left(\frac{1}{m} \operatorname{grad} m - \sigma \operatorname{grad} T\right) \tag{4–27.7}$$

or

$$_w\mathbf{J}_2 = -D[\operatorname{grad} c + c\beta \operatorname{grad} T - \sigma c(1 - cV_2) \operatorname{grad} T]. \tag{4–27.8}$$

The term containing β in Eq. (8) has nothing to do with thermal diffusion but refers to thermal expansion. Equation (8) is the starting point for the evaluation of thermal diffusion experiments in electrolyte solutions[1].

(b) Stationary State

Equation (4–4.23) is the fundamental formula for the convection-free stationary state. When applied to our special case, this leads to the following equations:†

$$\left.\begin{aligned} \operatorname{div} {}_w\mathbf{J}_u &= -b, \\ \operatorname{div} {}_w\mathbf{J}_+ &= \nu_+ b, \\ \operatorname{div} {}_w\mathbf{J}_- &= \nu_- b, \end{aligned}\right\} \tag{4–27.9}$$

where $_w\mathbf{J}_i$ is the diffusion current density of species i in the Fick reference system and b the rate of the dissociation reaction. If the first equation in (9) is multiplied by ν and the three equations are added then we find with Eq. (2):

$$\nu \operatorname{div} {}_w\mathbf{J}_2 = 0. \tag{4–27.10}$$

If the concentration and temperature gradients lie in one direction (a one-dimensional problem) and if the experimental vessel is closed (cf. p. 227) then there follows from Eq. (10):

$$_w\mathbf{J}_2 = 0, \tag{4–27.11}$$

hence

$$\mathbf{J}_2 = 0. \tag{4–27.12}$$

From this with Eq. (1) there again results

$$\mathbf{J}_i = 0 \quad (i = +, -, u). \tag{4–27.13}$$

Equation (4–24.22) is thus satisfied. Therefore Eq. (4–24.24) and especially (4–24.31) are also applicable.

---

† Though formally, for local dissociation equilibrium, the relation $b = 0$ holds, we must not apply this to the stationary state, since here the time-independent concentrations in each volume element result from the combined action of diffusion, thermal diffusion, and chemical reaction.

If we allow $*Q_i$ $(i = +, -, u)$ or $*Q_2$ to denote the heat of transport (in the Hittorf reference system) of species i or of the electrolyte respectively then we obtain from Eqs. (4-24.30) and (4-24.31):

$$*Q_2 = *Q_u = \nu_+ \, *Q_+ + \nu_- \, *Q_- \equiv *Q, \qquad (4\text{-}27.14)$$

$$\left(\frac{\partial \mu_2}{\partial m}\right)_{T,P} \text{grad } m = -\frac{*Q}{T} \text{grad } T. \qquad (4\text{-}27.15)$$

Here μ_2 is the chemical potential of the electrolyte. For the derivation of Eq. (15), we have used the methods of the thermodynamics of irreversible processes and particularly Onsager's reciprocity law.

From Eqs. (7) and (11) we find for the stationary state:

$$\frac{1}{m} \text{grad } m = \sigma \text{ grad } T. \qquad (4\text{-}27.16)$$

This relation is often used as the definition of the Soret coefficient σ.

Comparison of Eq. (15) with Eq. (16) leads to the relation between the heat of transport $*Q$ of the electrolyte and the Soret coefficient σ:

$$*Q = -\sigma T m \left(\frac{\partial \mu_2}{\partial m}\right)_{T,P}. \qquad (4\text{-}27.17)$$

From this with Eq. (4-19.21) there follows:

$$*Q = -\nu R T^2 \left[1 + m\left(\frac{\partial \ln \gamma / \gamma^\dagger}{\partial m}\right)_{T,P}\right]\sigma, \qquad (4\text{-}27.18)$$

where R denotes the gas constant and γ the conventional activity coefficient $(\gamma^\dagger = 1 \text{ kg mol}^{-1})$. If we have measured values for σ and γ available then $*Q$ can be determined from Eq. (18). The expression in parentheses in Eq. (18) assumes the value 1 in the limiting case $m \to 0$ (cf. Eq. 4-18.42).

The relations (14), (17), and (18) are general, i.e. not restricted to the stationary state.

(c) Experimental Values

The Soret coefficients of aqueous electrolyte solutions† are of order of magnitude 10^{-3} K^{-1}. They usually have a negative sign. (This means according to Eq. (16) an enrichment of the electrolyte in the colder region.) They are strongly concentration and temperature dependent. Typical examples are found in Figs. 24 and 25. (For high dilutions, the value of σ decreases with increasing concentration, cf. Figs. 26 and 27.) In a few cases, a sign reversal results at a definite concentration for a given temperature. One

---

† Of the older measurements, those of Tanner[138a, b] should be brought to our attention. A more recent summary and discussion of measured data is found in Tyrrell[108].

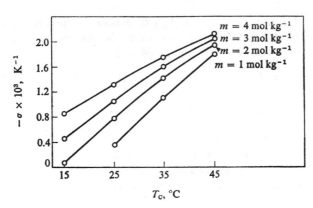

Fig. 24. Soret coefficient σ as a function of the Celsius temperature T_c for aqueous solutions of potassium chloride. After Longsworth[137].

example is potassium iodide for which, at small concentrations, the Soret coefficient σ is negative (the heat of transport $^*Q$ therefore positive), while at higher concentrations σ shows positive values (and $^*Q$ accordingly negative values) (see Fig. 26).

The quantity of most theoretical interest is the heat of transport $^*Q$ for dilute solutions. Here simple laws are to be expected. Investigations relating to this have been recently carried out by Agar and others[139–144].

From the data in Figs. 26 and 27 the following rules can be derived:

1. The limiting values for infinite dilution (see Eq. 18)

$$\lim_{m \to 0} {}^*Q \equiv {}^*Q^0 = -\nu RT^2 \sigma^0 \quad (\sigma^0 = \lim_{m \to 0} \sigma) \qquad (4\text{–}27.19)$$

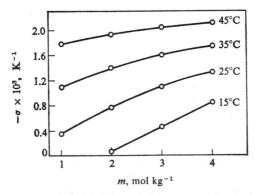

Fig. 25. Soret coefficient σ as a function of the molality m for aqueous solutions of potassium chloride. After Longsworth[137].

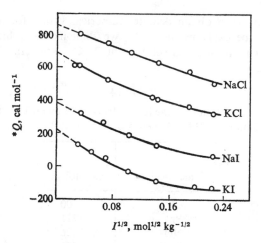

Fig. 26. Heats of transport $*Q$ of alkali halides in aqueous solution at 25.3°C as a function of the ionic strength I. After Snowdon and Turner[141].

for a given solvent and at a fixed temperature are additive in their ion constituents. In the relation (see Eq. 14)

$$*Q^0 = \nu_+ \, *Q^0_+ + \nu_- \, *Q^0_-, \qquad (4\text{–}27.20)$$

the limiting values $*Q^0_+$ and $*Q^0_-$ of the (individually unmeasurable) heats of transport of the ions must therefore, under these conditions, be individual quantities and thus be independent of the nature of the gegenion. We can thus assign an arbitrary value to $*Q^0_i$ for an arbitrarily selected ionic species i

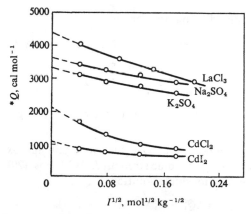

Fig. 27. Heats of transport $*Q$ of several salts in aqueous solution at 25.3°C as a function of the ionic strength I. After Snowdon and Turner[141].

(here: Cl^-), and then obtain definite numerical values for all other ions. For reasons to be explained in §4–31, we shall fix the value $*Q_i^0 \equiv 1700$ cal mol^{-1} for Cl^- in aqueous solution at 25°C. Then the $*Q_i^0$-values of Table 14 result.†

Table 14

Conventional Values of the Ionic Heats of Transport at Infinite Dilution $*Q_i^0$ for Aqueous Solutions at 25°C[146]‡

Ionic species i	$*Q_i^0$ cal mol^{-1}
Cl^-	1700
Br^-	1721
NO_3^-	1426
ClO_4^-	1481
H^+	1550
Na^+	−850
K^+	−1040
Rb^+	−915
Cs^+	−884
Ag^+	−300
Ba^{2+}	−730
La^{3+}	−800

‡ For lack of data, some of these values refer to $m = 0.01$ mol kg^{-1} (not to $m = 0$).

2. For very high dilution, the formula

$$*Q = v_+ *Q_+ + v_- *Q_- = *Q^0 + q^0\sqrt{I} \qquad (4\text{-}27.21)$$

is valid. In this, q^0 denotes a constant and

$$I \equiv \tfrac{1}{2}(z_+^2 v_+ + z_-^2 v_-)m \qquad (4\text{-}27.22)$$

denotes the "ionic strength" of the electrolyte (z_i = charge number of ionic species i).

3. For dilute solutions, the $*Q$-values are again approximately additive in ion constituents at a given ionic strength I. With Eqs. (20) and (21) we can thus write an equation for dilute solutions

$$*Q_i = *Q_i^0 + q_i^0\sqrt{I} \quad (i = +, -). \qquad (4\text{-}27.23)$$

† For a theoretical interpretation of $*Q^0$, the reader is referred to Hoch[145].

Here q_i^0 like $*Q_i^0$ is a characteristic constant for the ion i for a given solvent and a fixed temperature. Comparison with Eq. (20) and (21) produces:

$$q^0 = \nu_+ q_+^0 + \nu_- q_-^0. \tag{4-27.24}$$

We return to these regularities in §4-31.

(d) Thermal Diffusion Potential

In an electrolyte solution which is homogeneous with reference to concentrations and is in a temperature field, an internal electric field occurs almost instantaneously through the migration of ions. To this electrostatic field there corresponds an electric potential φ_0, which we—by analogy with the diffusion potential and sedimentation potential—will designate as the *thermal diffusion potential*. After a time, concentration gradients also arise in the solution, due to the Soret effect, and these lead to a diffusion potential. We wish to designate the sum of the thermal diffusion potential (in the narrower sense) and the diffusion potential as the "thermal diffusion potential in the wider sense" (symbol: φ_{Th}). In the stationary state, the quantity φ_{Th} assumes the special value φ_∞.

From Eq. (4-24.58) there follows directly for the thermal diffusion potential φ_{Th} in the wider sense, excluding pressure gradients and external force fields ($\mathbf{g} = 0$) but allowing any number of species (N),

$$F \operatorname{grad} \varphi_{Th} = -\sum_{k=2}^{N} \tau_k \left[(\operatorname{grad} \mu_k)_{T,P} + \frac{*Q_k}{T} \operatorname{grad} T \right]. \tag{4-27.25}$$

In this, particle type 1 is the solvent. Furthermore, F denotes the Faraday constant, τ_k, μ_k, and $*Q_k$ the reduced transport number, chemical potential, and heat of transport respectively of species k.

The expression for the thermal diffusion potential φ_0 in the narrower sense results immediately from Eq. (25) with $(\operatorname{grad} \mu_k)_{T,P} = 0$:

$$F \operatorname{grad} \varphi_\theta = -\sum_{k=2}^{N} \frac{\tau_k \, *Q_k}{T} \operatorname{grad} T. \tag{4-27.26}$$

The relation for the stationary value φ_∞ of the thermal diffusion potential results from Eq. (4-24.59) with $\mathbf{g} = 0$:

$$z_i F \operatorname{grad} \varphi_\infty = -(\operatorname{grad} \mu_i)_{T,P} - \frac{*Q_i}{T} \operatorname{grad} T \quad (i = 2, 3, \ldots, N). \tag{4-27.27}$$

The inferior i or k in the preceding formulas refers to any arbitrary species except the solvent. For neutral molecules, the condition $z_i = 0$ should be noted.

We again restrict discussion to solutions of a single electrolyte with two ionic species. Then Eq. (27) leads to Eqs. (4–24.26) through (4–24.31) which we have already used. The simplifications, however, which result from Eq. (25) need a more careful investigation.

From Eq. (25) and on consideration of Eq. (4–16.42) we derive for the special case considered here:

$$-F \operatorname{grad} \varphi_{\mathrm{Th}} = \frac{t_+}{z_+} \left[(\operatorname{grad} \mu_+)_{T,P} + \frac{^*Q_+}{T} \operatorname{grad} T \right]$$

$$+ \frac{t_-}{z_-} \left[(\operatorname{grad} \mu_-)_{T,P} + \frac{^*Q_-}{T} \operatorname{grad} T \right]$$

$$+ \tau_u \left[(\operatorname{grad} \mu_u)_{T,P} + \frac{^*Q_u}{T} \operatorname{grad} T \right]. \quad (4\text{–}27.28)$$

Here t_+ or t_- is the Hittorf transport number of the cations or anions and τ_u the reduced transport number of the undissociated electrolyte molecules. On the basis of Eqs. (4–16.16), (4–16.52), (4–24.28), (4–24.29), and (14) there is valid:

$$t_+ + t_- = 1, \quad (4\text{–}27.29)$$

$$z_+ \nu_+ + z_- \nu_- = 0, \quad (4\text{–}27.30)$$

$$\nu_+ \mu_+ + \nu_- \mu_- = \mu_u = \mu_2, \quad (4\text{–}27.31)$$

$$\nu_+ \,^*Q_+ + \nu_- \,^*Q_- = \,^*Q_u = \,^*Q, \quad (4\text{–}27.32)$$

$$t_+ + z_+ \nu_+ \tau_u = \vartheta_+, \qquad t_- + z_- \nu_- \tau_u = \vartheta_-, \quad (4\text{–}27.33)$$

where ϑ_+ or ϑ_- is the stoichiometric transport number of the cation or anion constituent of the electrolyte. From Eqs. (28) to (33) there follows:

$$-F \operatorname{grad} \varphi_{\mathrm{Th}} = \frac{1}{z_-} \left[(\operatorname{grad} \mu_-)_{T,P} + \frac{^*Q_-}{T} \operatorname{grad} T \right]$$

$$+ \frac{\vartheta_+}{z_+ \nu_+} \left[(\operatorname{grad} \mu_2)_{T,P} + \frac{^*Q}{T} \operatorname{grad} T \right]$$

$$= \frac{1}{z_+} \left[(\operatorname{grad} \mu_+)_{T,P} + \frac{^*Q_+}{T} \operatorname{grad} T \right]$$

$$+ \frac{\vartheta_-}{z_- \nu_-} \left[(\operatorname{grad} \mu_2)_{T,P} + \frac{^*Q}{T} \operatorname{grad} T \right]. \quad (4\text{–}27.34)$$

In this, we set

$$(\operatorname{grad} \mu_i)_{T,P} = \left(\frac{\partial \mu_i}{\partial m} \right)_{T,P} \operatorname{grad} m \quad (i = +, -), \quad (4\text{–}27.35)$$

$$(\operatorname{grad} \mu_2)_{T,P} = \left(\frac{\partial \mu_2}{\partial m} \right)_{T,P} \operatorname{grad} m. \quad (4\text{–}27.36)$$

Thermal diffusion potentials cannot be measured. The e.m.f. of a non-isothermal galvanic cell, which can be experimentally determined, arises from the combination of the thermal diffusion potential differences with other electric potential differences (cf. §4–29 and §4–30). Equation (34) shows this clearly: apart from the measurable quantities ϑ_+, ϑ_-, $(\partial\mu_2/\partial m)_{T,P}$ and $*Q$, the gradient of the thermal diffusion potential contains the experimentally undeterminable quantities $(\partial\mu_i/\partial m)_{T,P}$ and $*Q_i$ $(i = +, -)$.

Thermal diffusion potentials can also appear in solid or molten electrolytes. In this case, we choose as particle type 1, i.e. as the reference species of the Hittorf reference system, a suitable ionic species. If we have ion crystals with two ionic types (e.g. solid AgCl) or molten salts with two ionic species without neutral molecules (e.g. molten $AgNO_3$) then we consider either the cations or the anions as the reference species. If the second alternative is chosen then with Eq. (4–16.10) we find $t_+ = 1$, $t_- = 0$, $\tau_u = 0$. Here Eq. (27) and Eq. (28) produce the same result, since terms for the concentration gradients may be dropped:

$$-z_+ F\,\mathrm{grad}\ \varphi_{\mathrm{Th}} = \frac{*Q_+}{T}\,\mathrm{grad}\ T. \qquad (4\text{–}27.37)$$

This expression, in which $*Q_+$ denotes the heat of transport of the cations measured relative to the anions, is completely analogous to the formula (4–25.19) for the thermoelectric potential in electronic conductors (without concentration gradients). Thermocells with solid or molten electrolytes (salts) will be dealt with in §4–29 (p. 397).

4–28 HEAT CONDUCTION IN REACTING MEDIA

So far we have considered heat conduction only in the absence of chemical reactions (cf. §4–24). If reactions are allowed then the heat transfer becomes a complicated process; for, seen qualitatively, a transport of reaction heat in the concentration and temperature field must also result and overlap with the normal heat conduction mechanism. We now wish to treat this problem quantitatively with the methods of the thermodynamics of irreversible processes.

In the discussion of the cases of practical interest, we make the following assumptions:

1. Fluid media with creeping motions.
2. No external force fields.
3. Absence of space charges.
4. No passage of electric current.
5. No change of pressure.

Then, according to Eqs. (4–5.18), (4–6.12), (4–6.15), and (4–6.18), with the approximation $c_k \mathbf{v}_k \approx {}_v\mathbf{J}_k$ (cf. below) we find:

$$\frac{\bar{C}_P}{\bar{V}} \frac{\partial T}{\partial t} = -\mathrm{div}\left(\mathbf{J}_Q + \sum_{k=1}^{N} H_k\,{}_v\mathbf{J}_k\right) - \sum_{k=1}^{N} H_k \frac{\partial c_k}{\partial t}. \qquad (4\text{–}28.1)$$

Here $\bar{C}_P$ denotes the molar heat capacity for constant pressure, $\bar{V}$ the molar volume, T the absolute temperature, t the time, $\mathbf{J}_Q$ the heat current density, H_k the partial molar enthalpy of species k, ${}_v\mathbf{J}_k$ the diffusion current density of species k in the barycentric system, c_k or $\mathbf{v}_k$ the molarity or velocity of species k respectively, and N the number of chemical species.

At this point it is already obvious that the role of the heat current density $\mathbf{J}_Q$ in the normal heat conduction equation is now taken over by the quantity (see Eqs. 4–3.14, 4–3.15, and 4–7.5)

$$\mathbf{J}_Q^* \equiv \mathbf{J}_Q + \sum_{k=1}^{N} H_k\,{}_v\mathbf{J}_k$$

$$= \mathbf{J}_Q + \sum_{k=1}^{N} \tilde{H}_k \mathbf{J}_k^* = \mathbf{J}_Q + \sum_{i=1}^{N-1} (\tilde{H}_i - \tilde{H}_N)\mathbf{J}_i^*. \qquad (4\text{–}28.2)$$

Here $\mathbf{J}_k^*$ is the diffusion current density defined by Eq. (4–3.14) and $\tilde{H}_k$ the partial specific enthalpy of species k.

In practice, the thermal conductivity is usually measured for the stationary state. Then the following conditions hold:

$$\frac{\partial c_k}{\partial t} = 0 \quad (k = 1, 2, \ldots, N), \qquad \frac{\partial T}{\partial t} = 0 \quad \text{(stationary state).} \qquad (4\text{–}28.3)$$

From Eqs. (1) and (2) there thus follows:

$$\frac{\bar{C}_P}{\bar{V}} \frac{\partial T}{\partial t} = -\mathrm{div}\,\mathbf{J}_Q^* = 0 \quad \text{(stationary state).} \qquad (4\text{–}28.4)$$

Furthermore, there results from Eq. (4–4.1)

$$\mathrm{div}\,(c_k \mathbf{v}_k) = \sum_r \nu_{kr} b_r \quad \text{(stationary state).} \qquad (4\text{–}28.5)$$

In this, b_r denotes the reaction rate of the chemical reaction r and ν_{kr} the stoichiometric number of species k in reaction r.

In further discussion, we take the simplest case as a basis: a fluid system with concentration and temperature gradients containing two uncharged species between which a single reaction occurs. This reaction may be a rearrangement, an excitation, a dissociation, or an association. We write as the reaction scheme:

$$[1] \rightleftharpoons \nu[2]. \qquad (4\text{–}28.6)$$

For rearrangement or excitation reactions, we have: $\nu = 1$. As examples for dissociations or associations we quote

$$N_2O_4 \rightleftharpoons 2\,NO_2$$

$$HF \rightleftharpoons \tfrac{1}{6}(HF)_6$$

with $\nu = 2$ or $\nu = \tfrac{1}{6}$ respectively. Obviously, the relation between the molar masses M_1 and M_2 of the two species is:

$$M_1 = \nu M_2. \tag{4-28.7}$$

Furthermore, for the mean molar mass we obtain:

$$\overline{M} = M_1 x_1 + M_2 x_2 = M_2(\nu x_1 + x_2), \tag{4-28.8}$$

where x_i is the mole fraction of species i. We denote by α the degree of dissociation (degree of association, degree of rearrangement, etc.). We then find:

$$\alpha = \frac{x_2}{\nu x_1 + x_2} = \frac{M_2 x_2}{\overline{M}}. \tag{4-28.9}$$

Finally, there follows from Eqs. (1–12.8), (1–14.2), (1–14.22), and (1–22.34):

$$x_1\left(\frac{\partial \mu_1}{\partial x_1}\right)_{T,P} = x_2\left(\frac{\partial \mu_2}{\partial x_2}\right)_{T,P}, \tag{4-28.10}$$

$$A = \mu_1 - \nu \mu_2, \tag{4-28.11}$$

$$-T^2\left(\frac{\partial (A/T)}{\partial T}\right)_{P,x_1} = -h_{TP} = H_1 - \nu H_2. \tag{4-28.12}$$

Here μ_i denotes the chemical potential of species i, A the affinity of the chemical reaction, P the pressure, and h_{TP} the enthalpy of reaction ("heat of reaction at constant pressure"). From Eqs. (10) to (12) there results finally with grad $P = 0$:

$$\operatorname{grad}\left(\frac{A}{T}\right) = \frac{h_{TP}}{T^2}\operatorname{grad} T + \frac{\nu x_1 + x_2}{x_2 T}\left(\frac{\partial \mu_1}{\partial x_1}\right)_{T,P}\operatorname{grad} x_1. \tag{4-28.13}$$

For $N = 2$, Eq. (2) is reduced with Eqs. (7) and (12) to the expression:

$$\mathbf{J}_Q^* = \mathbf{J}_Q - h_{TP}\,{}_\nu\mathbf{J}_1. \tag{4-28.14}$$

With the help of Eqs. (4–26.6), (4–26.12), (4–26.14), and (4–26.35) as well as Eqs. (8) and (10) with grad $P = 0$ there results:

$$\mathbf{J}_Q^* = -\left(\lambda_0 - \frac{M_2 h_{TP} D_T}{\overline{M}\overline{V}T}\right)\operatorname{grad} T$$

$$+ \frac{M_2 x_2 h_{TP} D - \overline{M}(\partial \mu_1/\partial x_1)_{T,P} D_T}{x_2 \overline{M}\overline{V}}\operatorname{grad} x_1. \tag{4-28.15}$$

Here, according to §4–26, λ_0 is the thermal conductivity for the uniform state (for frozen reaction), D_T the thermal diffusion coefficient, D the diffusion coefficient, and $\overline{V}$ the molar volume. The Onsager reciprocity law is contained in Eq. (15).†

We now assume, for the sake of simplicity, that the reaction occurs so rapidly that there is local chemical equilibrium everywhere.‡ According to Eq. (1–21.19), there is, thus:

$$A = 0. \qquad (4\text{–}28.16)$$

Since this condition should be satisfied for each volume element, we can also write:

$$\text{grad} \left(\frac{A}{T}\right) = 0. \qquad (4\text{–}28.17)$$

From Eqs. (13) and (17) there immediately follows:

$$\frac{h_{TP}}{T} \text{ grad } T = -\frac{\nu x_1 + x_2}{x_2} \left(\frac{\partial \mu_1}{\partial x_1}\right)_{T,P} \text{ grad } x_1 \quad \text{(chemical equilibrium)}.$$
$$(4\text{–}28.18)$$

With this a relation between the temperature and the concentration gradients is provided.

From Eqs. (8), (15), and (18) we derive:

$$\mathbf{J_Q^*} = -\lambda_A \text{ grad } T \quad \text{(chemical equilibrium)} \qquad (4\text{–}28.19)$$

with

$$\lambda_A \equiv \lambda_0 + \frac{x_2 h_{TP}^2 D}{(\nu x_1 + x_2)^2 \overline{V} T (\partial \mu_1 / \partial x_1)_{T,P}} - \frac{2 h_{TP} D_T}{(\nu x_1 + x_2) \overline{V} T}. \qquad (4\text{–}28.20)$$

---

† Qualitatively, Eq. (15) states that the transport of heat in a reacting system is due to the following four effects:

(a) Normal heat conduction (term $-\lambda_0 \text{ grad } T$).

(b) Dufour effect (term with $-D_T \text{ grad } x_1$).

(c) Transport of heat of reaction in the temperature gradient (term with $h_{TP} D_T \text{ grad } T$).

(d) Transport of heat of reaction in the concentration gradient (term with $h_{TP} D \text{ grad } x_1$).

That the Dufour effect is essentially described by the thermal diffusion coefficient D_T, is required by Onsager's reciprocity law (§4–26).

‡ If we wish to omit this assumption and thus treat heat conduction for the delayed establishment of equilibrium (cf. below) then we must proceed from Eq. (4–24.7) in place of Eq. (16), and this leads, in our case, to a linear relation between the rate of reaction b and the affinity A, which holds near equilibrium

$$b = aA,$$

where a is a phenomenological coefficient. The calculation of the problem then becomes much more complicated[147].

By insertion of Eq. (19) in Eq. (4) there results:

$$\frac{\bar{C}_P}{\bar{V}}\frac{\partial T}{\partial t} = \text{div}\,(\lambda_A\,\text{grad}\,T) = 0 \quad \text{(stationary state at chemical equilibrium)}.$$

$$(4\text{-}28.21)$$

We denote the quantity λ_A, which can be measured using Eq. (21), as the *thermal conductivity at chemical equilibrium*. This quantity is, according to Eq. (20), related to the thermal conductivity λ_0 for the uniform state (for frozen reaction), to the diffusion coefficient D, to the thermal diffusion coefficient D_T, and to the thermodynamic properties of the mixture[147–150].

For the stationary state, by application of Eq. (5) to the reaction (6), there follows:

$$\text{div}\,(c_1\mathbf{v}_1) = -b, \qquad \text{div}\,(c_2\mathbf{v}_2) = \nu b,$$

where b is the rate of the reaction considered. From the two equations we find with the help of Eq. (4–3.5), (4–3.12), and Eq. (7):

$$\text{div}\,(\nu c_1\mathbf{v}_1 + c_2\mathbf{v}_2) = \frac{1}{M_2}\,\text{div}\,(\rho_1\mathbf{v}_1 + \rho_2\mathbf{v}_2) = \frac{1}{M_2}\,\text{div}\,(\rho\mathbf{v}) = 0,$$

that is,

$$\text{div}\,(\rho\mathbf{v}) = \rho\,\text{div}\,\mathbf{v} + \mathbf{v}\,\text{grad}\,\rho = 0.$$

In this, ρ_i denotes the mass concentration or partial density of species i, ρ the density, and $\mathbf{v}$ the barycentric velocity. We also obtain the same result directly from Eq. (4–4.4) with $\partial\rho/\partial t = 0$. If we set $\mathbf{v}\,\text{grad}\,\rho \approx 0$—that is, according to Eq. (4–4.6), $d\rho/dt \approx 0$—then there results

$$\text{div}\,\mathbf{v} \approx 0 \quad \text{(stationary state)}.$$

Accordingly, it is permissible in this case to describe freedom from convection by the condition $\mathbf{v} \approx 0$ and to proceed from Eq. (1) instead of from Eq. (4–24.48) which is more usefully applied to other problems (cf. §4–26).

If the reacting medium represents an ideal gas mixture then we have (R = gas constant)

$$\bar{V} = \frac{RT}{P}, \qquad \left(\frac{\partial\mu_1}{\partial x_1}\right)_{T,P} = \frac{RT}{x_1}. \tag{4–28.22}$$

With Eq. (9) and Eqs. (20) and (22) there follows:

$$\lambda_A = \lambda_0 + \frac{\alpha(1-\alpha)Ph_{TP}^2 D}{\nu R^2 T^3} - \frac{2\alpha Ph_{TP}D_T}{x_2 RT^2} \quad \text{(ideal gas mixture)}.$$

$$(4\text{-}28.23)$$

We can calculate or at least estimate the quantities λ_0, D, D_T for ideal gas mixtures using the kinetic theory of gases and experimental data for molecules which are similar to the particle types under consideration at present. There results, first, that the term with D_T lies within the uncertainty

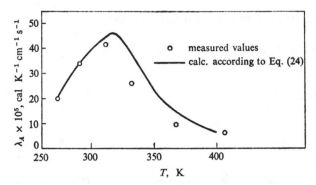

Fig. 28. Thermal conductivity λ_A of dissociating nitrogen tetroxide as a function of the absolute temperature T at a pressure of 257.5 torr. After Franck and Spalthoff[156].

of error limits for the numerical evaluation of λ_0 and the measurement of λ_A. Thus we write in place of Eq. (23):

$$\lambda_A = \lambda_0 + \frac{\alpha(1 - \alpha)Ph_{TP}^2 D}{\nu R^2 T^3}. \tag{4–28.24}$$

This formula[150] corresponds to an equation derived by Nernst[151].

The Nernst relation is obtained if we set $\nu = 2$ (dissociation reaction of the type $N_2O_4 \rightleftharpoons 2NO_2$) and take account of the fact that the diffusion coefficient D' used by Nernst is related to our diffusion coefficient D as follows:

$$D' = \frac{D}{\nu x_1 + x_2} = \frac{D}{2x_1 + x_2} = \tfrac{1}{2}(1 + \alpha)D.$$

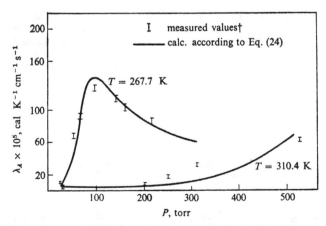

Fig. 29. Thermal conductivity λ_A of associating hydrogen fluoride as a function of the pressure P at two temperatures. According to Butler and Brokaw[159].
† Measurements according to Franck and Spalthoff[160].

With this we find from Eq. (24):

$$\lambda_A = \lambda_0 + \frac{\alpha(1 - \alpha)Ph_{TP}^2 D'}{(1 + \alpha)R^2T^3},$$

the Nernst formula.

In the application of Eq. (24) to concrete cases, the quantities α (as a function of T and P) and h_{TP}, in addition to λ_0 and D (cf. above), must be known. They are obtained from the measured equilibrium constants of the reaction in question (h_{TP} can also be obtained from the calorimetric determination of the heat of reaction). The most striking feature of Eq. (24) is the strong temperature and pressure dependence of λ_A in contrast to the normal thermal conductivity λ_0.

In Figs. 28 and 29, we compare the values of λ_A calculated according to Eq. (24) to measured thermal conductivities for nitrogen tetroxide ($N_2O_4 \rightleftharpoons 2NO_2$) and hydrogen fluoride ($HF \rightleftharpoons \frac{1}{6}(HF)_6$). It can be seen that the agreement is quite good.†

In other cases, we must take account of delayed equilibration in heat conduction measurements. This problem can also be treated thermodynamically[147,156].‡ Similarly, the relations between the thermal conductivity of polyatomic gases and the rotational and vibrational degrees of freedom can be examined in this way[158].

Finally, comparison of the thermal conductivity against temperature plot for a normal gas (Fig. 13, p. 334) with that for a dissociating or associating gas (Fig. 28 or Fig. 29) suggests an explanation for the abnormal behavior of water as compared to that of other liquids (Fig. 14, p. 335) on the basis of chemical reactions (associations)[161].

4–29 THERMOCELLS

(a) General

A *thermocell* is a nonisothermal galvanic cell represented by the diagram

		a		b		c		d		
	Terminal		Electrode		Electrolyte solution		Electrode		Terminal	
	(A)		(B)				(B)		(A)	
Temperature:	T_0		T_1		T_1		T_2		T_2	T_0
Potential:		φ_I φ_{II}		ψ_{III}		φ_{IV} φ_V		φ_{VI}	φ_{VII} φ_{VIII}	

Here T_0 is the temperature of the measuring instruments, T_1 or T_2 the temperature of the first or second electrode with the directly adjoining layer of

---

† Further material relating to the experimental examination of Eq. (24) or a generalization of this formula to reacting gaseous multi-component systems is found, for example, in the following papers[152–155].

‡ A very thorough investigation of the problem of heat conduction with the delayed establishment of equilibrium is found in Brokaw[157].

the electrolyte solution. In the solution, there are, in the most general
case, concentration and temperature gradients. The solution can consist of
any substances (electrolytes and nonelectrolytes) but should always contain
that ionic species for which the electrodes are reversible, e.g. Ag^+ ions in the
case of Ag electrodes or Cl^- ions in the case of Ag/AgCl electrodes. The
terminal in which a temperature gradient is present is the metallic lead,
e.g. a wire of silver, platinum, or copper. Pressure gradients and external
force fields will be excluded here.

Thermocells which, in place of the electrolyte solution, contain a solid
or molten electrolyte have also been investigated. Thus many of the earliest
experimental works relate to molten salts (Andrews, 1837; Hankel, 1858;
Glasstone and Tribe, 1881; Poincaré, 1890) and to solid salts (Chaperon,
1886; Thiele, 1925; Reinhold, 1928–1933). Below we give some references
to these types of thermocells.

The first measurements on thermocells with electrolyte solutions are
very old too (Wild, 1858; Bouty, 1879–1881; Gockel, 1885–1893; Eberling,
1887; G. Meyer, 1888; Hagenbach, 1894–1896; Richards, 1897). The presentation given below is specifically for such thermocells.

For all types of thermocells which are irreversible galvanic cells, the more
exact experimental and theoretical investigation took place relatively
recently, because an adequate description of the experimental results was first
made possible by the thermodynamics of irreversible processes [64, 92, 108,
146, 162–171]. Let us also mention the Peltier effect in thermocells[172a,b].
We will not treat this extensively but there results for this an equation
analogous to the Thomson relation (4–25.60)[64, 92].

The e.m.f. of the thermocell which we denote by Φ is by convention the
difference between the electric potential in the right-hand terminal and that
in the left-hand terminal, with no electric current flowing and the local
heterogeneous equilibria being established for all phase boundaries. According to our diagram, we thus have:

$$\Phi = \varphi_{VIII} - \varphi_I = (\varphi_{VIII} - \varphi_{VII}) + (\varphi_{VII} - \varphi_{VI})$$
$$+ (\varphi_{VI} - \varphi_V) + (\varphi_V - \varphi_{IV}) + (\varphi_{IV} - \varphi_{III})$$
$$+ (\varphi_{III} - \varphi_{II}) + (\varphi_{II} - \varphi_I). \tag{4–29.1}$$

Here the potential differences $\varphi_{VIII} - \varphi_{VII}$ and $\varphi_{II} - \varphi_I$ depend on the thermoelectric potentials (§4–25), and the quantity $(\varphi_V - \varphi_{IV})$ on a thermal diffusion
potential in the wider sense (§4–27). All other potential differences are—
since we stipulate local electrochemical equilibrium at the phase boundaries
a, b, c, and d—equilibrium potential differences ("contact potentials" or
"electrode potentials") and thus can be determined from the equilibrium
conditions of classical thermodynamics.

If we start with a homogeneous (with reference to concentrations)
electrolyte solution then we measure initially (at time $t = 0$) a definite

e.m.f. of the cell which we designate as the "initial value" Φ_0. In the course of time, concentration gradients build up due to thermal diffusion in the solution where a diffusion potential overlaps the thermal diffusion potential in the narrower sense (§4–27). The e.m.f. of the cell is thus a function of time. In the stationary final state (at the time $t \to \infty$) in which diffusion and thermal diffusion are balanced in the solution, we measure the "final value" Φ_∞ of the e.m.f. of the cell.

With

$$\Delta T \equiv T_2 - T_1 \tag{4–29.2}$$

we define the following limiting value:

$$\varepsilon \equiv \lim_{\Delta T \to 0} \frac{\Phi}{\Delta T}, \tag{4–29.3}$$

which we call the *thermoelectric power* of the thermocell (cf. p. 347). Accordingly, the expression

$$\varepsilon_0 \equiv \lim_{\Delta T \to 0} \frac{\Phi_0}{\Delta T}$$

or

$$\varepsilon_\infty \equiv \lim_{\Delta T \to 0} \frac{\Phi_\infty}{\Delta T}$$

is called the "initial value" or the "final value" of the thermoelectric power of the thermocell. According to our diagram and to Eq. (3), a positive or negative value of ε means that the warmer or colder electrode respectively corresponds to the positive pole. For the e.m.f. of the thermocell, we conclude from Eqs. (2) and (3) that

$$\Phi = \varepsilon \Delta T + \text{higher terms} \tag{4–29.4}$$

or

$$\Phi = \int_{T_1}^{T_2} \varepsilon \, dT, \tag{4–29.5}$$

where ε is to be regarded as a function of the temperature T.

The phase boundaries a and d are reversible for the electrons (indicated by inferior $\ominus$); the phase boundaries b and c are reversible for a definite ionic species of the solution (indicated by inferior i). We designate quantities which relate to the terminal or electrode substances by inferior A or B respectively, while quantities without an inferior label refer to the solution. Furthermore, let F be the Faraday constant, z_i the charge number of ionic species i, and $\mu_\ominus$ or μ_i the chemical potential of the electrons or of the ion species i. Then, for the contact and electrode potentials at the phase boundaries a, b, c, and d, we obtain the following expressions from the general condition (1–20.13) of electrochemical equilibrium:

$$F(\varphi_{\text{VII}} - \varphi_{\text{VI}}) = (\mu_{\ominus})_A - (\mu_{\ominus})_B \quad \text{(temperature } T_2\text{)}, \qquad \text{(4–29.6)}$$

$$F(\varphi_{\text{III}} - \varphi_{\text{II}}) = (\mu_{\ominus})_B - (\mu_{\ominus})_A \quad \text{(temperature } T_1\text{)}, \qquad \text{(4–29.7)}$$

$$F(\varphi_{\text{VI}} - \varphi_{\text{V}}) = \frac{1}{z_i} [\mu_i - (\mu_i)_B] \quad \text{(temperature } T_2\text{)}, \qquad \text{(4–29.8)}$$

$$F(\varphi_{\text{IV}} - \varphi_{\text{III}}) = \frac{1}{z_i} [(\mu_i)_B - \mu_i] \quad \text{(temperature } T_1\text{)}. \qquad \text{(4–29.9)}$$

Let m_k be the molality of substance k in the solution at the phase boundary b (insofar as m_k can be chosen as an independent composition variable), $m_k + \Delta m_k$ the corresponding quantity at the phase boundary c, and let $S_{\ominus}$, $(S_i)_B$, and S_i be the partial molar entropy of the electrons, of the ionic species i in the electrode and in the solution respectively, and P the pressure. With the abbreviations

$$\Delta\varphi_{\text{het}} \equiv (\varphi_{\text{VII}} - \varphi_{\text{VI}}) + (\varphi_{\text{III}} - \varphi_{\text{II}}) + (\varphi_{\text{VI}} - \varphi_{\text{V}}) + (\varphi_{\text{IV}} - \varphi_{\text{III}}),$$

$$\text{(4–29.10)}$$

$$\frac{dm_k}{dT} \equiv \lim_{\Delta T \to 0} \frac{\Delta m_k}{\Delta T}, \qquad \text{(4–29.11)}$$

$$\mu_{ik} \equiv \left(\frac{\partial \mu_i}{\partial m_k}\right)_{T, P, m_j}, \qquad \text{(4–29.12)}$$

where m_j stands for all independent molalities except m_k, and with the thermodynamic relations (cf. Eq. 1–12.7)

$$\left(\frac{\partial \mu_{\ominus}}{\partial T}\right)_P = -S_{\ominus}, \qquad \left(\frac{\partial (\mu_i)_B}{\partial T}\right)_P = -(S_i)_B, \qquad \text{(4–29.13)}$$

$$d\mu_i = -S_i \, dT + \sum_k \mu_{ik} \, dm_k \quad (P = \text{const}), \qquad \text{(4–29.14)}$$

we find from Eqs. (2) and (6) to (9):

$$z_i F \lim_{\Delta T \to 0} \frac{\Delta\varphi_{\text{het}}}{\Delta T} = (S_i)_B + z_i(S_{\ominus})_B - z_i(S_{\ominus})_A - S_i + \sum_k \mu_{ik} \frac{dm_k}{dT}.$$

$$\text{(4–29.15)}$$

We also have (cf. p. 318):

$$(S_i)_B + z_i(S_{\ominus})_B = S_a. \qquad \text{(4–29.16)}$$

Here S_a denotes a certain linear combination of the partial molar entropies of substances in the electrodes (in the simplest case, i.e. of electrodes of the first kind, the partial molar entropy of the appropriate neutral atomic species in the electrode). It can, for example, be shown[92] that the following statements hold:

$S_a = \bar{S}_{\text{Ag}}$ (molar entropy of solid silver)
for Ag electrodes (reversible to Ag^+),

$S_a = S_{\text{Cd}}$ (partial molar entropy of cadmium in the mixture)
for Cd amalgam electrodes (reversible to Cd^{2+}),

$S_a = \frac{1}{2}\bar{S}_{\text{H}_2}$ ($\bar{S}_{\text{H}_2}$ = molar entropy of gaseous hydrogen)
for H_2 electrodes (reversible to H^+),

$S_a = \bar{S}_{\text{AgCl}} - \bar{S}_{\text{Ag}}$ ($\bar{S}_{\text{AgCl}}$ = molar entropy of solid silver chloride)
for Ag/AgCl electrodes (reversible to Cl^-).

From Eqs. (15) and (16) there results:

$$z_i F \lim_{\Delta T \to 0} \frac{\Delta \varphi_{\text{het}}}{\Delta T} = S_a - z_i(S_\ominus)_A - S_i + \sum_k \mu_{ik} \frac{dm_k}{dT}. \quad (4\text{-}29.17)$$

According to Eqs. (1), (3), and (10), this expression, after division by $z_i F$, represents the contribution of the "heterogeneous effects" (contact and electrode potentials) to the thermoelectric power of the thermocell.

According to Eqs. (2) and (4–25.19) (cf. also Eq. 4–25.37), the potential differences depending on the thermoelectric potentials yield:

$$F[(\varphi_{\text{VIII}} - \varphi_{\text{VII}}) + (\varphi_{\text{II}} - \varphi_{\text{I}})] = [(S_\ominus)_A - (*S_\ominus)_A]\Delta T + \text{higher terms,}$$

$$(4\text{-}29.18)$$

where $(*S_\ominus)_A$ denotes the transported entropy of the electrons in the terminal. Finally, Eq. (4–27.25) with Eqs. (2) and (12) for the potential difference determined by the thermal diffusion potential in the electrolyte solution becomes:

$$F(\varphi_{\text{V}} - \varphi_{\text{IV}}) = -\sum_j \frac{\tau_j *Q_j}{T}\Delta T - \sum_{j,k} \tau_j \mu_{jk} \Delta m_k + \text{higher terms,}$$

$$(4\text{-}29.19)$$

where j denotes any (charged or uncharged) species of the solution (apart from the solvent), τ_j or $*Q_j$ the reduced transport number or heat of transport of this particle type, and μ_{jk} the quantity analogous to μ_{ik} in Eq. (12). With the abbreviation

$$\Delta\varphi_{\text{hom}} \equiv (\varphi_{\text{VIII}} - \varphi_{\text{VII}}) + (\varphi_{\text{II}} - \varphi_{\text{I}}) + (\varphi_{\text{V}} - \varphi_{\text{IV}}) \quad (4\text{-}29.20)$$

there follows from Eqs. (11), (18), and (19):

$$F \lim_{\Delta T \to 0} \frac{\Delta\varphi_{\text{hom}}}{\Delta T} = (S_\ominus)_A - (*S_\ominus)_A - \sum_j \frac{\tau_j *Q_j}{T} - \sum_{j,k} \tau_j \mu_{jk} \frac{dm_k}{dT}. \quad (4\text{-}29.21)$$

After division by F and following Eqs. (1), (3), and (20) this represents the contribution of the "homogeneous effects" (thermoelectric potentials and thermal diffusion potential in the wider sense) to the thermoelectric power of the thermocell.

From the combination of Eqs. (1), (3), (10), (17), (20), and (21) we arrive at the following formula:

$$F\varepsilon = \frac{S_a}{z_i} - (*S_\ominus)_A - \frac{S_i}{z_i} - \sum_j \frac{\tau_j {}^* Q_j}{T} + \sum_k \left(\frac{\mu_{ik}}{z_i} - \sum_j \tau_j \mu_{jk} \right) \frac{dm_k}{dT}.$$

(4–29.22)

This is the general equation for the thermoelectric power of a thermocell[92]. We repeat that species i is that ion type for which the electrodes are reversible, and that the inferior j denotes any species of the solution except the solvent, and the inferior k all substances in the solution whose concentrations are independent.

The initial value ε_0 of the thermoelectric power results from Eq. (22) with $dm_k/dT = 0$:

$$F\varepsilon_0 = \frac{S_a}{z_i} - (*S_\ominus)_A - \frac{S_i}{z_i} - \sum_j \frac{\tau_j {}^* Q_j}{T}.$$

(4–29.23)

The final value ε_∞ of the thermoelectric power results, according to Eqs. (19), (4–27.25), and (4–27.27), if the terms

$$\sum_j \frac{\tau_j {}^* Q_j}{T} + \sum_{j,k} \tau_j \mu_{jk} \frac{dm_k}{dT}$$

are substituted in Eq. (22) by the expression

$$\frac{1}{z_i} \left(\sum_k \mu_{ik} \frac{dm_k}{dT} + \frac{{}^* Q_i}{T} \right).$$

With the help of the relation (4–24.20)

$$*S_i = \frac{{}^* Q_i}{T} + S_i,$$

(4–29.24)

where $*S_i$ denotes the transported entropy of the ionic species i in the solution, we obtain

$$F\varepsilon_\infty = \frac{S_a}{z_i} - (*S_\ominus)_A - \frac{*S_i}{z_i}.$$

(4–29.25)

In spite of its simple appearance, Eq. (25), like Eqs. (22) and (23), is valid for any solution. The specialization to solutions of a single electrolyte with two ion species in a single neutral solvent brings simplifications only for Eqs. (22) and (23) but not for Eq. (25) (see below).

The quantity S_a can be obtained—up to an arbitrary additive constant—from equilibrium data. The arbitrariness in the numerical value is eliminated by the Planck normalization (p. 394). Thus, from the measurement of ε_∞ using Eq. (25), we find conventional values of

$$*S_i + z_i(*S_\ominus)_A. \tag{4-29.26}$$

If the transported entropy of electrons $*S_\ominus$ is also fixed by a normalization (p. 354) then we arrive at conventional values of the transported entropies of ions $*S_i$ (cf. below and §4-31).

(b) Special Case of a Single Electrolyte with Two Ionic Species

We now assume that the solution in the thermocell contains only one neutral solvent (component 1) and one single electrolyte (component 2) with two ionic species (cations = species $+$, anions = species $-$). There are, thus, besides solvent molecules and undissociated electrolyte molecules (species u), no other neutral particle types present. As the independent composition variable we choose the molality m of the electrolyte whose chemical potential we denote by μ_2. The inferior i again characterizes that ion for which the electrodes are reversible, so that the species i corresponds either to the cations (indicated by inferior $+$) or to the anions (indicated by inferior $-$). We write as an abbreviation:

$$\mu'_j \equiv \left(\frac{\partial \mu_j}{\partial m}\right)_{T,P} \quad (j = i, +, -, u), \qquad \mu'_2 \equiv \left(\frac{\partial \mu_2}{\partial m}\right)_{T,P}. \tag{4-29.27}$$

Equation (4-16.42) and Eq. (22) with Eq. (27) produce for the special case considered:

$$F\varepsilon = \frac{S_a}{z_i} - (*S_\ominus)_A - \frac{S_i}{z_i} - \frac{1}{T}\left(\frac{t_+}{z_+}*Q_+ + \frac{t_-}{z_-}*Q_- + \tau_u*Q_u\right)$$

$$+ \left(\frac{1}{z_i}\mu'_i - \frac{t_+}{z_+}\mu'_+ - \frac{t_-}{z_-}\mu'_- - \tau_u\mu'_u\right)\frac{dm}{dT}. \tag{4-29.28}$$

Here t_+ or t_- denotes the Hittorf transport number of the cations or anions respectively.

From comparison of Eq. (4-27.28) with Eq. (4-27.34) and on noting Eqs. (4-27.35), (4-27.36), and Eq. (27) there follows:

$$\frac{1}{T}\left(\frac{t_+}{z_+}*Q_+ + \frac{t_-}{z_-}*Q_- + \tau_u*Q_u\right) + \left(\frac{t_+}{z_+}\mu'_+ + \frac{t_-}{z_-}\mu'_- + \tau_u\mu'_u\right)\frac{dm}{dT}$$

$$= \frac{1}{z_-}\left(\mu'_- \frac{dm}{dT} + \frac{*Q_-}{T}\right) + \frac{\vartheta_+}{z_+\nu_+}\left(\mu'_2 \frac{dm}{dT} + \frac{*Q}{T}\right)$$

$$= \frac{1}{z_+}\left(\mu'_+ \frac{dm}{dT} + \frac{*Q_+}{T}\right) + \frac{\vartheta_-}{z_-\nu_-}\left(\mu'_2 \frac{dm}{dT} + \frac{*Q}{T}\right). \tag{4-29.29}$$

Here ν_+ or ν_- is the dissociation number of the cations or anions respectively, ϑ_+ or ϑ_- the stoichiometric transport number of the cation or anion constituent of the electrolyte, and $*Q$ the heat of transport of the electrolyte.

On consideration of Eq. (24), by combining Eqs. (28) and (29), we find:

$$F\varepsilon = \frac{S_a}{z_i} - (*S_\ominus)_A - \frac{*S_i}{z_i} + \frac{1 - \vartheta_i}{z_i\nu_i}\left(\frac{*Q}{T} + \mu_2'\frac{dm}{dT}\right). \quad (4\text{–}29.30)$$

This equation holds for the thermoelectric power ε at an arbitrary instant (with any concentration distribution in the solution). As in the corresponding formula (4–19.19) or (4–22.21) for a concentration cell or for gravitational or centrifugal cells, the stoichiometric transport number ϑ_i appears in Eq. (30).

The initial value ε_0 of the thermoelectric power results from Eq. (30) with $dm/dT = 0$ (homogeneous solution):

$$F\varepsilon_0 = \frac{S_a}{z_i} - (*S_\ominus)_A - \frac{*S_i}{z_i} + \frac{1 - \vartheta_i}{z_i\nu_i}\frac{*Q}{T}. \quad (4\text{–}29.31)$$

For the difference of the thermoelectric powers $(\varepsilon - \varepsilon_0)$, there follows from Eqs. (30) and (31):

$$F(\varepsilon - \varepsilon_0) = \frac{1 - \vartheta_i}{z_i\nu_i}\mu_2'\frac{dm}{dT}. \quad (4\text{–}29.32)$$

It is obvious from Eq. (32) with Eqs. (5) and (27) that the difference of the e.m.f. values $(\Phi - \Phi_0)$ is equal to the e.m.f. of the corresponding isothermal concentration cell, given by Eq. (4–19.19).

According to Eqs. (4–19.21), (4–19.22), and Eq. (27), we have:

$$\mu_2' = \nu RT\left[\frac{1}{m} + \left(\frac{\partial \ln \gamma/\gamma^\dagger}{\partial m}\right)_{T,P}\right],$$

$$\frac{\mu_2'}{z_i\nu_i} = \pm\left(\frac{1}{z_+} - \frac{1}{z_-}\right)RT\left[\frac{1}{m} + \left(\frac{\partial \ln \gamma/\gamma^\dagger}{\partial m}\right)_{T,P}\right], \quad (4\text{–}29.33)$$

where $\nu = \nu_+ + \nu_-$, R is the gas constant, γ is the conventional activity coefficient $(\gamma^\dagger = 1 \text{ kg mol}^{-1})$, and where the upper or lower sign is valid if the ionic species i represents the cation or anion respectively.

For the stationary state, we derive from Eq. (4–27.15) and Eq. (27):

$$\mu_2'\frac{dm}{dT} = -\frac{*Q}{T} \quad \text{(stationary state).} \quad (4\text{–}29.34)$$

This relation with Eq. (30) produces the final value ε_∞ of the thermoelectric power:

$$F\varepsilon_\infty = \frac{S_a}{z_i} - (*S_\ominus)_A - \frac{*S_i}{z_i}, \quad (4\text{–}29.35)$$

in agreement with Eq. (25).

According to Eq. (4-27.16), there is valid for the Soret coefficient σ:

$$\sigma = \frac{1}{m}\frac{dm}{dT} \quad \text{(stationary state).} \tag{4-29.36}$$

From this with Eqs. (32) and (33) we find:

$$\varepsilon_\infty - \varepsilon_0 = \pm\left(\frac{1}{z_+} - \frac{1}{z_-}\right)\frac{RT}{F}\left[1 + m\left(\frac{\partial \ln \gamma/\gamma^\dagger}{\partial m}\right)_{T,P}\right](1 - \vartheta_i)\sigma, \tag{4-29.37}$$

that is, a relation[64,168b] between the three measurable quantities σ, ε_0, and ε_∞.

From Eqs. (14), (24), (33), (34), and (36) with (4-27.31) and (4-27.32) there follows:

$$^*Q = -\nu RT^2\left[1 + m\left(\frac{\partial \ln \gamma/\gamma^\dagger}{\partial m}\right)_{T,P}\right]\sigma, \tag{4-29.38}$$

$$\nu_+ {^*S_+} + \nu_- {^*S_-} = \frac{^*Q}{T} + S_2 = S_2 - \nu RT\left[1 + m\left(\frac{\partial \ln \gamma/\gamma^\dagger}{\partial m}\right)_{T,P}\right]\sigma. \tag{4-29.39}$$

Equation (38) is identical to Eq. (4-27.18).† As is obvious from Eq. (39), a linear combination of the transported entropies of the ions can be obtained experimentally with the knowledge of the partial molar entropy S_2 of the electrolyte from the Soret coefficient.‡

---

† If Eq. (37) is combined with Eq. (38) and Eq. (4-19.22) then we obtain the relation[64, 168b]:

$$^*Q = \frac{z_i \nu_i FT(\varepsilon_0 - \varepsilon_\infty)}{1 - \vartheta_i}. \tag{4-29.38a}$$

An examination of the consistency of Soret effect data σ with e.m.f. measurements on thermocells ε_0, ε_∞ can thus follow with the knowledge of the activity coefficients γ and transport numbers ϑ_i, either directly according to Eq. (37), or by comparison of the heats of transport $^*Q$ calculated from Eq. (38) with the $^*Q$ values determined according to Eq. (38a)[173].

Conversely, if the Soret coefficients and the e.m.f. measurements are assumed to be reliable, either ϑ_i (with the knowledge of γ) or γ (with the knowledge of ϑ_i) can be determined, in principle, from experimental data with the help of Eq. (37). As in all earlier considerations of solutions of a single electrolyte with two ionic species, we have obtained the result that: the quantity accessible directly from measurement is the stoichiometric transport number ϑ_i (not the Hittorf transport number $t_i = \vartheta_i - z_i\nu_i\tau_u$) or the conventional activity coefficient γ (not the mean ionic activity coefficient $\gamma_\pm = \gamma/\alpha$, see p. 56).

‡ For the partial molar entropy S_2 of the electrolyte in the solution, we obtain according to Eqs. (1-12.7) and (1-19.26):

$$S_2 = S_2^\ominus - \nu R\left[\ln\left(\nu_\pm m\gamma\right) + T\left(\frac{\partial \ln \gamma/\gamma^\dagger}{\partial T}\right)_{P,m}\right]$$

[Continued on page 394]

The quantities S_a, S_2, $(*S_\ominus)_A$, γ, and ϑ_i are accessible from measured data. In this, the values for S_a and S_2 are numerically fixed in the usual fashion (Planck's normalization, see below) and the value for $(*S_\ominus)_A$ by the convention described earlier (cf. Table 13, p. 355). Accordingly, *conventional transported entropies of ions* $*S_+$ and $*S_-$ are obtained in the following ways[146, 168a, b]:

1. From measurements of ε_0 and σ according to Eqs. (31), (38), and (39).
2. From measurements of ε_0 and ε_∞ according to Eqs. (31), (35), and (39).

It follows from Eq. (37) that only two of the three measured quantities σ, ε_0, and ε_∞ are independent.

In §4–31, we will discuss in more detail the conventional transported entropies of ions so determined. The expression "conventional transported entropy" is not to be taken in the sense of an arbitrary division of a transport quantity, valid for the electrolyte, into individual ion parts—such as for the conventional heats of transport of ions in Table 14, p. 376—but has a deeper physical significance: the zero-point value of the entropy of any pure condensed substance in internal equilibrium or of the transported entropy of the electrons in any metal is set equal to zero because this is formally the simplest way to assess the validity of the (modernized) Nernst heat theorem or the disappearance of thermoelectric powers of any thermocouple at the absolute zero of temperature (p. 354). The convention relating to the zero-point entropy of pure phases is known as "Planck's normalization."

(c) Measurements

The primary measured quantities are the e.m.f. of the thermocell Φ and the (absolute) electrode temperatures (T_1 and T_2). If the thermoelectric power ε is constant in the temperature interval considered then according to Eqs. (2) and (5) we find:

$$\frac{\Phi}{T_2 - T_1} = \frac{\Phi}{\Delta T} = \varepsilon.$$

The supposition that $\varepsilon = $ const is a good approximation for temperature differences which do not exceed 5°C[169].

It can be shown[108] that such a simple determination of ε is not bound to the restricted assumption $\varepsilon = $ const. From measurements over larger

---

[*Continued from page 393*]
with

$$\nu_\pm^\nu \equiv \nu_+^{\nu+} \nu_-^{\nu-}.$$

Here S_2^θ is a standard value dependent only on T and P. For the numerical determination of this standard value, the Planck normalization is used. Furthermore, as we recognize, the knowledge of the activity coefficient γ as a function of concentration *and* temperature is necessary for the evaluation of the measurements.

temperature regions[174] ε is often found to have the form (which is similar to that for thermocouples):

$$\varepsilon = a + bT, \tag{4-29.40}$$

with a and b constant for given electrolyte concentrations. This equation with Eqs. (2) and (5) leads to the expression

$$\frac{\Phi}{\Delta T} = a + bT_m = \varepsilon(T_m) \tag{4-29.41}$$

with

$$T_m \equiv \tfrac{1}{2}(T_1 + T_2). \tag{4-29.42}$$

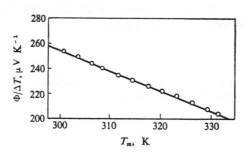

Fig. 30. Quotient of the initial value Φ_0 of the e.m.f. and the temperature difference ΔT of the electrodes as a function of the average (absolute) temperature T_m. According to measurements of Levin and Bonilla[174] on the thermocell (43) for 1.0014 N aqueous solution. After Tyrrell[108].

If, therefore, $\Phi/\Delta T$ is a linear function of the average temperature T_m, $\Phi/\Delta T$ may be identified with the thermoelectric power $\varepsilon(T_m)$ at the mean temperature T_m. In Fig. 30, we show one example as a test of Eq. (41). The measurements refer to the initial values of the e.m.f. Φ_0 of the thermocell

$$\begin{array}{cccccc} | \text{ Ag } | \text{ AgCl } | \text{ KCl(aq) } | \text{ AgCl } | \text{ Ag } | \\ T_0 \quad T_1 \qquad T_1 \qquad\quad T_2 \qquad T_2 \quad T_0 \end{array} \tag{4-29.43}$$

The results are not always as reasonable as in the case of Fig. 30. Equation (41) may, however, be applicable up to temperature differences of about 20°C.

The normal experimental arrangement for a thermocell with Ag/AgCl electrodes and aqueous potassium chloride solution appears in the following way:

$$\begin{array}{cccccccccc} | \text{ Cu } | \text{ Pt } | \text{ Ag } | \text{ AgCl } | \text{ KCl(aq) } | \text{ AgCl } | \text{ Ag } | \text{ Pt } | \text{ Cu } | \\ T_0 \quad T_1 \ T_1 \ T_1 \qquad T_1 \qquad T_2 \qquad T_2 \ T_2 \ T_2 \ T_0 \end{array}$$

$$\tag{4-29.43a}$$

Here Cu (copper wire) is the terminal (metallic lead) in which a temperature gradient is present. The insertion of Pt wires between terminals and electrodes is without influence on the e.m.f. of the cell. Equally, the temperature T_0 is unimportant. The e.m.f. of this thermocell is, however, not identical to that of the cell (43), since there Ag (silver wire) represents the terminal with the temperature drop. According to Eqs. (30) and (4–25.46), the difference in the thermoelectric powers of the thermocells (43) and (43a) for a given electrolyte concentration is equal to the value of the thermoelectric power of a thermocouple consisting of Cu and Ag.

The concentration dependence of ε_0 or ε_∞ in the case of a single electrolyte with two ion species results from Eq. (31) or (35). At high dilution, ϑ_i and $*Q$ are of the form (cf. Eqs. 4–16.35, 4–16.40, and Eq. 4–27.21): $\alpha + \beta\sqrt{I}$, where α and β represent constants at a given temperature and I the ionic strength of the electrolyte. As will be shown in §4–31, for small concentrations $*S_i$ has the form ($m^\dagger \equiv 1$ mol kg^{-1}):

$$*S_i = *S_i^{\ominus} - R\ln\frac{\nu_i m}{m^\dagger} + \alpha_i\sqrt{I},$$

where $*S_i^{\ominus}$ and α_i are constant for a fixed temperature. From this it follows from Eqs. (31) or (35) that the expression:

$$F\varepsilon_0 - \frac{R}{z_i}\ln\frac{\nu_i m}{m^\dagger} \equiv \eta_0 \qquad (4\text{–}29.44)$$

or

$$F\varepsilon_\infty - \frac{R}{z_i}\ln\frac{\nu_i m}{m^\dagger} \equiv \eta_\infty \qquad (4\text{–}29.45)$$

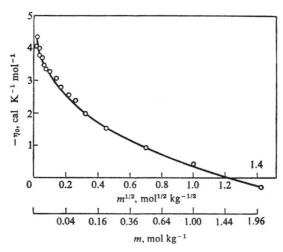

Fig. 31. The quantity η_0 according to Eq. (44) as a function of the molality m of the electrolyte for the thermocell (43a) with HCl (instead of KCl) at an average temperature of 25°C. After Haase and Schoenert[169].

must, for sufficiently small concentrations, be a linear function of $\sqrt{I}$. This will be shown with the use of ε_0 measurements for the thermocell (43a) with HCl (instead of KCl) in Fig. 31. (For this thermocell, there is valid: $z_i = -1, \nu_i = 1, I = m$).

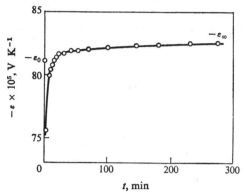

Fig. 32. Thermoelectric power ε of the thermocell (46) as a function of the time t at an average temperature of 25°C and a molality of the electrolyte of 0.001 mol kg^{-1}. After Haase and Behrend[171].

In Fig. 32, we show the change with time in thermoelectric power of the thermocell

$$| \text{Pt} \,|\, \text{Ag} \,|\, \text{AgNO}_3(\text{aq}) \,|\, \text{Ag} \,|\, \text{Pt} \,| \qquad (4\text{-}29.46)$$
$$T_0 \quad T_1 \quad T_1 \qquad\qquad T_2 \quad T_2 \quad T_0$$

at given values of the mean temperature and of the electrolyte concentration. (The $\varepsilon(t)$ curve does not begin at ε_0, since the temperature difference effective in the solution is set up only after some time.) From curves of this type the ε_∞-values for various temperatures and compositions can be determined.

(d) Thermocells with Solid or Molten Electrolytes

If we wish to apply Eq. (22) to thermocells with solid or molten electrolytes then the question of the reference system for the definition of the transport quantities (transport numbers, heats of transport, transported entropies) must be carefully investigated.†

In the equations developed thus far, the Hittorf reference system has been used: the velocity of a selected species is regarded as the reference velocity. This reference species corresponds to ions in the crystal lattice for electronic conductors and to the solvent for electrolyte solutions. The entire

---

† The misunderstandings possible in this have led to different interpretations of the experimental results[175–180].

formalism can be retained for solid or molten electrolytes if a suitable ion type is defined as a reference species (cf. p. 379). We explain this procedure in the following for the simplest case.[37]

We consider an ionic crystal with two ion types (such as solid silver chloride) or an ionized melt with two ionic species without neutral particles (such as molten silver nitrate). Moreover, we assume for simplicity that, for the corresponding thermocell, the electrodes consist of that metal M in a solid form whose ions appear in the solid or molten salt MX_n. We have thus a thermocell of the type

$$M \mid MX_n \mid M \qquad (4\text{–}29.47)$$
$$T_1 \quad\quad T_2$$

Examples are:

$$Ag \mid AgCl \text{ (solid)} \mid Ag \qquad (4\text{–}29.48)$$
$$T_1 \quad\quad\quad\quad T_2$$

$$Pb \mid PbCl_2 \text{ (solid)} \mid Pb \qquad (4\text{–}29.49)$$
$$T_1 \quad\quad\quad\quad T_2$$

$$Ag \mid AgNO_3 \text{ (liquid)} \mid Ag. \qquad (4\text{–}29.50)$$
$$T_1 \quad\quad\quad\quad T_2$$

In such thermocells, the Soret effect is not possible. In Eq. (22), we have therefore: $dm_k/dT = 0$. Furthermore, in Eq. (22) the inferior i must refer to the cations (metal ions). If we now choose the anions (indicated by inferior $-$) or cations (indicated by inferior $+$) as the reference species, we find with Eqs. (4–16.10) and (4–16.42) for the summation remaining in Eq. (22):

$$\sum_j \tau_j {}^*Q_j = \tau_+ {}^*Q_+ = \frac{t_+}{z_+} {}^*Q_+ = \frac{{}^*Q_+}{z_+}$$

or

$$\sum_j \tau_j {}^*Q_j = \tau_- {}^*Q_- = \frac{t_-}{z_-} {}^*Q_- = \frac{{}^*Q_-}{z_-}.$$

Here ${}^*Q_+$ or ${}^*Q_-$ is the heat of transport of the cations or anions in the solid or molten salt measured relative to the anions or cations respectively. As is obvious, the Hittorf transport number t_+ or t_- which, for our choice of reference system, always has the value 1, has disappeared from the formula. Finally, if we note the relations

$$\nu_+ z_+ + \nu_- z_- = 0,$$
$$\nu_+ S_+ + \nu_- S_- = \bar{S}_{MX_n},$$
$$S_a = \bar{S}_M,$$

where $\bar{S}_M$ or $\bar{S}_{MX_n}$ denotes the molar entropy of the solid electrode metal or of the solid or molten salt respectively, then the simple formula (cf. Eq. 25):

$$F\varepsilon = \frac{1}{z_+} \bar{S}_M - (^*S_\ominus)_A - \frac{1}{z_+} {}^*S_+ \qquad (4–29.51)$$

or

$$F\varepsilon = \frac{1}{z_+ \nu_+} (\nu_+ \bar{S}_M - \bar{S}_{MX_n}) - (^*S_\ominus)_A - \frac{1}{z_-} {}^*S_- \qquad (4–29.52)$$

results from Eqs. (22) and (24). Here $^*S_+$ or $^*S_-$ denotes the transported entropy of the cations or anions in the solid or molten salt measured relative to the anions or cations respectively. Comparison of Eq. (51) with Eq. (52) leads to the relation

$$\nu_+ {}^*S_+ + \nu_- {}^*S_- = \bar{S}_{MX_n}. \qquad (4–29.53)$$

From the last equation with Eq. (24) there follows:

$$^*Q \equiv \nu_+ {}^*Q_+ + \nu_- {}^*Q_- = 0. \qquad (4–29.54)$$

Equations (51) and (54) are in agreement with the statements of Pitzer[181].†
With the help of Eqs. (51) to (53) the conventional transported entropies of ions $^*S_+$ and $^*S_-$ in solid or molten salts can be obtained from e.m.f. measurements on thermocells of the type (48) and (49) or (50). Examples illustrating this are found in §4–31 (Tables 16 and 17, pp. 407 and 408).

4–30 ELECTROLYTIC THERMOCOUPLES

An *electrolytic thermocouple* is a nonisothermal galvanic cell represented by the diagram

$$
\begin{array}{ccccccc}
 & & & a & & b & & c & & \\
| \text{Terminal} | & \text{Electrode} | & \text{Solution I} & || & \text{Solution II} & || & \text{Solution I} & | & \text{Electrode} | & \text{Terminal} | \\
T_0 & T_1 & T_1 & \uparrow T_1 \uparrow & & \uparrow T_2 \uparrow & & T_1 & T_1 & T_0 \\
 & & & \alpha \quad \beta & & \gamma \quad \delta & & & &
\end{array}
$$

The terminals and electrodes are always similar to one another. Solutions I and II contain different electrolytes in the same solvent. In solution I, that ion species should be present for which the electrodes are reversible. At the sites denoted by $||$ there are the "bridge solutions". Room temperature is denoted by T_0, while T_1 and T_2 represent the temperatures of the two liquid junctions. Pressure gradients, external force fields, concentration gradients, and, thus, the Soret effect will be excluded. "Initial values" must therefore be determined in the e.m.f. measurements.

---

† Nevertheless, the considerations of Pitzer are still based on the quasithermodynamic theory of Eastman and Wagner.

The measurable e.m.f. of the above cell is composed of the diffusion potentials at the phase boundaries a and b as well as thermal diffusion potentials inside the solution II (between a and b) and inside the solution I (between b and c). All other potential differences (contact and electrode potentials as well as thermoelectric potentials) mutually compensate for each other. If suitable electrodes are considered to be inserted at the positions α, β, γ, and δ then we recognize that our electrolytic thermocouple corresponds to a combination of two thermocells with two concentration cells (with transference).

As an example of an experimentally investigated electrolytic thermocouple, we consider the cell[182]†

$$| \text{Cu} | \text{Pt} | \text{Ag} | \text{AgCl} | \text{HCl(aq)} \| \text{KCl(aq)} \| \text{HCl(aq)} | \text{AgCl} | \text{Ag} | \text{Pt} | \text{Cu} | \quad \text{(A)}$$

with temperature markings T_0, T_1 ... T_1, T_1, T_2, T_1 ... T_1, T_0 and positions α, β, γ, δ.

Here the aqueous solutions of HCl and KCl have equal molalities. Let the e.m.f. of this nonisothermal cell be Φ and the thermoelectric power ε ($\approx \Phi/\Delta T$, $\Delta T = T_2 - T_1$).

We now consider the isothermal concentration cells

Temperature T_1: $\text{Ag} | \text{AgCl} | \text{HCl(aq)} \| \text{KCl(aq)} | \text{AgCl} | \text{Ag}$ (B)

Temperature T_2: $\text{Ag} | \text{AgCl} | \text{KCl(aq)} \| \text{HCl(aq)} | \text{AgCl} | \text{Ag}$ (C)

and the thermocells

$$| \text{Cu} | \text{Pt} | \text{Ag} | \text{AgCl} | \text{KCl(aq)} | \text{AgCl} | \text{Ag} | \text{Pt} | \text{Cu} | \quad \text{(I)}$$

with temperatures T_0, T_1 ... T_1, T_2 ... T_2, T_0

$$| \text{Cu} | \text{Pt} | \text{Ag} | \text{AgCl} | \text{HCl(aq)} | \text{AgCl} | \text{Ag} | \text{Pt} | \text{Cu} | \quad \text{(II)}$$

with temperatures T_0, T_1 ... T_1, T_2 ... T_2, T_0

and call the e.m.f. values of these cells in that sequence Φ_1, Φ_2, Φ_I, Φ_II. Let the thermoelectric powers (initial values) of the two thermocells be ε_I ($\approx \Phi_\text{I}/\Delta T$) and ε_II ($\approx \Phi_\text{II}/\Delta T$).

If we consider Ag/AgCl electrodes to be introduced into the electrolytic thermocouple (A) at α and β (temperature T_1) as well as at γ and δ (temperature T_2) then we arrive at the following relation:

$$\Phi = \Phi_1 + \Phi_2 + \Phi_\text{I} - \Phi_\text{II}$$

or

$$\varepsilon = \varepsilon' + \varepsilon_\text{I} - \varepsilon_\text{II}, \tag{4–30.1}$$

where

$$\varepsilon' \equiv \frac{\Phi_1 + \Phi_2}{T_2 - T_1}. \tag{4–30.2}$$

---

† There are only few earlier investigations (Wild, 1858; Duane, 1898; Podszus, 1908).

Thus the thermoelectric power ε of the electrolytic thermocouple can be related to the thermoelectric powers ε_I and ε_{II} of the two thermocells and to the expression ε'. The quantity ε' must, strictly speaking, be obtained from e.m.f. measurements on the isothermal concentration cells (B) and (C) where the liquid junctions should be set up in exactly the same (reproducible) way as in the electrolytic thermocouple (A). The rigorous verification of Eq. (1) requires awkward experimental methods without leading to new information. Therefore, we restrict the following discussion to some simple considerations on available data.

If sufficiently dilute solutions are assumed then the electrode potential is independent of the nature of the cation (H^+ or K^+) in the aqueous chloride solutions. The e.m.f. of the isothermal concentration cells is thus, in this approximation, determined only by the diffusion potential at the liquid junctions.

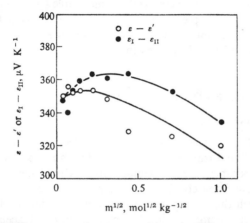

Fig. 33. The difference $\varepsilon - \varepsilon'$ or $\varepsilon_I - \varepsilon_{II}$ determined from measurements on the electrolytic thermocouple (A) and Eq. (3) or from measurements on the thermocells (I) and (II) as a function of the molality m of the electrolyte at an average temperature of 25°C. According to Haase and Sauermann[182].

As has been known for a long time (cf. p. 300), diffusion potentials in solutions of two or more electrolytes can, in general, be determined neither experimentally nor theoretically; indeed they can often not even be reproduced. If, however, two ideal dilute solutions† of uni-univalent electrolytes of the same concentration containing a common ionic species adjoin one another in a definite way, we have a particularly simple case. Here the calculations of Planck[184a,b] and Henderson[185] produce the same result, so that we may

_____

† Recently, a more general analysis has been carried out by Spiro[183].

assume that the value of the diffusion potential in this special case is independent of the structure of the transition layer between the two electrolyte solutions.† Thus, if we assume ideal dilute solutions in the isothermal concentration cells (B) and (C) then the Planck–Henderson formula for Eq. (2) produces:

$$\varepsilon' = \frac{R}{F}\left(\frac{T_1}{T_2 - T_1} \ln \frac{(\Lambda_1)_{\text{HCl}}}{(\Lambda_2)_{\text{HCl}}} + \frac{T_1}{T_2 - T_1} \ln \frac{(\Lambda_2)_{\text{KCl}}}{(\Lambda_1)_{\text{KCl}}} + \ln \frac{(\Lambda_2)_{\text{KCl}}}{(\Lambda_2)_{\text{HCl}}}\right).$$

$$(4\text{–}30.3)$$

Here R denotes the gas constant, F the Faraday constant, and Λ_1 or Λ_2 the equivalent conductance of the aqueous solution of HCl and KCl respectively at the temperature T_1 or T_2 in the limiting case of infinite dilution.

According to the above, the expression $\varepsilon - \varepsilon'$ determined from the thermoelectric power ε of the electrolytic thermocouple (A) and from Eq. (3) approximately agrees with the quantity $\varepsilon_{\text{I}} - \varepsilon_{\text{II}}$ derived from the thermoelectric powers of the thermocells (I) and (II). Agreement is better the more dilute the solutions and Fig. 33 proves this statement.

4–31 TRANSPORTED ENTROPIES OF IONS

(a) Electrolyte Solutions

According to §4–29, for solutions of a single electrolyte with two ionic species, the conventional transported entropies $*S_+$ and $*S_-$ of cations and anions can be experimentally determined from e.m.f. measurements on thermocells and from Soret coefficients. We now wish to concern ourselves in greater detail with these quantities, which are characteristic for the non-isothermal transport phenomena in electrolyte solutions (except pure heat conduction) in a similar way as are the ion mobilities for isothermal processes (electric conduction and diffusion).

The definition (4–24.20) produces the relation:

$$*S_i = \frac{*Q_i}{T} + S_i \quad (i = +, -),$$

$$(4\text{–}31.1)$$

where T denotes the absolute temperature, $*Q_i$ the heat of transport of species i, and S_i the partial molar entropy of species i. Both $*S_i$ and $*Q_i$ are measured in the Hittorf reference system, where the solvent represents the reference species.

---

† In the case of the electrolytic thermocouple A, reproducible "flowing junctions" were used.

According to classical thermodynamics (cf. Eqs. 1–12.7, 1–19.15, and 1–19.16),

$$S_i = S_i^\ominus - R\left[\ln\left(m_i\gamma_i\right) + T\left(\frac{\partial \ln \gamma_i/\gamma^\dagger}{\partial T}\right)_{P,m}\right] \qquad (4\text{–}31.2)$$

with

$$\lim_{m\to 0} \gamma_i = \gamma^\dagger \quad (\equiv 1 \text{ kg mol}^{-1}). \qquad (4\text{–}31.3)$$

Here $S_i^\ominus$ is the standard value of the partial molar entropy of the ionic species i, which, at a fixed temperature† and for a given solvent, depends only on the nature of the ion i,‡ R the gas constant, m_i the molality of the ion type i, γ_i the practical activity coefficient of ionic species i, P the pressure, and m the molality of the electrolyte.

On the basis of our knowledge about Soret coefficients (p. 376) we can set up for the heat of transport $*Q_i$ of the ionic species i:

$$*Q_i = *Q_i^0 + q_i \qquad (4\text{–}31.4)$$

and

$$\lim_{m\to 0} q_i = 0. \qquad (4\text{–}31.5)$$

$*Q_i^0$, the limiting value of $*Q_i$ at infinite dilution, is, like $S_i^\ominus$, at constant temperature and for given solvent, an individual characteristic constant of the ionic type i, that is, independent of the nature of the ion of opposite charge. The quantity q_i is a concentration function which vanishes for $m = 0$.

From Eqs. (1) to (5) there results[146, 168a, 170] ($m^\dagger \equiv 1 \text{ mol kg}^{-1}$):

$$*S_i = *S_i^\ominus - R\ln\frac{m_i}{m^\dagger} - R\left[\ln\frac{\gamma_i}{\gamma^\dagger} + T\left(\frac{\partial \ln \gamma_i/\gamma^\dagger}{\partial T}\right)_{P,m}\right] + \frac{q_i}{T} \qquad (4\text{–}31.6)$$

with

$$*S_i^\ominus \equiv S_i^\ominus + \frac{*Q_i^0}{T}, \qquad (4\text{–}31.7)$$

$$*S_i^\ominus = \lim_{m\to 0}\left(*S_i + R\ln\frac{m_i}{m^\dagger}\right). \qquad (4\text{–}31.8)$$

---

† We always ignore the insignificant effect of the pressure for electrolyte solutions.
‡ This means that the standard value $S_2^\ominus$ of the partial molar entropy of the electrolyte in the solution (cf. footnote to p. 393)

$$S_2^\ominus = \nu_+ S_+^\ominus + \nu_- S_-^\ominus$$

is additive in its ion contributions. Thus we can, for the purpose of tabulation, set the $S_i^\ominus$ value for a definite ion species arbitrarily equal to zero in a similar way as for the heats of transport (Table 14, p. 376), and thus obtain numerically determined values for the $S_i^\ominus$ of the other ion types. If we proceed from the usual normalization $S_{H^+}^\ominus = 0$ (for aqueous solutions at 25°C) then we find the $S_i^\ominus$ values which are given in Table 15, p. 406.

The quantity $*S_i^{\ominus}$ is the *standard value* of the (conventional) transported entropy. According to Eq. (7), at a fixed temperature and for a given solvent it represents a constant characteristic for the ion i.

Simple regularities are to be expected only at sufficiently high dilution. Since in this case the electrolyte is completely dissociated, we have, according to Eq. (1–19.13),

$$m_i = \nu_i m, \tag{4–31.9}$$

where ν_i denotes the dissociation number of the ion species i. Moreover, we obtain from the Debye–Hueckel limiting law:

$$\ln \frac{\gamma_i}{\gamma^\dagger} = -az_i^2 \sqrt{I}. \tag{4–31.10}$$

Here z_i is the charge number of ion i, I is the ionic strength of the electrolyte (Eq. 4–27.22), and a is a quantity which depends only on the temperature and on the nature of the solvent and is also predicted by theory (Eq. 4–16.40e). Eventually, there follows from Soret effect data and Eqs. (4–27.23) and (4):

$$q_i = q_i^0 \sqrt{I}, \tag{4–31.11}$$

where q_i^0 represents a constant characteristic for the ion i at a fixed temperature and a given solvent.†

Using the abbreviation

$$\beta_i \equiv *S_i + R \ln \frac{\nu_i m}{m^\dagger} \tag{4–31.12}$$

we find from Eqs. (6) to (11):

$$\beta_i = *S_i^{\ominus} + \alpha_i \sqrt{I} \tag{4–31.13}$$

with

$$\alpha_i \equiv z_i^2 R \left[a + T \left(\frac{\partial a}{\partial T} \right)_P \right] + \frac{q_i^0}{T}, \tag{4–31.14}$$

$$*S_i^{\ominus} = \lim_{m \to 0} \beta_i. \tag{4–31.15}$$

---

† For the (measurable) conventional activity coefficient γ, there results from Eq. (10) as well as from Eqs. (1–19.8), (1–19.9), (1–19.17), and (1–19.18) with $\alpha = 1$ (complete dissociation):

$$\ln \frac{\gamma}{\gamma^\dagger} = az_+ z_- \sqrt{I}.$$

In this form, the Debye–Hueckel limiting law is experimentally verified as the equation of the limiting tangent for the curve $\ln \gamma/\gamma^\dagger (\sqrt{I})$ at $I = 0$.

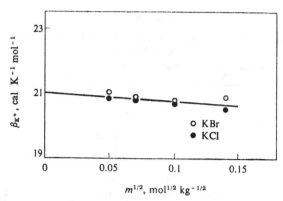

Fig. 34. The quantity β_{K^+} as a function of the molality m of the electrolyte for aqueous solutions of KCl and KBr at 25°C. According to Haase, Hoch, and Schoenert[170].†

If, therefore, β_i is plotted against $\sqrt{I}$ then α_i is obtained from the slope. Here the intersection with the ordinate axis $*S_i^\ominus$ and the value of the slope α_i depend at constant temperature and for a given solvent only on the nature of the ion i. Recent measurements on very dilute solutions verify this statement[146, 170, 171] and an example is presented in Fig. 34. Table 15 gives a survey of the constants $*S_i^\ominus$ so determined for ions in aqueous solution at 25°C. The experimental data are still not accurate enough to make a tabulation of the α_i and q_i^0 values worthwhile.‡ These can be calculated from Eq. (14).∥

Figure 35 gives an example of a plot of the function $\beta_i(\sqrt{I})$ over a larger concentration range. As we may see, Eq. (13) becomes invalid for concentrated solutions; the formula represents only, as does the Debye–Hueckel limiting law, the equation of the limiting tangent.

It should be noted that, in the region of validity of Eq. (13), unknown Soret coefficients can be calculated from Eq. (4–29.39) and Eq. (12) using the constants assembled in Table 15[170].

The splitting of $*S_i^\ominus$ into $S_i^\ominus$ and $*Q_i^0/T$ using Eq. (7) is arbitrary. If, however, the $S_i^\ominus$-values for aqueous solutions at 25°C are fixed by convention, usually by $S_{H^+}^\ominus = 0$, as we shall do, then $*Q_i^0$ is also numerically determined for aqueous solutions at 25°C from the relation

$$*Q_i^0 = T(*S_i^\ominus - S_i^\ominus) \qquad (4\text{–}31.16)$$

† More recent data are to be found in Haase and Hoch[146].
‡ The molecular statistical theory of Helfand and Kirkwood[186] also leads to an expression of the form (13).
∥ The first published values of $*S_i^\ominus$ stem from Choroschin and Temkin[187], whose method of evaluation, however, amounts to the supposition $\alpha_i = 0$.

Table 15

Standard Values of the Transported Entropies $*S_i^{\ominus}$, Standard Values of the Partial Molar Entropies $S_i^{\ominus}$, and Conventional Values of the Heats of Transport at Infinite Dilution $*Q_i^0$ for 16 Ionic Species in Aqueous Solution at 25°C[146]

Ionic species i	$*S_i^{\ominus}$ cal K^{-1} mol^{-1}	$S_i^{\ominus}$ cal K^{-1} mol^{-1}	$*Q_i^0$ cal mol^{-1}
Cl$^-$	18.9	13.2	1700
Br$^-$	24.8	19.3	1640
NO$_3^-$	39.9	35.0	1460
ClO$_4^-$	47.8	43.2	1370
H$^+$	5.2	0.0	1550
Li$^+$	−0.8	3.4	−1250
Na$^+$	11.0	14.4	−1010
K$^+$	20.0	24.5	−1340
NH$_4^+$	22.0	27.0	−1490
Rb$^+$	27.0	29.7	− 810
Cs$^+$	29.2	31.8	− 780
Ag$^+$	17.0	17.7	− 210
Ca^{2+}	−16.6	−13.2	−1010
Sr^{2+}	−11.9	−9.4	− 750
Ba^{2+}	2.1	3.0	− 270
La^{3+}	−42.5	−39.0	−1040

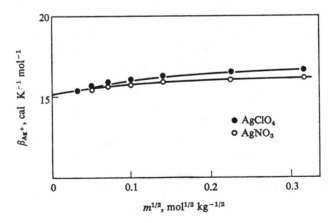

Fig. 35. The quantity β_{Ag^+} as a function of the molality m of the electrolyte for aqueous solutions of AgNO$_3$ and AgClO$_4$ at 25°C. According to Haase and Behrend[171].†

---

† More recent data are to be found in Haase and Hoch[146].

Conventional Transported Entropies of the Cations and Anions ($*S_+$ and $*S_-$) in Solid Salts ($\bar{S}_{MX_n}$ = molar entropy of the solid salt)

Salt	Temperature K	$\bar{S}_{MX_n}$ cal K^{-1} mol^{-1}	$*S_+$ cal K^{-1} mol^{-1}	$*S_-$ cal K^{-1} mol^{-1}	Remarks
AgCl	400	26.8	44.0	− 17.2	Cation conductor
	500	29.9	41.0	− 11.1	
	600	32.5	37.6	− 5.1	
	700	34.8	31.8	+ 3.0	
AgBr	400	29.5	46.0	− 16.5	Cation conductor
	500	32.8	39.2	− 6.4	
	600	35.8	32.8	+ 3.0	
AgI (II)	400	31.5	40.6	− 9.1	Cation conductor
AgI (I)	500	38.1	26.3	+11.8	
	600	40.5	27.4	+13.1	
CuCl	400	25.6	33.0	− 7.4	Cation conductor
	500	28.8	32.6	− 3.8	
	600	31.8	32.0	− 0.2	
CuBr	400	^{a}26.9	46.0	− 19.1	Cation conductor
	500	^{a}29.9	40.6	− 10.7	
	600	^{a}32.4	35.0	− 2.6	
CuI (III)	400	^{a}26.9	33.0	− 6.1	Cation conductor
	500	^{a}29.9	32.6	− 2.7	
	600	^{a}32.4	32.0	+ 0.4	
CuI (I)	750		22.0		Cation conductor
PbCl$_2$	500	42.4	46.0	− 1.8	Cation conductor
PbBr$_2$	500	48.6	40.0	+ 4.3	Anion conductor
PbI$_2$	550	54.4	2.0	+26.2	Anion and cation conductor
	650	57.9	12.5	+22.7	

^{a} Estimated.

following from Eq. (7). If we now determine the quantity $*Q_i^0$ for the ion Cl^- in this way then there results a definite numerical value $*Q_{Cl^-}^0$ which is different from zero. The $*Q_i^0$-values thus fixed for the other ions are shown in Table 14 where, except for the value for Cl^-, all quantities have been derived from Soret effect measurements. These must then agree with the values for $*Q_i^0$ calculated from Eq. (16) on the basis of measurements on thermocells. The differences which can be seen from Tables 14 and 15 represent a criterion for the quality of the measurements.[38]

(b) Solid and Molten Electrolytes

According to Eqs. (4–29.51) and (4–29.53) the transported entropies of the cations and anions ($*S_+$ and $*S_-$) can be determined for solid and molten electrolytes from e.m.f. measurements on thermocells of the type shown in (4–29.47). Here again we are dealing with conventional values since the molar entropies ($\bar{S}_M$ and $\bar{S}_{MX_n}$) appearing in Eqs. (4–29.51) and (4–29.53) or electronic transported entropies $(*S_\ominus)_A$ are numerically fixed by the Planck normalization or by the method described previously (p. 354). We use the Hittorf reference system, choosing the anion as the reference species for $*S_+$ and the cation as the reference species for $*S_-$.

Table 17

Conventional Transported Entropies of the Cations and Anions ($*S_+$ and $*S_-$) in Molten Salts ($\bar{S}_{MX_n}$ = molar entropy of the molten salt)

Salt	Temperature K	$\bar{S}_{MX_n}$ cal K^{-1} mol^{-1}	$*S_+$ cal K^{-1} mol^{-1}	$*S_-$ cal K^{-1} mol^{-1}
$AgNO_3$	500	54.0	21.0	33
AgCl	800	41.2	26	15.2
AgBr	750	42.6	27	15.6
AgI	850	45.2	27	18.2
CuCl	735	39.1	23.7	15.4
	798	40.4	24.2	16.2
	861	41.6	24.7	16.9
$ZnCl_2$	600		8	
$SnCl_2$	600		22	

In Table 16, we show the cation transported entropies $*S_+$ determined for solid salts by Pitzer[181] from Eq. (4–29.51) and literature data and the anion transported entropies $*S_-$ derived with the help of Eq. (4–29.53). The values used in this† for the molar entropies of the solid salt $\bar{S}_{MX_n}$ are also found there. Finally, we have enlarged the table with the statement of whether

---

† The values of $\bar{S}_{MX_n}$ were taken (also for the molten salts) from the tabulated work of Landolt and Boernstein[188], as well as from the compilation of Kelley[189].

we are dealing with a cation conductor, an anion conductor, or a mixed conductor. For the salts considered here, the question of electronic conduction does not arise in the temperature range investigated. For pure cation conductors or pure anion conductors, the chosen reference system coincides with the crystal lattice if we consider $*S_+$ or $*S_-$ respectively. As is obvious, $*S_+$ is always positive by a considerable amount, while $*S_-$ can be negative and can exhibit a change of sign at a definite temperature.

In Table 17, the corresponding figures for molten salts are shown. The $*S_+$ values for CuCl stem from Nichols and Langford[178]; other values are from Pitzer[181]. Obviously, molten salts always simultaneously show cation and anion conduction.[39]

D. Complicated Processes

4-32 VISCOUS FLOW

(a) General

Viscous flow is accompanied by the formation of viscous pressures by velocity gradients in a streaming fluid. Here we only consider isotropic systems without rotational movements and without electromagnetic fields.†

If we use a rectangular coordinate system (position-coordinates z_1, z_2, z_3) then we have, at first, nine viscous pressures P_{ij} ($i, j = 1, 2, 3$) and nine velocity gradients $\partial v_i / \partial z_j$ (v_i = component of barycentric velocity $\mathbf{v}$, $\partial / \partial z_j$ = operator of the differentiation with respect to a space-coordinate at a fixed time). Because of the symmetry relations (cf. p. 230)

$$P_{ij} = P_{ji} \quad (i, j = 1, 2, 3) \tag{4-32.1}$$

and

$$X_{ij} = X_{ji} \quad (i, j = 1, 2, 3) \tag{4-32.2}$$

with (cf. Eq. 4-12.5)

$$X_{ij} \equiv -\frac{1}{2}\left(\frac{\partial v_i}{\partial z_j} + \frac{\partial v_j}{\partial z_i}\right) \quad (i, j = 1, 2, 3) \tag{4-32.3}$$

there remain, however, only six independent viscous pressures (e.g. the normal pressures P_{11}, P_{22}, P_{33}, and the tangential or shear pressures P_{12}, P_{13}, P_{23}) and six independent combinations of the velocity gradients (X_{11}, etc.).

---

† The case of a flowing medium in a centrifuge or in a magnetic field, in which the reciprocity relations should be taken account of in a particularly complicated form (§4-35), was investigated by Hooyman[44]. The hydrodynamics of liquid helium (II) developed in the light of the thermodynamics of irreversible processes was dealt with by Prigogine and Mazur[190, 191].

With Eqs. (1) and (2) we write the last term in the dissipation function (4–12.2) in the following form:

$$\Psi_\eta \equiv \sum_{i=1}^{3} \sum_{j=1}^{3} P_{ij} X_{ij}$$

$$= P_{11}X_{11} + P_{22}X_{22} + P_{33}X_{33} + 2P_{12}X_{12} + 2P_{13}X_{13} + 2P_{23}X_{23}.$$
$$(4\text{–}32.4)$$

In this, there thus appear six independent generalized flows (viscous pressures or "momentum flux densities") and six independent generalized forces.

The flowing medium can support arbitrary gradients of temperature, pressure, and concentrations and can also contain chemical species that react. Thus, besides the viscous flow, other transport phenomena (diffusion, heat conduction, thermal diffusion, etc.) are possible as well as chemical reactions. We need not take account of the transport processes when setting up the phenomenological laws for the viscous flow because of the symmetry principle of Curie (p. 249). The chemical reactions, on the other hand, must be considered since, in principle, a coupling between these and the viscous flow may arise. Thus we keep the last two terms in the general expression (4–12.2) for the dissipation function Ψ in isotropic media without electrification and magnetization:

$$\Psi = \sum_r b_r A_r + \Psi_\eta \geqslant 0, \qquad (4\text{–}32.5)$$

where b_r or A_r denotes the reaction rate or the affinity of the reaction r respectively, and the summation is over all independent chemical reactions.

(b) Shear Viscosity and Bulk Viscosity

First we exclude chemical reactions. The decisive part of the dissipation function is then, according to Eqs. (4) and (5),

$$\Psi = \Psi_\eta = P_{11}X_{11} + P_{22}X_{22} + P_{33}X_{33}$$
$$+ 2P_{12}X_{12} + 2P_{13}X_{13} + 2P_{23}X_{23} \geqslant 0. \qquad (4\text{–}32.6)$$

According to the general methods of the thermodynamics of irreversible processes, we would expect that for near-equilibrium conditions, i.e. for not too large values of the velocity gradients, phenomenological equations are valid which, according to Eq. (6), lead to six linear equations with 36 coefficients. However, a coordinate transformation, analogous to a corresponding transformation of the stress–strain relation (the generalized Hooke's law) for isotropic elastic solid bodies, shows that, due to the isotropy of the fluid medium, there remain[192] 6 independent equations with only 2 independent coefficients:

$$P_{ij} = P_{ji} = 2\eta X_{ij} + (\zeta - \tfrac{2}{3}\eta)\delta_{ij}(X_{11} + X_{22} + X_{33}) \quad (i, j = 1, 2, 3).$$
$$(4\text{–}32.7)$$

Here δ_{ij} denotes the Kronecker symbol ($\delta_{ij} = 1$ for $i = j$, $\delta_{ij} = 0$ for $i \neq j$). We call η the *shear viscosity* (or simply *viscosity*) and ζ the *bulk viscosity* (or *pressure viscosity*). Equation (7) is in agreement with experience.

Insertion of Eq. (7) in Eq. (6) yields $\eta \geqslant 0$, $\zeta \geqslant 0$. With $\zeta = 0$ there follows from Eqs. (6) and (7) an expression which is known as the "Rayleigh dissipation function" and from which the name "dissipation function" originated for Ψ in general.

From consideration of Eq. (3) and the relation

$$\text{div } \mathbf{v} = \frac{\partial v_1}{\partial z_1} + \frac{\partial v_2}{\partial z_2} + \frac{\partial v_3}{\partial z_3}, \tag{4–32.8}$$

Eq. (7) can also be formulated thus:

$$P_{ij} = P_{ji} = -\eta\left(\frac{\partial v_i}{\partial z_j} + \frac{\partial v_j}{\partial z_i}\right) + (\tfrac{2}{3}\eta - \zeta)\,\delta_{ij}\,\text{div } \mathbf{v} \quad (i, j = 1, 2, 3). \tag{4–32.9}$$

Equation (7) or (9) goes back in principle to St. Venant (1843) and Stokes (1845). Indeed, in Eq. (7) or (9) nothing more is assumed than a linear dependence of the viscous pressures on the velocity gradients, and the isotropy of the medium is taken into account.

In an *incompressible liquid*, as may be assumed in many problems of hydrodynamics, the density is constant in position and time. Thus, according to Eq. (4–4.4), there is:

$$\text{div } \mathbf{v} = 0 \quad \text{(incompressibility condition).} \tag{4–32.10}$$

Here the second term in Eq. (9) is dropped for $i = j$, too. With the insertion of Eqs. (9) and (10) in Eq. (4–5.11) the equations of Navier (1822) and Stokes (1845) are obtained. However, in the following we will always consider the more general case of a compressible fluid.

If only *one* velocity component v_1 appears, the gradient of which lies in the z_2-axis, then Eq. (9) is simplified to the well-known elementary expression:

$$P_{12} = -\eta \frac{\partial v_1}{\partial z_2}. \tag{4–32.11}$$

This is Newton's law (1687). [40]

If a volume element of a fluid is uniformly compressed or expanded, we have

$$\frac{\partial v_1}{\partial z_1} = \frac{\partial v_2}{\partial z_2} = \frac{\partial v_3}{\partial z_3} = \tfrac{1}{3}\,\text{div } \mathbf{v}.$$

From this and from Eq. (9) we find

$$P_{ii} = -\zeta \operatorname{div} \mathbf{v} \quad (i = 1, 2, 3). \tag{4-32.12}$$

The significance of the bulk viscosity is thus exposed: ζ is a measure of the additional normal pressure which is equal on all sides and which arises for a uniform deformation in a fluid. Here, because $\zeta > 0$ (cf. above), P_{ii} is positive if a compression (div $\mathbf{v} < 0$) is present.

While the quantity η is accessible as "viscosity" in all tables (cf. Table 18), measurements of the bulk viscosity have until recently been unavailable. Such values have first become known from the experimental investigations which were suggested by Eckart[193] and carried out by Liebermann[194] on the propagation of sound in liquids.

Table 18

Viscosity (Shear Viscosity) η for Fluid One-component Systems at Atmospheric Pressure[21c, 195] (P = poise)

Substance	Temperature	$\eta \times 10^4$ P
Argon (gas)	100 K	0.839
	200 K	1.594
	300 K	2.269
Water (liquid)	0°C	178.7
	25°C	89.03
	50°C	54.67
	75°C	37.88
	100°C	28.29
Hydrogen cyanide (liquid)	18°C	20.6
Benzene (liquid)	25°C	60.28
Ethanol (liquid)	25°C	107.8
Sulfuric acid (liquid)	25°C	2454
Glycerine (liquid)	25°C	94,500

The shear viscosity η of gases and liquids depends strongly on the temperature, a little on the pressure, and (for mixtures) noticeably on the composition. According to the measurements of Liebermann on water, the pressure viscosity ζ appears to show practically the same temperature dependence as η. Moreover, according to Liebermann, the numerical value for the ratio ζ/η is of importance for liquids. Thus η for water amounts to about 10^{-2} P around 17°C, while the corresponding value of ζ is approximately 3×10^{-2} P. For benzene and carbon disulfide at around 17°C, the quotient ζ/η is actually larger than 100. Recent investigations[196] on liquid $ZnCl_2$ lead to the following values at 311°C: $\eta = 44.5$ P, $\zeta = 22.9$ P.

The kinetic theory of gases[105, 197] produces for monatomic ideal gases: $\zeta = 0$. For polyatomic ideal gases as well as for real gases, the statement

$\zeta > 0$ is to be expected according to the kinetic theory. The corresponding molecular theory for liquids is still undeveloped. In almost all cases, nevertheless, the bulk viscosity is related to molecular relaxation processes.[†]

(c) Chemical Viscosity

With chemical reactions occurring in flowing media, Eq. (7) or (9) is not generally valid. This arises from consideration of the dissipation function (5). We must take account of a coupling between reactions and viscous flow on establishing the phenomenological equations. The tensor elements P_{ij} will thus depend linearly on the scalar affinities A_r (after multiplication by the unit tensor δ_{ij}). Accordingly, the scalar rates of reaction, b_r must be set up as linear functions not only of the affinities but also of $X_{11} + X_{22} + X_{33}$ ($= -\operatorname{div} \mathbf{v}$). (The last quantity represents the "scalar invariant" or "trace" of the tensor X_{ij}.) Accordingly, on consideration of Eqs. (3), (4), and (5), we obtain the following relations[4b]:

$$P_{ij} = P_{ji} = -\eta\left(\frac{\partial v_i}{\partial z_j} + \frac{\partial v_j}{\partial z_i}\right) + \left(\tfrac{2}{3}\eta - \zeta\right)\delta_{ij}\operatorname{div}\mathbf{v} + \delta_{ij}\sum_{r=1}^{R}\lambda_r A_r$$

$$(i, j = 1, 2, 3), \quad (4\text{–}32.13)$$

$$b_r = \sum_{s=1}^{R} a_{rs}A_s - \lambda_r'\operatorname{div}\mathbf{v} \quad (r = 1, 2, \ldots, R). \qquad (4\text{–}32.14)$$

Here R denotes the number of independent reactions. The a_{rs}, λ_r, and λ_r' are phenomenological coefficients. Equation (13) generalizes Eq. (9), while Eq. (14) represents the generalization of the linear equation (4–24.7) for chemical reactions near to equilibrium.

For the coefficients a_{rs}, the Onsager reciprocity law holds in the usual form (4–24.8):

$$a_{rs} = a_{sr} \quad (r, s = 1, 2, \ldots, R). \qquad (4\text{–}32.15)$$

The quantities λ_r and λ_r', however, according to §1–26 (p. 91) are "Casimir coefficients" since the affinities are "forces of the α-type," the velocity gradients (here $-\operatorname{div}\mathbf{v}$) are "forces of the β-type."[‡] Thus the reciprocity relation (1–26.3) should be applied, and this produces:

$$\lambda_r = -\lambda_r' \quad (r = 1, 2, \ldots, R). \qquad (4\text{–}32.16)$$

We denote λ_r as the *chemical viscosity* assigned to the reaction r.

[†] An extensive bibliography on all papers which relate to the bulk viscosity is given by Karim and Rosenhead[198]. The quantity $\zeta - \tfrac{2}{3}\eta$ is often denoted as "second viscosity."

[‡] With reversal of time the A_r retain their values, while $-\operatorname{div}\mathbf{v}$, as in fact any velocity gradient, reverses its sign.

On specialization to a single reaction (reaction rate b, affinity A) the system of equations:

$$P_{ij} = P_{ji} = -\eta\left(\frac{\partial v_i}{\partial z_j} + \frac{\partial v_j}{\partial z_i}\right) + (\tfrac{2}{3}\eta - \zeta)\,\delta_{ij}\,\mathrm{div}\,\mathbf{v} = \delta_{ij}\lambda A \quad (i,j = 1, 2, 3),$$

$$(4\text{–}32.17)$$

$$b = aA + \lambda\,\mathrm{div}\,\mathbf{v} \qquad\qquad (4\text{–}32.18)$$

follows from Eqs. (13), (14), and (16) with the four coefficients η (shear viscosity), ζ (bulk viscosity), λ (chemical viscosity), and a. Here the concept of the "chemical reaction" need not be understood literally: it may refer to an arbitrary transformation (relaxation process).

Since experimental verification of the above relations is possible in practice only from measurements of the propagation of sound, as is the bulk viscosity, we introduce the specific entropy $\tilde{S}$ and the specific volume $\tilde{V}$ (the reciprocal density) in place of the temperature and pressure (cf. §2–14) as independent variables for the description of the thermodynamic state of a volume element. In the following, we wish to neglect transport effects like diffusion, heat conduction, thermal diffusion, etc., i.e. we set the diffusion flows and the heat flow equal to zero. Then a further variable ξ is sufficient for the determination of the intensive state variables other than $\tilde{S}$ and $\tilde{V}$, and this "extent of reaction" is defined most suitably for the present case by the differential relation

$$d\xi = \frac{\rho}{v_k M_k}\,d\chi_k$$

(ρ = density, v_k = stoichiometric number, M_k = molar mass, χ_k = mass fraction of species k). Accordingly, there follows from Eq. (4–4.20) with $J_k^* = 0$ (no diffusion flows),

$$\frac{d\xi}{dt} = b.$$

Here the operator d/dt denotes the total (substantial) derivative with respect to the time (cf. Eq. 4–4.5). If we introduce the relaxation time τ of the reaction (for an adiabatic–isochoric course) with Eq. (2–9.19) then we find from Eq. (18) and the last relation:†

---

† We see that Eq. (2–9.19) can be taken from homogeneous systems and applied to the present case in the following way: the affinity A of the chemical reaction in any volume element is a function of the chosen independent variables:

$$A = A(\tilde{S},\ \tilde{V}\ \xi).$$

Thus there is valid for adiabatic–isochoric state changes not far from equilibrium (expansion about the equilibrium state $A = 0$, $\xi = \bar{\xi}$ for $\tilde{S}$ = const, $\tilde{V}$ = const):

$$A = \left(\frac{\partial A}{\partial \xi}\right)_{S,\tilde{V}}(\xi - \bar{\xi}).$$

$$\frac{d\xi}{dt} = \lambda \operatorname{div} \mathbf{v} - \frac{1}{\tau}\left(\frac{\partial \xi}{\partial A}\right)_{\tilde{S},\tilde{V}} A. \qquad (4\text{-}32.19)$$

Furthermore, near equilibrium we can set:

$$\frac{d\xi}{dt} \approx \frac{d\tilde{\xi}}{dt}, \qquad (4\text{-}32.20)$$

where $\tilde{\xi}$ denotes the equilibrium value of the extent of reaction ξ. Since $\tilde{\xi}$ depends only on $\tilde{S}$ and $\tilde{V}$ we furthermore write:

$$\frac{d\tilde{\xi}}{dt} = \left(\frac{\partial \tilde{\xi}}{\partial \tilde{S}}\right)_{\tilde{V}} \frac{d\tilde{S}}{dt} + \left(\frac{\partial \tilde{\xi}}{\partial \tilde{V}}\right)_{\tilde{S}} \frac{d\tilde{V}}{dt}. \qquad (4\text{-}32.21)$$

With $1/\tilde{V} = \rho$ there follows from Eq. (4–4.8):

$$\frac{d\tilde{V}}{dt} = \tilde{V} \operatorname{div} \mathbf{v}. \qquad (4\text{-}32.22)$$

On the basis of Eqs. (4–9.7) and (4–9.8), and on neglecting the diffusion fluxes and heat flux as previously assumed, and on noting the fact that the local entropy production and thus the dissipation function contains only terms of the second order in div v and A (cf. Eqs. 5, 17, and 18), we can, to the approximation considered here, set:

$$\frac{d\tilde{S}}{dt} \approx 0. \qquad (4\text{-}32.23)$$

From Eqs. (19) to (23) there results:

$$A = \tau\left(\frac{\partial A}{\partial \xi}\right)_{\tilde{S},\tilde{V}}\left[\lambda - \tilde{V}\left(\frac{\partial \tilde{\xi}}{\partial \tilde{V}}\right)_{\tilde{S}}\right] \operatorname{div} \mathbf{v}. \qquad (4\text{-}32.24)$$

Here the affinity A is related to the divergence of the barycentric velocity $\mathbf{v}$.

If no velocity gradients are present and if we consequently consider the chemical reaction as the only irreversible process there then follows from Eq. (18) with div v = 0

$$b = \frac{d\xi}{dt} = aA = a\left(\frac{\partial A}{\partial \xi}\right)_{\tilde{S},\tilde{V}}(\xi - \tilde{\xi})$$

or with $a = \text{const}$ (cf. §2–9):

$$\xi - \tilde{\xi} = Ce^{-t/\tau},$$

where

$$\tau \equiv -\frac{1}{a}\left(\frac{\partial \xi}{\partial A}\right)_{\tilde{S},\tilde{V}}$$

is the relaxation time and C an integration constant.

On inserting Eq. (24) into Eq. (17) we obtain finally for the viscous pressures:

$$P_{ij} = P_{ji} = -\eta\left(\frac{\partial v_i}{\partial z_j} + \frac{\partial v_j}{\partial z_i}\right) + (\tfrac{2}{3}\eta - \zeta - \zeta')\,\delta_{ij}\,\mathrm{div}\,\mathbf{v} \quad (4\text{–}32.25)$$

with

$$\zeta' \equiv \tau\left(\frac{\partial A}{\partial \xi}\right)_{s,\bar{v}} \lambda\left[\bar{V}\left(\frac{\partial \xi}{\partial \bar{V}}\right)_s - \lambda\right]. \quad (4\text{–}32.26)$$

The chemical reaction thus formally produces a contribution to the bulk viscosity. If we set $\zeta = 0$ then, due to Eqs. (25) and (26), we would have a bulk viscosity which experimentally is not distinguishable from the "normal case."

This result[199] suggests that *any* measured bulk viscosity is due to a delayed equilibrium among the internal transformations (thermal relaxations, etc.), that the bulk viscosity thus represents only the phenomenological substitute for the explicit consideration of molecular relaxation processes. This concept[199] is identical to most kinetic interpretations (cf. above).

4–33 ROTATING SYSTEMS

Let there be a system in a centrifugal field. Let the angular velocity be denoted by $\boldsymbol{\Omega}$. Then on species k at a distance $\mathbf{r}$ from the axis of rotation there acts the molar force (sum of molar centrifugal force and of molar Coriolis force)

$$\mathbf{K}_k = M_k(\Omega^2 \mathbf{r} + 2\mathbf{v}_k \times \boldsymbol{\Omega}), \quad (4\text{–}33.1)$$

where M_k denotes the molar mass, $\mathbf{v}_k$ the average velocity of the particle type k (relative to the rotating system), and $\mathbf{v}_k \times \boldsymbol{\Omega}$ the vector product of the vectors $\mathbf{v}_k$ and $\boldsymbol{\Omega}$. If we exclude gravitational and electromagnetic fields then Eq. (1) represents the general expression for the molar force with reference to species k.

As is obvious from our statements in §1–21, the Coriolis force plays no role in the formulation of the equilibrium conditions since it does not produce a contribution to the potential of the external force fields. We now wish to investigate whether the Coriolis force should be considered in the course of irreversible processes in rotating systems (sedimentation, etc.). (In our treatment in §4–14, §4–15, and §4–20, we had explicitly or implicitly excluded transverse effects.)

If, for the sake of simplicity, we assume a fluid medium in creeping motion then the force law is given by Eq. (4–5.13):

$$\sum_k c_k \mathbf{K}_k = \mathrm{grad}\,P. \quad (4\text{–}33.2)$$

Here c_k is the molarity of species k and P the pressure.

Obviously, $\mathbf{K}_k$ in Eqs. (1) and (2) with the conditions assumed here has the same meaning. Accordingly, there results:

$$\sum_k M_k c_k (\Omega^2 \mathbf{r} + 2\mathbf{v}_k \times \mathbf{\Omega}) = \operatorname{grad} P. \tag{4–33.3}$$

Now, due to Eqs. (4–3.5), (4–3.6), and (4–3.12), there is valid:

$$\sum_k M_k c_k = \rho, \qquad \sum_k M_k c_k \mathbf{v}_k = \rho \mathbf{v}, \tag{4–33.4}$$

where ρ is the density and $\mathbf{v}$ the barycentric velocity. With this and Eq. (3) we find:

$$\rho(\Omega^2 \mathbf{r} + 2\mathbf{v} \times \mathbf{\Omega}) = \operatorname{grad} P. \tag{4–33.5}$$

This is the explicit formulation of Eq. (2) for the present case.

In the energy balance (4–6.4), the Coriolis forces drop out because they do not contribute to the potential energy in the external force fields (cf. above). This can be formally confirmed by substitution of Eq. (1).

It is seen from Eqs. (4–3.5), (4–3.11), (4–3.12), (4–3.13), (4–12.4), and Eq. (1) that the dissipation function in the formulation (4–12.2) contains no term which relates explicitly to the centrifugal field. This is because the barycentric system is used in Eq. (4–12.2). If another reference system is used then centrifugal and Coriolis forces arise in the formula for the dissipation function. This is seen, for example, in that the transition from Eq. (4–8.15) to Eq. (4–10.4) requires Eq. (4–10.2) which again relates back to the force law (2). In the following, we treat these formulations somewhat more precisely.

Again we assume fluid media in creeping motion as in Eq. (2). Furthermore, we define the diffusion current density $\mathbf{J}_k$ of species k to be:

$$\mathbf{J}_k \equiv c_k(\mathbf{v}_k - \mathbf{\omega}). \tag{4–33.6}$$

Here $\mathbf{\omega}$ is an arbitrary reference velocity, so that $\mathbf{J}_k$ represents the vector of the diffusion flux density of the particle type k for an arbitrary reference system. Furthermore, we exclude chemical reactions. The dissipation function Ψ can then be written in the following form which results from Eqs. (4–12.3), (4–12.4), and (4–12.14):

$$\Psi = -\mathbf{J}_Q \frac{\operatorname{grad} T}{T} + \sum_k \mathbf{J}_k [\mathbf{K}_k - (\operatorname{grad} \mu_k)_T]. \tag{4–33.7}$$

Here $\mathbf{J}_Q$ denotes the heat current density, T the absolute temperature, and μ_k the chemical potential of species k. The summation is over all particle types present.

From Eqs. (1), (4), and (6) there follows:

$$\sum_k \mathbf{J}_k \mathbf{K}_k = \sum_k M_k \mathbf{J}_k \Omega^2 \mathbf{r} + 2 \sum_k M_k \mathbf{J}_k (\mathbf{v}_k \times \mathbf{\Omega}),$$

$$\sum_k M_k \mathbf{J}_k (\mathbf{v}_k \times \mathbf{\Omega}) = \sum_k M_k c_k (\mathbf{v}_k - \mathbf{\omega})(\mathbf{v}_k \times \mathbf{\Omega})$$

$$= -\rho \mathbf{\omega}(\mathbf{v} \times \mathbf{\Omega}) = \rho(\mathbf{v} - \mathbf{\omega})(\mathbf{v} \times \mathbf{\Omega})$$

$$= \sum_k M_k \mathbf{J}_k (\mathbf{v} \times \mathbf{\Omega}),$$

that is,

$$\sum_k \mathbf{J}_k \mathbf{K}_k = \sum_k M_k \mathbf{J}_k (\Omega^2 \mathbf{r} + 2\mathbf{v} \times \mathbf{\Omega}). \qquad (4\text{–}33.8)$$

Furthermore, according to Eq. (4–15.2),

$$(\text{grad } \mu_k)_T = V_k \text{ grad } P + (\text{grad } \mu_k)_{T,P}. \qquad (4\text{–}33.9)$$

Here V_k is the partial molar volume of species k. From Eqs. (7) to (9) with Eq. (5) there results

$$\Psi = -\mathbf{J}_Q \frac{\text{grad } T}{T}$$

$$+ \sum_k \mathbf{J}_k [(M_k - \rho V_k)(\Omega^2 \mathbf{r} + 2\mathbf{v} \times \mathbf{\Omega}) - (\text{grad } \mu_k)_{T,P}]. \qquad (4\text{–}33.10)$$

This is the general expression[34] for the dissipation function in a rotating fluid system with creeping motion excluding chemical reactions and all force fields except the centrifugal field. (Since we also exclude electric fields, the formula holds only for uncharged species. The extension of the relation to systems with charged particle types does not present any difficulties.)

In Eq. (10), the "fluxes" $\mathbf{J}_k$ and the "forces"

$$\mathbf{X}_k \equiv (M_k - \rho V_k)(\Omega^2 \mathbf{r} + 2\mathbf{v} \times \mathbf{\Omega}) - (\text{grad } \mu_k)_{T,P} \qquad (4\text{–}33.11)$$

are not all independent; according to Eqs. (4–3.4), (4–20.17), (4–20.18), and Eq. (4) the identities

$$\sum_k \frac{\omega_k}{c_k} \mathbf{J}_k = 0, \qquad (4\text{–}33.12)$$

$$\sum_k c_k \mathbf{X}_k = 0 \qquad (4\text{–}33.13)$$

are valid. Here the ω_k represent the weight factors for averaging the velocities. If we introduce the diffusion current density in the Hittorf reference system (4–3.30)

$$_1\mathbf{J}_k \equiv c_k (\mathbf{v}_k - \mathbf{v}_1) \qquad (4\text{–}33.14)$$

in place of J_k then we obtain from Eqs. (10) and (11), by analogy with Eq. (4–12.17),

$$\Psi = -\mathbf{J}_Q \frac{\operatorname{grad} T}{T} + \sum_{k=2}^{N} {}_1\mathbf{J}_k\mathbf{X}_k. \qquad (4\text{–}33.15)$$

Here N denotes the number of species. We recognize that in Eq. (15) only independent "fluxes" and "forces" appear.

We further restrict the discussion to isothermal systems ($\operatorname{grad} T = 0$), and from Eq. (15) find for the dissipation function:

$$\Psi = \sum_{k=2}^{N} {}_1\mathbf{J}_k\mathbf{X}_k. \qquad (4\text{–}33.16)$$

We must proceed from this expression if we wish to set up the phenomenological equations for diffusion and sedimentation in their most general form.

In experiments with centrifuges or ultracentrifuges, the condition

$$|\mathbf{v} \times \boldsymbol{\Omega}| \ll |\Omega^2\mathbf{r}| \qquad (4\text{–}33.17)$$

is always satisfied. Thus, in view of Eq. (11), Eq. (16) is reduced to the expression

$$\Psi = \sum_{k=2}^{N} {}_1\mathbf{J}_k[(M_k - \rho V_k)\Omega^2\mathbf{r} - (\operatorname{grad} \mu_k)_{T,P}]. \qquad (4\text{–}33.18)$$

Even if the Coriolis force no longer appears explicitly in this relation it is still effective indirectly: the motions of the individual particles are influenced by forces which are perpendicular to their direction of motion and to the axis of rotation and are proportional to the particle velocities. This determines preferential directions in the medium and a modification of the "microscopic reversibility." Accordingly, the components of the vectors in Eq. (18) must be treated separately and the reciprocity relations written in a special form. We defer temporarily the treatment of these complications (§4–35) and exclude here all "transverse effects." The phenomenological equations then assume the usual form (cf. Eq. 4–15.10):

$${}_1\mathbf{J}_i = \sum_{k=2}^{N} a_{ik}[(M_k - \rho V_k)\Omega^2\mathbf{r} - (\operatorname{grad} \mu_k)_{T,P}] \quad (i = 2, 3, \ldots, N),$$
$$(4\text{–}33.19)$$

where a_{ik} are the phenomenological coefficients.

With $\Omega = 0$ (no centrifugal field) we obtain from Eq. (19) the equations (4–17.47) used in §4–17 for diffusion in multi-component systems. With $(\operatorname{grad} \mu_k)_{T,P} = 0$ (no concentration gradients) we find:

$${}_1\mathbf{J}_i = \sum_{k=2}^{N} a_{ik}(M_k - \rho V_k)\Omega^2\mathbf{r} \quad (i = 2, 3, \ldots, N). \qquad (4\text{–}33.20)$$

These equations obviously describe sedimentation in multi-component systems.

Because of the relation (4–17.50)

$$_1\mathbf{J}_i = \sum_{k=2}^{N} \varepsilon_{ik}\ _w\mathbf{J}_k \quad (i = 2, 3, \ldots, N) \tag{4–33.21}$$

with

$$\varepsilon_{ik} \equiv \delta_{ik} + \frac{c_i V_k}{c_1 V_1}, \tag{4–33.22}$$

where δ_{ik} denotes the Kronecker delta ($\delta_{ik} = 1$ for $i = k$, $\delta_{ik} = 0$ for $i \neq k$) and $_w\mathbf{J}_k$ denotes the diffusion current density of species k in the Fick reference system, we can formulate Eqs. (4–20.2)

$$_w\mathbf{J}_k = c_k s_k \Omega^2 \mathbf{r} \quad (k = 2, 3, \ldots, N), \tag{4–33.23}$$

with which the sedimentation coefficients s_k are introduced as follows:

$$_1\mathbf{J}_i = \sum_{k=2}^{N} c_k \varepsilon_{ik} s_k \Omega^2 \mathbf{r} \quad (i = 2, 3, \ldots, N). \tag{4–33.24}$$

From this we recognize that the phenomenological equations (20) are in agreement with the empirical equations (23) for sedimentation.

Comparison of coefficients in Eqs. (20) and (24) leads to the formula

$$\sum_{k=2}^{N} a_{ik}(M_k - \rho V_k) = \sum_{k=2}^{N} c_k \varepsilon_{ik} s_k \quad (i = 2, 3, \ldots, N), \tag{4–33.25}$$

that is, to a relation between the phenomenological coefficients and the sedimentation coefficients.

The Onsager reciprocity law (4–24.12)

$$a_{ik} = a_{ki} \quad (i, k = 2, 3, \ldots, N) \tag{4–33.26}$$

with Eq. (25) produces a relation between the sedimentation coefficients of a multi-component system. This equation, whose explicit formulation we will not discuss because of the lack of appropriate experimental data, also results from Eqs. (4–17.60) and (4–20.25).

4–34 MATTER IN AN ELECTROMAGNETIC FIELD

(a) Definitions and Basic Laws

Consider an electromagnetic field in a vacuum or in a space filled with matter. The medium here can be isotropic or anisotropic.

In the formulation of the fundamental laws, we make use of a "rational" system of definitions.[41] In this system, the fundamental differential equations of electrodynamics do not contain the factor 4π, while in the integrated relations for the spherically symmetric case (e.g. point charges or charged spheres) the number 4π often appears. In the "irrational" system, which we do not use here, the situation is reversed.†

The *electric field strength* $\mathfrak{E}$ is determined at each point in an electromagnetic field according to its magnitude and direction such that the force acting on a small stationary test charge q placed at this point is $q\mathfrak{E}$. Here q is positive or negative according to the sign of the charge.

The *magnetic induction* $\mathfrak{B}$ is defined by the following statement: the force acting on a small test element of length $d\mathbf{s}$ of a linear conductor which is brought into a magnetic field and through which an electric current i flows is given by the vector product $i\,d\mathbf{s} \times \mathfrak{B}$. The magnetic vector $\mathfrak{B}$ thus corresponds to the electric vector $\mathfrak{E}$.

If a charge q moves with a velocity $\mathbf{v}$ relative to an electromagnetic field then the product $q\mathbf{v}$ represents the analog to the "current element" $i\,d\mathbf{s}$. According to the above, the force $\mathfrak{K}$ on the moving charge is:

$$\mathfrak{K} = q(\mathfrak{E} + \mathbf{v} \times \mathfrak{B}). \qquad (4\text{–}34.1)$$

If the charges are continuously distributed in a medium then a charge number z_k (positive or negative integer for positive or negative particles) is assigned to each charged species k. Let F be the Faraday constant. Then $z_k F$ denotes the molar charge of species k. If the particles of type k have the average velocity $\mathbf{v}_k$ (relative to the electromagnetic field) then, from Eq. (1), the molar force acting on species k and denoted by $\mathbf{L}_k$ becomes:

$$\mathbf{L}_k = z_k F(\mathfrak{E} + \mathbf{v}_k \times \mathfrak{B}). \qquad (4\text{–}34.2)$$

Equation (1) or (2) is the expression for the "Lorentz force." (Sometimes, as previously in our text, only the second term on the right-hand side of Eq. (1) or (2), relating to the magnetic field, is denoted as the "Lorentz force.") For matter which can be polarized, there are still further macroscopic forces, which will be treated later (p. 428). Equation (1) or (2) refers to the force acting on the "true" charges.‡

---

† The difference between the rational and the irrational systems has nothing to do with the choice of units but is a difference in the definition of certain basic quantities[200]. Arbitrary units can be used in both systems. We will later return to the question of dimensions and units.

‡ In the microphysical sense, i.e. when considering a single charged particle like an electron or an ion, no forces exist other than those given by Eq. (2), which is then to be divided by the Avogadro constant so that, in place of the Faraday constant, the elementary charge appears. Nevertheless, $\mathfrak{E}$ and $\mathfrak{B}$ are here the "effective" values, not the macroscopic values of the electric field strength and magnetic induction.

From the basic vectors $\mathfrak{E}$ and $\mathfrak{B}$ we can derive: the *electric potential* φ (a scalar quantity) and the *magnetic vector potential* $\mathfrak{A}$ (a vector). These are defined by the following equations:†

$$\mathfrak{E} = -\operatorname{grad} \varphi \quad \text{(electrostatic field)}, \tag{4–34.3}$$

$$\mathfrak{B} = \operatorname{curl} \mathfrak{A}. \tag{4–34.4}$$

Here Eq. (3) holds for an electrostatic field, Eq. (4) for an arbitrary electromagnetic field.

Let us consider a parallel plate capacitor small enough to serve as a test element. If this capacitor is brought into an electrostatic field, we can measure the induced charge for that orientation of the parallel plates in which the greatest induced charge arises, and this maximum charge divided by the area of the capacitor is said to be the *electric displacement* $\mathfrak{D}$. The direction of this vector is from the negative to the positive induced charge and is perpendicular to the plane of the capacitor.

Finally, we define a vector for the magnetic field analogous to $\mathfrak{D}$. To do this we introduce a small uniform cylindrical test coil (number of turns w) whose length l is large in comparison to its diameter, and we send an electric current through the coil and vary the current i and the orientation of the coil until the magnetic field is compensated for by the coil field. Then the expression iw/l defines the magnitude of the *magnetic field strength* $\mathfrak{H}$. The direction of this vector is from the north pole to the south pole of the coil.

In isotropic media, the directions of $\mathfrak{E}$ and $\mathfrak{D}$ or of $\mathfrak{B}$ and $\mathfrak{H}$ coincide. For anisotropic bodies, this is no longer the case.

The four vectors $\mathfrak{E}$, $\mathfrak{B}$, $\mathfrak{D}$, and $\mathfrak{H}$ are interrelated by the *Maxwell equations* (Maxwell, 1864):

$$\operatorname{curl} \mathfrak{E} + \dot{\mathfrak{B}} = 0, \tag{4–34.5}$$

$$\operatorname{curl} \mathfrak{H} - \dot{\mathfrak{D}} = \mathbf{I}. \tag{4–34.6}$$

Here a dot above a symbol denotes the derivative with respect to time at a fixed position. The vector $\mathbf{I}$ is, as previously, the electric current density.

---

† The quantity curl $\mathbf{a}$ (where $\mathbf{a}$ is any vector) represents a vector. If x, y, and z are space-coordinates, then we have for the cartesian components of curl $\mathbf{a}$:

$$(\operatorname{curl} \mathbf{a})_x = \frac{\partial a_z}{\partial y} - \frac{\partial a_y}{\partial z},$$

$$(\operatorname{curl} \mathbf{a})_y = \frac{\partial a_x}{\partial z} - \frac{\partial a_z}{\partial x},$$

$$(\operatorname{curl} \mathbf{a})_z = \frac{\partial a_y}{\partial x} - \frac{\partial a_x}{\partial y},$$

where a_x, a_y, and a_z denote the cartesian components of $\mathbf{a}$ and all differentiations are carried out at fixed time.

Besides Eqs. (5) and (6) we need the additional relations

$$\text{div } \mathfrak{D} = \rho_{\text{el}}, \tag{4–37.7}$$

$$\text{div } \mathfrak{B} = 0. \tag{4–34.8}$$

Here ρ_{el} denotes the charge density (space density of the electric charge). If c_k is the molarity of species k then we have:

$$\rho_{\text{el}} = \sum_k c_k z_k F. \tag{4–34.9}$$

Equation (8) justifies the introduction of the vector potential $\mathfrak{A}$ in Eq. (4), because div curl $\mathfrak{A} = 0$.

From Eqs. (3), (4), and (5) with the vector identity curl grad $\varphi = 0$ there results:

$$\mathfrak{E} = -\text{grad } \varphi - \dot{\mathfrak{A}}. \tag{4–34.10}$$

For magnetic fields which are constant in time or vanishing ($\dot{\mathfrak{A}} = 0$), Eq. (10) reduces to Eq. (3).

So far all relations are equally valid for vacuum and for any (isotropic or anisotropic) media. The "matter equations" now to be discussed are more restricted.

(b) Matter Equations

We begin with the relations for a *vacuum*. Here:

$$\mathfrak{D} = \varepsilon_0 \mathfrak{E}, \tag{4–34.11}$$

$$\mathfrak{B} = \mu_0 \mathfrak{H}, \tag{4–34.12}$$

where ε_0 and μ_0 are universal constants. The quantity ε_0 is called *permittivity of vacuum*, the quantity μ_0 *permeability of vacuum*. From the Maxwell equations (5), (6) and Eqs. (11) and (12), the speed of light c_0 *in vacuo* can be shown to be

$$c_0 = (\varepsilon_0 \mu_0)^{-1/2}. \tag{4–34.13}$$

For space filled with matter, Eqs. (11) and (12) do not apply. For an arbitrary (isotropic or anisotropic) medium, we define two new vectors $\mathfrak{P}$ and $\mathfrak{J}$:

$$\mathfrak{D} \equiv \varepsilon_0 \mathfrak{E} + \mathfrak{P}, \tag{4–34.14}$$

$$\mathfrak{B} \equiv \mu_0 \mathfrak{H} + \mathfrak{J}. \tag{4–34.15}$$

$\mathfrak{P}$ is called *electric polarization*, $\mathfrak{J}$ *magnetic polarization*. According to Eqs. (11) and (12), these two quantities vanish in a vacuum.

The quantity

$$\mathfrak{M} \equiv \frac{\mathfrak{B}}{\mu_0} - \mathfrak{H} = \frac{\mathfrak{J}}{\mu_0}$$

is often denoted as the "magnetization." Both $\mathfrak{J}V$ and $\mathfrak{M}V$ may be considered to be the "magnetic moment" of polarized matter of volume V.

If we restrict further discussion to *isotropic media* and exclude ferro-magnetic (and "ferroelectric") substances then we obtain the following relations in place of Eqs. (11) and (12)

$$\mathfrak{D} = \varepsilon_m \mathfrak{E}, \tag{4–34.16}$$

$$\mathfrak{B} = \mu_m \mathfrak{H}. \tag{4–34.17}$$

Here ε_m and μ_m are "material constants," i.e. coefficients dependent on the type and state (temperature, pressure, composition) of the medium. The quantity ε_m is called *permittivity*, the quantity μ_m *permeability*. The dimension-less expression

$$\varepsilon \equiv \frac{\varepsilon_m}{\varepsilon_0} \tag{4–34.18}$$

is designated as *dielectric constant* (or "relative permittivity"). The dimension-less quantity

$$\mu \equiv \frac{\mu_m}{\mu_0} \tag{4–34.19}$$

is called *magnetic constant* (or "relative permeability"). It is these quantities which are found in tables (besides the "magnetic susceptibility," cf. below).

From Eqs. (14) to (19) there follows:

$$\mathfrak{P} = \psi \mathfrak{E}, \tag{4–34.20}$$

$$\mathfrak{J} = \chi \mathfrak{H}, \tag{4–34.21}$$

where

$$\psi \equiv \varepsilon_0(\varepsilon - 1), \qquad \chi \equiv \mu_0(\mu - 1). \tag{4–34.22}$$

On historical grounds, the dimensionless quantity

$$\frac{\varepsilon - 1}{4\pi} = \frac{\psi}{4\pi\varepsilon_0}$$

is called the "electric susceptibility." Analogously, the dimensionless expression

$$\frac{\mu - 1}{4\pi} = \frac{\chi}{4\pi\mu_0}$$

is called the "magnetic susceptibility." This is often tabulated in place of μ.

We will apply Eqs. (16) and (17) or (20) and (21), because they are limited in their validity, at a single place (Eq. 83) without further use in the fundamental equations.

(c) Dimensions and Units

On the question of the *dimensions* we proceed in such a way that we introduce the four independent dimensions, length l, time t, energy E, and electric charge q. With this there results the following summary of dimensions for the most important electromagnetic quantities:[42]

Table 19

Symbol	Name	Dimension
q	Charge	q
i	Electric current	$t^{-1}q$
$\mathfrak{I}$	Electric current density	$l^{-2}t^{-1}q$
$\mathfrak{E}$	Electric field strength	$l^{-1}Eq^{-1}$
φ	Electric potential	Eq^{-1}
$\mathfrak{D}$	Electric displacement	$l^{-2}q$
$\mathfrak{P}$	Electric polarization	$l^{-2}q$
ϵ_0	Permittivity of vacuum	$l^{-1}E^{-1}q^2$
ϵ_m	Permittivity	$l^{-1}E^{-1}q^2$
$\mathfrak{B}$	Magnetic induction	$l^{-2}tEq^{-1}$
$\mathfrak{A}$	Magnetic vector potential	$l^{-1}tEq^{-1}$
$\mathfrak{H}$	Magnetic field strength	$l^{-1}t^{-1}q$
$\mathfrak{J}$	Magnetic polarization	$l^{-2}tEq^{-1}$
μ_0	Permeability of vacuum	$l^{-1}t^2Eq^{-2}$
μ_m	Permeability	$l^{-1}t^2Eq^{-2}$

The basic *units* are now:

length   1 cm or 1 m $\equiv 10^2$ cm
time   1 s
energy   1 erg or 1 J (joule) $\equiv 10^7$ erg

charge   1 Fr (franklin) or 1 C (coulomb) $\equiv \dfrac{b}{10}$ Fr

Here the unit Fr is determined by the equation[201]

$$\varepsilon_0 = \frac{1}{4\pi}\frac{\text{Fr}^2}{\text{erg cm}} \qquad (4\text{–}34.23)$$

and the numerical factor b by the equation

$$c_0 = b\,\frac{\text{cm}}{\text{s}} \qquad (4\text{–}34.24)$$

($b \approx 3 \times 10^{10}$).

As derived units we mention[28]:

1 dyn $\equiv$ 1 erg cm^{-1}
1 N (newton) $\equiv$ 1 J m^{-1} = 10^5 dyn
1 A (ampère) $\equiv$ 1 C s^{-1}
1 Bi (biot) $\equiv$ 10 A = b Fr s^{-1}
1 V (volt) $\equiv$ 1 J C^{-1}
1 Ω (ohm) $\equiv$ 1 V A^{-1}
1 F (farad) $\equiv$ 1 C V^{-1}

1 H (henry) $\equiv$ 1 J A^{-2}

1 Oe (oersted) $\equiv \dfrac{10^3}{4\pi}$ A m^{-1}

1 T (tesla) $\equiv$ 1 V s m^{-2}

1 G (gauss) $\equiv 10^{-4}$ T

From Eqs. (13), (23), and (24) the quantities ε_0 and μ_0 become

$$\varepsilon_0 = \frac{10^{11}}{4\pi\, b^2}\frac{C^2}{J\, m} = \frac{10^{11}}{4\pi\, b^2}\frac{A\, s}{V\, m} = \frac{10^{11}}{4\pi\, b^2}\frac{F}{m}, \qquad (4\text{-}34.25)$$

$$\mu_0 = \frac{4\pi}{b^2}\frac{erg\, s^2}{Fr^2\, cm}, \qquad (4\text{-}34.26)$$

$$\mu_0 = \frac{4\pi}{10^7}\frac{J\, s^2}{C^2\, m} = \frac{4\pi}{10^7}\frac{V\, s}{A\, m} = \frac{4\pi}{10^7}\frac{H}{m} = 1\,\frac{G}{Oe}. \qquad (4\text{-}34.27)$$

In the irrational system, the permittivity ε_0' and the permeability μ_0' of vacuum are given by the following relations:

$$\varepsilon_0' \equiv 4\pi\varepsilon_0, \qquad \mu_0' \equiv \frac{\mu_0}{4\pi}.$$

From this we derive by means of Eqs. (23) and (26):

$$\varepsilon_0' = 1\,\frac{Fr^2}{erg\, cm}, \qquad \mu_0' = 1\,\frac{erg}{Bi^2\, cm}.$$

Indeed, the two units of charge, Fr (franklin) and Bi s (biot $\times$ second), corresponding to the "absolute electrostatic unit of charge" and to the "absolute electromagnetic unit of charge" respectively, were chosen to give the numerical value unity to both ε_0' and μ_0'.

The unit "faraday" referred to in §4–16 and §4–19 has not been accepted universally, but it is much used in electrochemistry. This unit of charge can be defined as follows:

$$1\ \text{faraday} \equiv n^\dagger F \approx 96500\ C$$

where

$$n^\dagger \equiv 1\ \text{mol}.$$

(d) Mass and Charge Balance

Let the average velocity of a particle of type k in any volume element be $\mathbf{v}_k$. If $\mathbf{v}$ is the barycentric velocity of the volume element then the vector

$$_v\mathbf{J}_k \equiv c_k(\mathbf{v}_k - \mathbf{v}) \qquad (4\text{-}34.28)$$

represents the diffusion current density of species k in the barycentric system (see Eq. 4-3.11).

The electric current density $\mathbf{I}$ which appears in Eq. (6) is given by the expression

$$\mathbf{I} = F \sum_k z_k c_k \mathbf{v}_k. \tag{4–34.29}$$

According to Eqs. (9) and (28), the vector $\mathbf{I}$ in Eq. (29) can be split into two parts:

$$\mathbf{I} = \mathbf{I}_v + \mathbf{i} \tag{4–34.30}$$

with

$$\mathbf{I}_v \equiv F \sum_k z_k c_k \mathbf{v} = \rho_{\mathrm{el}} \mathbf{v}, \tag{4–34.31}$$

$$\mathbf{i} \equiv F \sum_k z_k \,_v \mathbf{J}_k. \tag{4–34.32}$$

Here $\mathbf{I}_v$ can be regarded as the density of the "convective electric flux" and $\mathbf{i}$ as the density of the "conductive electric flux." For an insulator, there is valid for any charged species k ($z_k \neq 0$): $_v\mathbf{J}_k = 0$, that is, $\mathbf{i} = 0$, $\mathbf{I} = \mathbf{I}_v$, while for an electric conductor the electroneutrality condition (condition for the disappearance of the space charge)

$$\rho_{\mathrm{el}} = F \sum_k z_k c_k = 0 \quad \text{(conductor)} \tag{4–34.33}$$

must be satisfied, so that $\mathbf{I}_v = 0$, $\mathbf{I} = \mathbf{i}$. In the last case, $\mathbf{i}$ is independent of the reference system (see p. 224). According to Eqs. (4–4.1) and (4–4.12), the local mass balance has the following form:

$$\frac{\partial c_k}{\partial t} = -\operatorname{div}(c_k \mathbf{v}_k) + \sum_r \nu_{kr} b_r = -\operatorname{div}(c_k \mathbf{v})$$

$$-\operatorname{div} \,_v\mathbf{J}_k + \sum_r \nu_{kr} b_r. \tag{4–34.34}$$

Here the operator $\partial/\partial t$ denotes the derivative with respect to time t at a fixed position, ν_{kr} the stoichiometric number of species k in the chemical reaction r, and b_r the reaction rate of reaction r. The summations are over all reactions.

The law of conservation of charge, applied to chemical reactions, requires:

$$\sum_k z_k \nu_{kr} = 0, \tag{4–34.35}$$

where the summation is over all reacting species and the equation holds for each individual reaction.

If we multiply Eq. (34) by $z_k F$, sum over all species, and take note of Eqs. (9), (29), and (35), we find:

$$\frac{\partial \rho_{el}}{\partial t} = -\operatorname{div} \mathbf{I}. \qquad (4\text{-}34.36)$$

For conductors, according to Eqs. (30) to (33) from Eq. (36), there follows:

$$\operatorname{div} \mathbf{I} = \operatorname{div} \mathbf{i} = 0 \quad \text{(conductor)}. \qquad (4\text{-}34.37)$$

Equation (36) is the analog of the local balance (4-4.4) for the total mass. Equation (37) corresponds to condition (4-32.10) for incompressible liquids.

In the following, we consider the most general case as was done for the Maxwell equations: the medium, which is in an electromagnetic field, contains both space charges and movable charged species and can be polarized. It thus represents a transition between a "conductor" and an "insulator" ("dielectric").

(e) Momentum Balance

The momentum balance in the form (4-5.1) and the relation (4-5.8) following from it can be applied to any isotropic medium without viscous pressures. We thus write ($v^2 \equiv \mathbf{v} \cdot \mathbf{v}$):

$$\rho \frac{d\mathbf{v}}{dt} = \mathfrak{K}_V - \operatorname{grad} P, \qquad (4\text{-}34.38)$$

$$\frac{\partial}{\partial t}\left(\frac{\rho}{2} v^2\right) + \operatorname{div}\left(\frac{\rho}{2} v^2 \mathbf{v}\right) = \mathbf{v}(\mathfrak{K}_V - \operatorname{grad} P). \qquad (4\text{-}34.39)$$

Here ρ denotes the density, $d\mathbf{v}/dt$ the barycentric acceleration, $\mathfrak{K}_V$ the resultant of the force densities of the external forces, and P the pressure.

We allow no other external fields besides the electromagnetic field. Then $\mathfrak{K}_V$ is the resultant of the force densities of the "ponderomotive forces." We thus have:

$$\mathfrak{K}_V = \mathfrak{K}_V' + \mathfrak{K}_V'' + \mathfrak{K}_V'''. \qquad (4\text{-}34.40)$$

In this,

$$\mathfrak{K}_V' = \sum_k c_k \mathbf{L}_k$$

is the Lorentz force per unit volume, which can be represented in the following form owing to Eqs. (2), (9), and (29):

$$\mathfrak{K}_V' = \rho_{el}\mathfrak{E} + \mathbf{I} \times \mathfrak{B}. \qquad (4\text{-}34.41)$$

The vector $\mathfrak{K}_V''$ is the force per unit volume which arises from the electric polarization and also appears for an uncharged insulator. It can be formulated in various ways. The formulas in the literature ("Helmholtz force,"

"Kelvin force," etc.), which apparently deviate from one another, arise because the pressure P in Eq. (38) is of different significance in the individual cases. As can be shown[202], the formulation as the "Kelvin force" is the simplest from the macrophysical standpoint:

$$\mathfrak{K}_V'' = (\text{Grad } \mathfrak{E})\mathfrak{P}. \tag{4-34.42}$$

Here Grad $\mathfrak{E}$ denotes the vector gradient of $\mathfrak{E}$. The cartesian components of the vector (Grad **b**) **a** are

$$\sum_i a_i \frac{\partial b_i}{\partial z_k} \quad (i, k = 1, 2, 3),$$

where z_1, z_2, z_3 are the space-coordinates and the operator $\partial/\partial z_k$ describes the derivatives with respect to position at a fixed time. Equation (42) is analogous to the well-known formula for the force on a dipole in an inhomogeneous electric field.† The quantity P in Eq. (38) now means the pressure we would measure with an electrically inert manometer in a direction perpendicular to the electric field.

Finally, the vector $\mathfrak{K}_V'''$ is the force per unit volume insofar as it arises from the magnetic polarization. Since in the equations for the electrification and magnetization work the vectors $\mathfrak{E}$ and $\mathfrak{P}$ correspond to the vectors $\mathfrak{H}$ and $\mathfrak{I}$ (p. 5), we write by analogy with Eq. (42):

$$\mathfrak{K}_V''' = (\text{Grad } \mathfrak{H})\mathfrak{I}. \tag{4-34.43}$$

According to Eqs. (1), (41), and (43), the force resulting from a magnetic field is due to the magnetic induction $\mathfrak{B}$ or to the magnetic field strength $\mathfrak{H}$ if we are dealing with a moving charge or with a magnetized body respectively.

We obtain from Eqs. (40) to (43) the resulting force density:

$$\mathfrak{K}_V = \rho_{\text{el}}\mathfrak{E} + \mathbf{I} \times \mathfrak{B} + (\text{Grad } \mathfrak{E})\mathfrak{P} + (\text{Grad } \mathfrak{H})\mathfrak{I}. \tag{4-34.44}$$

If we introduce Eq. (44) into Eq. (38) then we have the formulation for the momentum balance for a volume element in an isotropic medium in the electromagnetic field excluding viscous flow and gravitational and centrifugal fields.

---

† From the molecular point of view electric polarization is due to partial orientation of molecules in the external electric field, provided there are natural dipoles, and to the induction of dipoles in the molecules by the field. The force acting on these molecular dipoles can then be calculated according to Eq. (2) (cf. footnote to p. 421), where $\mathfrak{E}$ means the effective electric field strength of the resulting inhomogeneous field in the environment of the molecule considered. Macroscopically, Eq. (42) results, since the "internal" field strength can naturally not be present in a macrophysical expression.

(f) Energy Balance

The Maxwell equations (5) and (6), on multiplication of the first and the second equation by $\mathfrak{H}$ and $\mathfrak{E}$ respectively, and then subtracting and using the identity

$$\mathfrak{H} \operatorname{curl} \mathfrak{E} - \mathfrak{E} \operatorname{curl} \mathfrak{H} = \operatorname{div} (\mathfrak{E} \times \mathfrak{H}) \qquad (4\text{-}34.45)$$

produce the following relation:

$$\mathfrak{E} \dot{\mathfrak{D}} + \mathfrak{H} \dot{\mathfrak{B}} + \operatorname{div} (\mathfrak{E} \times \mathfrak{H}) = - \mathfrak{E} \mathbf{I}.$$

The vector product $\mathfrak{E} \times \mathfrak{H}$ is called "Poynting's vector." On noting Eqs. (14) and (15) there results from this:

$$\frac{\partial}{\partial t} \left(\frac{\varepsilon_0}{2} \mathfrak{E}^2 + \frac{\mu_0}{2} \mathfrak{H}^2 \right) = -\operatorname{div} (\mathfrak{E} \times \mathfrak{H}) - \mathfrak{E} \mathbf{I} - \mathfrak{E} \frac{\partial \mathfrak{P}}{\partial t} - \mathfrak{H} \frac{\partial \mathfrak{J}}{\partial t}. \qquad (4\text{-}34.46)$$

Now it is suggested that we denote the quantity

$$E_V^* \equiv \frac{\varepsilon_0}{2} \mathfrak{E}^2 + \frac{\mu_0}{2} \mathfrak{H}^2 \qquad (4\text{-}34.47)$$

as the density of the "electromagnetic energy" on the basis of its dimension and in view of the form of Eq. (46) which is that of a balance equation.† Correspondingly, the vector $\mathfrak{E} \times \mathfrak{H}$ may be interpreted as the "electromagnetic energy current density." After integration over volume and time Eq. (46) may be interpreted thus: the increase in the electromagnetic energy of the system localized in the field is equal to the electromagnetic energy flowing through the boundary surfaces of the volume considered in the system (described by the corresponding surface integral, which arises from the term with the Poynting vector), less the work done by the field, namely the electrical work on regions which have electric currents flowing through them, plus the (reversible) electrification and magnetization work on regions which can be polarized (cf. Eq. 1–3.5). This statement is to be expanded further since a complete "energy balance" must also contain terms like the "kinetic energy," "mechanical work," "internal energy," and "heat."

Consider a first extension resulting from the momentum balance (38). From Eqs. (39) and (44) there follows for isotropic systems without gravitational and centrifugal fields and excluding viscous flow:

$$\frac{\partial}{\partial t} \left(\frac{\rho}{2} v^2 \right) = -\operatorname{div} \left(\frac{\rho}{2} v^2 \mathbf{v} \right) - \mathbf{v} \operatorname{grad} P$$

$$+ \mathbf{v}[\rho_{\mathrm{el}} \mathfrak{E} + \mathfrak{J} \times \mathfrak{B} + (\operatorname{Grad} \mathfrak{E})\mathfrak{P} + (\operatorname{Grad} \mathfrak{H})\mathfrak{J}]. \qquad (4\text{-}34.48)$$

---

† The Maxwell definition of the density of the electromagnetic energy

$$(E_V^*)_{\text{Maxwell}} \equiv \tfrac{1}{2}(\mathfrak{E}\mathfrak{D} + \mathfrak{H}\mathfrak{B})$$

leads to complications as soon as permittivity and permeability are not constant in time.

Since $\rho/2v^2$ represents the density of the macroscopic kinetic energy of the volume element considered (referred to the centre of gravity), Eq. (48) is to be regarded as a local balance of the kinetic energy.

Using Eqs. (30) and (31) as well as the abbreviation

$$L \equiv [\mathbf{i} \times \mathfrak{B} + (\text{Grad } \mathfrak{E})\mathfrak{P} + (\text{Grad } \mathfrak{H})\mathfrak{J}]\mathbf{v} \qquad (4\text{-}34.49)$$

we derive from the addition of Eq. (46) to Eq. (48) and noting Eq. (47):

$$\frac{\partial}{\partial t}\left(E_V^* + \frac{\rho}{2}v^2\right) = -\text{div}\left(\mathfrak{E} \times \mathfrak{H} + \frac{\rho}{2}v^2\mathbf{v}\right) - \mathbf{v}\,\text{grad}\,P$$

$$-\mathbf{i}\mathfrak{E} - \mathfrak{E}\frac{\partial \mathfrak{P}}{\partial t} - \mathfrak{H}\frac{\partial \mathfrak{J}}{\partial t} + L. \qquad (4\text{-}34.50)$$

This relation is the local balance for the sum of electromagnetic and kinetic energy for isotropic systems without gravitational and centrifugal fields excluding viscous flow.

Next, for the internal energy, we formulate a general local balance, following from Eq. (4-2.4):

$$\frac{\partial U_V}{\partial t} = -\text{div}\,\mathbf{J}_U + q(U). \qquad (4\text{-}34.51)$$

Here U_V denotes the density of the internal energy of the region, the vector $\mathbf{J}_U$ the current density of the internal energy, and $q(U)$ the local production of the internal energy.

We interpret the quantity

$$E_V \equiv E_V^* + \frac{\rho}{2}v^2 + U_V$$

as the density of the total energy. With Eq. (50) and (51) we will thus write:

$$\frac{\partial E_V}{\partial t} = \frac{\partial}{\partial t}\left(E_V^* + \frac{\rho}{2}v^2 + U_V\right)$$

$$= -\text{div}\left(\mathfrak{E} \times \mathfrak{H} + \frac{\rho}{2}v^2\mathbf{v} + \mathbf{J}_U\right)$$

$$- \mathbf{v}\,\text{grad}\,P - \mathbf{i}\mathfrak{E} - \mathfrak{E}\frac{\partial \mathfrak{P}}{\partial t} - \mathfrak{H}\frac{\partial \mathfrak{J}}{\partial t} + L + q(U). \qquad (4\text{-}34.52)$$

This balance equation should be valid for isotropic systems without viscous flow, gravitational or centrifugal fields.

In order that the above interpretation be meaningful, we must write a conservation law for the total energy. Equation (52) must therefore have the form

$$\frac{\partial E_V}{\partial t} = -\text{div}\,\mathbf{J}_E, \qquad (4\text{-}34.53)$$

where the vector $\mathbf{J}_E$ denotes the flux density of the total energy. From Eqs. (52) and (53) there follows immediately:

$$\mathbf{J}_E = \mathfrak{E} \times \mathfrak{H} + \frac{\rho}{2} v^2 \mathbf{v} + \mathbf{J}_U, \tag{4-34.54}$$

$$q(U) = \mathbf{v} \operatorname{grad} P + i\mathfrak{E} + \mathfrak{E} \frac{\partial \mathfrak{P}}{\partial t} + \mathfrak{H} \frac{\partial \mathfrak{J}}{\partial t} - L. \tag{4-34.55}$$

The expression for the vector $\mathbf{J}_U$ is, within certain limits, arbitrary, since the definition of the "heat flow" for systems in electromagnetic fields is not fixed. If, however, we desire a connection with our previous conventions, we must require that, for systems in stationary electromagnetic fields without electric and magnetic polarization, Eq. (52) be identical to Eq. (4-6.4). We eliminate the uncertainty which remains by a transformation so that Eq. (4-6.10) holds for systems in arbitrary electromagnetic fields. For the flux density of the internal energy in Eq. (4-6.10), we use Eqs. (4-6.11) and (4-6.13) to write

$$\mathbf{J}_U = H_V \mathbf{v} + \mathbf{J}_Q + \sum_k H_k \, {}_v\mathbf{J}_k, \tag{4-34.56}$$

where H_V denotes the enthalpy density ($= U_V + P$), H_k the partial molar enthalpy of species k, and $\mathbf{J}_Q$ the heat current density. We also retain Eq. (56), but we must now generalize the term "enthalpy" and, with Eq. (1-12.26), set the enthalpy density equal to:

$$H_V = U_V + P - \mathfrak{E}\mathfrak{P} - \mathfrak{H}\mathfrak{J}. \tag{4-34.57}$$

Thus the "heat flow" is unambiguously determined by Eq. (56) for systems in arbitrary electromagnetic fields.

From Eqs. (52), (55), (56), and (57) there results the final form of the local balance of the total energy:

$$\frac{\partial E_V}{\partial t} = \frac{\partial}{\partial t} \left(E_V^* + \frac{\rho}{2} v^2 + U_V \right)$$

$$= -\operatorname{div} \left(\mathfrak{E} \times \mathfrak{H} + \frac{\rho}{2} v^2 \mathbf{v} + U_V \mathbf{v} + P\mathbf{v} - (\mathfrak{E}\mathfrak{P})\mathbf{v} \right.$$

$$\left. - (\mathfrak{H}\mathfrak{J})\mathbf{v} + \mathbf{J}_Q + \sum_k H_k \, {}_v\mathbf{J}_k \right). \tag{4-34.58}$$

We can now see that Eq. (58) under certain conditions is reduced to Eq. (4-6.4).

For a system in a stationary electromagnetic field without electric and magnetic polarization, we have:

$$\dot{\mathfrak{P}} = 0, \quad \dot{\mathfrak{D}} = 0, \quad \mathfrak{P} = 0, \quad \mathfrak{J} = 0, \tag{4-34.59}$$

that is, on account of Eqs. (5), (6), (14), (15), (45), and (47):

$$\text{curl } \mathfrak{E} = 0, \quad \text{curl } \mathfrak{H} = \mathbf{I}, \quad -\mathfrak{E}\mathbf{I} = \text{div} (\mathfrak{E} \times \mathfrak{H}), \quad \frac{\partial E_V^*}{\partial t} = 0. \quad (4\text{-}34.60)$$

From Eqs. (58) to (60) we find:

$$\frac{\partial}{\partial t}\left(\frac{\rho}{2} v^2 + U_V\right) = -\text{div}\left(\frac{\rho}{2} v^2 \mathbf{v} + U_V\mathbf{v} + P\mathbf{v} + \mathbf{J}_Q + \sum_k H_k \,_V\mathbf{J}_k\right) + \mathfrak{E}\mathbf{I}.$$
$$(4\text{-}34.61)$$

This relation is identical to Eq. (4–6.4) if we take account of the assumed absence of gravitational and centrifugal fields and accordingly insert the relation following from Eqs. (2) and (29) into Eq. (4–6.4):

$$\sum_k \mathbf{K}_k c_k \mathbf{v}_k = \sum_k \mathbf{L}_k c_k \mathbf{v}_k = \mathbf{I}\mathfrak{E}.$$

If Eq. (58) is compared with Eq. (61) then we recognize again that our definitions are meaningful: in Eq. (58) the density of the electromagnetic energy is added on the left-hand side to the density of the kinetic and internal energy, while the Poynting vector appears on the right-hand side in addition to the convective flux densities of the kinetic and internal energy. The negative divergence of this becomes the Joule term ($\mathfrak{E}\mathbf{I}$) as may be seen from Eq. (60) when Eq. (61) is valid. Moreover, the expression div ($\mathfrak{E}\mathfrak{P}$)$\mathbf{v}$ + ($\mathfrak{H}\mathfrak{J}$)$\mathbf{v}$ must be added on the right-hand side of Eq. (58) to the term $-\text{div} (P\mathbf{v})$ which describes the mechanical work (reversible volume work) and which appears in Eq. (61), and that expression refers to the (reversible) electrification and magnetization work.

With the help of the identity

$$H_V = \sum_k c_k H_k \qquad (4\text{-}34.62)$$

which follows from Eq. (1–6.10), we derive from Eqs. (28), (49), (51), (55), and (56) the explicit form of the local balance for the internal energy:

$$\frac{\partial U_V}{\partial t} = -\text{div}\left(\mathbf{J}_Q + \sum_k H_k c_k \mathbf{v}_k\right) + \mathbf{v}[\text{grad } P - (\text{Grad } \mathfrak{E})\mathfrak{P} - (\text{Grad } \mathfrak{H})\mathfrak{J}]$$

$$+ \mathfrak{E}\frac{\partial \mathfrak{P}}{\partial t} + \mathfrak{H}\frac{\partial \mathfrak{J}}{\partial t} + \mathbf{i}(\mathfrak{E} + \mathbf{v} \times \mathfrak{B}). \qquad (4\text{-}34.63)$$

This equation corresponds to Eq. (4–8.3) which was used in the determination of the entropy balance for the systems considered in §4–8. We will thus use Eq. (63) to set up the entropy balance for isotropic systems in electromagnetic fields (without viscous flow and without gravitational and centrifugal fields). Equation (63) already contains the momentum balance.

Equations (58) and (63) generalize the equations of Mazur and Prigogine[202],† in which the electric field is regarded as varying slowly and in which the influence of a magnetic field is neglected.

As is obvious from our general discussion on p. 11, the internal energy is a function of the internal state variables of the region in question. For the polarization of matter in an electromagnetic field, the electric field strength $\mathfrak{E}$ and the magnetic field strength $\mathfrak{H}$ may be regarded as internal state variables along with temperature, pressure, and amount. The electromagnetic energy depends on the first variables as does the internal energy of a polarizable medium.

At this point we already recognize clearly that a general form of continuum physics is not produced by mechanics alone nor by electrodynamics alone but that only thermodynamics offers the possibility of the most general macroscopic description of an arbitrary system.

(g) Entropy Balance

For any volume element of an isotropic system with electrification and magnetization, the generalized Gibbs equation in the form (1–12.25) is valid for sufficiently slow irreversible processes (cf. §1–23 and §4–1). If we replace the differentials in Eq. (1–12.25) by the partial derivatives with respect to time at a fixed position (operator $\partial/\partial t$) then we obtain for the local rate of increase of the entropy density S_V in a volume element:

$$\frac{\partial S_V}{\partial t} = \frac{1}{T}\left(\frac{\partial U_V}{\partial t} - \mathfrak{E}\frac{\partial \mathfrak{P}}{\partial t} - \mathfrak{H}\cdot\frac{\partial \mathfrak{I}}{\partial t} - \sum_k \mu_k \frac{\partial c_k}{\partial t}\right). \qquad (4–34.64)$$

Here T is the absolute temperature and μ_k the chemical potential of species k.

The generalized Gibbs–Duhem relation in the form (1–13.7) has the same range of validity as Eq. (64) and from the former we derive directly:

$$\operatorname{grad} P - (\operatorname{Grad} \mathfrak{E})\mathfrak{P} - (\operatorname{Grad} \mathfrak{H})\mathfrak{I} = \sum_k c_k(\operatorname{grad} \mu_k)_T. \qquad (4–34.65)$$

Here $(\operatorname{grad} \mu_k)_T$ denotes the gradient of μ_k after subtraction of the term containing grad T.

---

† The other differences in the presentation of these authors compared to ours are due to another definition of the "heat flow" and to the use of the irrational system of electrodynamics.

Moreover, according to Eqs. (1–12.6), (1–12.8), and (1–14.2), we have:

$$\mu_k = H_k - TS_k, \tag{4–34.66}$$

$$\text{grad}\left(\frac{\mu_k}{T}\right) = -\frac{H_k}{T^2}\,\text{grad}\,T + \frac{1}{T}(\text{grad}\,\mu_k)_T, \tag{4–34.67}$$

$$A_r = -\sum_k \nu_{kr}\mu_k, \tag{4–34.68}$$

where S_k denotes the partial molar entropy of species k and A_r the affinity of chemical reaction r.

Furthermore, we use the vector identities

$$\frac{1}{T}\,\text{div}\,\mathbf{J}_Q = \text{div}\left(\frac{\mathbf{J}_Q}{T}\right) + \frac{\mathbf{J}_Q}{T^2}\,\text{grad}\,T, \tag{4–34.69}$$

$$\frac{1}{T}\,\text{div}\sum_k H_k c_k \mathbf{v}_k = \text{div}\frac{\sum_k H_k c_k \mathbf{v}_k}{T} + \frac{\sum_k H_k c_k \mathbf{v}_k}{T^2}\,\text{grad}\,T, \tag{4–34.70}$$

$$\sum_k \frac{\mu_k}{T}\,\text{div}\,(c_k \mathbf{v}_k) = \text{div}\left(\sum_k \frac{\mu_k}{T}\,c_k \mathbf{v}_k\right) - \sum_k c_k \mathbf{v}_k\,\text{grad}\left(\frac{\mu_k}{T}\right). \tag{4–34.71}$$

If we insert Eqs. (34) and (63) into Eq. (64) and take account of Eq. (28) as well as of Eqs. (65) to (71) then we find the equation for the local balance of the entropy (cf. Eqs. 4–8.13 to 4–8.15):

$$\frac{\partial S_V}{\partial t} = -\text{div}\,\mathbf{J}_S + \vartheta \tag{4–34.72}$$

with

$$\mathbf{J}_S \equiv \frac{\mathbf{J}_Q}{T} + \sum_k c_k S_k \mathbf{v}_k, \tag{4–34.73}$$

$$\vartheta \equiv -\frac{1}{T^2}\,\mathbf{J}_Q\,\text{grad}\,T - \frac{1}{T}\sum_k {_v}\mathbf{J}_k(\text{grad}\,\mu_k)_T$$

$$+ \frac{1}{T}\sum_r b_r A_r + \frac{1}{T}\mathbf{i}(\mathfrak{E} + \mathbf{v}\times\mathfrak{B}). \tag{4–34.74}$$

The vector $\mathbf{J}_S$ is the *entropy current density* and ϑ the *local entropy production*. For this, we again require, as in (4–8.16),

$$\vartheta \geqslant 0, \tag{4–34.75}$$

where the inequality sign holds for actual (irreversible) processes inside the volume element and the equality sign for the reversible limiting case (in particular for equilibrium).

The above form of the entropy balance can, by our assumptions in the derivation of Eq. (63), be applied to isotropic systems in arbitrary electromagnetic fields provided gravitational and centrifugal fields as well as viscous flow are excluded. On the other hand, Eqs. (4–8.13) to (4–8.15) are valid for isotropic systems with viscous flow which are in stationary external force fields (gravitational and centrifugal fields or stationary electromagnetic fields) but which show no electric or magnetic polarization. Accordingly, the equations for the entropy flux (4–8.14) and (73) are in fact the same but the formulas for the local entropy production (4–8.15) and (74) are different from each other.†

(h) Dissipation Function

The occurrence of irreversible processes is represented by the local entropy production or by the *dissipation function* (cf. Eq. 4–12.1)

$$\Psi \equiv T\vartheta. \tag{4–34.76}$$

If we note Eq. (32), which connects the "conductive electric current density" $\mathbf{i}$ to the diffusion current densities $_v\mathbf{J}_k$, then from Eqs. (74) to (76) we obtain for the dissipation function in our case[203]:

$$\Psi = \mathbf{J}_Q\mathbf{X}_Q + \sum_k {}_v\mathbf{J}_k\mathbf{X}_k + \sum_r b_r A_r \geqslant 0 \tag{4–34.77}$$

with

$$\mathbf{X}_Q \equiv -\frac{1}{T}\operatorname{grad} T, \tag{4–34.78}$$

$$\mathbf{X}_k \equiv z_k F(\mathfrak{E} + \mathbf{v} \times \mathfrak{B}) - (\operatorname{grad} \mu_k)_T. \tag{4–34.79}$$

The quantities $\mathbf{J}_Q$, $_v\mathbf{J}_k$, and b_r are the *generalized fluxes*, the expressions $\mathbf{X}_Q$, $\mathbf{X}_k$, and A_r the *generalized forces*.

An apparent formal difference between Eq. (77) and Eq. (4–12.2) consists of the fact that in Eq. (4–12.2) the generalized force conjugate to the diffusion flux density $_v\mathbf{J}_k$ has the form (4–12.4)

$$\mathbf{X}_k = \mathbf{K}_k - (\operatorname{grad} \mu_k)_T,$$

---

† In the recent work of de Groot and Mazur[5], the entropy balance for polarizable systems in electromagnetic fields is formulated for another case: viscous pressures are taken account of, and in place of the chemical reactions—which in the most general sense include relaxation phenomena—the electric and magnetic relaxation are explicitly introduced, by distinguishing "instantaneous values" and "equilibrium values" of the electric and magnetic field. The latter procedure corresponds to the distinction mentioned previously (p. 43) between the "internal" and "external" tensile force for elastic media. That kind of difference already appears for homogeneous systems in the "dissipated work" and thus in the dissipation function (p. 85).

where K_k is the external molar force acting on species k. Excluding gravitational and centrifugal fields (as always here) as well as electrification and magnetization (as in Eq. 4–12.2) we have, according to Eq. (2),

$$\mathbf{K}_k = z_k F(\mathfrak{E} + \mathbf{v}_k \times \mathfrak{B}).$$

Now, if we form the summation (cf. Eq. 28)

$$\sum_k {}_v\mathbf{J}_k\mathbf{K}_k = \sum_k c_k(\mathbf{v}_k - \mathbf{v})\mathbf{K}_k$$

then with $\mathbf{v}_k(\mathbf{v}_k \times \mathfrak{B}) = 0$, $\mathbf{v}(\mathbf{v} \times \mathfrak{B}) = 0$ we obtain

$$\sum_k {}_v\mathbf{J}_k\mathbf{K}_k = \sum_k {}_v\mathbf{J}_k z_k F(\mathfrak{E} + \mathbf{v} \times \mathfrak{B}),$$

that is, the same expression as in Eq. (79).

For polarizable media in arbitrary electromagnetic fields, the dissipation function has *formally* the same form as for systems without electrification and magnetization in stationary electromagnetic fields, if for the latter systems we introduce the quantity

$$z_k F(\mathfrak{E} + \mathbf{v} \times \mathfrak{B})$$

in place of K_k. Actually, on closer examination, substantial differences arise as will be shown in the following.

With Eq. (10) and the definition (1–20.15) of the electrochemical potential

$$\eta_k \equiv \mu_k + z_k F\varphi \tag{4–34.80}$$

the generalized force (79) can be written:

$$\mathbf{X}_k = -(\text{grad } \eta_k)_T + z_k F\left(\mathbf{v} \times \mathfrak{B} - \frac{\partial \mathfrak{A}}{\partial t}\right). \tag{4–34.81}$$

If only stationary fields are considered then the term with $\partial \mathfrak{A}/\partial t$ drops out. Furthermore, if the discussion is restricted to electrostatic fields, the term with $\mathbf{v} \times \mathfrak{B}$ also disappears. In the latter case, the generalized force conjugate to the diffusion flux density is reduced to the expression

$$\mathbf{X}_k = -(\text{grad } \eta_k)_T,$$

just as it appears in normal electrochemical problems (cf. Eq. 4–15.11 and Eq. 4–24.2).

For systems in arbitrary electromagnetic fields, the general form (79) or (81) must be retained. Accordingly, the electrochemical potential plays a fundamental role[203–205] only for systems in electrostatic fields.

According to the results of electrodynamics of moving media[206a], the electric field strength $\mathfrak{E}$ and the magnetic induction $\mathfrak{B}$ depend on the reference system of the observer. The sum

$$\mathfrak{E} + \boldsymbol{\omega} \times \mathfrak{B},$$

where ω denotes the reference velocity, represents, however, an invariant quantity.† It is, therefore, satisfactory for the general theory that such an expression appears in Eq. (79).

So far we have considered only the difference between stationary and nonstationary fields. There remains a discussion of how a polarizable medium can be distinguished from a nonpolarizable medium.

For systems without electrification and magnetization, the chemical potentials μ_k are only functions of the temperature, of the pressure, and of the composition. If, on the other hand, electric or magnetic polarization appears then the μ_k depend also on the electric or magnetic field strength. We can, in fact, state immediately how this dependence appears. It follows from Eqs. (1–12.34) and (1–12.35):

$$d\mu_k = -\left(\frac{\partial \mathfrak{P}}{\partial c_k}\right)_{T,\mathfrak{E},\mathfrak{H},c_j} d\mathfrak{E} - \left(\frac{\partial \mathfrak{J}}{\partial c_k}\right)_{T,\mathfrak{E},\mathfrak{H},c_j} d\mathfrak{H}$$

$$+ \sum_{i=1}^{N} \left(\frac{\partial \mu_k}{\partial c_i}\right)_{T,\mathfrak{E},\mathfrak{H},c_j} dc_i \quad (T = \text{const}). \quad (4\text{–}34.82)$$

Here T, $\mathfrak{E}$, $\mathfrak{H}$, and $c_1, c_2, \ldots, c_N$ (N = number of species) have been chosen as independent variables. The inferior c_j signifies the constancy of all concentrations except that with respect to which the quantities are differentiated. On noting Eqs. (20) and (21) there results from integration of Eq. (82):

$$\mu_k = \mu_k^0 - \frac{1}{2}\left(\frac{\partial \psi}{\partial c_k}\right)_{T,c_j} \mathfrak{E}^2 \quad (T = \text{const}, \ \mathfrak{H} = \text{const}, \ c_i = \text{const}), \quad (4\text{–}34.83a)$$

$$\mu_k = \mu_k^0 - \frac{1}{2}\left(\frac{\partial \chi}{\partial c_k}\right)_{T,c_j} \mathfrak{H}^2 \quad (T = \text{const}, \ \mathfrak{E} = \text{const}, \ c_i = \text{const}). \quad (4\text{–}34.83b)$$

The quantity μ_k^0 in Eq. (83a) or (83b) denotes that value of the chemical potential of species k which is found at the temperature and composition in question without an electric or magnetic field. As may be seen with the help of Eq. (22), μ_k^0 is different from μ_k if the permittivity or permeability of the medium is concentration dependent (pressure or density dependent for one-component systems).‡ Let it be noted that, except in Eq. (83) of our thermodynamic treatment in this section, the "matter equations" (16) and (17) have not been used.

---

† See Eqs. (89) and (90).
‡ A derivation of Eq. (83a) which is similar in principle is found in Mazur and Prigogine[202].

The diffusion current densities in the dissipation function (77) are related to one another by the identity (4–3.13)

$$\sum_k M_k \,_v\mathbf{J}_k = 0 \qquad (4\text{–}34.84)$$

(M_k = molar mass of species k). Instead of eliminating a diffusion flux in the dissipation function with the help of Eq. (84) (cf. Eq. 4–12.12), we immediately take the simpler path following the procedure on p. 248 which leads to the relation (4–12.17).

Thus we assume "mechanical equilibrium", in particular, for a fluid medium, "creeping motion."† Then the barycentric acceleration ($d\mathbf{v}/dt$) in Eq. (38) disappears and we obtain with Eqs. (44) and (65):

$$\rho_{\text{el}}\mathfrak{E} + \mathbf{I} \times \mathfrak{B} + (\text{Grad } \mathfrak{E})\mathfrak{P} + (\text{Grad } \mathfrak{H})\mathfrak{J} = \text{grad } P, \quad (4\text{–}34.85)$$

$$\rho_{\text{el}}\mathfrak{E} + \mathbf{I} \times \mathfrak{B} = \sum_k c_k(\text{grad } \mu_k)_T. \qquad (4\text{–}34.86)$$

With Eqs. (9), (28), (29), and (79) we derive for the second term on the right-hand side of Eq. (77):

$$\sum_k \,_v\mathbf{J}_k\mathbf{X}_k = \sum_k c_k(\mathbf{v}_k - \mathbf{v})[z_k F(\mathfrak{E} + \mathbf{v} \times \mathfrak{B}) - (\text{grad } \mu_k)_T]$$

$$= \mathbf{I}(\mathfrak{E} + \mathbf{v} \times \mathfrak{B}) - \sum_k c_k\mathbf{v}_k(\text{grad } \mu_k)_T$$

$$- \mathbf{v}\left[\rho_{\text{el}}\mathfrak{E} - \sum_k c_k(\text{grad } \mu_k)_T\right].$$

Equation (86) thus leads to the formula:

$$\sum_k \,_v\mathbf{J}_k\mathbf{X}_k = \mathbf{I}\mathfrak{E} - \sum_k c_k\mathbf{v}_k(\text{grad } \mu_k)_T$$

$$= \sum_k c_k\mathbf{v}_k[z_k F\mathfrak{E} - (\text{grad } \mu_k)_T]. \qquad (4\text{–}34.87)$$

For an arbitrary reference velocity $\boldsymbol{\omega}$ and from Eqs. (9), (29), and (86), we write:

$$\boldsymbol{\omega} \sum_k c_k[z_k F\mathfrak{E} - (\text{grad } \mu_k)_T]$$

$$= -\boldsymbol{\omega}(\mathbf{I} \times \mathfrak{B}) = \mathbf{I}(\boldsymbol{\omega} \times \mathfrak{B}) = F \sum_k z_k c_k(\mathbf{v}_k - \boldsymbol{\omega})(\boldsymbol{\omega} \times \mathfrak{B}).$$

---

† Since we use the following equations only for the formulation of the equilibrium conditions, there arise no considerations with reference to the mechanical equilibrium in solid bodies (cf. p. 315).

Then we obtain, with the definition (cf. Eq. 4–3.1)

$$\mathbf{J}_k \equiv c_k(\mathbf{v}_k - \boldsymbol{\omega}) \tag{4–34.88}$$

and with Eq. (87), the relation:

$$\sum_k {}_v\mathbf{J}_k\mathbf{X}_k = \sum_k \mathbf{J}_k\mathbf{X}'_k \tag{4–34.89}$$

with

$$\mathbf{X}'_k \equiv z_k F(\mathfrak{E} + \boldsymbol{\omega} \times \mathfrak{B}) - (\text{grad } \mu_k)_T. \tag{4–34.90}$$

In Eqs. (89) and (90), a linear relation between the "fluxes" $\mathbf{J}_k$ or between the "forces" $\mathbf{X}'_k$ arises because of Eq. (4–3.4) or Eq. (86). We eliminate the dependent quantities in that we use the Hittorf reference system (cf. Eq. 4–3.30):

$$\boldsymbol{\omega} \equiv \mathbf{v}_1, \quad {}_1\mathbf{J}_k \equiv c_k(\mathbf{v}_k - \mathbf{v}_1), \quad {}_1\mathbf{J}_1 \equiv 0. \tag{4–34.91}$$

With Eqs. (89) and (91) the dissipation function (77) for a system with N species and R linearly independent chemical reactions can be written:

$$\Psi = \mathbf{J}_Q\mathbf{X}_Q + \sum_{k=2}^{N} {}_1\mathbf{J}_k\mathbf{X}'_k + \sum_{r=1}^{R} b_r A_r \geqslant 0, \tag{4–34.92}$$

by analogy with Eq. (4–12.17). In Eq. (92), which is valid for mechanical equilibrium, all "fluxes" $(J_Q, {}_1J_2, {}_1J_3, \ldots, {}_1J_N; b_1, b_2, \ldots, b_R)$ or all "forces" $(\mathbf{X}_Q, \mathbf{X}'_2, \mathbf{X}'_3, \ldots, \mathbf{X}'_N; A_1, A_2, \ldots, A_R)$ are independent of one another.†

(i) Equilibrium Conditions

If there is *local equilibrium*, the independent "fluxes" and the independent "forces" must disappear in each volume element of the system, so that the dissipation function becomes zero everywhere.

From the disappearance of the "forces" there follows with Eqs. (10), (78), (80), (90), (91), and (92):

$$\text{grad } T = 0, \quad T = \text{const}, \tag{4–34.93}$$

$$z_k F(\mathfrak{E} + \mathbf{v}_1 \times \mathfrak{B}) = (\text{grad } \mu_k)_T = \text{grad } \mu_k \quad (k = 2, 3, \ldots, N) \tag{4–34.94}$$

---

† Although we use Eq. (92) in the following only for the formulation of the equilibrium conditions, it is evident that the dissipation function in this form represents the starting point for the description of many kinds of irreversible processes. All problems in whose phenomenological treatment the assumption of mechanical equilibrium is allowed are included in this; among others, electric conduction, diffusion, heat conduction, and thermal diffusion in fluid media in the presence of arbitrary electromagnetic fields with the inclusion of electric and magnetic polarization. In connection with this, there are the effects which were experimentally and theoretically investigated by Hanssen[207], and which appear when galvanic cells are brought into an external magnetic field.

or

$$z_k F \left(\mathbf{v}_1 \times \mathfrak{B} - \frac{d\mathfrak{A}}{\partial t} \right) = (\text{grad } \eta_k)_T = \text{grad } \eta_k \quad (k = 2, 3, \ldots, N) \quad (4\text{–}34.95)$$

$$A_r = 0 \quad (r = 1, 2, \ldots, R). \quad (4\text{–}34.96)$$

Equilibrium conditions (93) and (96) appear obvious. Equations (94) and (95), on the other hand, require comment.

If we assume an *electrostatic field* ($\mathfrak{B} = 0$, $\partial\mathfrak{A}/\partial t = 0$) then from Eqs. (94) and (95) we derive

$$\text{grad } \mu_k = z_k F \mathfrak{E}, \quad (4\text{–}34.97)$$

$$\text{grad } \eta_k = 0, \quad \eta_k = \text{const.} \quad (4\text{–}34.98)$$

Equations (97) and (98) are in agreement with (1–21.20) to (1–21.22) which were obtained with the gross methods of classical thermodynamics. However, it is apparent here that these simple equilibrium conditions also hold for electrified media. Then the dependence of μ_k or η_k on the electric field strength will be described by Eq. (82).

Comparison of Eq. (95) with Eq. (98) indicates that the electrochemical potential η_k plays the role of a quantity which determines equilibrium only for systems in electrostatic fields. In the most general case, the conditions for the equilibrium distribution of the various species in an electromagnetic field cannot be expressed by a constant "potential."

Here we have permitted fields which can change in time, and macroscopic motions in the system. Thus, in the conditions for *local* equilibrium which hold for each volume element, quantities like $\mathbf{v}_1$ and $\partial\mathfrak{A}/dt$ can appear. These quantities vanish from the conditional equations for *gross* equilibrium which relate to the total system and which can be defined only for systems at rest in stationary fields (§1–17 to §1–21).

According to Eq. (91), since here all independent "fluxes" vanish, we have for local equilibrium:

$$_1\mathbf{J}_k = c_k(\mathbf{v}_k - \mathbf{v}_1) = 0 \quad (k = 1, 2, \ldots, N),$$

that is,

$$\mathbf{v}_1 = \mathbf{v}_2 = \cdots = \mathbf{v}_N.$$

From this with Eqs. (4–3.2) and (4–3.3) there follows for any arbitrary reference velocity $\boldsymbol{\omega}$ and thus also for the barycentric velocity $\mathbf{v}$:

$$\boldsymbol{\omega} = \mathbf{v} = \mathbf{v}_1 = \mathbf{v}_2 = \cdots = \mathbf{v}_N. \quad (4\text{–}34.99)$$

We may therefore insert into the equilibrium conditions, in place of $\mathbf{v}_1$, the velocity of any species (in particular, the velocity of the species in consideration at that time) or any arbitrary mean velocity. Once more this confirms that the vector sum

$$\mathfrak{E} + \omega \times \mathfrak{B}$$

plays the role of an invariant for changes of the reference system.

With the help of the last statement we can show, finally, that Eqs. (94) and (95) are valid not only for $k = 2, 3, \ldots, N$ but also for $k = 1$. According to Eq. (99), we may write in place of Eq. (94):

$$z_k F(\mathfrak{E} + \mathbf{v}_k \times \mathfrak{B}) = (\text{grad } \mu_k)_T \quad (k = 2, 3, \ldots, N). \quad (4\text{–}34.100)$$

Equation (86), with Eqs. (9) and (29), produces:

$$F \sum_{k=1}^{N} z_k c_k (\mathfrak{E} + \mathbf{v}_k \times \mathfrak{B}) = \sum_{k=1}^{N} c_k (\text{grad } \mu_k)_T. \quad (4\text{–}34.101)$$

From the last two equations there results:

$$z_1 F(\mathfrak{E} + \mathbf{v}_1 \times \mathfrak{B}) = (\text{grad } \mu_1)_T. \quad (4\text{–}34.102)$$

Thus the assertion above is proved.

If we compare Eqs. (100) and (102) with Eq. (2) we see that we can also write these equilibrium conditions in the shorter form (cf. Eq. 1–21.24)

$$\text{grad } \mu_k = (\text{grad } \mu_k)_T = \mathbf{L}_k \quad (k = 1, 2, \ldots, N), \quad (4\text{–}34.103)$$

where $\mathbf{L}_k$ denotes the molar Lorentz force for species k. Equation (103) is, together with the relations (93) and (96)

$$\text{grad } T = 0, \qquad A_r = 0 \quad (r = 1, 2, \ldots, R), \quad (4\text{–}34.104)$$

without doubt the most concise formulation of the conditions for local equilibrium for systems in arbitrary electromagnetic fields.

According to Eqs. (68) and (104), the equilibrium condition

$$\sum_{k=1}^{N} \nu_{kr} \, \text{grad } \mu_k = 0 \quad (r = 1, 2, \ldots, R) \quad (4\text{–}34.105)$$

must be satisfied. At first sight, this relation appears to contradict the equation

$$\sum_{k=1}^{N} \nu_{kr} \, \text{grad } \mu_k = \sum_{k=1}^{N} \nu_{kr} \mathbf{L}_k \quad (r = 1, 2, \ldots, R)$$

which can be derived from Eq. (103). However, it now follows from Eqs. (2), (35), and (99):

$$\sum_{k=1}^{N} \nu_{kr} \mathbf{L}_k = 0 \quad (r = 1, 2, \ldots, R),$$

with which the condition (105) is satisfied.

(j) Generalization of Ohm's Law

We consider systems without chemical reactions and without gradients in the molarities which, moreover, either cannot be polarized or which have fields which are constant in position. Then there results from Eqs. (74) to (76) with Eq. (82) for the dissipation function:

$$\Psi = -\mathbf{J}_Q \frac{\text{grad } T}{T} + \mathbf{i}(\mathfrak{E} + \mathbf{v} \times \mathfrak{B}) \geqslant 0. \qquad (4\text{-}34.106)$$

This relation will, in the following, be applied especially to electronic conductors but will at first be investigated for the more general case.

We here exclude the complications which are caused by the galvanomagnetic effects and thermomagnetic effects (§4–36) and which require the splitting of the vectors in Eq. (106) into components. Also we assume, for the sake of simplicity, a temperature which is constant in position (grad $T = 0$). Then, for states "near equilibrium", we have the simple phenomenological equation

$$\mathbf{i} = \kappa(\mathfrak{E} + \mathbf{v} \times \mathfrak{B}). \qquad (4\text{-}34.107)$$

The phenomenological coefficient κ is obviously a generalized "specific conductance." Equation (107) represents a generalization of Ohm's law (4–16.24) to systems in arbitrary electromagnetic fields.

It is interesting that relativistic electrodynamics also produces Ohm's law in the form (107) (Minkowski, 1908) provided $\mathbf{v}$ is small compared to the velocity of light[206b].

(k) Electronic Conductors

Under normal experimental conditions the assumptions which lead to Eq. (106) are satisfied for electronic conductors. Moreover, we can usually assume systems at rest, i.e. neglect the barycentric velocity $\mathbf{v}$ according to Eq. (4–24.51). Thus we obtain for the dissipation function for electronic conductors at rest from Eq. (106):

$$\Psi = -\mathbf{J}_Q \frac{\text{grad } T}{T} + \mathbf{I}\mathfrak{E} \geqslant 0, \qquad (4\text{-}34.108)$$

since the conductive current density $\mathbf{i}$ coincides with the total current density $\mathbf{I}$ for all electric conductors according to Eqs. (30) to (33).

Equation (108), in which the magnetic field no longer appears explicitly, represents the starting point for our discussion of the galvanomagnetic and thermomagnetic effects in electron conductors (§4–36).

If no magnetic field is present then Eq. (108) leads to the phenomenological equations (4–25.1) and (4–25.2) which are known to us and which describe the electric conduction, heat conduction, and thermoelectric effects in electronic conductors.

4-35 RECIPROCITY RELATIONS FOR SYSTEMS IN CENTRIFUGAL AND MAGNETIC FIELDS

On particles of a system in a centrifugal field (§4–33) or in a magnetic field (§4–34), in addition to the centrifugal force or the electrostatic force, there are the Coriolis force or the magnetic part of the Lorentz force respectively. The last mentioned forces are perpendicular to the instantaneous direction of particle motion and perpendicular to the axis of rotation or direction of the magnetic field.   They are proportional to the magnitude of the particle velocity. These facts yield three conclusions:

1. Even an originally isotropic system has preferential directions, if it is brought into a centrifuge or into a magnetic field. Accordingly, the "fluxes" and "forces" contained in the dissipation function must split up into their components, and the phenomenological equations for these components must be formulated individually just as for anisotropic media (§4–37).

2. The phenomenological coefficients α_{ij} are functions of the angular velocity Ω or of the magnetic induction $\mathfrak{B}$. This is the case even if, under the existing experimental conditions, the Coriolis force or the magnetic part of the Lorentz force no longer appears explicitly in the dissipation function (cf. Eq. 4–33.18 or Eq. 4–34.108).

3. The reciprocity relations must be written in the form (1–26.4). This form of the reciprocity law results from the fact that, considering the "principle of microscopic reversibility," the direction of the centrifugal or magnetic field must be reversed when time is reversed.†

If we write the phenomenological equations in the form

$$J_i = \sum_j \alpha_{ij} X_j \qquad (4\text{–}35.1)$$

(J_i = generalized fluxes, X_j = generalized forces) then the Onsager–Casimir reciprocity law (1–26.4) states:

$$\alpha_{ij}(\Omega) = \varepsilon_i \varepsilon_j \alpha_{ji}(-\Omega) \qquad (4\text{–}35.2)$$

or

$$\alpha_{ij}(\mathfrak{B}) = \varepsilon_i \varepsilon_j \alpha_{ji}(-\mathfrak{B}). \qquad (4\text{–}35.3)$$

Here $\varepsilon_i = 1$ or $\varepsilon_j = 1$, if X_i or X_j is even with respect to time reversal ("force of the α-type"), while $\varepsilon_i = -1$ or $\varepsilon_j = -1$, if X_i or X_j is uneven with respect to time reversal ("force of the β-type").

For $i = j$ ($\alpha_{ij} = \alpha_{ji} = \alpha_{ii}$, $\varepsilon_i \varepsilon_j = \varepsilon_i^2 = 1$), Eq. (2) or (3) indicates that the phenomenological coefficient α_{ii} must be an even function of Ω or $\mathfrak{B}$. In the simplest case, α_{ii} is independent of Ω or $\mathfrak{B}$.

---

† A modernized and complete molecular–statistical derivation of the Onsager–Casimir reciprocity relations (1–26.4) for systems in magnetic fields is found in de Groot and others[208–210].

The application of the above relations to rotating systems is not of practical interest at present. Statement (3), which is valid for magnetic fields, is significant for the treatment of the galvanomagnetic and thermomagnetic effects (§4–36).

4–36 GALVANOMAGNETIC AND THERMOMAGNETIC EFFECTS

(a) General

Galvanomagnetic or thermomagnetic effects arise in electric conductors in an external magnetic field if an electric current flows or a temperature gradient is present respectively.† The best-known example is the "Hall effect": in an isothermal conductor, through which an electric current flows perpendicular to the magnetic field, there is an electric field produced which is perpendicular to the plane of the current and magnetic field directions ("galvanomagnetic transverse effect").

Here we consider for simplicity only isotropic electronic conductors at rest without concentration gradients and without chemical reactions in homogeneous fields. With this we exclude, among other things, "heterogeneous effects" which take place through the adjoining of different materials or of different magnetic regions and which play a role in the experimental confirmation of certain effects (cf. §4–25).

Under these conditions, the dissipation function Ψ has the simple form (4–34.108):

$$\Psi = -\mathbf{J}_Q \frac{\text{grad } T}{T} + \mathbf{I}\mathfrak{E}. \qquad (4\text{–}36.1)$$

Here T denotes the absolute temperature, $\mathbf{J}_Q$ the heat current density, $\mathbf{I}$ the electric current density, and $\mathfrak{E}$ the electric field strength. Because of the magnetic field we must now split the vectors in Eqs. (1) into their spatial components (§4–35).

We use a rectangular coordinate system (position-coordinates x, y, z) and denote the spatial (cartesian) components of $\mathbf{J}_Q$ by J_x, J_y, J_z, those of

$$\mathbf{X}_Q \equiv -\frac{1}{T} \text{grad } T$$

by X_x, X_y, X_z, those of $\mathbf{I}$ by I_x, I_y, I_z, and those of $\mathfrak{E}$ by $\mathfrak{E}_x$, $\mathfrak{E}_y$, $\mathfrak{E}_z$. Thus, according to Eq. (1),

$$\Psi = J_x X_x + J_y X_y + J_z X_z + I_x \mathfrak{E}_x + I_y \mathfrak{E}_y + I_z \mathfrak{E}_z \qquad (4\text{–}36.2)$$

---

† Comprehensive presentations of these effects can be found from, for example, Meissner[211] and Jan[212]. A general treatment from the standpoint of electron theory of metals is given by Meixner[109, 213].

with

$$X_x \equiv -\frac{1}{T}\frac{\partial T}{\partial x}, \quad X_y \equiv -\frac{1}{T}\frac{\partial T}{\partial y}, \quad X_z \equiv -\frac{1}{T}\frac{\partial T}{\partial z}, \quad (4\text{-}36.3)$$

where, for example, the operator $\partial/\partial x$ signifies the derivative with respect to the position-coordinate x at a fixed time.

The irreversible processes will now be described as a first approximation ("near equilibrium") by the "phenomenological equations," i.e. by homogeneous linear relations between the six "fluxes" (J_x, etc.) and the six "forces" (X_x, etc.) in Eq. (2), where the "phenomenological coefficients", which we denote by α_{11}, α_{12}, etc., are functions of the magnetic induction $\mathfrak{B}$ (cf. Eq. 4-35.1). For practical reasons, it is expedient to treat the components of the electric current density (I_x, I_y, I_z) and the quantities X_x, X_y, X_z as independent variables, that is, to exchange the role of the "fluxes" and "forces" in the last three terms of Eq. (2). We then obtain the following linear equations:

$$\left.\begin{array}{l}
\mathfrak{E}_x = \alpha_{11}I_x + \alpha_{12}I_y + \alpha_{13}I_z + \alpha_{14}X_x + \alpha_{15}X_y + \alpha_{16}X_z, \\
\mathfrak{E}_y = \alpha_{21}I_x + \alpha_{22}I_y + \alpha_{23}I_z + \alpha_{24}X_x + \alpha_{25}X_y + \alpha_{26}X_z, \\
\mathfrak{E}_z = \alpha_{31}I_x + \alpha_{32}I_y + \alpha_{33}I_z + \alpha_{34}X_x + \alpha_{35}X_y + \alpha_{36}X_z, \\
J_x = \alpha_{41}I_x + \alpha_{42}I_y + \alpha_{43}I_z + \alpha_{44}X_x + \alpha_{45}X_y + \alpha_{46}X_z, \\
J_y = \alpha_{51}I_x + \alpha_{52}I_y + \alpha_{53}I_z + \alpha_{54}X_x + \alpha_{55}X_y + \alpha_{56}X_z, \\
J_z = \alpha_{61}I_x + \alpha_{62}I_y + \alpha_{63}I_z + \alpha_{64}X_x + \alpha_{65}X_y + \alpha_{66}X_z.
\end{array}\right\} \quad (4\text{-}36.4)$$

Not all of the 36 coefficients of this system of equations are independent. From the assumption of isotropy (in the absence of the magnetic field) and from the Onsager–Casimir reciprocity law (§4–35) certain relations between the coefficients follow, as we will see below.

It can be shown (cf. §4–37) that the above equations also describe the corresponding processes in anisotropic media.

If we introduce, to abbreviate the notation, the tensors

$$\rho \equiv \begin{pmatrix} \alpha_{11} & \alpha_{12} & \alpha_{13} \\ \alpha_{21} & \alpha_{22} & \alpha_{23} \\ \alpha_{31} & \alpha_{32} & \alpha_{33} \end{pmatrix}, \quad (4\text{-}36.5)$$

$$\sigma \equiv \begin{pmatrix} \alpha_{14} & \alpha_{15} & \alpha_{16} \\ \alpha_{24} & \alpha_{25} & \alpha_{26} \\ \alpha_{34} & \alpha_{35} & \alpha_{36} \end{pmatrix}, \quad (4\text{-}36.6)$$

$$-\frac{{}^{*}Q}{F} \equiv \begin{pmatrix} \alpha_{41} & \alpha_{42} & \alpha_{43} \\ \alpha_{51} & \alpha_{52} & \alpha_{53} \\ \alpha_{61} & \alpha_{62} & \alpha_{63} \end{pmatrix}, \quad (4\text{-}36.7)$$

$$T\lambda \equiv \begin{pmatrix} \alpha_{44} & \alpha_{45} & \alpha_{46} \\ \alpha_{54} & \alpha_{55} & \alpha_{56} \\ \alpha_{64} & \alpha_{65} & \alpha_{66} \end{pmatrix} \quad (4\text{-}36.8)$$

(F = Faraday constant) then we can replace the six scalar relations (4) by two vector equations:

$$\mathfrak{E} = \rho \mathbf{I} + \sigma \mathbf{X}_Q, \tag{4–36.9}$$

$$\mathbf{J}_Q = -\frac{*Q}{F} \mathbf{I} + T\lambda \mathbf{X}_Q. \tag{4–36.10}$$

Here ρ is the tensor of the resistivity (cf. §4–38), λ the thermal conductivity tensor (cf. §4–39), σ the tensor of the coefficients of the thermoelectric potential, and $*Q$ the tensor of the heat of transport of electrons (cf. below).

If no magnetic field is present and if the medium is isotropic then the tensors (5) to (8) degenerate to scalars. Then we write Eqs. (9) and (10) in the form

$$\mathfrak{E} = a_{11}\mathbf{I} + a_{12}\mathbf{X}_Q, \tag{4–36.11}$$

$$\mathbf{J}_Q = a_{21}\mathbf{I} + a_{22}\mathbf{X}_Q \tag{4–36.12}$$

with the four scalar coefficients a_{11}, a_{12}, a_{21}, and a_{22}. It should now be noted that, with reference to reversal of time, $\mathbf{X}_Q$ is even but $\mathbf{I}$ is odd. (For a time transformation $t \to -t$, the current reverses its direction while the temperature gradient remains.) Thus the reciprocity law is valid in the form (1–26.3):

$$a_{12} = -a_{21}. \tag{4–36.13}$$

By comparing Eqs. (9) and (10) with Eqs. (11) and (12) and on taking account of Eq. (13) there follows:

$$\mathfrak{E} = \rho \mathbf{I} - \frac{*Q}{FT} \operatorname{grad} T, \tag{4–36.14}$$

$$\mathbf{J}_Q = -\frac{*Q}{F} \mathbf{I} - \lambda \operatorname{grad} T. \tag{4–36.15}$$

This system of equations agrees exactly with Eqs. (4–25.16) and (4–25.17), where ρ is the resistivity (the reciprocal value of the specific conductance), $*Q$ the heat of transport of electrons, and λ the thermal conductivity (for the state without electric current). The quantity $*Q/FT$ in Eq. (14) can be called the "coefficient of the thermoelectric potential." From this our notation in Eqs. (9) and (10) is clear in the general case. Equations (14) and (15) describe electric conduction, heat conduction, and thermoelectric phenomena in isotropic electronic conductors without magnetic fields (cf. §4–25).

As has been indicated, the presentation of the following irreversible processes is included in Eq. (4) or (9) and (10) in the most general case: electric conduction, heat conduction, thermoelectric, galvanomagnetic, and thermomagnetic effects in anisotropic electron conductors. Insofar as the classes of crystals in question do not require symmetry relations, only the Onsager–Casimir reciprocity law (4–35.3) permits statements about the coefficients. Since Eq. (4–35.3) requires that the 36 coefficients of our system of equations yield 15 independent relations, there remain 21 independent coefficients in the general case[109].

After the general discussion of processes in anisotropic media (§4–37) we will treat electric conduction (§4–38) and heat conduction (§4–39) in anisotropic systems without a magnetic field. Here we are concerned only with effects in isotropic media with a magnetic field.

Let an isotropic electronic conductor (e.g. a liquid metal, a metal unit crystal of the cubic system, or a polycrystalline metal without preferential directions of the crystallites) be placed in an external magnetic field (magnetic induction $\mathfrak{B}$). Let the vector $\mathfrak{B}$ have the direction of the z-axis. Isotropy of the medium (in the x–y-plane) then requires that, for an arbitrary rotation of the coordinate system about the z-axis, Eqs. (4) or (9) and (10) remain unchanged. This invariance condition has mathematically the following effect: in tensors (5) to (8) the first two elements of the last row and of the last column disappear; the second element of the first row becomes the negative value of the second element of the first column; the first two elements of the main diagonal become identical. Thus we find from the assumption of isotropy of the medium:

$$\left.\begin{aligned}
\alpha_{11} &= \alpha_{22}, & \alpha_{12} &= -\alpha_{21}, & \alpha_{13} &= \alpha_{31} = \alpha_{23} = \alpha_{32} = 0, \\
\alpha_{14} &= \alpha_{25}, & \alpha_{15} &= -\alpha_{24}, & \alpha_{16} &= \alpha_{34} = \alpha_{26} = \alpha_{35} = 0, \\
\alpha_{41} &= \alpha_{52}, & \alpha_{42} &= -\alpha_{51}, & \alpha_{43} &= \alpha_{61} = \alpha_{53} = \alpha_{62} = 0, \\
\alpha_{44} &= \alpha_{55}, & \alpha_{45} &= -\alpha_{54}, & \alpha_{46} &= \alpha_{64} = \alpha_{56} = \alpha_{65} = 0
\end{aligned}\right\} \quad (4\text{–}36.16)$$

and the system of equations (4) can be written:

$$\left.\begin{aligned}
\mathfrak{E}_x &= \alpha_{11}I_x + \alpha_{12}I_y + \alpha_{14}X_x + \alpha_{15}X_y, \\
\mathfrak{E}_y &= -\alpha_{12}I_x + \alpha_{11}I_y - \alpha_{15}X_x + \alpha_{14}X_y, \\
\mathfrak{E}_z &= \alpha_{33}I_z + \alpha_{36}X_z, \\
J_x &= \alpha_{41}I_x + \alpha_{42}I_y + \alpha_{44}X_x + \alpha_{45}X_y, \\
J_y &= -\alpha_{42}I_x + \alpha_{41}I_y - \alpha_{45}X_x + \alpha_{44}X_y, \\
J_z &= \alpha_{63}I_z + \alpha_{66}X_z.
\end{aligned}\right\} \quad (4\text{–}36.17)$$

These are six linear equations with 12 different scalar coefficients. In fact, for isotropic electronic conductors there exist 12 elementary effects.

For reasons of symmetry, the following statements are valid for isotropic systems:

$\alpha_{11}, \alpha_{14}, \alpha_{33}, \alpha_{36}, \alpha_{41}, \alpha_{44}, \alpha_{63},$ and α_{66} are even functions of $\mathfrak{B}$;   (4–36.18)

$\alpha_{12}, \alpha_{15}, \alpha_{42},$ and α_{45} are odd functions of $\mathfrak{B}$.   (4–36.19)

According to Eq. (4–35.3), the Onsager–Casimir reciprocity law leads to the following relations (cf. Eqs. 16, 18, and 19):

$$\alpha_{12}(\mathfrak{B}) = \quad \alpha_{21}(-\mathfrak{B}) = -\alpha_{12}(-\mathfrak{B}), \qquad (4\text{–}36.20)$$

$$\alpha_{14}(\mathfrak{B}) = -\alpha_{41}(-\mathfrak{B}) = -\alpha_{41}(\mathfrak{B}), \qquad (4\text{–}36.21)$$

$$\alpha_{15}(\mathfrak{B}) = -\alpha_{51}(-\mathfrak{B}) = \quad \alpha_{42}(-\mathfrak{B}) = -\alpha_{42}(\mathfrak{B}), \qquad (4\text{–}36.22)$$

$$\alpha_{36}(\mathfrak{B}) = -\alpha_{63}(-\mathfrak{B}) = -\alpha_{63}(\mathfrak{B}), \qquad (4\text{–}36.23)$$

$$\alpha_{45}(\mathfrak{B}) = \quad \alpha_{54}(-\mathfrak{B}) = -\alpha_{45}(-\mathfrak{B}). \qquad (4\text{–}36.24)$$

Here Eqs. (20) and (24) are contained in (16) and (19). There thus remain 3 "true" reciprocity relations and there are only 9 independent effects[109].

For clarity, we again write the 3 independent statements for the coefficients in Eq. (17) resulting from the reciprocity law according to Eqs. (21) to (23):

$$\alpha_{14} = -\alpha_{41}, \qquad (4\text{–}36.25)$$

$$\alpha_{15} = -\alpha_{42}, \qquad (4\text{–}36.26)$$

$$\alpha_{36} = -\alpha_{63}. \qquad (4\text{–}36.27)$$

These 3 relations correspond to 3 independent relations between the 12 elementary effects.

(b) Classification of Effects

In the assignment of the measurable effects to the 12 elementary effects, i.e. to the 12 coefficients of the system of equations (17), it must be noted that certain effects split up into "isothermal" and "adiabatic" phenomena, that is, they represent "double effects," and furthermore that certain phenomena only become measurable if several conductors are combined into a heterogeneous system which corresponds to a thermocouple (cf. §4–25). For the sake of conciseness, we exclude heterogeneous effects and adiabatic phenomena in the double effects and discuss only the simple homogeneous effects (local effects).

Under these conditions, Eqs. (3) and (17) yield the classification scheme of Table 20. It should be remembered that of the 20 terms on the right-hand side of the linear system of equations (17) only 12 are actually, from the beginning, different from one another. Thus the relation

$$\mathfrak{E}_x = \alpha_{12}I_y \quad (I_x = 0, \, X_x = 0, \, X_y = 0)$$

describes in principle the same phenomenon as the equation

$$\mathfrak{E}_y = -\alpha_{12}I_x \quad (I_y = 0, \, X_x = 0, \, X_y = 0).$$

Many effects bear the names of their discoverers (Ettingshausen, Hall, Leduc, Nernst, Righi). The phenomenon represented by the last two formulas, for example, is the "Hall effect" which was mentioned earlier (Fig. 36).

The two thermomagnetic longitudinal effects in the transverse field (effects 9 and 10 in Table 20) are sometimes also called "2nd Righi–Leduc

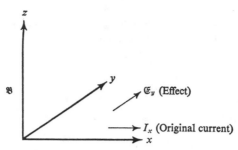

Fig. 36. Hall effect (Hall, 1879): an electric current in the x-direction (current density I_x) produces, in the presence of a magnetic field in the z-direction (magnetic induction $\mathfrak{B}$), an electric field in the y-direction (field strength $\mathfrak{E}_y$).

effect" and "2nd Ettingshausen–Nernst effect." While these designations are superfluous, a short title for effect 6, which in the pertinent literature has no name, should be introduced. The designation "Kelvin effect" is suggested since William Thomson (Lord Kelvin) experimentally examined thermoelectric phenomena in magnetic fields as early as 1856. The "Kelvin effect," like the Nernst effect (effect 4), is related to the Peltier effect in magnetic fields. Correspondingly, there is a relation analogous to the Thomson relation (4–25.69) to be expected between these phenomena and effects 10 and 12 (the thermoelectric potentials in magnetic fields) (cf. below).

According to Eqs. (18) and (19), the coefficients which describe the transverse effects (Hall effect, Ettingshausen effect, Righi–Leduc effect, and Ettingshausen–Nernst effect) represent odd functions of the magnetic induction $\mathfrak{B}$ while all other coefficients represent even functions of $\mathfrak{B}$. As a first approximation, we may thus expect that the coefficients of the transverse effects will be proportional to $\mathfrak{B}$ (= magnitude of the magnetic induction). This has been confirmed experimentally. On the other hand, experience shows that the coefficients of the other effects usually have the form $a + b\mathfrak{B}^2$, where a and b are (temperature-dependent) material constants.

For the "material constants" of the transverse effects (see Table 20)

$$\frac{\alpha_{12}}{\mathfrak{B}} = a_{\mathrm{H}} \quad \text{(Hall coefficient)}, \tag{4–36.28}$$

$$-T\frac{\alpha_{42}}{\alpha_{44}\mathfrak{B}} = \frac{\chi}{\mathfrak{B}} = a_{\mathrm{E}} \quad \text{(Ettingshausen coefficient)}, \tag{4–36.29}$$

$$-\frac{\alpha_{45}}{\alpha_{44}\mathfrak{B}} = a_{\mathrm{RL}} \quad \text{(Righi–Leduc coefficient)}, \tag{4–36.30}$$

$$\frac{\alpha_{15}}{T\mathfrak{B}} = \frac{\eta}{\mathfrak{B}} = a_{\mathrm{EN}} \quad \text{(Ettingshausen–Nernst coefficient)}, \tag{4–36.31}$$

we give some examples in Table 21 (copper at three temperatures).

Table 20

Classification of the Galvanomagnetic and Thermomagnetic Elementary Effects for Isotropic Electronic Conductors (x, y, z = cartesian coordinates; T = absolute temperature; $\mathfrak{E}_x, \mathfrak{E}_y, \mathfrak{E}_z$ = components of the electric field strength; I_x, I_y, I_z = components of the electric current density; J_x, J_y, J_z = components of the heat current density. The magnetic induction $\mathfrak{B}$ lies in the z-direction)

Class	Effect	Conditions	Description of the effect
Galvanomagnetic transverse effects	1. Hall effect (1879)	$I_y = 0$, $\partial T/\partial x = 0$, $\partial T/\partial y = 0$	$\dfrac{\mathfrak{E}_y}{I_x} = -\alpha_{12}$ $\equiv -a_{\mathrm{H}}\mathfrak{B}$
	2. Ettingshausen effect (1887)	$I_y = 0$, $\partial T/\partial x = 0$, $J_y = 0$	$\dfrac{\partial T/\partial y}{I_x} = -T\dfrac{\alpha_{42}}{\alpha_{44}}$ $\equiv \chi \equiv a_{\mathrm{E}}\mathfrak{B}$
Galvanomagnetic longitudinal effects in the transverse field	3. Electric conduction in the transverse magnetic field	$I_y = 0$, $\partial T/\partial x = 0$, $\partial T/\partial y = 0$	$\dfrac{\mathfrak{E}_x}{I_x} = \alpha_{11} \equiv \rho_{\mathrm{tr}}$
	4. Nernst effect (1887)	$I_y = 0$, $\partial T/\partial y = 0$, $J_x = 0$	$\dfrac{\partial T/\partial x}{I_x} = T\dfrac{\alpha_{41}}{\alpha_{44}} \equiv \nu$
Galvanomagnetic longitudinal effects in the longitudinal field	5. Electric conduction in the longitudinal magnetic field	$\partial T/\partial z = 0$	$\dfrac{\mathfrak{E}_z}{I_z} = \alpha_{33}$
	6. "Kelvin effect"	$J_z = 0$	$\dfrac{\partial T/\partial z}{I_z} = T\dfrac{\alpha_{63}}{\alpha_{66}} \equiv \beta$
Thermomagnetic transverse effects	7. Righi–Leduc effect (1887)	$I_x = 0$, $I_y = 0$, $J_y = 0$	$\dfrac{\partial T/\partial y}{\partial T/\partial x} = \dfrac{\alpha_{45}}{\alpha_{44}}$ $\equiv -a_{\mathrm{RL}}\mathfrak{B}$
	8. Ettingshausen–Nernst effect (1886)	$I_x = 0$, $I_y = 0$, $\partial T/\partial y = 0$	$\dfrac{E_y}{\partial T/\partial x} = \dfrac{\alpha_{15}}{T} \equiv \eta$ $\equiv a_{\mathrm{EN}}\mathfrak{B}$
Thermomagnetic longitudinal effects in the transverse field	9. Heat conduction in the transverse magnetic field	$I_x = 0$, $I_y = 0$, $\partial T/\partial y = 0$	$-\dfrac{J_x}{\partial T/\partial x} = \dfrac{\alpha_{44}}{T}$ $\equiv \lambda_{\mathrm{tr}}$
	10. Thermoelectric potential in the transverse magnetic field	$I_x = 0$, $I_y = 0$, $\partial T/\partial y = 0$	$\dfrac{\mathfrak{E}_x}{\partial T/\partial x} = -\dfrac{\alpha_{14}}{T}$ $\equiv \zeta$
Thermomagnetic longitudinal effects in the longitudinal field	11. Heat conduction in the longitudinal magnetic field	$I_z = 0$	$-\dfrac{J_z}{\partial T/\partial z} = \dfrac{\alpha_{66}}{T}$ $\equiv \lambda_{\mathrm{l}}$
	12. Thermoelectric potential in the longitudinal magnetic field	$I_z = 0$	$\dfrac{\mathfrak{E}_z}{\partial T/\partial z} = -\dfrac{\alpha_{36}}{T}$ $\equiv \gamma$

(c) Relations Between the Effects

By applying the definitions of Table 20 to the reciprocity relations (25) to (27) we obtain the following three explicit interrelations:

$$\lambda_{tr}\chi = T\eta, \tag{4–36.32}$$

$$\lambda_{tr}\nu = T\zeta, \tag{4–36.33}$$

$$\lambda_1\beta = T\gamma. \tag{4–36.34}$$

Here λ_{tr} or λ_1 denotes the thermal conductivity in the transverse or longitudinal magnetic field respectively.

Table 21

Coefficients of the Transverse Effects for Copper. After Hall[214]

Temperature °C	$a_H \times 10^{11}$ m^3 s^{-1} A^{-1}	$a_E \times 10^8$ m^3 s^{-1} K A^{-1} V^{-1}	$a_{RL} \times 10^3$ m^2 s^{-1} V^{-1}	$a_{EN} \times 10^8$ m^2 s^{-1} K^{-1}
25	5.36	1.446	2.698	2.155
55	5.37	1.544	2.448	2.104
85	5.38	1.674	2.272	2.014

Equation (32) formulates a relation between the Ettingshausen effect χ and the Ettingshausen–Nernst effect η. This relation was stated earlier by Bridgman (1929).

Equation (33) represents a relation between the Nernst effect ν and the thermoelectric potential in the transverse magnetic field ζ[215]. It corresponds to an extension of the Thomson relation (4–25.69) to systems in transverse magnetic fields.

Finally, Eq. (34) describes the relation between the Kelvin effect β and the thermoelectric potential in a longitudinal magnetic field γ[209b]. It corresponds to an extension of the Thomson relation (4–25.69) to systems in longitudinal magnetic fields.

There are only sufficient experimental data for the Bridgman relation (32) which, according to Eqs. (29) and (31), may be written in the form

$$\lambda_{tr}a_E = Ta_{EN}. \tag{4–36.35}$$

The measured data are assembled in Table 22. The confirmation of Eq. (35) is not too impressive. It must be noted, however, that the measurements are very difficult and reproducibility is often poor. In that single case in which all investigations were carried out on the same sample, namely on arsenic[216], Eq. (35) holds almost quantitatively. Good agreement is also obtained with bismuth for which all effects are unusually large and for which many experimental results are available.

Table 22

Experimental Verification of the Bridgman Relation ($\lambda_{tr}a_E = Ta_{EN}$) for Several Metals at Temperatures around 20°C. After Miller[55]

Metal	$\dfrac{\lambda_{tr}a_E}{T} \times 10^8$ m² s⁻¹ K⁻¹	$a_{EN} \times 10^8$ m² s⁻¹ K⁻¹	$\dfrac{\lambda_{tr}a_E}{Ta_{EN}}$
Ag	2.2	1.8	1.22
Al	0.60	0.42	1.43
As	22.0	22.5	0.98
Au	1.2	1.8	0.66
Bi	2200	2340	0.94
Cd	0.9	1.2	0.75
Co	22.0	21.9	1.00
Cu	2	1.9	1.05
Fe	8.6	9.5	0.91
Ni	22.0	30.4	0.72
Pd	4.00	3.26	1.23
Sb	220	176	1.25
Zn	1.00	0.73	1.37

Finally, it should be noted that other relations between the effects are often given in the literature. These are not consequences of the Onsager–Casimir reciprocity law but are based on certain models from electron theory. The best-known relation of this kind relates the Hall coefficient a_H to the resistivity (the reciprocal specific conductance) in the transverse magnetic field ρ_{tr} and the Righi–Leduc coefficient a_{RL}:

$$a_H = \rho_{tr}a_{RL}. \tag{4–36.36}$$

Equation (36) is well confirmed experimentally. By combination of this relation with Eq. (35) we find:

$$\frac{a_H a_{EN}}{a_{RL} a_E} = \frac{\rho_{tr}\lambda_{tr}}{T}. \tag{4–36.37}$$

Application of the Wiedemann–Franz–Sommerfeld law to metals in magnetic fields would produce for the "Lorenz constant," i.e.

$$L \equiv \frac{\rho_{tr}\lambda_{tr}}{T}, \tag{4–36.38}$$

the universal value (see p. 336)

$$L = \frac{\pi^2}{3}\left(\frac{R}{F}\right)^2 \approx 2.44 \times 10^{-8} \text{ V}^2 \text{ K}^{-2} \tag{4–36.39}$$

(R = gas constant). The data in Table 21 with Eqs. (37) and (38) yield the following values for copper:

$$L = 2.96 \times 10^{-8} \, V^2 \, K^{-2} \quad (25°C),$$
$$L = 2.99 \times 10^{-8} \, V^2 \, K^{-2} \quad (55°C),$$
$$L = 2.38 \times 10^{-8} \, V^2 \, K^{-2} \quad (85°C).$$

These are in tolerable agreement with Eq. (39). The question of the validity of the Wiedemann–Franz law for metals in magnetic fields is quite complicated[212].

4-37 PROCESSES IN ANISOTROPIC SYSTEMS

Having shown in §4–34 how the transfer from stationary to arbitrary electromagnetic fields is handled in our basic equations (mass balance, momentum balance, energy balance, entropy balance, etc.) we will now ask how the above-mentioned equations could be modified if we give up the assumption of isotropy of the medium. However, in order not to be forced to take account of all complications simultaneously, we will again assume stationary external force fields and we will ignore electric and magnetic polarization. Besides elastic stresses, we allow viscous pressures and therefore include "viscoelastic bodies" in our considerations. Thus we treat processes in arbitrary anisotropic systems insofar as electrification and magnetization do not play a role and external fields, if present, are constant in time.

The calculations are considerably simplified if we do not proceed, as in §4–8 and §4–34, from the balance equations with the local time derivatives (operator $\partial/\partial t$), but from the corresponding relations with the substantial derivatives (operator d/dt) (see Eq. 4–4.5).

We write the local *mass balance* in the form (4–4.20):

$$\rho \frac{d\chi_k}{dt} = -\operatorname{div} \mathbf{J}_k^* + M_k \sum_r \nu_{kr} b_r. \qquad (4\text{--}37.1)$$

Here ρ denotes the density, χ_k the mass fraction of species k, $\mathbf{J}_k^*$ the quantity defined by Eq. (4–3.14), i.e. the diffusion current density in the barycentric system, M_k the molar mass of species k, ν_{kr} the stoichiometric number of species k in the chemical reaction r, and b_r the rate of reaction of the chemical reaction r. According to Eqs. (4–3.11) and (4–3.14), the relation

$$\mathbf{J}_k^* = M_k \,_v\mathbf{J}_k = M_k c_k (\mathbf{v}_k - \mathbf{v}) \qquad (4\text{--}37.2)$$

is valid, where $\mathbf{v}$ is the barycentric velocity, $_v\mathbf{J}_k$ the diffusion flux density defined by Eq. (4–3.11), c_k the molarity, and $\mathbf{v}_k$ the velocity of species k.

In addition, the *momentum balance equation* in the form (4–5.11) and the relation (4–5.12) following from it can be used provided that by Π ("pressure tensor") we understand the following tensor:

$$\Pi = \begin{pmatrix} P_{11} - \tau_{11} & P_{12} - \tau_{12} & P_{13} - \tau_{13} \\ P_{21} - \tau_{21} & P_{22} - \tau_{22} & P_{23} - \tau_{23} \\ P_{31} - \tau_{31} & P_{32} - \tau_{32} & P_{33} - \tau_{33} \end{pmatrix}. \tag{4–37.3}$$

Here the P_{ij} $(= P_{ji})$ are the viscous pressures and the τ_{ij} $(= \tau_{ji})$ the stress components $(i, j = 1, 2, 3)$. For negligible viscous flow or for purely elastic media, Π is reduced—after reversal of sign—to the stress tensor in the sense of the theory of elasticity.

Equation (4–5.12) now states:

$$\frac{\partial}{\partial t} \left(\frac{\rho}{2} v^2 \right) + \operatorname{div} \left(\frac{\rho}{2} v^2 \mathbf{v} \right) = \mathbf{v} \left(\sum_k c_k \mathbf{K}_k - \operatorname{Div} \Pi \right). \tag{4–37.4}$$

Here $\mathbf{K}_k$ is the molar external force acting on species k and $\operatorname{Div} \Pi$ the tensor divergence of Π, i.e. a vector with the cartesian components

$$\sum_{j=1}^{3} \frac{\partial (P_{ij} - \tau_{ij})}{\partial z_j} \quad (i = 1, 2, 3), \tag{4–37.5}$$

where z_1, z_2, z_3 are the space-coordinates and the operator $\partial / \partial z_j$ is the derivative with respect to the position-coordinate at a fixed time.

Let $\Pi \mathbf{v}$ be a vector with the cartesian components

$$\sum_{j=1}^{3} (P_{ij} - \tau_{ij}) v_j \quad (i = 1, 2, 3), \tag{4–37.6}$$

where v_1, v_2, v_3 are the cartesian components of $\mathbf{v}$. Then $\operatorname{div} (\Pi \mathbf{v})$ is that part of Eq. (4–6.4) corresponding to the term $\operatorname{div} (P\mathbf{v})$ which describes the deformation work per unit volume and unit time. Thus we can simply extend Eq. (4–6.4), which holds for isotropic media without viscous flow, to anisotropic systems with viscous flow if we substitute $\operatorname{div} (\Pi \mathbf{v})$ for $\operatorname{div} (P\mathbf{v})$. We immediately simplify the local *energy balance* so obtained by taking account of Eqs. (2) and (4) as well as of the identity

$$\operatorname{div} (\Pi \mathbf{v}) = \mathbf{v} \operatorname{Div} \Pi + \Pi : \operatorname{Grad} \mathbf{v}, \tag{4–37.7}$$

where $\operatorname{Grad} \mathbf{v}$ is the vector gradient of $\mathbf{v}$ (that is, a tensor) and the sign : denotes the inner product of the two tensors after double contraction, and thus leads to a scalar quantity. Thus on noting Eq. (4–6.16) we find:

$$\rho \frac{d\tilde{U}}{dt} = \frac{\partial U_V}{\partial t} + \operatorname{div} (U_V \mathbf{v})$$

$$= -\operatorname{div} \left(\mathbf{J}_Q + \sum_k H_{k \, V} \mathbf{J}_k \right) + \sum_k \mathbf{K}_{k \, V} \mathbf{J}_k - \Pi : \operatorname{Grad} \mathbf{v}. \tag{4–37.8}$$

Here $\tilde{U}$ is the specific internal energy, U_V the density of the internal energy, $\mathbf{J}_Q$ the heat current density, and H_k the partial molar enthalpy of species k. Equation (8) combines the momentum and energy balances.

Expansion of the last term in Eq. (8) into its components shows†

$$\Pi : \mathrm{Grad}\ \mathbf{v} = \sum_{i=1}^{3} \sum_{j=1}^{3} (P_{ij} - \tau_{ij}) \frac{\partial v_i}{\partial z_j} \equiv \sum_{i,j} (P_{ij} - \tau_{ij}) \frac{\partial v_i}{\partial z_j}$$

$$= \sum_{i,j} P_{ij} \frac{\partial v_i}{\partial z_j} - \sum_{i,j} \tau_{ij} \frac{\partial e_{ij}}{\partial t}. \tag{4-37.9}$$

Here the e_{ij} ($= e_{ji}$) are the components of the strain tensor.

For isotropic media, we have

$$\tau_{12} = \tau_{21} = \tau_{13} = \tau_{31} = \tau_{23} = \tau_{32} = 0, \qquad \tau_{11} = \tau_{22} = \tau_{33} = -P,$$

that is, according to Eq. (9),

$$\sum_{i,j} \tau_{ij} \frac{de_{ij}}{dt} = \sum_{i,j} \tau_{ij} \frac{\partial v_i}{\partial z_j} = -P\ \mathrm{div}\ \mathbf{v},$$

where P is the static pressure. Accordingly, Eq. (8) is reduced to Eq. (4–6.9) for isotropy, as it must be.

For sufficiently slow irreversible processes, the generalized Gibbs equation holds for each volume element. This equation has the form (1–9.14) for an anisotropic medium without electrification and magnetization (see Eq. 1–3.3):

$$T \frac{d\tilde{S}}{dt} = \frac{d\tilde{U}}{dt} - \frac{1}{\rho} \sum_{i,j} \tau_{ij} \frac{de_{ij}}{dt} - \sum_{k} \tilde{\mu}_k \frac{d\chi_k}{dt}, \tag{4-37.10}$$

where (see Eq. 1–9.15)

$$\tilde{\mu}_k = \frac{\mu_k}{M_k}. \tag{4-37.11}$$

Here T is the absolute temperature, $\tilde{S}$ the specific entropy, and μ_k or $\tilde{\mu}_k$ the molar (as used thus far) or the specific chemical potential of species k respectively.

From Eqs. (1) and (8) we substitute the derivatives $d\chi_k/dt$ and $d\tilde{U}/dt$ in Eq. (10), taking account of Eqs. (2), (9), and (11), and we set

$$\mathbf{J}_Q^* \equiv \mathbf{J}_Q + \sum_{k} H_k\,{}_v\mathbf{J}_k. \tag{4-37.12}$$

---

† For the substantial derivative d/dt of the components of the symmetric strain tensor, there is valid:

$$\frac{de_{ij}}{dt} = \frac{\partial e_{ij}}{\partial t} + \mathbf{v}\ \mathrm{grad}\ e_{ij} = \frac{1}{2}\left(\frac{\partial v_i}{\partial z_j} + \frac{\partial v_j}{\partial z_i}\right) \quad (i, j = 1, 2, 3).$$

Then we obtain

$$\rho \frac{d\tilde{S}}{dt} = \frac{1}{T}\sum_k \mathbf{K}_k \, _v\mathbf{J}_k - \frac{1}{T}\sum_{i,j} P_{ij}\frac{\partial v_i}{\partial z_j} - \frac{1}{T}\sum_k\sum_r \nu_{kr}\mu_k b_r$$

$$- \frac{1}{T}\,\text{div}\,\mathbf{J}_Q^* + \frac{1}{T}\sum_k \mu_k\,\text{div}\,_v\mathbf{J}_k. \qquad (4\text{-}37.13)$$

We now note the vector identities

$$\frac{1}{T}\,\text{div}\,\mathbf{J}_Q^* = \text{div}\left(\frac{\mathbf{J}_Q^*}{T}\right) + \mathbf{J}_Q^*\frac{\text{grad}\,T}{T^2}, \qquad (4\text{-}37.14)$$

$$\frac{1}{T}\sum_k \mu_k\,\text{div}\,_v\mathbf{J}_k = \text{div}\sum_k \frac{1}{T}(\mu_k\,_v\mathbf{J}_k) - \sum_k\,_v\mathbf{J}_k\,\text{grad}\,\frac{\mu_k}{T}, \quad (4\text{-}37.15)$$

and the general thermodynamic relations (cf. Eqs. 4–34.66 to 4–34.68 as well as Eq. 4–9.3)

$$A_r = -\sum_k \nu_{kr}\mu_k, \qquad (4\text{-}37.16)$$

$$\text{grad}\left(\frac{\mu_k}{T}\right) = -\frac{H_k}{T^2}\,\text{grad}\,T + \frac{1}{T}(\text{grad}\,\mu_k)_T, \qquad (4\text{-}37.17)$$

$$\mu_k = H_k - TS_k, \qquad (4\text{-}37.18)$$

$$S_V = \sum_k c_k S_k, \qquad (4\text{-}37.19)$$

where A_r denotes the affinity of reaction r, S_k the partial molar entropy of species k, and S_V the entropy density. With the identity (cf. Eq. 4–4.19)

$$\rho\frac{d\tilde{S}}{dt} = \frac{\partial S_V}{\partial t} + \text{div}\,(S_V\mathbf{v}) \qquad (4\text{-}37.20)$$

we derive from Eqs. (12) to (19):

$$\frac{\partial S_V}{\partial t} = -\text{div}\,\mathbf{J}_S + \vartheta \qquad (4\text{-}37.21)$$

with

$$\mathbf{J}_S \equiv \frac{\mathbf{J}_Q}{T} + \sum_k c_k S_k \mathbf{v}_k, \qquad (4\text{-}37.22)$$

$$\vartheta \equiv -\frac{1}{T^2}\mathbf{J}_Q\,\text{grad}\,T + \frac{1}{T}\sum_k\,_v\mathbf{J}_k[\mathbf{K}_k - (\text{grad}\,\mu_k)_T]$$

$$+ \frac{1}{T}\sum_r b_r A_r - \frac{1}{T}\sum_{i,j} P_{ij}\frac{\partial v_i}{\partial z_j}. \qquad (4\text{-}37.23)$$

Equation (21) is the local *entropy balance* with the *entropy current density* $\mathbf{J}_S$ and the *local entropy production* ϑ given by Eqs. (22) and (23). The quantity ϑ must be positive for the actual (irreversible) course of the processes, while it disappears for the reversible limiting case (in particular at equilibrium) (cf. p. 241):

$$\vartheta \geqslant 0. \tag{4–37.24}$$

Equations (21) to (23) formally agree with Eqs. (4–8.13) to (4–8.15). The anisotropy of the medium is only seen from the fact that the chemical potentials μ_k depend on the stress components τ_{ij} (or the strain components e_{ij}) as well as on temperature and composition. For isotropic media, we need only consider the pressure (or the density) in place of the τ_{ij} or e_{ij}.

For the *dissipation function*

$$\Psi \equiv T\vartheta, \tag{4–37.25}$$

we can write from Eqs. (23) and (24) (cf. Eqs. 4–12.2 to 4–12.5):

$$\Psi = \mathbf{J}_Q \mathbf{X}_Q + \sum_k {}_v\mathbf{J}_k\mathbf{X}_k + \sum_r b_r A_r + \sum_{i,j} P_{ij}X_{ij} \geqslant 0 \tag{4–37.26}$$

with

$$\mathbf{X}_Q \equiv -\frac{1}{T}\,\mathrm{grad}\,T, \tag{4–37.27}$$

$$\mathbf{X}_k \equiv \mathbf{K}_k - (\mathrm{grad}\,\mu_k)_T, \tag{4–37.28}$$

$$X_{ij} \equiv -\frac{1}{2}\left(\frac{\partial v_i}{\partial z_j} + \frac{\partial v_j}{\partial z_i}\right) \quad (i,j = 1, 2, 3). \tag{4–37.29}$$

The quantities $\mathbf{J}_Q$, ${}_v\mathbf{J}_k$, b_r, and P_{ij} are the *generalized fluxes*; the expressions $\mathbf{X}_Q$, $\mathbf{X}_k$, A_r, and X_{ij} are the *generalized forces*.

If relaxation processes are interpreted as "internal transformations" and if they are formally described like chemical reactions (cf. §2–11) then Eq. (26) can serve as the starting point for a thermodynamic–phenomenological theory of transport phenomena in anisotropic media with "internal damping" which can be related to the theory of "after effects"[217].

If no velocity gradients and no gradients of the concentrations and stress components are present then:

$$X_{ij} = 0, \qquad (\mathrm{grad}\,\mu_k)_T = 0 \tag{4–37.30}$$

and the dissipation function (26) is reduced to the expression

$$\Psi = \mathbf{J}_Q \mathbf{X}_Q + \sum_k {}_v\mathbf{J}_k\mathbf{K}_k + \sum_r b_r A_r \geqslant 0. \tag{4–37.31}$$

This includes two important special cases which we will now discuss.

If, on the one hand, temperature gradients and external forces are excluded then we have

$$X_Q = 0, \qquad K_k = 0 \qquad\qquad (4\text{–}37.32)$$

and according to Eq. (31):

$$\Psi = \sum_r b_r A_r \geqslant 0, \qquad\qquad (4\text{–}37.33)$$

just as for chemical reactions in homogeneous isotropic media (see Eq. 2–2.13). Since the affinities A_r depend on the temperature, on the composition, and on the stress or strain components, the explicit evaluation of the formulas following from Eq. (33) is more complicated than for isotropic systems. In this way, a thermodynamic theory of elastic relaxation can be developed[218, 219].

If, on the other hand, chemical reactions as well as internal transformations are excluded and only electromagnetic fields considered as external force fields, which must, according to our basic assumptions, be stationary and which may not lead to polarization of the material, then according to Eq. (2) and Eqs. (4–34.2) and (4–34.29) for systems at rest:

$$b_r = 0, \qquad \sum_k {}_v J_k K_k = I\mathfrak{E}, \qquad\qquad (4\text{–}37.34)$$

where I denotes the electric current density and $\mathfrak{E}$ the electric field strength. With this there follows from Eq. (31):

$$\Psi = J_Q X_Q + I\mathfrak{E} \geqslant 0, \qquad\qquad (4\text{–}37.35)$$

in agreement with Eq. (4–34.108).

We again formulate *phenomenological equations* as a first approximation to a description of irreversible processes in anisotropic systems. We must now consider the scalar fluxes and all components of the vectorial and tensorial fluxes in Eq. (26) as homogeneous linear functions of the scalar forces as well as of all components of the vectorial and tensorial forces.† This process is familiar from §4–35 and §4–36 since isotropic systems in magnetic fields also exhibit preferential directions.

In order to set up the phenomenological equations for anisotropic electronic conductors at rest in stationary electromagnetic fields, we must proceed from Eq. (35). Then we recover the equations (4–36.4). Thus the assertion formulated in §4–36 is proved, namely, that this system of equations describes

---

† The symmetry principle of P. Curie (p. 249) does not hold here. The only restrictions, which are valid from the beginning, follow from possible symmetry relations of the crystal type, because the phenomenological equations must be covariant with regard to crystallographic symmetry transformations of the lattice (cf. Table 23, p. 466).

electric conduction and heat conduction as well as the thermoelectric, galvanomagnetic, and thermomagnetic effects in anisotropic electronic conductors.

The *reciprocity law* of Onsager and Casimir is, in the most general case, to be applied in the form (4–35.3). Our examples in §4–38 and §4–39 relate, however, to systems without magnetic fields in which only "Onsager coefficients" appear (§1–26). The phenomenological coefficients α_{ik} thus obey here the simple condition:

$$\alpha_{ik} = \alpha_{ki}. \tag{4-37.36}$$

In general, it should be noted that symmetry statements for anisotropic media can have four different origins. If we consider a relation of the type

$$a_{ik} = a_{ki}, \tag{4-37.37}$$

where the a_{ik} are arbitrary quantities (that is, not necessarily phenomenological coefficients), then this may follow from

1. The basic laws of mechanics (e.g. symmetry of the tensors of strain, stress, or viscous pressure, cf. above).
2. The basic laws of classical thermodynamics (e.g. the symmetry of the matrix of the elasticity constants or of the tensor of the dielectric constants, cf. §4–38).
3. The geometric symmetry properties of the crystal class in question (cf. §4–39).
4. The Onsager reciprocity relations (e.g. the symmetry of the viscosity matrix, of the electric conductivity tensor, or of the thermal conductivity tensor, cf. §4–38 and §4–39).

4–38 ELECTRIC CONDUCTION AND ELECTRIC POLARIZATION IN ANISOTROPIC MEDIA

(a) Electric Conduction

As the simplest example of electric conduction in anisotropic media, we consider the transport of charge in anisotropic electronic conductors such as in metals with a noncubic crystal lattice. Temperature gradients, concentration gradients (for solid solutions), gradients of the elastic stresses, plastic processes, and internal transformations will be excluded. Also, in addition to an electrostatic field, there shall be no other external force field effective. Then, according to Eq. (4–37.35), the dissipation function becomes:

$$\Psi = \mathbf{I}\mathfrak{E} = I_x \mathfrak{E}_x + I_y \mathfrak{E}_y + I_z \mathfrak{E}_z. \tag{4-38.1}$$

Here $\mathbf{I}$ is the electric current density with the cartesian components I_x, I_y, I_z, and $\mathfrak{E}$ the electric field strength with the cartesian components $\mathfrak{E}_x$, $\mathfrak{E}_y$, $\mathfrak{E}_z$.

According to the discussion in §4–37, irreversible processes are described to a first approximation by the phenomenological equations which, in the case of electric conduction, following Eq. (1), yield a system of three linear equations with nine coefficients:

$$\left.\begin{array}{l} I_x = \kappa_{11}\mathfrak{E}_x + \kappa_{12}\mathfrak{E}_y + \kappa_{13}\mathfrak{E}_z, \\ I_y = \kappa_{21}\mathfrak{E}_x + \kappa_{22}\mathfrak{E}_y + \kappa_{23}\mathfrak{E}_z, \\ I_z = \kappa_{31}\mathfrak{E}_x + \kappa_{32}\mathfrak{E}_y + \kappa_{33}\mathfrak{E}_z. \end{array}\right\} \qquad (4\text{-}38.2)$$

The scalar quantities κ_{ij} ($i, j = 1, 2, 3$) are the phenomenological coefficients. In the most general case, they depend on the local state variables (temperature, stress components, concentrations).

Equation (2) is the generalization of Ohm's law (4-16.24) to anisotropic media. Accordingly, we designate the matrix

$$\kappa = \begin{pmatrix} \kappa_{11} & \kappa_{12} & \kappa_{13} \\ \kappa_{21} & \kappa_{22} & \kappa_{23} \\ \kappa_{31} & \kappa_{32} & \kappa_{33} \end{pmatrix} \qquad (4\text{-}38.3)$$

as the tensor of the specific conductance. The three scalar relations (2) can be combined symbolically into a vector equation using Eq. (3) (cf. Eq. 4-36.9):

$$\mathbf{I} = \kappa\mathfrak{E}. \qquad (4\text{-}38.4)$$

As long as the crystal class in question does not require symmetry relations leading to a reduction of the independent coefficients (cf. Table 23, p. 466), only the Onsager reciprocity law (4-37.36) produces a general statement:

$$\kappa_{ij} = \kappa_{ji} \quad (i, j = 1, 2, 3). \qquad (4\text{-}38.5)$$

The conductance tensor is thus always symmetric.

Although early workers[220] already conjectured the symmetry of the conductance tensor for anisotropic conductors, there is so far no adequate experimental evidence to verify Eq. (5). This is explained partly by the lack of metal unit crystals with the required symmetry properties (almost all crystal lattices of metals have a symmetry which is too high), and partly by the great sensitivity of resistance measurements on crystals to traces of impurities. Since in the analogous case of the thermal conductivity (§4–39) the situation is more favorable, we delay further discussion of the electric conductivity in crystalline media.

(b) Electric Polarization

We now consider the electric polarization in an anisotropic insulator.

The equations for the cartesian components $\mathfrak{P}_x$, $\mathfrak{P}_y$, $\mathfrak{P}_z$ of the electric polarization $\mathfrak{P}$ as functions of the corresponding components of the electric field strength for electrified anisotropic media are:†

---

† Here substances like "Rochelle" salt (sodium potassium tartrate) are excluded, since these are "ferroelectric," i.e. they exhibit saturation and hysteresis phenomena, so that the ψ_{ij} depend on the field strength.

$$\left.\begin{array}{l} \mathfrak{P}_x = \psi_{11}\mathfrak{E}_x + \psi_{12}\mathfrak{E}_y + \psi_{13}\mathfrak{E}_z, \\ \mathfrak{P}_y = \psi_{21}\mathfrak{E}_x + \psi_{22}\mathfrak{E}_y + \psi_{23}\mathfrak{E}_z, \\ \mathfrak{P}_z = \psi_{31}\mathfrak{E}_x + \psi_{32}\mathfrak{E}_y + \psi_{33}\mathfrak{E}_z \end{array}\right\} \qquad (4\text{-}38.6)$$

with

$$\psi_{ij} \equiv \varepsilon_0(\varepsilon_{ij} - 1) \quad (i, j = 1, 2, 3). \qquad (4\text{-}38.7)$$

Here ε_0 denotes the permittivity of vacuum and the ε_{ij} the components of the tensor of the dielectric constant (cf. Eqs. 4–34.20 and 4–34.22). The ψ_{ij} or ε_{ij} are "material constants," i.e. dependent on the type of the medium and on the local values of the intensive state variables. They can therefore be represented as functions of the temperature T, of the stress components τ_{ij}, and of the concentrations (molarities) c_k of the species contained in the region:

$$\psi_{ij} = \psi_{ij}(T, \tau_{ij}, c_k) \quad (i, j = 1, 2, 3). \qquad (4\text{-}38.8)$$

This system of equations (6) formally corresponds to Eqs. (2) but has the character of an equation of state.

Now the generalized Gibbs equation in the form (1–9.9) states:†

$$T\, dS_V = dU_V - \sum_{i=1}^{3} \sum_{j=1}^{3} \tau_{ij}\, de_{ij} - \mathfrak{E}_x\, d\mathfrak{P}_x - \mathfrak{E}_y\, d\mathfrak{P}_y - \mathfrak{E}_z\, d\mathfrak{P}_z - \sum_{k} \mu_k\, dc_k.$$
$$(4\text{-}38.9)$$

Here S_V or U_V is the entropy density or the density of the internal energy respectively and μ_k the chemical potential of species k. If we introduce the density of the Gibbs function using Eqs. (1–6.7) and (1–11.2):

$$G_V = U_V - TS_V - \sum_{i=1}^{3} \sum_{j=1}^{3} \tau_{ij}e_{ij} - \mathfrak{E}_x\mathfrak{P}_x - \mathfrak{E}_y\mathfrak{P}_y - \mathfrak{E}_z\mathfrak{P}_z \qquad (4\text{-}38.10)$$

then with the help of Eq. (9) we obtain:

$$dG_V = -S_V\, dT - \sum_{i=1}^{3} \sum_{j=1}^{3} e_{ij}\, d\tau_{ij} - \mathfrak{P}_x\, d\mathfrak{E}_x - \mathfrak{P}_y\, d\mathfrak{E}_y$$
$$- \mathfrak{P}_z\, d\mathfrak{E}_z + \sum_{k} \mu_k\, dc_k. \qquad (4\text{-}38.11)$$

---

† See Eq. (1–3.6). We set the volume V of the region in question equal to the reference volume V_0. Equation (9) is equivalent to the relation (4–37.10) for $\mathfrak{P}_x = \mathfrak{P}_y = \mathfrak{P}_z = 0$.

Since G_V is a function of state—in contrast to Ψ in Eq. (1)—we can make a statement about the coefficients ψ_{ij} of Eq. (6) with the help of *classical* thermodynamics alone. Namely, it follows from Eq. (11) that

$$\frac{\partial^2 G_V}{\partial \mathfrak{E}_i \, \partial \mathfrak{E}_j} = -\frac{\partial \mathfrak{P}_i}{\partial \mathfrak{E}_j} = \frac{\partial^2 G_V}{\partial \mathfrak{E}_j \, \partial \mathfrak{E}_i} = -\frac{\partial \mathfrak{P}_j}{\partial \mathfrak{E}_i}$$

or

$$\frac{\partial \mathfrak{P}_i}{\partial \mathfrak{E}_j} = \frac{\partial \mathfrak{P}_j}{\partial \mathfrak{E}_i} \quad (i, j = x, y, z). \tag{4–38.12}$$

With this there results from Eqs. (6) to (8):

$$\psi_{ij} = \psi_{ji}, \qquad \varepsilon_{ij} = \varepsilon_{ji} \quad (i, j = 1, 2, 3). \tag{4–38.13}$$

Thus the tensor of the dielectric constant is always symmetric. This statement holds without regard to the crystal system and also has nothing to do with the Onsager reciprocity law.†

For sufficiently small deformations, we can, by analogy with Eq. (6), set up a linear equation between the six independent strain components and the six independent stress components. This "generalized Hooke's law" (Cauchy, 1829) is thus a system of six linear equations with 36 temperature- and concentration-dependent coefficients (the "elasticity constants" ζ). From Eq. (11) the symmetry statements (W. Thomson, 1855) follow in an analogous way to the above, with the help of classical thermodynamic considerations:

$$\zeta_{ij} = \zeta_{ji} \quad (i, j = 1, 2, \ldots, 6).$$

Thus the number of the independent elasticity constants is reduced from 36 to 21.

The above-mentioned Hooke–Cauchy law is, like Eq. (6), an "equation of state", while Eq. (2) represents an equation for irreversible processes.

The equation for viscous flow in anisotropic media is essentially analogous to the generalized Ohm's law (2) but formally similar to the Hooke–Cauchy law. According to Eq. (4–37.26), the phenomenological equations lead to homogeneous linear relations between the 6 independent viscous pressures and the 6 independent combinations (4–37.29) of the velocity gradients. The 36 coefficients of the system of equations form the viscosity matrix. This matrix is, due to the Onsager reciprocity law (4–37.36), symmetric. Thus, in the general case, there are 21 independent coefficients again.

4–39 HEAT CONDUCTION IN ANISOTROPIC MEDIA

Let there be temperature gradients in an anisotropic medium. Let velocity gradients, gradients of the stress components, and of the concentrations, as

---

† The reduction of the 9 coefficients of the linear equations (6) to 6 independent coefficients by classical thermodynamic considerations is due to W. Thomson (Lord Kelvin) and has been proved in principle in the same way as above in the outstanding book by Voigt[220].

well as reactions and external force fields be excluded. Then heat conduction appears in each volume element as the only irreversible process and according to Eq. (4–37.35) the dissipation function becomes:

$$\Psi = \mathbf{J}_Q \mathbf{X}_Q = J_x X_x + J_y X_y + J_z X_z. \tag{4–39.1}$$

Here $\mathbf{J}_Q$ is the heat current density with the cartesian components J_x, J_y, J_z, and $\mathbf{X}_Q$ is the vector (see Eq. 4–37.27)

$$\mathbf{X}_Q \equiv -\frac{1}{T}\,\mathrm{grad}\,T \tag{4–39.2}$$

with the cartesian components

$$X_x = -\frac{1}{T}\frac{\partial T}{\partial x}, \quad X_y = -\frac{1}{T}\frac{\partial T}{\partial y}, \quad X_z = -\frac{1}{T}\frac{\partial T}{\partial z}, \tag{4–39.3}$$

where T denotes the absolute temperature, x, y, z the space-coordinates, and the operators $\partial/\partial x$, $\partial/\partial y$, and $\partial/\partial z$ the space derivatives at a fixed time.

The phenomenological equations, which describe the heat conduction to a first approximation, according to §4–37 and Eq. (1) state:

$$\left.\begin{aligned} J_x &= \alpha_{11} X_x + \alpha_{12} X_y + \alpha_{13} X_z, \\ J_y &= \alpha_{21} X_x + \alpha_{22} X_y + \alpha_{23} X_z, \\ J_z &= \alpha_{31} X_x + \alpha_{32} X_y + \alpha_{33} X_z. \end{aligned}\right\} \tag{4–39.4}$$

The nine phenomenological coefficients α_{ij} are, in general, functions of the local state variables (temperature, stress components, concentrations). They are connected to the thermal conductivities λ_{ij} as follows from Eq. (3):

$$\alpha_{ij} = T\lambda_{ij} \quad (i, j = 1, 2, 3). \tag{4–39.5}$$

The matrix

$$\lambda = \begin{pmatrix} \lambda_{11} & \lambda_{12} & \lambda_{13} \\ \lambda_{21} & \lambda_{22} & \lambda_{23} \\ \lambda_{31} & \lambda_{32} & \lambda_{33} \end{pmatrix} \tag{4–39.6}$$

is called the thermal conductivity tensor.

Equation (4) is the generalization of Fourier's law (4–24.32) to anisotropic media and was proposed in 1832 by Duhamel (derived on a molecular physical basis with some hypotheses) and then in a more general way in 1851 by Stokes. Equation (4) is analogous to Ohm's law (4–38.2) for anisotropic media and has a wide region of validity. Failure of Eq. (4) is to be expected only at extremely large temperature gradients, corresponding to the fact that Ohm's law becomes invalid only at very high field strengths.

The Onsager reciprocity law (4–37.36) requires that

$$\alpha_{ij} = \alpha_{ji} \quad (i, j = 1, 2, 3) \tag{4–39.7}$$

or according to Eq. (5):

$$\lambda_{ij} = \lambda_{ji} \quad (i, j = 1, 2, 3). \tag{4–39.8}$$

The thermal conductivity tensor (6) is thus always symmetric.

Equation (8) was suspected quite early (Duhamel, 1832) and was made probable by measurements (De Senarmont, 1848). However, Soret (1893) and Voigt (1903) were the first to refine the experimental methods, so that it is possible to speak of an experimental confirmation of Eq. (8). This finding is one of those facts which led Onsager (1931) to his reciprocity law.

Before experiments in this field are discussed, we must be clear about which statements follow *only from the geometric symmetry relations* for different crystal types. The general equation (4) with 9 coefficients holds only for crystals of the triclinic system, while regular crystals behave isotropically, i.e. they exhibit only a single independent coefficient.

We obtain the summary shown in Table 23[220] which, moreover, is valid for the coefficients of any system of linear equations of the type (4). For example, it holds for the coefficients in Eqs. (4–38.2) and (4–38.6).†

The simplest nontrivial case for which the validity of the Onsager reciprocity relations (8) can be tested relates to crystal classes 12 and 13 of the trigonal system, crystal classes 17, 18, and 20 of the tetragonal system, as well as crystal classes 24, 25, and 27 of the hexagonal system (see Table 23). In all these cases, the system of equations (4) is reduced with (3) and (5) to the following equations

$$\left. \begin{aligned} J_x &= -\lambda_{11} \frac{\partial T}{\partial x} - \lambda_{12} \frac{\partial T}{\partial y}, \\ J_y &= \lambda_{12} \frac{\partial T}{\partial x} - \lambda_{11} \frac{\partial T}{\partial y}, \\ J_z &= -\lambda_{33} \frac{\partial T}{\partial z}. \end{aligned} \right\} \tag{4–39.9}$$

Here the Onsager reciprocity law (8) states:

$$\lambda_{12} = \lambda_{21}.$$

This relation is consistent with the statement that

$$\lambda_{12} = -\lambda_{21},$$

which follows from Eqs. (6) and (9), only under the condition

$$\lambda_{12} = 0. \tag{4–39.10}$$

---

† Here the coefficients are presented in the principal coordinate system. That is, the coordinate axes coincide as far as possible with the crystallographic principal axes.

A test of the reciprocity law simply amounts to an experimental verification of Eq. (10) for suitable crystals (e.g. dolomite, wolframic limestone, or apatite).

Table 23

Crystal system	Crystal classes (with examples)	Scheme of coefficients (thermal conductivity tensor)		
Triclinic	1, 2 (1: copper sulfate)	λ_{11} λ_{21} λ_{31}	λ_{12} λ_{22} λ_{32}	λ_{13} λ_{23} λ_{33}
Monoclinic	3, 4, 5 (3: gypsum, 5: sugar)	λ_{11} λ_{21} 0	λ_{12} λ_{22} 0	0 0 λ_{33}
Rhombic	6, 7, 8 (6: topaz, 7: rochelle salt, 8: willemite)	λ_{11} 0 0	0 λ_{22} 0	0 0 λ_{33}
Trigonal	9, 10, 11 (9: calcite, 10: quartz, 11: tourmaline)	λ_{11} 0 0	0 λ_{22} 0	0 0 λ_{33}
Trigonal	12, 13 (12: dolomite, 13: sodium periodate)	λ_{11} $-\lambda_{12}$ 0	λ_{12} λ_{11} 0	0 0 λ_{33}
Tetragonal	14, 15, 16, 19 (14: rutile, 15: nickel sulfate, 19: copper pyrite)	λ_{11} 0 0	0 λ_{11} 0	0 0 λ_{33}
Tetragonal	17, 18, 20 (17: wolframic limestone)	λ_{11} $-\lambda_{12}$ 0	λ_{12} λ_{11} 0	0 0 λ_{33}
Hexagonal	21, 22, 23, 26 (21: beryl, 23: silver iodide)	λ_{11} 0 0	0 λ_{11} 0	0 0 λ_{33}
Hexagonal	24, 25, 27 (24: apatite, 25: nepheline)	λ_{11} $-\lambda_{12}$ 0	λ_{12} λ_{11} 0	0 0 λ_{33}
Regular (cubic)	28, 29, 30, 31, 32 (28: rock salt, 29: sylvite, 30: zincblende, 31: pyrite, 32: sodium chlorate)	λ_{11} 0 0	0 λ_{11} 0	0 0 λ_{11}

The most sensitive experimental method of the various arrangements[55, 220] was suggested by P. Curie (1893) and later carried out by Voigt (1903).

Let a long thin slab be cut from a crystal of the above-mentioned type perpendicularly to the principal crystallographic axis lying in the z-direction, so that the x–y-plane falls on a principal plane. Now the slab ends are brought into contact with two thermostats of temperatures T_A and T_B (Fig. 37).

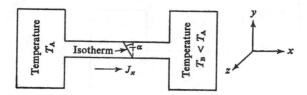

Fig. 37. Scheme of the experimental arrangement. After Curie and Voigt.

By an artifice (see below) we find that the heat flow in the y-direction practically disappears. Then from Eq. (9) we find that

$$J_x = -\lambda_{11} \frac{\partial T}{\partial x} - \lambda_{12} \frac{\partial T}{\partial y}, \qquad (4\text{–}39.11\text{a})$$

$$0 = \lambda_{12} \frac{\partial T}{\partial x} - \lambda_{11} \frac{\partial T}{\partial y}, \qquad (4\text{–}39.11\text{b})$$

in which the heat flow in the z-direction is irrelevant. From Eq. (11b) there follows:

$$\frac{\partial T}{\partial y} \bigg/ \frac{\partial T}{\partial x} = \frac{\lambda_{12}}{\lambda_{11}}. \qquad (4\text{–}39.12)$$

If the isotherms (i.e. the curves $T = $ const) on the surface of the slab form an angle α with the y-direction (see Fig. 37) then it can be shown (Fig. 38)

$$\frac{\lambda_{12}}{\lambda_{11}} = \tan \alpha. \qquad (4\text{–}39.13)$$

For $\lambda_{12} = 0$ (Eq. 10), α must be equal to zero.

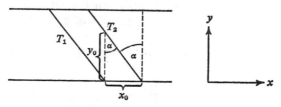

Fig. 38. Enlarged section of the slab surface in Fig. 37 (T_1 and T_2 are two neighboring isotherms). We have

$$\frac{\partial T}{\partial x} = \frac{T_2 - T_1}{x_0} = \frac{T_2 - T_1}{y_0 \tan \alpha}, \qquad \frac{\partial T}{\partial y} = \frac{T_2 - T_1}{y_0},$$

that is,

$$\frac{\partial T/\partial y}{\partial T/\partial x} = \tan \alpha.$$

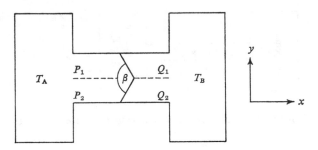

Fig. 39. Scheme of the twin-slabs method (after Voigt): the cut surface (dotted line) divides the two halves of the slab such that the upper half is in the initial position (Fig. 37) and the lower half is so turned round the x-axis that the originally neighboring points P_2 and P_1 or Q_1 and Q_2 no longer lie together. The isotherms of the two halves enclose the angle $\beta = \pi - 2\alpha$.

During the experiments Voigt (1903) used the "method of the twin slabs": he halved the crystal slab along the x-axis, turned the one half about the x-axis and cemented the two halves together (Fig. 39). In this arrangement, which leads to V-formed isotherms, we may assume that in the neighborhood of the cut surface there is no heat flow in the y-direction (see Fig. 39). The isotherms are made visible by a wax of suitable melting point which is spread in a thin layer on the slab. (The molten wax leaves behind clear traces on solidification.) A thick heated copper slab takes over the role of the "thermostat" of temperature T_A, while the other end of the crystal slab (temperature T_B) is at room temperature. From Eq. (13) and Fig. 39 there results:

$$\frac{\lambda_{12}}{\lambda_{11}} = \tan\left(\frac{\pi}{2} - \frac{\beta}{2}\right) \qquad (4\text{--}39.14)$$

with β measured in the middle of the slab near the cut area (Fig. 39).

Voigt observed with several crystal samples of dolomite and apatite that the angle β had the value of 180° with an experimental error of not more than 4', that is, that the isotherms lay practically in the y-direction. From this there follows with Eq. (14):

$$\frac{\lambda_{12}}{\lambda_{11}} = 0$$

with an uncertainty of less than 0.05%. This is an impressive verification of Eq. (10) which follows from the Onsager reciprocity law.

4–40 FURTHER PROBLEMS

In our presentation of irreversible processes in continuous systems, we have restricted ourselves to the general development of the macroscopic theory, and have chosen as examples such problems which are either educational as illustrations or which have been thoroughly investigated from an experimental viewpoint. We have either considered briefly or left out discussion of theories which are not yet concluded and of experimental applications which, for the present, are of little interest. Also, following the intention of this monograph, we have not gone into molecular–statistical or kinetic questions.

Now a few short references to more recent theoretical developments will be given. These have not been discussed so far for the above reasons but remain worthy of mention because they will probably achieve great significance in the future.

First, it should be mentioned that there is a region to which the methods of the thermodynamics of irreversible processes have not yet been systematically applied, although the formalism is in principle clear: the thermodynamic–phenomenological description of irreversible processes in *interfaces*. Difficulties appear due, among other things, to the fact that interfacial layers in nonequilibrium states are in general not "autonomous"[221].

In a certain sense, a generalization of the present macroscopic theory of irreversible processes is the thermodynamic investigation of processes in systems with continuously varying *internal degrees of freedom* (cf. §1–16). Processes like the deformation of macromolecules in streaming fluids or the orientation of polar molecules in electromagnetic fields ("dielectric relaxation") are examples of this. The problem can be solved either by an extension of the basic equations of the theory (mass balance, momentum balance, energy and entropy balance, local entropy production, phenomenological equations, etc.) to the internal configuration space[222] or by an extension of the theory of relaxation phenomena (cf. §2–11) to systems with an infinite number of parameters[223,240]. In this way, the Debye formula for dipole relaxation can be derived from the thermodynamic–phenomenological theory of irreversible processes.

General investigations of the frequency dependence of thermodynamic quantities and of the presentation of transport phenomena as relaxation phenomena[4c,241] form the transition from the above problem to the more general questions of statistical mechanics of irreversibility.

Another extension of the macroscopic theory which is important in principle—in particular for systems in electromagnetic fields—is the *relativistic thermodynamics of irreversible processes*[4c, 6b, 224–226].

Finally, the *statistical mechanics of irreversible processes* has been developed remarkably in latter years. Its purpose, on the one hand, is to produce the fundamental statements of the thermodynamics of irreversible phenomena and also to delineate the range of validity of the theory (cf. §1–23); on the other hand, it permits calculation of the course of special processes. The most important and most interesting is the general question which was posed and partially answered by Boltzmann and Gibbs for classical thermodynamics: why does every actual macroscopic system exhibit irreversible processes, while the laws for the motion of the individual particles show the typical reversibility of classical mechanics? For the solution of this problem, there are so far only certain starting points. For the presentation of these, the reader is referred to the original literature [227–237,239,242]. With reference to other problems of the molecular statistics of irreversible processes we will only mention more recent reviews[4d, 5, 237–239].[43]

REFERENCES

1. J. N. Agar, *Trans. Faraday Soc.* **56**, 776 (1960).
2. I. Prigogine, *Etude thermodynamique des Phénomènes irréversibles*, Paris–Liége (1947).
3. J. Meixner, *Ann. Physik* (5) **41**, 409 (1942).
4. J. Meixner and H. G. Reik, *Handbuch der Physik*, Vol. III/2, Berlin–Goettingen–Heidelberg (1959), (a) p. 447; (b) p. 427; (c) pp. 483, 487, 494; (d) p. 505.
5. S. R. de Groot and P. Mazur, *Non-Equilibrium Thermodynamics*, Amsterdam (1962).
6. C. Eckart, *Phys. Rev.* (a) **58**, 267, 269 (1940); (b) **58**, 919 (1940).
7. J. Meixner, *Z. Physik. Chem.* (*Leipzig*) (B) **53**, 235 (1943).
8. H. Schoenert and C. Sinistri, *Z. Elektrochem.* **66**, 413 (1962).
9. R. W. Laity and C. T. Moynihan, *J. Phys. Chem.* **67**, 723 (1963).
10. E. W. Washburn, *Z. Physik. Chem.* (*Leipzig*) **66**, 513 (1909).
11. G. N. Lewis, *J. Am. Chem. Soc.* **32**, 862 (1910).
12. D. A. MacInnes and L. G. Longsworth, *Chem. Rev.* **11**, 171 (1932).
13. R. J. Bearman, *J. Chem. Phys.* **36**, 2432 (1962).
14. M. Spiro, *Trans. Faraday Soc.* **61**, 350 (1965).
15. M. Selvaratnam and M. Spiro, *Trans. Faraday Soc.* **61**, 360 (1965).
16. M. Spiro, *J. Chem. Phys.* **42**, 4060 (1965).
17. R. Haase, *Z. Physik. Chem.* (*Frankfurt*) **39**, 27 (1963).
18. C. Sinistri, *J. Phys. Chem.* **66**, 1600 (1962).
19. B. R. Sundheim, *J. Chem. Phys.* **40**, 27 (1964).
20. R. Haase, *Z. Elektrochem.* (a) **62**, 279 (1958); (b) **62**, 1043 (1958).
21. R. A. Robinson and R. H. Stokes, *Electrolyte Solutions*, 2nd ed., London (1959), (a) p. 463; (b) p. 289; (c) p. 457.
22. R. Haase, P.-F. Sauermann, and K. H. Dücker, *Z. Physik. Chem.* **43**, 218 (1964).
23. G. C. Hood and C. A. Reilly, *J. Chem. Phys.* **32**, 127 (1960).
24. R. Haase, G. Lehnert, and H.-J. Jansen, *Z. Physik. Chem.* (*Frankfurt*) **42**, 32 (1964).

25. R. Haase, P.-F. Sauermann, and K.-H. Dücker, *Z. Physik. Chem.* (a) **46**, 129, 140 (1965); (b) **47**, 224 (1965); (c) **48**, 206 (1966).

26. R. Haase and K.-H. Dücker, *Z. Physik. Chem.* (*Franfkurt*) **54**, 319 (1967).

27. R. M. Fuoss and F. Accascina, *Electrolytic Conductance*, New York–London (1959).

28. U. Stille, *Messen und Rechnen in der Physik*, 2nd ed., Braunschweig (1961).

29. J. N. Agar, in *The Structure of Electrolytic Solutions* (Ed. W. J. Hamer), New York (1959), p. 200.

30. A. A. Noyes and K. G. Falk, *J. Am. Chem. Soc.* **33**, 1436 (1911).

31. M. Spiro, *J. Chem. Educ.* **33**, 464 (1956).

32. L. G. Longsworth, *J. Am. Chem. Soc.* **69**, 1288 (1947).

33. W. Jost, *Diffusion* (Methoden der Messung und Auswertung) (Fortschritte der physikalischen Chemie, Vol. 1), Darmstadt (1957).

34. G. J. Hooyman, H. Holtan, Jr., P. Mazur, and S. R. de Groot, *Physica* **19**, 1095 (1953).

35. J. G. Kirkwood, R. L. Baldwin, P. J. Dunlop, L. J. Gosting, and G. Kegeles, *J. Chem. Phys.* **33**, 1505 (1960).

36. F. Sauer and V. Freise, *Z. Elektrochem.* **66**, 353 (1962).

37. R. Haase and M. Siry, *Z. Physik. Chem.* (*Frankfurt*) **57**, 56 (1968).

38. S. Claesson and L.-O. Sundelof, *J. Chim. Phys.* **54**, 914 (1957).

39. U. Dehlinger, *Z. Physik* **102**, 633 (1936).

40. U. Dehlinger, *Z. Metallk.* **29**, 401 (1937).

41. R. Becker, *Z. Metallk.* **29**, 245 (1937).

42. L. Onsager, *Ann. N.Y. Acad. Sci.* **46**, 241 (1945).

43. R. Haase, *Thermodynamik der Mischphasen*, Berlin–Goettingen–Heidelberg (1956), (a) p. 182; (b) p. 580; (c) p. 359 *et seq.*; (d) p. 524.

44. G. J. Hooyman, *Thesis* Leiden (1955).

45. J. Meixner, *Ann. Physik* (5) (a) **43**, 268 (1943); (b) **43**, 244 (1943).

46. P. J. Dunlop and L. J. Gosting, *J. Phys. Chem.* **63**, 86 (1959).

47. P. J. Dunlop, *J. Phys. Chem.* **63**, 612, 2089 (1959).

48. H. Fujita and L. J. Gosting, *J. Phys. Chem.* **64**, 1256 (1960).

49. R. P. Wendt, *J. Phys. Chem.* **66**, 1279 (1962).

50. L. A. Woolf, D. G. Miller, and L. J. Gosting, *J. Am. Chem. Soc.* **84**, 317 (1962).

51. R. P. Wendt, *J. Phys. Chem.* **66**, 1279 (1962), **69**, 1227 (1965).

52. P. J. Dunlop, *J. Phys. Chem.* **69**, 4276 (1965).

53. O. W. Edwards, R. L. Dunn, J. D. Hatfield, E. O. Huffman, and K. L. Elmore, *J. Phys. Chem.* **70**, 217 (1966).

54. D. G. Miller, *J. Phys. Chem.* (a) **62**, 767 (1958); (b) **63**, 570, 2089 (1959); (c) **69**, 3374 (1965).

55. D. G. Miller, *Chem. Rev.* **60**, 15 (1960).

56. P. F. MIJNLIEFF and H. A. VREEDENBERG, *J. Phys. Chem.* **70**, 2158 (1966).

57. Y. L. YAO, *J. Chem. Phys.* **45**, 110 (1966).

58. J. RASTAS, *Acta Polytech. Scand.*, Ch. 50, Helsinki (1966).

59. C. H. BYERS and C. J. KING, *J. Phys. Chem.* **70**, 2499 (1966).

60. J. K. BURCHARD and H. L. TOOR, *J. Phys. Chem.* **66**, 2015 (1962).

61. F. O. SHUCK and H. L. TOOR, *J. Phys. Chem.* **67**, 540 (1963).

62. H. T. CULLINAN, JR., and H. L. TOOR, *J. Phys. Chem.* **69**, 3941 (1965).

63. H. SCHOENERT, *J. Phys. Chem.* **64**, 733 (1960).

64. R. HAASE, *Trans. Faraday Soc.* **49**, 724, footnote p. 728 (1953).

65. W. H. STOCKMAYER, *J. Chem. Phys.* **33**, 1291 (1960).

66. D. G. MILLER, *J. Phys. Chem.* **64**, 1598 (1960).

67. P. B. LORENZ, *J. Phys. Chem.* **65**, 704 (1961).

68. C. W. GARLAND, S. TONG, and W. H. STOCKMAYER, *J. Phys. Chem.* **69**, 1718 (1965).

69. D. G. MILLER, *J. Phys. Chem.* **70**, 2639 (1966); (a) R. HAASE and J. RICHTER, *Z. Naturforsch.* **22a**, 1761 (1967).

70. L. ONSAGER and R. M. FUOSS, *J. Phys. Chem.* **36**, 2689 (1932).

71. G. S. HARTLEY, *Phil. Mag.* **12**, 473 (1931).

72. E. A. GUGGENHEIM, *Trans. Faraday Soc.* **50**, 1048 (1954).

73. E. A. GUGGENHEIM, *Thermodynamics*, 3rd ed., Amsterdam (1957), p. 400.

74. M. SPIRO, *Trans. Faraday Soc.* **55**, 1207 (1959).

75. R. HAASE, *Z. Elektrochem.* **57**, 87, 448 (1953).

76. R. HAASE, *Z. Physik. Chem.* (*Frankfurt*) **25**, 26 (1960).

77. R. HAASE and H. SCHOENERT, *Z. Elektrochem.* **64**, 1155 (1960).

78. R. HAASE, in *Ultracentrifugal Analysis in Theory and Experiment* (Ed. J. W. WILLIAMS), Academic Press, New York–London (1963), p. 13. The reader should perhaps ignore the editor's comments on this paper.

79. L. PELLER, *J. Chem. Phys.* **29**, 415 (1958).

80. R. HAASE, *Angew. Chem.* **77**, 517 (1965).

81. T. SVEDBERG, *Kolloid-Z.* (Zsigmondy Festschrift) **36**, 53 (1925).

82. G. V. SCHULZ, *Z. Physik. Chem.* (*Leipzig*) **193**, 168 (1944).

83. O. LAMM, *Acta Chem. Scand.* **7**, 173 (1953).

84. O. LAMM, *Trans. Roy. Inst. Technol.*, Stockholm No. **134** (1959).

85. R. HAASE, *Kolloid-Z.* (a) **138**, 105 (1954); (b) **147**, 141 (1956).

86. J. M. CREETH, *J. Phys. Chem.* **66**, 1228 (1962).

87. F. E. LA BAR and R. L. BALDWIN, *J. Am. Chem. Soc.* **85**, 3105 (1963).

88. G. REHAGE, Private communication.

89. O. ERNST, Private communication.

90. H. FUJITA, *Mathematical Theory of Sedimentation Analysis*, New York–London (1962).

91. H. SCHOENERT, *Z. Physik. Chem.* (*Frankfurt*) **30**, 52 (1961).

92. R. Haase, *Z. Physik. Chem. (Frankfurt)* **11**, 379 (1957).
93. Th. des Coudres, *Wiedemanns Ann. Physik Chem.* **57**, 232 (1896).
94. S. W. Grinnell and F. O. Koenig, *J. Am. Chem. Soc.* **64**, 682 (1942).
95. Th. des Coudres, *Wiedemanns Ann. Physik Chem.* **49**, 284 (1893).
96. R. C. Tolman, *Proc. Am. Acad. Arts Sci.* **46**, 109 (1910).
97. R. C. Tolman, *J. Am. Chem. Soc.* **33**, 121 (1911).
98. D. A. MacInnes and B. R. Ray, *J. Am. Chem. Soc.* **71**, 2987 (1949).
99. D. A. MacInnes and M. O. Dayhoff, *J. Chem. Phys.* **20**, 1034 (1952).
100. B. R. Ray, D. M. Beeson, and H. F. Crandall, *J. Am. Chem. Soc.* **80**, 1029 (1958).
101. D. G. Miller, *Am. J. Phys.* **24**, 595 (1956).
102. H. S. Harned and B. B. Owen, *The Physical Chemistry of Electrolytic Solutions*, 3rd ed., New York (1958).
103. Landolt–Boernstein, Vol. II, Part 7, Berlin–Goettingen–Heidelberg (1960).
104. R. Haase, *Z. Elektrochem.* **54**, 450 (1950).
105. S. Chapman and T. G. Cowling, *The Mathematical Theory of Non-Uniform Gases*, Cambridge (1953).
106. R. Haase, *Z. Naturforsch.* **6a**, 420 (1951).
107. M. W. Zemansky, *Heat and Thermodynamics*, London–New York–Toronto (1951), (a) pp. 85–88; (b) p. 301–302; (c) p. 303.
108. H. J. V. Tyrrell, *Diffusion and Heat Flow in Liquids*, London (1961), p. 301–302.
109. J. Meixner, *Ann. Physik* (5) **40**, 165 (1941).
110. L. Onsager, *Phys. Rev.* (a) **37**, 405 (1931); (b) **38**, 2265 (1931).
111. H. B. Callen, *Phys. Rev.* **73**, 1349 (1948).
112. C. A. Domenicali, *Rev. Mod. Phys.* **26**, 237 (1954).
113. R. Haase, *Z. Naturforsch.* **11a**, 681 (1956).
114. M. I. Temkin and A. W. Choroschin, *Russ. J. Phys. Chem.* (*English Transl.*) **26**, 500 (1952).
115. Landolt–Boernstein, Vol. IV, Part 3, Berlin–Goettingen–Heidelberg (1957).
116. L. Waldmann, *Z. Physik* (a) **121**, 501 (1943); (b) **124**, 2. 30, 175 (1948).
117. L. Waldmann, *Z. Naturforsch.* **4a**, 105 (1949).
118. K. Clusius, *Helv. Phys. Acta* **22**, 135 (1949).
119. R. Haase, *Z. Physik. Chem. (Leipzig)* **196**, 219 (1950).
120. R. P. Rastogi and G. L. Madan, *J. Chem. Phys.* **43**, 4179 (1965).
121. K. E. Grew and T. L. Ibbs, *Thermal Diffusion in Gases*, Cambridge (1952).
122. J. Meixner, *Ann. Physik* (5) **39**, 333 (1941).
123. R. Haase, *Z. Physik* **127**, 1 (1950).
124. E. W. Becker, *Z. Naturforsch.* **5a**, 457 (1950).
125. E. W. Becker, *J. Chem. Phys.* **19**, 131 (1951).

126. W. M. RUTHERFORD and J. G. ROOF, *J. Phys. Chem.* **63**, 1506 (1959).

127. W. M. RUTHERFORD, *A.I.Ch.E.J.* **9**, 841 (1963).

128. HAASE, BORGMANN, and LEE. To be published.

129. I. PRIGOGINE, L. DE BROUCKÈRE, and R. AMAND, *Physica* **16**, 577, 851 (1950).

130. I. PRIGOGINE, L. DE BROUCKÈRE, and R. BUESS, *Physica* **18**, 915 (1952).

131. G. THOMAES, *Physica* **17**, 885 (1951).

132. L. J. TICHACEK and H. G. DRICKAMER, *J. Phys. Chem.* **60**, 820 (1956).

133. F. H. HORNE and R. J. BEARMAN, *J. Chem. Phys.* **37**, 2842, 2857 (1962).

134. G. THOMAES, *J. Chem. Phys.* **25**, 32 (1956).

135. R. HAASE and B. K. BIENERT, *Ber. Bunsenges. Physik. Chem.* **71**, 392 (1967).

136. L. J. TICHACEK, W. S. KMAK, and H. G. DRICKAMER, *J. Phys. Chem.* **60**, 660 (1956).

137. L. G. LONGSWORTH, *J. Phys. Chem.* **61**, 1557 (1957).

138. C. C. TANNER, *Trans. Faraday Soc.* (a) **23**, 75 (1927); (b) **49**, 611 (1953).

139. J. N. AGAR and J. C. R. TURNER, *Proc. Roy. Soc.* (*London*), *Ser. A* **255**, 307 (1960).

140. J. N. AGAR, *J. Phys. Chem.* **64**, 1000 (1960).

141. P. N. SNOWDON and J. C. R. TURNER, *Trans. Faraday Soc.* **56**, 1409, 1812 (1960).

142. N. H. SAGERT and W. G. BRECK, *Trans. Faraday Soc.* **57**, 436 (1961).

143. A. D. PAYTON and J. C. R. TURNER, *Trans. Faraday Soc.* **58**, 55 (1962).

144. B. D. BUTLER and J. C. R. TURNER, *J. Phys. Chem.* **69**, 3598 (1965).

145. K. HOCH, *Z. Physik. Chem.* (*Frankfurt*) **56**, 30 (1967).

146. R. HAASE and K. HOCH, *Z. Physik. Chem.* (*Frankfurt*) **46**, 63 (1965).

147. J. MEIXNER, *Z. Naturforsch.* **7a**, 553 (1952).

148. E. U. FRANCK, *Z. Physik. Chem.* (*Leipzig*) **201**, 16 (1952).

149. I. PRIGOGINE and R. BUESS, *Bull. Classe Sci., Acad. Roy. Belg.* **38**, 711 (1952).

150. R. HAASE, *Z. Naturforsch.* **8a**, 729 (1953).

151. W. NERNST, *Festschrift Ludwig Boltzmann*, Leipzig (1904), p. 904.

152. K. P. COFFIN, *J. Chem. Phys.* **31**, 1290 (1959).

153. R. S. BROKAW, *J. Chem. Phys.* **32**, 1005 (1960).

154. B. N. SRIVASTAVA and A. K. BARUA, *J. Chem. Phys.* **35**, 329 (1961).

155. P. K. CHAKRABORTI, *J. Chem. Phys.* **38**, 575 (1963).

156. E. U. FRANCK and W. SPALTHOFF, *Z. Elektrochem.* **58**, 374 (1954).

157. R. S. BROKAW, *J. Chem. Phys.* **35**, 1569 (1961).

158. J. MEIXNER, *Z. Naturforsch.* **8a**, 69 (1953).

159. J. N. BUTLER and R. S. BROKAW, *J. Chem. Phys.* **26**, 1636 (1957).

160. E. U. FRANCK and W. SPALTHOFF, *Naturwissenschaften* **40**, 580 (1953).

161. M. EIGEN, *Z. Elektrochem.* **56**, 176 (1952).

162. J. N. AGAR and W. G. BRECK, *Trans. Faraday Soc.* **53**, 167 (1957).

163. W. G. Breck and J. N. Agar, *Trans. Faraday Soc.* **53**, 179 (1957).

164. J. N. Agar, *Rev. Pure Appl. Chem.* **8**, 1 (1958).

165. J. N. Agar, in *Advances in Electrochemistry and Electrochemical Engineering* (Ed. P. Delahay), Vol. 3, *Electrochemistry*, New York–London (1963), p. 31.

166. W. G. Breck, G. Cadenhead, and M. Hammerli, *Trans. Faraday Soc.* **61**, 37 (1965).

167. W. G. Breck and J. Lin, *Trans. Faraday Soc.* **61**, 1511, 2223 (1965).

168. R. Haase, *Z. Physik. Chem. (Frankfurt)* (a) **13**, 21 (1957); (b) **14**, 292 (1958).

169. R. Haase and H. Schoenert, *Z. Physik. Chem. (Frankfurt)* **25**, 193 (1960).

170. R. Haase, K. Hoch, and H. Schoenert, *Z. Physik. Chem. (Frankfurt)* **27**, 421 (1961).

171. R. Haase and G. Behrend, *Z. Physik. Chem. (Frankfurt)* **31**, 375 (1962).

172. E. Lange and T. Hesse, *Z. Elektrochem.* (a) **38**, 428 (1932); (b) **39**, 374 (1933).

173. L. G. Longsworth, in *The Structure of Electrolytic Solutions* (Ed. W. J. Hamer), New York (1959), p. 183.

174. H. Levin and C. F. Bonilla, *J. Electrochem. Soc.* **98**, 388 (1951).

175. H. Holtan, Jr., "Electric Potentials in Thermocouples and Thermocells," *Thesis* Utrecht (1953).

176. H. Holtan, Jr., *Proc. Koninkl. Ned. Akad. Wetenschap.* (B) **56**, 498 (1953).

177. B. R. Sundheim and J. Rosenstreich, *J. Phys. Chem.* **63**, 419 (1959).

178. A. R. Nichols and C. T. Langford, *J. Electrochem. Soc.* **107**, 842 (1960).

179. R. Schneebaum and B. R. Sundheim, *Discussions Faraday Soc.* **32**, 197 (1961).

180. B. R. Sundheim and J. D. Kellner, *J. Phys. Chem.* **69**, 1204 (1965).

181. K. S. Pitzer, *J. Phys. Chem.* **65**, 147 (1961).

182. R. Haase and P.-F. Sauermann, *Z. Physik. Chem. (Frankfurt)* **27**, 42 (1961).

183. M. Spiro, *Electrochim. Acta* **11**, 569 (1966).

184. M. Planck, *Wied. Ann.* (a) **39**, 161 (1890); (b) **40**, 561 (1890).

185. P. Henderson, *Z. Physik. Chem. (Leipzig)* (a) **59**, 118 (1907); (b) **63**, 325 (1908).

186. E. Helfand and J. G. Kirkwood, *J. Chem. Phys.* **32**, 857 (1960).

187. A. W. Choroschin and M. I. Temkin, *Russ. J. Phys. Chem. (English Transl.)* **26**, 773 (1952).

188. Landolt–Boernstein, Vol. II, Part 4, Berlin–Goettingen–Heidelberg (1961).

189. K. K. Kelley, *U.S. Bur. Mines, Bull.* **584** (1960).

190. I. Prigogine and P. Mazur, *Physica* **17**, 661 (1951).

191. P. Mazur and I. Prigogine, *Physica* **17**, 680 (1951).

192. A. Sommerfeld, *Vorlesungen ueber Theoretische Physik*, Vol. II: *Mechanik der deformierbaren Medien*, 4th ed., Leizpig (1957).

193. C. Eckart, *Phys. Rev.* **73**, 68 (1948).

194. L. N. LIEBERMANN, *Phys. Rev.* **75**, 1415 (1949).

195. E. A. GUGGENHEIM and J. E. PRUE, *Physicochemical Calculations*, Amsterdam (1955), p. 27.

196. G. J. GRUBER and T. A. LITOVITZ, *J. Chem. Phys.* **40**, 13 (1964).

197. J. O. HIRSCHFELDER, C. F. CURTISS, and R. B. BIRD, *Molecular Theory of Gases and Liquids*, New York–London (1954).

198. S. M. KARIN and L. ROSENHEAD, *Rev. Mod. Phys.* **24**, 108 (1952).

199. J. MEIXNER, *Z. Physik* **131**, 456 (1952).

200. E. A. GUGGENHEIM, *Phil. Mag.* **33**, 487 (1942).

201. E. A. GUGGENHEIM, *Nature* **148**, 751 (1941).

202. P. MAZUR and I. PRIGOGINE, *Mem. Classe Sci., Acad. Roy. Belg.* **28**, Part 1 (1953).

203. I. PRIGOGINE, *J. Chim. Phys.* **49**, 79 (1952).

204. I. PRIGOGINE, P. MAZUR, and R. DEFAY, *J. Chim. Phys.* **50**, 146 (1953).

205. S. R. DE GROOT, P. MAZUR, and H. A. TOLHOEK, *Physica* **19**, 549 (1953).

206. A. SOMMERFELD, *Vorlesungen ueber Theoretische Physik*, Vol. III: *Elektrodynamik*, Leipzig (1949), (a) p. 248; (b) p. 291.

207. K.-J. HANSSEN, *Z. Naturforsch.* **9a**, 323, 919 (1954).

208. P. MAZUR and S. R. DE GROOT, *Physica* **19**, 961 (1953).

209. R. FIESCHI, S. R. DE GROOT, and P. MAZUR, *Physica* (a) **20**, 67 (1954); (b) **20**, 259 (1954).

210. R. FIESCHI, S. R. DE GROOT, P. MAZUR, and J. VLIEGER, *Physica* **20**, 245 (1954).

211. W. MEISSNER, *Handbuch der Experimentalphysik*, Vol. II, Part 2, Leipzig (1935), p. 311.

212. J.-P. JAN, *Solid State Phys.* **5**, 1 (1957).

213. J. MEIXNER, *Ann. Physik* (5) **35**, 701 (1939).

214. E. H. HALL, *Proc. Am. Acad. Arts Sci.* **72**, 301 (1938).

215. P. MAZUR and I. PRIGOGINE, *J. Phys. Radium* **12**, 616 (1951).

216. N. C. LITTLE, *Phys. Rev.* **28**, 418 (1926).

217. J. MEIXNER, *Proc. Roy. Soc. (London), Ser. A* **226**, 57 (1954).

218. J. MEIXNER, *Z. Naturforsch.* **9a**, 654 (1954).

219. G. FALK and J. MEIXNER, *Z. Naturforsch.* **11a**, 782 (1956).

220. W. VOIGT, *Lehrbuch der Kristallphysik*, Leipzig and Berlin (1910).

221. R. DEFAY and I. PRIGOGINE, *Tension superficielle et Adsorption*, Liège (1951).

222. I. PRIGOGINE and P. MAZUR, *Physica* **19**, 241 (1953).

223. J. MEIXNER, *Kolloid-Z.* **134**, 3 (1953).

224. G. A. KLUITENBERG, *Thesis* Leiden (1954).

225. G. A. KLUITENBERG, S. R. DE GROOT, and P. MAZUR, *Physica* **19**, 698, 1079 (1953).

226. G. A. KLUITENBERG and S. R. DE GROOT, *Physica* **20**, 199 (1954).

227. N. G. VAN KAMPEN, *Physica* (a) **20**, 603 (1954); (b) **23**, 641, 707 (1957).

228. L. VAN HOVE, *Physica* (a) **21**, 517, 901 (1955); (b) **22**, 343 (1956); (c) **23**, 341 (1957).

229. H. N. V. TEMPERLEY, *Proc. Cambridge Phil. Soc.* **52**, No. 4, 712 (1956).

230. W. KOHN and J. M. LUTTINGER, *Phys. Rev.* **108**, 590 (1957).

231. E. W. MONTROLL and J. WARD, *Physica* **25**, 423 (1959).

232. I. PRIGOGINE and F. HENIN, *J. Math. Phys.* **1**, 349 (1960).

233. J. MEIXNER and H. KOENIG, *Rheol. Acta* **1**, 190 (1958).

234. J. MEIXNER, *Z. Naturforsch.* **16a**, 721 (1961).

235. J. MEIXNER, *J. Math. Phys.* **4**, 154 (1963).

236. J. MEIXNER, *Acta Phys. Polon.* **27**, 113 (1965).

237. H. GRAD, *Commun. Pure Appl. Math.* **5**, 455 (1952).

238. R. EISENSCHITZ, *Statistical Theory of Irreversible Processes*, Oxford (1958).

239. I. PRIGOGINE, *Non-Equilibrium Statistical Mechanics*, New York–London (1962).

240. B. D. COLEMAN and M. E. GURTIN, *J. Chem. Phys.* **47**, 597 (1967).

241. J. MEIXNER, *Z. Physik* **193**, 366 (1966).

242. F. SCHLÖGL, *Z. Physik* **191**, 81 (1966); **195**, 98 (1966); **198**, 559 (1967).

CHAPTER 5

STATIONARY STATES

5-1 INTRODUCTION

In §1–17, a *stationary state* was defined as a nonequilibrium state in which the intensive state variables of the system are constant in time. If the system is homogeneous then the above variables have the same value everywhere, that is, they are invariant with respect to time and position. If we are dealing with a heterogeneous (discontinuous) system then the intensive variables inside one phase of the system are again constant, but can exhibit a discontinuity on passing through the phase boundary. Finally, in continuous systems, the intensive quantities generally vary from place to place.

If the amounts of the individual regions of the system, that is, for instance, the mass of a homogeneous system, or the amounts of the phases of a heterogeneous system, or the masses of the volume elements of a continuous system, are invariant with respect to time then the extensive quantities of the system no longer depend on time. If the above condition is not satisfied then this can, for example, mean that individual regions increase their mass at the expense of other regions, although the intensive variables are constant with respect to time. (Example: a phase of a two-phase system may, at given values of pressure, temperature, and composition, be consumed, while the other phase is continuously increased with respect to its amount.)

Thus we can, in general, write

$$\frac{\partial \zeta}{\partial t} = 0 \quad \text{(stationary state)}, \tag{5-1.1}$$

where ζ denotes any intensive state variable of an arbitrary system, t the time, and $\partial/\partial t$ the operator of the time derivative at a fixed position. For an extensive state function Z, which should refer to the total system, there is, according to the above,

$$\frac{dZ}{dt} = 0 \quad \text{(stationary state with the masses of all regions held constant)}, \tag{5-1.2}$$

where the total derivative d/dt is written because the position as a variable drops out for the total system. With this we obtain for the total entropy S of a system:

$$\frac{dS}{dt} = 0 \quad \text{(stationary state with the masses of all regions held constant).} \tag{5–1.3}$$

From Eq. (1–24.3) and Eq. (3) there follows:

$$\frac{dS}{dt} = \frac{d_aS}{dt} + \frac{d_iS}{dt} = 0 \quad \text{(stationary state with the masses of all regions held constant),} \tag{5–1.4}$$

where d_aS/dt is the entropy flow and d_iS/dt the entropy production. Since, for stationary states, irreversible processes inside the system occur continuously, there results

$$\frac{d_iS}{dt} > 0, \qquad \frac{d_aS}{dt} = -\frac{d_iS}{dt} < 0. \tag{5–1.5}$$

For *equilibrium*, on the other hand, we find

$$\frac{d_iS}{dt} = 0 \quad \text{(equilibrium).} \tag{5–1.6}$$

Since Eq. (4) remains valid in this case, there results

$$\frac{d_aS}{dt} = 0 \quad \text{(equilibrium).} \tag{5–1.7}$$

Equations (5), (6), and (7) clearly show the difference between equilibrium states and stationary nonequilibrium states.

Equations (5) state that for a stationary state of the above-mentioned type the positive entropy production inside the system is compensated for by a negative entropy flow. This negative entropy flow can come about by heat or mass exchange of a system with the surroundings or by both. If we are dealing with heat exchange only then, obviously, heat must flow out of the system into the surroundings (see Eqs. 2–2.11, 3–4.6, and 4–11.2).† From (5) there follows, in general, that stationary states of the type considered are impossible for thermally insulated systems ($d_aS/dt = 0$).

Quite generally, a stationary state depends, as was discussed in §1–17, on a continuous mutual exchange between system and surroundings, while an equilibrium state is determined by the conditions inside the system alone.

Now it is interesting to ask how the entropy production

$$\Theta \equiv \frac{d_iS}{dt} > 0 \tag{5–1.8}$$

---

† For nonhomogeneous systems, it is possible, according to Eqs. (3–4.6) and (4–11.2), that heat is taken out of the surroundings into certain system parts, while other parts of the system give up heat to the surroundings.

of a system, inside of which irreversible processes occur, changes in the course of time. This problem has been investigated in various ways during recent years and has been discussed with the use of numerous examples[1–10]. Here we shall restrict the discussion to the most important general results and to some examples.[44] We begin with the simplest cases (chemical reactions in homogeneous systems) in the next section and then gradually go on to more complicated cases.

5-2 HOMOGENEOUS SYSTEMS

In a homogeneous system, in which chemical reactions occur, a stationary state can be achieved if temperature and pressure are held constant in time by the external conditions and a suitable exchange of matter with the surroundings makes it possible that the concentrations of all substances are time independent. Such a reaction system thus represents an open phase. As an example, consider a homogeneous system in which the reaction M → N occurs continuously at a given temperature and at given pressure and to which the species M is delivered from a reservoir while the species N is extracted from the reaction mixture. Thus, under suitable experimental conditions, all concentrations remain constant (Fig. 40).

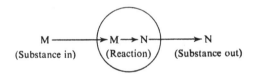

M ———————► M → N ———————► N
(Substance in)   (Reaction)   (Substance out)

Fig. 40. Open phase with chemical reaction inside.

We will not yet consider the stationary state, but first imagine an open phase with chemical reactions at constant temperature and pressure at an arbitrary instant and we wish to know the change in time of the entropy production of this homogeneous system. The variability of the entropy production with time must, thus, be determined here by a change with time of the concentrations of the various substances.

According to Eq. (2–2.12), the entropy production Θ in our system is

$$\Theta = \frac{1}{T} \sum_r w_r A_r > 0. \tag{5-2.1}$$

Here T is the absolute temperature, w_r the reaction rate and A_r the affinity of the reaction r. From this there follows for $T = \text{const}$:

$$\frac{d\Theta}{dt} = \Lambda + \Xi \tag{5-2.2}$$

with

$$\Lambda \equiv \frac{1}{T} \sum_r A_r \frac{dw_r}{dt}, \tag{5–2.3}$$

$$\Xi \equiv \frac{1}{T} \sum_r w_r \frac{dA_r}{dt}, \tag{5–2.4}$$

where t denotes the time.

We use as independent variables the temperature T (T = const), the pressure P (P = const), and the amounts of substance n_i of the species present in the reaction system. We find from Eqs. (2–2.3) and (2–2.6) that

$$\frac{dn_i}{dt} = \sum_r \nu_{ir} w_r + \frac{d_a n_i}{dt}, \tag{5–2.5}$$

$$\frac{dA_r}{dt} = -\sum_i \nu_{ir} \frac{d\mu_i}{dt}, \tag{5–2.6}$$

$$\frac{d\mu_i}{dt} = \sum_k \mu_{ik} \frac{dn_k}{dt} \tag{5–2.7}$$

with (see Eqs. 1–12.5 and 1–12.17)

$$\mu_{ik} \equiv \left(\frac{\partial \mu_i}{\partial n_k}\right)_{T,P,n_j} = \left(\frac{\partial^2 G}{\partial n_i \, \partial n_k}\right)_{T,P,n_j}. \tag{5–2.8}$$

Here ν_{ir} is the stoichiometric number of species i in reaction r, $d_a n_i$ the increase in the amount of substance of species i by supply from the surroundings during the time element dt, μ_i the chemical potential of particle type i, and G the Gibbs function of the phase. The inferior n_j in the derivatives in Eq. (8) indicates that all amounts apart from that with respect to which the derivative is taken are held constant.

Inserting Eqs. (5) to (7) into Eq. (4) we obtain:

$$\Xi = \frac{1}{T} \sum_i \sum_k \mu_{ik} \left(\frac{d_a n_i}{dt} - \frac{dn_i}{dt}\right) \frac{dn_k}{dt}. \tag{5–2.9}$$

The general stability condition (1–22.7) in connection with Eq. (8) says:

$$\sum_i \sum_k \mu_{ik} \alpha_i \alpha_k \geqslant 0, \tag{5–2.10}$$

where the α_i and α_k are arbitrary variables. The equality sign in (10) holds if:

1. Pure changes in the quantity of phase at constant composition are considered (p. 71).
2. Stability limits or critical phases are present (p. 74).
3. The variables α_i or α_k vanish.

The first two cases are uninteresting here. In the following discussion, only the third possibility will be considered.

We recognize that with the help of (10) alone no statement about the sign of (9) can be derived. A clear statement results if the following assumptions are made:

(a) The exchange of matter with the surroundings is so adjusted that the amounts of certain species are constant from the start, while the boundaries are barred to all other substances.

(b) The total amount of substance of the phase is constant.

(c) The chemical potentials of those species (indicated by inferior i), for which a supply or removal takes place, do not depend on the concentrations of the other substances, so that $\mu_{ik} = 0$ for $i \neq k$.†

An ideal system, in which the reaction

$$M \to \cdots \to N$$

takes place, leading from the initial species M over all possible intermediate steps to the final product N and for which the amounts of the substances M and N are held constant by a suitable exchange of mass, represents an example. Then the amounts of the intermediate products are first variable in time due to the continuous course of the reaction, until they become constant in a possible stationary final state.

As is obvious, all terms in Eq. (9) which relate to the supplied or removed species (in the above example M and N) disappear under the above assumptions. Furthermore, the term $d_a n_i/dt$ in Eq. (9) vanishes for all other substances. Accordingly, we find from Eqs. (9) and (10)

$$\Xi = -\frac{1}{T} \sum_i \sum_k \mu_{ik} \frac{dn_i}{dt} \frac{dn_k}{dt} \leqslant 0. \qquad (5\text{-}2.11)$$

Here the equality sign is valid for

$$\frac{dn_i}{dt} = 0, \qquad \frac{dn_k}{dt} = 0,$$

that is, for the stationary state (at constant total amount of the phase). The quantity Ξ is, accordingly, always negative, as long as time changes of the concentrations appear. In the stationary final state, it disappears and it thus reaches its maximum value at the stationary state.

---

† Consider, for example, the equation (valid for ideal gas mixtures, ideal mixtures, and ideal dilute solutions, all denoted concisely in the following by "ideal systems"):

$$\mu_i = \mu_i^0(T, P) + RT \ln x_i,$$

where μ_i^0 is a standard value of the chemical potential, R the gas constant, and $x_i = n_i/n$ (n = total amount of substance) the mole fraction of species i. Then there follows with $n = $ const: $\mu_{ik} = 0$ for $i \neq k$.

Now we wish to assume that all chemical reactions occur not far removed from equilibrium. Then the phenomenological equations

$$w_r = \sum_s a_{rs} A_s \qquad (5\text{-}2.12)$$

hold, according to Eq. (2–4.1), and the Onsager reciprocity relations

$$a_{rs} = a_{sr} \qquad (5\text{-}2.13)$$

for the phenomenological coefficients a_{rs}, according to Eq. (2–4.2). Furthermore, we assume that the phenomenological coefficients are time independent. This means that besides the temperature and the pressure some more variables must possibly be held constant (cf. p. 112). We then obtain from Eqs. (2), (3), (4), (11), (12), and (13):

$$\Lambda = \Xi, \qquad \frac{d\Theta}{dt} = 2\,\Xi \leqslant 0. \qquad (5\text{-}2.14)$$

Accordingly, under the above-mentioned special conditions, the entropy production continuously decreases until it reaches a (positive) minimum value in the stationary state.

The two statements (11) and (14), as will be shown in the following sections, may be applied to more complicated systems. [45]

We can, using special reaction mechanisms, investigate under which conditions the theorem of the minimum entropy production can be applied to stationary nonequilibrium states which are far removed from equilibrium. For more details, the reader is referred to the original literature[11–13].

5-3 HETEROGENEOUS (DISCONTINUOUS) SYSTEMS

Now we consider only that type of heterogeneous (discontinuous) system which has been almost exclusively considered in Chapter 3: a closed "valve system," i.e. a system consisting of two homogeneous subsystems (phases) which, as a whole, exhibits no exchange of matter with the surroundings but in which mass, electricity, and heat exchange can occur between the two phases by a valve (capillary, membrane, etc.). The two individual phases are thus open systems, while the total system is closed. Also we assume, as in Chapter 3, small differences in temperature, pressure, potential, and concentrations between the two phases of the system. [46]

Such a system represents a limiting case of a continuous system. Thus we will not discuss here the variation in the entropy production with time for the general case, since this will be discussed in §5–4 for continuous systems. We restrict ourselves to consideration of a special case which is not covered in the discussion of §5–4. There we assume the pressure to be constant in time but now we are particularly interested in a system with pressures which are variable in time.†

This special case relates to a nonisothermal fluid system of one component without electric potential differences and in which the .alve allows matter and heat transport between two phases as well as the accumulation of a pressure difference (§3–10). Thus permeation, heat conduction (through the valve), and thermomechanical phenomena (Knudsen effect, fountain effect, thermal osmosis, mechanocaloric effects) can occur. The entropy production Θ according to Eqs. (3–5.2), (3–5.8), (3–5.9), (3–5.10), and (3–5.11) has the simple form

$$\Theta = \frac{1}{T}\left(J_1 \overline{V} \Delta P + J_Q \frac{\Delta T}{T}\right) > 0. \tag{5–3.1}$$

Here T is the average absolute temperature of the system, $\overline{V}$ the molar volume at the mean temperature and at the mean pressure of the system, J_1 the flow of matter ($= dn/dt$, n = amount of substance in the first phase, t = time), ΔP the difference between the pressure in the second phase and that in the first phase, ΔT the corresponding temperature difference, and J_Q the heat flow or, more precisely, the heat flowing per unit time from the second phase into the first phase (cf. Eq. 3–3.7 as well as Fig. 7, p. 191).

We assume T and ΔT to be constant, and thus imagine the temperatures of the two homogeneous subsystems to be given as fixed by the experimental conditions (thermostats) at each instant. Then the quantities ΔP, J_1, and J_Q will change until the flow of mass through the valve vanishes in the stationary final state ($J_1 = 0$) and, accordingly, ΔP and J_Q assume constant values. The stationary pressure difference ΔP_{stat} is the "thermomechanical pressure difference" (cf. §3–10 and §3–11). We can regard the molar volume $\overline{V}$ as time independent since the average temperature or the average pressure does not change or remains practically constant.

From Eq. (1) under these conditions we have

$$\frac{d\Theta}{dt} = \Lambda + \Xi \tag{5–3.2}$$

with

$$\Lambda \equiv \frac{1}{T}\left(\overline{V}\Delta P \frac{dJ_1}{dt} + \frac{\Delta T}{T}\frac{dJ_Q}{dt}\right), \tag{5–3.3}$$

$$\Xi \equiv \frac{\overline{V}}{T}J_1 \frac{d\Delta P}{dt}, \tag{5–3.4}$$

---

† The case of a liquid–vapor system, immediately excluded here, will also not be included in our calculations in §5–4. Prigogine and Defay[4] have derived a relation between the minimum of the entropy production and stationary states (azeotropic transformations) for a binary liquid–vapor system with an azeotropic point.

by analogy with Eqs. (5–2.2) to (5–2.4). Now, according to the above, J_1 and $d\Delta P/dt$ have either opposite signs (in the nonstationary case in which the pressure of that phase to which matter is transported increases) or they vanish simultaneously (in the stationary final state). Thus there results from Eq. (4):

$$\Xi \leqslant 0, \tag{5–3.5}$$

by analogy with (5–2.11). Here the equality sign is valid for the stationary state.

We can now prove that the inequality (5) may be considered in the same way as (5–2.11). While condition (5–2.10) applies to (5–2.11) and is intrinsic for *chemical stability*, the sign-determining criterion in (5) is the condition for *mechanical stability*. In order to understand this, we write the rate of increase of the volumes V' and V'' of the two phases, which exhibit constant temperatures T' and T'' but variable amounts n' and n'' and variable pressures P' and P'', in the form

$$\frac{dV'}{dt} = \bar{V}' \frac{dn'}{dt} - \kappa_T' V' \frac{dP'}{dt},$$

$$\frac{dV''}{dt} = \bar{V}'' \frac{dn''}{dt} - \kappa_T'' V'' \frac{dP''}{dt},$$

where $\bar{V}$ is the molar volume ($\bar{V} = V/n$) and κ_T the compressibility (see Eq. 1–22.12) of the phase in question. According to our definitions,

$$J_1 = \frac{dn'}{dt} = -\frac{dn''}{dt}, \qquad \frac{d\Delta P}{dt} = \frac{dP''}{dt} - \frac{dP'}{dt}.$$

Since we can always *in principle* set up experimental conditions so that the volumes V' and V'' remain constant in time, we derive

$$J_1 = n'\kappa_T' \frac{dP'}{dt} = -n''\kappa_T'' \frac{dP''}{dt}.$$

With the stability condition (1–22.13)

$$\kappa_T > 0$$

it thus follows that the quantities J_1 and $d\Delta P/dt$ always exhibit opposite signs provided they do not both disappear. This relates the statement about the signs, which was intuitively based above, back to the condition for mechanical stability.

In Eq. (1), small values of ΔT and ΔP were assumed. However, this means near-equilibrium conditions (cf. §3–5). Thus we can also apply the phenomenological equations (3–10.14) and (3–10.15):

$$\left.\begin{aligned}
J_1 &= \alpha_{11}\overline{V}\Delta P + \alpha_{1Q}\frac{\Delta T}{T}, \\[2mm]
J_Q &= \alpha_{Q1}\overline{V}\Delta P + \alpha_{QQ}\frac{\Delta T}{T}.
\end{aligned}\right\} \tag{5-3.6}$$

Due to the time independence of the average temperature and of the average pressure (see above) the phenomenological coefficients α_{11}, α_{1Q}, α_{Q1}, and α_{QQ} are constant. Furthermore, Onsager's reciprocity law:

$$\alpha_{1Q} = \alpha_{Q1} \tag{5-3.7}$$

holds. We thus obtain from Eqs. (2) to (7):

$$\Lambda = \Xi, \qquad \frac{d\Theta}{dt} = 2\Xi \leqslant 0, \tag{5-3.8}$$

by analogy with Eq. (5–2.14). Accordingly, the entropy production continuously decreases until it has reached a (positive) minimum value in the stationary final state.

The question relating to the change in time of the total entropy of our heterogeneous system is still of interest. This question can be answered exactly for the special case of the Knudsen effect (cf. §3–10 and §3–11) because here the stationary pressure difference is determined by the temperatures of the two subsystems alone and because the equation of state for ideal gases can be applied to each of the subsystems.

Let the volumes of the two gas-filled chambers (phases ' and ") in the Knudsen experiment be constant and of equal size ($\equiv V$). The (variable) pressures and amounts of substance in the two subsystems shall for the initial state be denoted by P_0' and P_0'' and by n_0' and n_0'' respectively, and for the stationary final state by P_∞' and P_∞'' and by n_∞' and n_∞''. Then, according to the equation of state (R = gas constant),

$$\left.\begin{aligned}
P_0'V &= n_0'RT', & P_0''V &= n_0''RT'', \\
P_\infty'V &= n_\infty'RT', & P_\infty''V &= n_\infty''RT''.
\end{aligned}\right\} \tag{5-3.9}$$

Moreover, the kinetic theory of gases produces, in agreement with experience, the relation (see Eq. 3–11.11):

$$\frac{P_\infty'}{P_\infty''} = \sqrt{\tau} \tag{5-3.10}$$

with

$$\tau \equiv \frac{T'}{T''}. \tag{5-3.11}$$

Here the approximation

$$|\Delta T| = |T'' - T'| \ll T \tag{5–3.12}$$

is again used as a basis and T denotes the average absolute temperature of the system. From Eqs. (11) and (12) we obtain

$$1 - \tau \approx \frac{\Delta T}{T}. \tag{5–3.13}$$

From Eqs. (9), (10), and (11) there follows:

$$\frac{n_0''}{n_0'} = \tau \frac{P_0''}{P_0'}, \tag{5–3.14}$$

$$\frac{n_\infty''}{n_\infty'} = \sqrt{\tau}. \tag{5–3.15}$$

Moreover,

$$n_0' + n_0'' = n_\infty' + n_\infty''. \tag{5–3.16}$$

With the help of classical thermodynamics and the assumption of ideal gases for the entropies S_0 and S_∞ of the total heterogeneous system in the initial and final states the following formulas may be derived:

$$S_0 = n_0'\left(\sigma' - R \ln \frac{n_0'}{V} \frac{V^\dagger}{n^\dagger}\right) + n_0''\left(\sigma'' - R \ln \frac{n_0''}{V} \frac{V^\dagger}{n^\dagger}\right), \tag{5–3.17}$$

$$S_\infty = n_\infty'\left(\sigma' - R \ln \frac{n_\infty'}{V} \frac{V^\dagger}{n^\dagger}\right) + n_\infty''\left(\sigma'' - R \ln \frac{n_\infty''}{V} \frac{V^\dagger}{n^\dagger}\right). \tag{5–3.18}$$

Here σ is a function dependent on the temperature ($n^\dagger \equiv 1$ mol, $V^\dagger \equiv 1$ l). With Eqs. (1–22.16) and (1–22.17) we can write

$$T \frac{d\sigma}{dT} = \bar{C}_V > 0, \tag{5–3.19}$$

where $\bar{C}_V$ represents the molar heat capacity for constant volume. Thus we find

$$\sigma'' - \sigma' \approx \bar{C}_V \frac{\Delta T}{T}, \tag{5–3.20}$$

where $\bar{C}_V$ refers to the average temperature of the system. From Eqs. (16) to (20) we obtain the approximate expression for the entropy difference:

$$S_\infty - S_0 \approx (n_\infty'' - n_0'')\bar{C}_V \frac{\Delta T}{T}$$

$$+ R\left(n_0' \ln \frac{n_0'}{n^\dagger} + n_0'' \ln \frac{n_0''}{n^\dagger} - n_\infty' \ln \frac{n_\infty'}{n^\dagger} - n_\infty'' \ln \frac{n_\infty''}{n^\dagger}\right). \tag{5–3.21}$$

The approximation $\bar{C}_V = \text{const}$ which was assumed is allowed because of assumption (12).

We must now specify the initial state. Consider two cases:

(a) The system is initially homogeneous with respect to the concentration n_0/V. Then the initial pressures in the two subsystems are different and the following is obvious from Eqs. (14) to (16):

$$n_0' = n_0'' \equiv n_0, \tag{5-3.22}$$

$$n_\infty' = \frac{2n_0}{1 + \sqrt{\tau}}, \tag{5-3.23}$$

$$n_\infty'' = \frac{2n_0\sqrt{\tau}}{1 + \sqrt{\tau}}, \tag{5-3.24}$$

$$\frac{P_0'}{P_0''} = \tau. \tag{5-3.25}$$

Inserting Eqs. (22) to (24) into Eq. (21) and expanding the logarithms while taking account of Eqs. (13) and (19) produces

$$S_\infty - S_0 \approx -\frac{n_0}{16}(4\bar{C}_V + R)\left(\frac{\Delta T}{T}\right)^2 < 0. \tag{5-3.26}$$

Thus the entropy of the system in the stationary final state is *smaller* than that in the initial state with uniform concentration distribution.†

(b) The system is initially homogeneous with respect to pressure. Then the initial concentrations (initial amounts) in the two subsystems are different from one another. It is obvious from Eqs. (14) to (16) that

$$P_0' = P_0'', \tag{5-3.27}$$

$$n_0'' = n_0'\tau \equiv n_0\tau, \tag{5-3.28}$$

$$n_\infty' = n_0\frac{1 + \tau}{1 + \sqrt{\tau}}, \tag{5-3.29}$$

$$n_\infty'' = n_0\frac{(1 + \tau)\sqrt{\tau}}{1 + \sqrt{\tau}}. \tag{5-3.30}$$

Inserting Eqs. (28) to (30) into Eq. (21) and expanding the logarithms while taking account of Eqs. (13) and (19) produces

$$S_\infty - S_0 \approx \frac{n_0}{16}(4\bar{C}_V + 3R)\left(\frac{\Delta T}{T}\right)^2 > 0. \tag{5-3.31}$$

---

† Prigogine[2, 3] has already obtained this result.

The entropy of the system in the stationary final state is thus *greater* than that in the initial state with uniform pressure distribution.

While, in our example, the *entropy production* always decreases according to Eq. (8), on approaching a stationary state the *entropy* of the system can increase or decrease according to (26) and (31) depending on the initial conditions chosen. If the degree of "heterogeneity" in the system is judged by the concentration distribution then Eqs. (26) and (31) with Eqs. (15), (22), and (28) can be summarized as follows: the entropy of the system decreases with increasing concentration difference. However, if the pressure distribution is regarded as a measure of the "heterogeneity" of the system then with Eqs. (10), (25), and (27) Eqs. (26) and (31) can be interpreted thus: the entropy of the system increases with increasing pressure difference.

5–4 CONTINUOUS SYSTEMS

For a continuous system, in which irreversible processes (electric conduction, diffusion, heat conduction, thermal diffusion, chemical reactions, etc.) occur, the change of entropy production with time is naturally much more complicated than for a homogeneous or heterogeneous (discontinuous) system. Nevertheless, the question can be answered† in principle using the same scheme as was developed in §5–2 and §5–3.

We split the change of the local entropy production into two parts (see Eqs. 1–24.8 and 1–24.23):

$$\vartheta = \frac{1}{T} \sum_i J_i X_i = \sum_i J_i X_i' > 0, \tag{5-4.1}$$

$$d\vartheta = d_J\vartheta + d_X\vartheta \tag{5-4.2}$$

with

$$d_J\vartheta \equiv \sum_i X_i'\, dJ_i, \qquad d_X\vartheta \equiv \sum_i J_i\, dX_i'. \tag{5-4.3}$$

Here T denotes the absolute temperature of the volume element considered, J_i a generalized flux, and X_i the appropriate generalized force. Furthermore,

$$X_i' \equiv \frac{1}{T} X_i. \tag{5-4.4}$$

For the total entropy production Θ of the continuous system,

$$\Theta = \int \vartheta\, dV, \tag{5-4.5}$$

---

† This scheme was in fact developed especially for continuous systems by Glansdorff and Prigogine[7, 8], whose work is, in its essentials, taken as a basis for the presentation here.

where dV denotes a volume element and the integration is over the entire space which the system fills. If we write

$$d_j\vartheta \equiv \lambda \, dt, \qquad d_x\vartheta \equiv \xi \, dt, \tag{5-4.6}$$

where t is the time, and if we imagine the system as one contained in a fixed volume but one which may be permeable to matter, then we obtain from Eqs. (1) to (5)

$$\frac{d\Theta}{dt} = \Lambda + \Xi \tag{5-4.7}$$

with

$$\Lambda \equiv \int \lambda \, dV = \int \left(\sum_i X_i' \frac{\partial J_i}{\partial t} \right) dV, \tag{5-4.8}$$

$$\Xi \equiv \int \xi \, dV = \int \left(\sum_i J_i \frac{\partial X_i'}{\partial t} \right) dV. \tag{5-4.9}$$

Here the operator $\partial/\partial t$ denotes the partial derivative with respect to time at a fixed position. Equations (7) to (9) represent a transfer of Eqs. (5-2.2) to (5-2.4) or of Eqs. (5-3.2) to (5-3.4) to continuous systems.

We now direct our attention to the quantity Ξ which is given by Eq. (9).

For simplicity, we restrict the discussion to isotropic systems without electric and magnetic polarization. Viscous flow will also be excluded. Then from Eqs. (4-8.15) with Eqs. (1), (2), (3), and (6) we find

$$\begin{aligned}
\xi &= \sum_i J_i \frac{\partial X_i'}{\partial t} \\
&= \mathbf{J}_Q \frac{\partial}{\partial t} \left(\operatorname{grad} \frac{1}{T} \right) + \sum_k {}_v\mathbf{J}_k \frac{\partial}{\partial t} \left[\frac{\mathbf{K}_k}{T} - \frac{(\operatorname{grad} \mu_k)_T}{T} \right] \\
&\quad + \sum_r b_r \frac{\partial}{\partial t} \left(\frac{A_r}{T} \right).
\end{aligned} \tag{5-4.10}$$

Here $\mathbf{J}_Q$ denotes the heat current density, ${}_v\mathbf{J}_k$ the diffusion current density of species k (in the barycentric system), $\mathbf{K}_k$ the molar external force which acts on particle type k, μ_k the chemical potential of species k, b_r the rate of reaction of chemical reaction r, and A_r the affinity of reaction r.

With the help of Eqs. (4-8.6), (4-8.7), and (4-8.9) we obtain

$$\mathbf{J}_Q \operatorname{grad} \frac{1}{T} = \operatorname{div} \frac{\mathbf{J}_Q}{T} - \frac{1}{T} \operatorname{div} \mathbf{J}_Q, \tag{5-4.11}$$

$$\frac{1}{T} (\operatorname{grad} \mu_k)_T = \operatorname{grad} \frac{\mu_k}{T} - H_k \operatorname{grad} \frac{1}{T}, \tag{5-4.12}$$

$${}_v\mathbf{J}_k \operatorname{grad} \frac{\mu_k}{T} = \operatorname{div} \left(\frac{\mu_k}{T} {}_v\mathbf{J}_k \right) - \frac{\mu_k}{T} \operatorname{div} {}_v\mathbf{J}_k, \tag{5-4.13}$$

where H_k is the partial molar enthalpy of species k. With the definition (4–7.5)

$$\mathbf{J}_Q^* \equiv \mathbf{J}_Q + \sum_k H_k \,_v\mathbf{J}_k \qquad (5\text{–}4.14)$$

there follows from Eqs. (11) to (13):

$$\mathbf{J}_Q \,\mathrm{grad}\, \frac{1}{T} - \sum_k \frac{1}{T} \,_v\mathbf{J}_k (\mathrm{grad}\, \mu_k)_T$$

$$= \mathbf{J}_Q^* \,\mathrm{grad}\, \frac{1}{T} - \sum_k \,_v\mathbf{J}_k \,\mathrm{grad}\, \frac{\mu_k}{T}$$

$$= \mathrm{div}\, \frac{\mathbf{J}_Q^*}{T} - \frac{1}{T}\,\mathrm{div}\, \mathbf{J}_Q^* - \sum_k \mathrm{div}\left(\frac{\mu_k}{T}\,_v\mathbf{J}_k\right) + \sum_k \frac{\mu_k}{T}\,\mathrm{div}\,_v\mathbf{J}_k. \quad (5\text{–}4.15)$$

With this we can eliminate the gradients in Eq. (10).

In the transition from ξ to Ξ, we take note of Eq. (13) and use Gauss's law. Moreover, we assume time invariance of the external forces:

$$\frac{\partial K_k}{\partial t} = 0. \qquad (5\text{–}4.16)$$

Then we derive from Eqs. (9) and (10):

$$\Xi = \Xi_\Omega = \Xi_V \qquad (5\text{–}4.17)$$

with the surface integral

$$\Xi_\Omega = \int \left[J_{Qn}^* \frac{\partial}{\partial t}\left(\frac{1}{T}\right) - \sum_k \,_vJ_{kn} \frac{\partial}{\partial t}\left(\frac{\mu_k}{T}\right) \right] d\Omega \qquad (5\text{–}4.18)$$

and the volume integral

$$\Xi_V = \int \left[-\frac{\partial}{\partial t}\left(\frac{1}{T}\right)\mathrm{div}\, \mathbf{J}_Q^* + \sum_k \,_v\mathbf{J}_k K_k \frac{\partial}{\partial t}\left(\frac{1}{T}\right) \right.$$

$$\left. + \sum_k \frac{\partial}{\partial t}\left(\frac{\mu_k}{T}\,\mathrm{div}\,_v\mathbf{J}_k\right) + \sum_r b_r \frac{\partial}{\partial t}\left(\frac{A_r}{T}\right) \right] dV, \qquad (5\text{–}4.19)$$

where $d\Omega$ is a surface element of the enclosing surface, $\mathbf{J}_{Qn}^*$ the normal component of $\mathbf{J}_Q^*$, and $_vJ_{kn}$ the normal component of $_v\mathbf{J}_k$.

Furthermore, we assume that all possible convective flow processes have died away. Then the pressure P is time independent and the barycentric velocity $\mathbf{v}$ is negligible:

$$\frac{\partial P}{\partial t} = 0, \qquad \mathbf{v} = 0. \qquad (5\text{–}4.20)$$

Then from Eqs. (4–4.12), (4–6.5), (4–6.12), (4–6.15), and (4–6.18) with Eq. (14) there results

$$\text{div } _v\mathbf{J}_k = -\frac{\partial c_k}{\partial t} + \sum_r \nu_{kr} b_r, \tag{5–4.21}$$

$$-\text{div } \mathbf{J}_Q^* = -\sum_k {}_v\mathbf{J}_k K_k + \sum_k H_k \frac{\partial c_k}{\partial t} + \frac{\bar{C}_P}{\bar{V}} \frac{\partial T}{\partial t}. \tag{5–4.22}$$

Here c_k denotes the molarity of species k, ν_{kr} the stoichiometric number of species k in reaction r, $\bar{V}$ the molar volume, and $\bar{C}_P$ the molar heat capacity at constant pressure. Furthermore, the general thermodynamic relations (cf. Eq. 5–2.6 and Eq. 12) hold:

$$\left(\frac{\partial(\mu_k/T)}{\partial T}\right)_{P,c} = -\frac{H_k}{T^2}, \tag{5–4.23}$$

$$A_r = -\sum_k \nu_{kr} \mu_k, \tag{5–4.24}$$

where the inferior c signifies constancy of all independent concentrations. From Eq. (23) there follows:

$$\frac{\partial}{\partial t}\left(\frac{\mu_k}{T}\right) = -\frac{H_k}{T^2}\frac{\partial T}{\partial t} + \frac{1}{T}\mu_{ki}\sum_i \frac{\partial c_i}{\partial t} \tag{5–4.25}$$

with

$$\mu_{ki} \equiv \left(\frac{\partial \mu_k}{\partial c_i}\right)_{T,P,c_j(j \neq i)}. \tag{5–4.26}$$

In Eq. (25), the summation is over all independent concentrations.

By bringing Eqs. (21), (22), (24), and (25) into Eq. (19) there finally results:

$$\Xi_V = -\int \left[\frac{1}{T^2}\frac{\bar{C}_P}{\bar{V}}\left(\frac{\partial T}{\partial t}\right)^2 + \frac{1}{T}\sum_i \sum_k \mu_{ki}\frac{\partial c_i}{\partial t}\frac{\partial c_k}{\partial t}\right] dV. \tag{5–4.27}$$

From the stability conditions (5–2.10) and (1–22.15) there results:†

$$\Xi_V \leqslant 0, \tag{5–4.28}$$

---

† In the substitution of the n_i by the c_i in (5–2.10), we have excluded only the case of changes of amounts at constant composition (cf. p. 481). The conditions (5–2.10) now denote chemical stability as well as stability with respect to phase separation. The relation (1–22.15) indicates thermal stability ($\bar{C}_P > 0$).

where the equality sign is valid only for the case

$$\frac{\partial T}{\partial t} = 0, \qquad \frac{\partial c_i}{\partial t} = 0, \qquad \frac{\partial c_k}{\partial t} = 0,$$

that is, for the stationary state.

We now require boundary conditions which result in the disappearance of the surface integral (18). For example, the system may be thermally insulated or the temperature and the concentrations of all species may have time-independent values everywhere at the bounding surfaces of the system, or the system surface may be impermeable to matter for a temperature which is constant in time, or parts of several conditions may be valid. Under these rather general subsidiary conditions from Eqs. (27) and (28) we find that [47]

$$\Xi \leqslant 0, \tag{5-4.29}$$

as in (5–2.11) and (5–3.5). Here also the stability conditions dictate the interpretation of the signs: in the transition from Eq. (27) to (28) we have used the conditions for thermal and material stability.

The inequality (29) does not assume near-equilibrium states. If, however, we wish to obtain a statement about the entropy production Θ itself, we must make this assumption. Then we may set up the phenomenological equations of the form (see Eq. 1)

$$J_i = \sum_k \alpha_{ik} X_k' \tag{5-4.30}$$

and apply Onsager's reciprocity law

$$\alpha_{ik} = \alpha_{ki}. \tag{5-4.31}$$

Moreover, if we regard the phenomenological coefficients α_{ik} as constant, then we find from Eqs. (3) and (6) to (9) for the conditions under which (29) is valid:

$$\Lambda = \Xi, \qquad \frac{d\Theta}{dt} = 2\Xi \leqslant 0, \tag{5-4.32}$$

by analogy with (5–2.14) and (5–3.8). Thus, in this case, the entropy production continuously decreases, until it has reached a (positive) minimum value in the stationary final state.

5-5 APPLICATIONS TO BIOLOGICAL SYSTEMS

A living thing or a part of an organism is, from the thermodynamic point of view, an open system inside of which irreversible processes (chemical reactions, osmosis, diffusion, etc.) continuously occur. The processes in the interior lead to a positive entropy production, the influences of the surroundings to an entropy flow with, at first, an undetermined sign. The total entropy can both increase and decrease in time, as is the case for any open system. Thus we will apply the concepts of thermodynamics of irreversible processes in open systems to biological systems.†

If anomalous states, short periodic vibrations, and rhythmic processes (illness, sleep, increased or decreased activity, etc.) are excluded then a *fully grown* organism is in a *stationary state* with constant masses of the individual regions.

According to Eq. (5-1.4), the entropy of a normally grown living object is thus constant in time, so that the entropy flow is negative and is equal to the entropy production in magnitude.

Leaving aside the heat exchange with the surroundings (since this can take place in both directions, according to the type of living thing and according to the surrounding temperature), the negative entropy flow occurs by an intake and output of matter of varying chemical nature, that is, it is connected with, among other things, the smaller "entropy value" of foodstuffs in comparison to that of the excretion products. The remark of Schroedinger[18] is to be understood in this sense: "the organism feeds on negative entropy."

In the stationary state, the internal energy of the organism must, by Eq. (5-1.2), be time independent. This is only possible if the decrease of the internal energy through the external work continually done by the living object (and through heat loss to the surroundings) is compensated for by an increase of the internal energy due to exchange of matter with the surroundings. In this exchange of matter (taking in of food, breathing, excreting of waste products), the substances which are taken in must, on the average, contain a higher "energy value" than the excretion products. We have thus extended the above quotation by the—wellnigh trivial—statement: "an organism also feeds on positive energy."

The two sentences quoted express the validity of the first and second laws for biological systems (here, especially, in the stationary final state).

During growing up and the individual (ontogenetic) development, a living thing corresponds to a nonstationary open system with irreversible processes in the interior, striving towards a stationary state (the final step of development). If the analogy to normal thermodynamic systems is carried still further then it will be supposed that the entropy production of the organism, if related to a given volume, decreases in the course of time and in the stationary final state reaches a (positive) minimum value, as is, accord-

† That there exists no contradiction between the "structure formation" in nature and the second law of thermodynamics can be confirmed in all particulars[10, 14, 15]. The application of the methods of the thermodynamics of irreversible processes to biological problems was first investigated by Prigogine and Wiame [16] and later by Jung[17].

ing to §5–2 to §5–4, the case under quite definite assumptions in the most varied of thermodynamic systems. This thesis is suggested by an observation which is often made, according to which the relative intensity of mass exchange (referred to unit mass or unit volume) of a living thing decreases in the course of the individual development and reaches a minimum for the fully grown organism. However, this biological minimum principle has as little general validity as the principle of minimum entropy production. Thus a certain hesitation in the application of such extreme principles to living things seems to be appropriate.

Stated more precisely, the situation according to experiments on many animal species is as follows[19, 20]: the intensity of metabolism v, for instance, measured as the rate of intake of oxygen or as the rate of output of carbon dioxide, depends on the mass m of the individual (which in the course of development continuously increases) according to the general law

$$v = am^b,$$

where a and b are characteristic positive constants for the animal species or animal group in question. For many animals (e.g. mammals, fishes, mussels), there is valid: $b = \frac{2}{3}$, so that the relative intensity of metabolism v/m decreases with increasing mass and exhibits a minimal value in the final state of development. Here the quantity v is obviously proportional to the surface ("surface rule"). For other animals (e.g. insects, snails of the family Helicidae), we find: $b = 1$, so that v/m is constant, in contradiction to the above-mentioned minimum principle. Finally, still a third type of metabolism (e.g. snails of the family Planorbidae), for which the exponent b lies between 1 and $\frac{2}{3}$, is known. Here again the law of the decrease of v/m, in the course of development holds.

REFERENCES

1. I. PRIGOGINE, *Bull. Classe Sci., Acad. Roy. Belg.* **31**, 600 (1945); **40**, 471 (1954).

2. I. PRIGOGINE, *Etude thermodynamique des Phénomènes irréversibles*, Paris–Liége (1947).

3. I. PRIGOGINE, *Introduction to Thermodynamics of Irreversible Processes*, Springfield, Illinois (1955).

4. I. PRIGOGINE and R. DEFAY, *Bull. Classe Sci., Acad. Roy. Belg.* **32**, 694 (1946); **33**, 48 (1947).

5. H. WERGELAND, *Kgl. Norske Videnskab. Selskabs.* **24**, 110 (1951).

6. P. MAZUR, *Bull. Classe Sci., Acad. Roy. Belg.* **38**, 182 (1952).

7. P. GLANSDORFF and I. PRIGOGINE, *Physica* **20**, 773 (1954).

8. P. GLANSDORFF and J. PASSELECQ, *Bull. Classe Sci., Acad. Roy. Belg.* **43**, 188 (1957).

9. R. HAASE, *Z. Naturforsch.* (a) **6a**, 522 (1951); (b) **8a**, 729 (1953).

10. R. Haase, *Med. Grundlagenforsch.* **II**, 717 (1959).
11. K. G. Denbigh, *Trans. Faraday Soc.* **48**, 389 (1952).
12. A. E. Nielsen, *Bull. Classe Sci., Acad. Roy. Belg.* **40**, 539 (1954).
13. H. C. Mel, *Bull. Classe Sci., Acad. Roy. Belg.* **40**, 834 (1954).
14. R. Haase, *Z. Elektrochem.* **55**, 566 (1951).
15. R. Haase, *Naturwissenschaften* **44**, 409 (1957).
16. I. Prigogine and J. M. Wiame, *Experientia* **2**, 451 (1946).
17. F. Jung, *Naturwissenschaften* **43**, 73 (1956).
18. E. Schroedinger, *Was ist Leben?* German translation, 2nd ed., Muenchen (1951).
19. L. von Bertalanffy, *Biophysik der Fliessgleichgewichte*, Braunschweig (1953).
20. L. von Bertalanffy and W. J. Pirozynski, *Biol. Bull.* **105**, 240 (1953).

APPENDIX

Notes Added to the Dover Edition (1990)

[1]The words "volume element" should be replaced by the expression "space element".

[2]The term "absolute temperature" is obsolete. The quantity T is called "thermodynamic temperature" or briefly "temperature" (unit: K).

[3]The problem of nonadditivity of work in a discontinuous system has been treated by R. Haase, *Electrochim. Acta* **32**, 1655 (1987).

[4]The energy balance for a discontinuous system in the most general case (including nonuniform pressure) has been given by R. Haase, *Electrochim. Acta* **32**, 1655 (1987); **34**, 387 (1989).

[5]The modern name for the quantity c_k is "(amount-of-substance) concentration of species k".

[6]A more thorough discussion may be consulted in the papers cited in note 4, above. See also R. Haase, *Electrochim. Acta* **31**, 545 (1986).

[7]See also R. Haase, "Survey of Fundamental Laws", in *Physical Chemistry* (ed. H. Eyring, D. Henderson, and W. Jost), New York (1971), Volume I (Thermodynamics), p. 90.

[8]For a heterogeneous system containing charged species (electrochemical system) Equations (2) and (4) have to be modified (compare Chapter 3). But the subsequent conclusions continue to hold. See again the papers cited in notes 4 and 6, above.

[9]It is evident that, since the volume element is a physical quantity, the general expression "volume element" as used on p. 52 (and previously) should be replaced by "space element".

[10]Now one has agreed to define the activity coefficients in such a way that they are dimensionless. Then expressions such as $m_i\gamma_i$ have to be replaced by $m_i\gamma_i/m^t$ with $m^t = 1$ mol kg^{-1}. The dissociation constants remain dimensionless.

[11]This formula continues to hold for incomplete dissociation if equilibrium prevails, as follows from Eq. (20).

[12]The reaction variable ζ_r is related to the extent of reaction ξ_r (p. 53) by $d\zeta_r = d\xi_r/V$ where V is the volume of the space element considered.

[13]Nor does the condition (29) hold for semiconductors.

[14]There are some more textbooks on nonequilibrium thermodynamics. For example: A. Katchalsky and P. F. Curran, *Nonequilibrium Thermodynamics in Biophysics*, Cambridge, Massachusetts (1965); P. Glansdorff and I. Prigogine, *Thermodynamic Theory of Structure, Stability and Fluctuations*, London–New York–Sydney–Toronto (1971); and K. S. Førland, T. Førland, and S. K. Ratkje, *Irreversible Thermodynamics*, Chichester–New York–Brisbane–Toronto–Singapore (1988).

[15]The quantity α is now called "electric conductance", the name "electric conductivity" being restricted to what used to be called "specific conductance" or "specific conductivity". Thus conductance and conductivity are the reciprocals of resistance and resistivity, respectively.

[16]In recent years the investigations have been extended to nonideal reacting systems, following the pioneer work of J. N. Brönsted, *Z. Physik. Chem.* **102**, 169 (1922); **115**, 337 (1925). See R. Haase, *Z. Physik. Chem. (Frankfurt)* **128**, 225 (1981); **131**, 127 (1982); **153**, 217 (1987). See also E. O. Timmermann, *An. Asoc. Quim. Argent.* **73**, 287 (1985).

[17]The extension of the formulas to nonideal systems has been carried through in the papers cited in note 16, above.

[18]A generalization of the relations including nonideal systems has been given by R. Haase, *Z. Physik. Chem. (Frankfurt)* **132**, 1 (1982).

[19]Since a system through which there passes an electric current is always an open system, the treatment should be extended to open discontinuous systems (see the papers cited in notes 4 and 6). The general thermodynamic theory of irreversible processes in open reacting multiphase systems has been developed by R. Haase, *Electrochim. Acta* **34**, 387 (1989).

[20]If the system is considered to be an open system, as it should in the case where there is nonzero current, the entropy flow includes terms relating to the matter exchanged with the surroundings, while the entropy production remains unchanged. See R. Haase, *Electrochim. Acta* **31**, 545 (1986).

[21]The connection between the measurable rate of volume change and the quantity (8) is also more complicated in the case where there is nonzero electric current. See R. Haase, *Z. Physik. Chem (Frankfurt)* **103**, 235 (1976). It is interesting to note that, after introduction of the measurable volume change in equations such as (9), the unmeasurable quantity $\Delta\varphi$ has to be replaced by the measurable electric potential difference so that only measurable quantities appear in the dissipation function. See R. Haase, *Z. Physik. Chem. (Frankfurt)* **159**, 219 (1988).

[22]In a general treatment of the isothermal membrane phenomena (see the last paper cited in note 21) we come to the conclusion that in the final form of the dissipation function the following "fluxes" and conjugate "forces" occur: measurable rate of volume change and pressure difference, relative flows of the species (as in the last formulas), and chemical potential differences due to composition differences, electric current, and measurable electric potential difference.

[23]The quantity κ is now called "electric conductivity" (compare note 15, above).

[24]See, however, the footnote on p. 243 of the paper by R. Haase, *Z. Physik. Chem. (Frankfurt)* **103**, 235 (1976).

[25]A rigorous treatment of electroosmosis and related phenomena and recent experimental data are to be found in the following papers: R. Haase and K. Harff, *J. Membr. Sci.* **12**, 279 (1983); R. Haase and N. Özlen, *Z. Physik. Chem. (Frankfurt)* **148**, 255 (1986). An interpretation of permeation through membranes is given by R. Haase and E. O. Timmermann, *J. Membr. Sci.* **10**, 57 (1982).

[26]A modern treatment of isothermal membrane phenomena with due consideration of the quantities actually measured is carried through by R. Haase, *Z. Physik. Chem. (Frankfurt)* **159**, 219 (1988). Compare note 22, above.

[27]More experimental data on thermoosmosis of water and methanol through cellophane membranes have been obtained by R. Haase, H. J. de Greiff, and H. J. Buchner, *Z. Naturforsch.* **25a**, 1080 (1970), and by R. Haase and H. J. de Greiff, *Z. Naturforsch.* **26a**, 1773 (1971). The theory of permeation and thermoosmosis is dealt with by R. Haase and E. O. Timmermann, *J. Membr. Sci.* **10**, 57 (1982).

[28]Permeation and osmosis through a semipermeable membrane (time dependence of phenomena) are treated by R. Haase, *Z. Physik. Chem. (Frankfurt)* **140**, 1 (1984); **141**, 251 (1984).

[29]Here again the term "volume element" should be replaced by "space element". See note 9, above.

[30]We recall that the name recommended for κ is (electric) conductivity. It also should be stressed that in ionic melt mixtures the reference system requires careful consideration. See R. Haase, *Z. Naturforsch.* **28a**, 1897 (1973); **29a**, 534 (1974).

[31]For modern data and a more convenient way of describing the composition and temperature dependence of the diffusion coefficient see R. Haase, *Z. Naturforsch.* **31a**, 1025 (1976); R. Haase und H. J. Jansen, *Z. Naturforsch.* **35a**, 1116 (1980); R. Haase and W. Engels, *Z. Naturforsch.* **38a**, 281 (1983); **41a**, 1337 (1986). Modern textbooks on diffusion are: E. L. Cussler, *Diffusion*, Cambridge (1984); H. J. V. Tyrrell and K. R. Harris, *Diffusion in Liquids*, London (1984).

[32]A more thorough discussion appears in R. Haase, *Transportvorgänge*, Darmstadt (1987).

[33]The general connection between the diffusion coefficient and the transport numbers in the range of validity of the Onsager-Fuoss law ($c^{1/2}$ range) has been derived by R. Haase, *Z. Naturforsch.* **30a**, 1211 (1975).

[34]The corresponding formula for ionic melts (with three ion constituents) has been given by R. Haase, *Electrochim. Acta* **23**, 391 (1978).

[35]The modern name of κ is "(electric) conductivity".

[36]See R. Haase, H.-W. Borgmann, K.-H. Dücker, and W.-P. Lee, *Z. Naturforsch.* **26a**, 1224 (1971). See also R. Haase, *Ber. Bunsenges. Physik. Chem.* **76**, 256 (1972).

[37]The evaluation of measurements on thermocells containing molten salt mixtures has been carried out by R. Haase, U. Prüser, and J. Richter, *Ber. Bunsenges. Physik. Chem.* **81**, 577 (1977).

[38]Experimental values for thermocells containing solutions of alkali halides in methanol and in N-methylformamide are due to R. Haase and H.-J. Jansen, *Z. Physik. Chem. (Frankfurt)* **61**, 310 (1968).

[39]Transported entropies for molten salt mixtures are given in the paper cited in note 37, above.

[40]Two standard textbooks on hydrodynamics should be mentioned here: H. Lamb, *Hydrodynamics,* 6th ed., New York (1945) [Dover reprint, 1984]; V. G. Levich, *Physicochemical Hydrodynamics,* Englewood Cliffs, New Jersey (1962).

[41]This is part of the International System of Units ("SI", for Système International d'Unités) now in general use.

[42]In the SI (see note 41) the base quantities are length, time, mass, amount of substance, (thermodynamic) temperature, electric current, and luminous intensity. The corresponding SI base units are the metre (symbol: m), the second (s), the kilogram (kg), the mole (mol), the kelvin (K), the ampère (A), and the candela (cd). All other units are "derived units", e.g. the joule (J) and the coulomb (C). Electric displacement and magnetic induction are also called electric flux density and magnetic flux density, respectively. Now the units franklin (Fr), biot (Bi), oerstedt (Oe), and gauss (G) are no longer used.

[43]See also, for example, H. J. Kreuzer, *Nonequilibrium Thermodynamics and Its Statistical Foundation,* Oxford (1981).

[44]See also, for example, P. Glansdorff and I. Prigogine, *Thermodynamic Theory of Structure, Stability and Fluctuations,* London (1971); G. Nicolis and I. Prigogine, *Self-Organization in Nonequilibrium Systems,* New York (1977).

[45]A thorough investigation of the time dependence of thermodynamic quantities for chemical reactions in open systems has been carried out by R. Haase, *Z. Physik. Chem. (Frankfurt)* **135**, 1 (1983).

[46]A more general analysis shows that the inequality (t: time, μ_i^α: chemical potential, n_i^α: amount of species i in phase α)

$$\sum_\alpha \sum_i (d\mu_i^\alpha/dt)(dn_i^\alpha/dt) \geqslant 0$$

holds for any discontinuous system with fixed temperature in each phase (although the temperature may vary from phase to phase). The proof is due to R. Haase, *Z. Physik. Chem. (Frankfurt),* **115**, 125 (1979). The Glansdorff-Prigogine evolution criterion (see (5–2.11) or (5–3.5)) is a special case of this inequality. The theorem of minimum entropy production (Prigogine) is again a special case of the evolution criterion. Compare R. Haase, *Z. Physik. Chem. (Frankfurt)* **103**, 247 (1976); **106**, 113 (1977); **107**, 254 (1977).

[47]The inequality (5–4.29) is the Glansdorff-Prigogine evolution criterion for continuous systems. See the books cited in note 44, above.

AUTHOR INDEX

SUBJECT INDEX

505

A CATALOG OF SELECTED
DOVER BOOKS
IN SCIENCE AND MATHEMATICS

A CATALOG OF SELECTED
DOVER BOOKS
IN SCIENCE AND MATHEMATICS

QUALITATIVE THEORY OF DIFFERENTIAL EQUATIONS, V.V. Nemytskii and V.V. Stepanov. Classic graduate-level text by two prominent Soviet mathematicians covers classical differential equations as well as topological dynamics and erqodic theory. Bibliographies. 523pp. 5⅜ × 8½. 65954-2 Pa. $10.95

MATRICES AND LINEAR ALGEBRA, Hans Schneider and George Phillip Barker. Basic textbook covers theory of matrices and its applications to systems of linear equations and related topics such as determinants, eigenvalues and differential equations. Numerous exercises. 432pp. 5⅜ × 8½. 66014-1 Pa. $8.95

QUANTUM THEORY, David Bohm. This advanced undergraduate-level text presents the quantum theory in terms of qualitative and imaginative concepts, followed by specific applications worked out in mathematical detail. Preface. Index. 655pp. 5⅜ × 8½. 65969-0 Pa. $10.95

ATOMIC PHYSICS (8th edition), Max Born. Nobel laureate's lucid treatment of kinetic theory of gases, elementary particles, nuclear atom, wave-corpuscles, atomic structure and spectral lines, much more. Over 40 appendices, bibliography. 495pp. 5⅜ × 8½. 65984-4 Pa. $11.95

ELECTRONIC STRUCTURE AND THE PROPERTIES OF SOLIDS: The Physics of the Chemical Bond, Walter A. Harrison. Innovative text offers basic understanding of the electronic structure of covalent and ionic solids, simple metals, transition metals and their compounds. Problems. 1980 edition. 582pp. 6⅛ × 9¼. 66021-4 Pa. $14.95

BOUNDARY VALUE PROBLEMS OF HEAT CONDUCTION, M. Necati Özisik. Systematic, comprehensive treatment of modern mathematical methods of solving problems in heat conduction and diffusion. Numerous examples and problems. Selected references. Appendices. 505pp. 5⅜ × 8½. 65990-9 Pa. $11.95

A SHORT HISTORY OF CHEMISTRY (3rd edition), J.R. Partington. Classic exposition explores origins of chemistry, alchemy, early medical chemistry, nature of atmosphere, theory of valency, laws and structure of atomic theory, much more. 428pp. 5⅜ × 8½. (Available in U.S. only) 65977-1 Pa. $10.95

A HISTORY OF ASTRONOMY, A. Pannekoek. Well-balanced, carefully reasoned study covers such topics as Ptolemaic theory, work of Copernicus, Kepler, Newton, Eddington's work on stars, much more. Illustrated. References. 521pp. 5⅜ × 8½. 65994-1 Pa. $11.95

PRINCIPLES OF METEOROLOGICAL ANALYSIS, Walter J. Saucier. Highly respected, abundantly illustrated classic reviews atmospheric variables, hydrostatics, static stability, various analyses (scalar, cross-section, isobaric, isentropic, more). For intermediate meteorology students. 454pp. 6⅛ × 9¼. 65979-8 Pa. $12.95

RELATIVITY, THERMODYNAMICS AND COSMOLOGY, Richard C. Tolman. Landmark study extends thermodynamics to special, general relativity; also applications of relativistic mechanics, thermodynamics to cosmological models. 501pp. 5⅜ × 8½. 65383-8 Pa. $11.95

APPLIED ANALYSIS, Cornelius Lanczos. Classic work on analysis and design of finite processes for approximating solution of analytical problems. Algebraic equations, matrices, harmonic analysis, quadrature methods, much more. 559pp. 5⅜ × 8½. 65656-X Pa. $11.95

SPECIAL RELATIVITY FOR PHYSICISTS, G. Stephenson and C.W. Kilmister. Concise elegant account for nonspecialists. Lorentz transformation, optical and dynamical applications, more. Bibliography. 108pp. 5⅜ × 8½. 65519-9 Pa. $3.95

INTRODUCTION TO ANALYSIS, Maxwell Rosenlicht. Unusually clear, accessible coverage of set theory, real number system, metric spaces, continuous functions, Riemann integration, multiple integrals, more. Wide range of problems. Undergraduate level. Bibliography. 254pp. 5⅜ × 8½. 65038-3 Pa. $7.00

INTRODUCTION TO QUANTUM MECHANICS With Applications to Chemistry, Linus Pauling & E. Bright Wilson, Jr. Classic undergraduate text by Nobel Prize winner applies quantum mechanics to chemical and physical problems. Numerous tables and figures enhance the text. Chapter bibliographies. Appendices. Index. 468pp. 5⅜ × 8½. 64871-0 Pa. $9.95

ASYMPTOTIC EXPANSIONS OF INTEGRALS, Norman Bleistein & Richard A. Handelsman. Best introduction to important field with applications in a variety of scientific disciplines. New preface. Problems. Diagrams. Tables. Bibliography. Index. 448pp. 5⅜ × 8½. 65082-0 Pa. $10.95

MATHEMATICS APPLIED TO CONTINUUM MECHANICS, Lee A. Segel. Analyzes models of fluid flow and solid deformation. For upper-level math, science and engineering students. 608pp. 5⅜ × 8½. 65369-2 Pa. $12.95

ELEMENTS OF REAL ANALYSIS, David A. Sprecher. Classic text covers fundamental concepts, real number system, point sets, functions of a real variable, Fourier series, much more. Over 500 exercises. 352pp. 5⅜ × 8½. 65385-4 Pa. $8.95

PHYSICAL PRINCIPLES OF THE QUANTUM THEORY, Werner Heisenberg. Nobel Laureate discusses quantum theory, uncertainty, wave mechanics, work of Dirac, Schroedinger, Compton, Wilson, Einstein, etc. 184pp. 5⅜ × 8½. 60113-7 Pa. $4.95

INTRODUCTORY REAL ANALYSIS, A.N. Kolmogorov, S.V. Fomin. Translated by Richard A. Silverman. Self-contained, evenly paced introduction to real and functional analysis. Some 350 problems. 403pp. 5⅜ × 8½. 61226-0 Pa. $7.95

PROBLEMS AND SOLUTIONS IN QUANTUM CHEMISTRY AND PHYSICS, Charles S. Johnson, Jr. and Lee G. Pedersen. Unusually varied problems, detailed solutions in coverage of quantum mechanics, wave mechanics, angular momentum, molecular spectroscopy, scattering theory, more. 280 problems plus 139 supplementary exercises. 430pp. 6½ × 9¼. 65236-X Pa. $10.95

ASYMPTOTIC METHODS IN ANALYSIS, N.G. de Bruijn. An inexpensive, comprehensive guide to asymptotic methods—the pioneering work that teaches by explaining worked examples in detail. Index. 224pp. 5⅜ × 8½.   64221-6 Pa. $5.95

OPTICAL RESONANCE AND TWO-LEVEL ATOMS, L. Allen and J.H. Eberly. Clear, comprehensive introduction to basic principles behind all quantum optical resonance phenomena. 53 illustrations. Preface. Index. 256pp. 5⅜ × 8½.
65533-4 Pa. $6.95

COMPLEX VARIABLES, Francis J. Flanigan. Unusual approach, delaying complex algebra till harmonic functions have been analyzed from real variable viewpoint. Includes problems with answers. 364pp. 5⅜ × 8½.   61388-7 Pa. $7.95

ATOMIC SPECTRA AND ATOMIC STRUCTURE, Gerhard Herzberg. One of best introductions; especially for specialist in other fields. Treatment is physical rather than mathematical. 80 illustrations. 257pp. 5⅜ × 8½.   60115-3 Pa. $4.95

APPLIED COMPLEX VARIABLES, John W. Dettman. Step-by-step coverage of fundamentals of analytic function theory—plus lucid exposition of 5 important applications: Potential Theory; Ordinary Differential Equations; Fourier Transforms; Laplace Transforms; Asymptotic Expansions. 66 figures. Exercises at chapter ends. 512pp. 5⅜ × 8½.   64670-X Pa. $10.95

ULTRASONIC ABSORPTION: An Introduction to the Theory of Sound Absorption and Dispersion in Gases, Liquids and Solids, A.B. Bhatia. Standard reference in the field provides a clear, systematically organized introductory review of fundamental concepts for advanced graduate students, research workers. Numerous diagrams. Bibliography. 440pp. 5⅜ × 8½.   64917-2 Pa. $8.95

UNBOUNDED LINEAR OPERATORS: Theory and Applications, Seymour Goldberg. Classic presents systematic treatment of the theory of unbounded linear operators in normed linear spaces with applications to differential equations. Bibliography. 199pp. 5⅜ × 8½.   64830-3 Pa. $7.00

LIGHT SCATTERING BY SMALL PARTICLES, H.C. van de Hulst. Comprehensive treatment including full range of useful approximation methods for researchers in chemistry, meteorology and astronomy. 44 illustrations. 470pp. 5⅜ × 8½.   64228-3 Pa. $9.95

CONFORMAL MAPPING ON RIEMANN SURFACES, Harvey Cohn. Lucid, insightful book presents ideal coverage of subject. 334 exercises make book perfect for self-study. 55 figures. 352pp. 5⅜ × 8¼.   64025-6 Pa. $8.95

OPTICKS, Sir Isaac Newton. Newton's own experiments with spectroscopy, colors, lenses, reflection, refraction, etc., in language the layman can follow. Foreword by Albert Einstein. 532pp. 5⅜ × 8½.   60205-2 Pa. $8.95

GENERALIZED INTEGRAL TRANSFORMATIONS, A.H. Zemanian. Graduate-level study of recent generalizations of the Laplace, Mellin, Hankel, K. Weierstrass, convolution and other simple transformations. Bibliography. 320pp. 5⅜ × 8½.   65375-7 Pa. $7.95

THE ELECTROMAGNETIC FIELD, Albert Shadowitz. Comprehensive undergraduate text covers basics of electric and magnetic fields, builds up to electromagnetic theory. Also related topics, including relativity. Over 900 problems. 768pp. 5⅜ × 8¼. 65660-8 Pa. $15.95

FOURIER SERIES, Georgi P. Tolstov. Translated by Richard A. Silverman. A valuable addition to the literature on the subject, moving clearly from subject to subject and theorem to theorem. 107 problems, answers. 336pp. 5⅜ × 8½. 63317-9 Pa. $7.95

THEORY OF ELECTROMAGNETIC WAVE PROPAGATION, Charles Herach Papas. Graduate-level study discusses the Maxwell field equations, radiation from wire antennas, the Doppler effect and more. xiii + 244pp. 5⅜ × 8½. 65678-0 Pa. $6.95

DISTRIBUTION THEORY AND TRANSFORM ANALYSIS: An Introduction to Generalized Functions, with Applications, A.H. Zemanian. Provides basics of distribution theory, describes generalized Fourier and Laplace transformations. Numerous problems. 384pp. 5⅜ × 8½. 65479-6 Pa. $8.95

THE PHYSICS OF WAVES, William C. Elmore and Mark A. Heald. Unique overview of classical wave theory. Acoustics, optics, electromagnetic radiation, more. Ideal as classroom text or for self-study. Problems. 477pp. 5⅜ × 8½. 64926-1 Pa. $10.95

CALCULUS OF VARIATIONS WITH APPLICATIONS, George M. Ewing. Applications-oriented introduction to variational theory develops insight and promotes understanding of specialized books, research papers. Suitable for advanced undergraduate/graduate students as primary, supplementary text. 352pp. 5⅜ × 8½. 64856-7 Pa. $8.50

A TREATISE ON ELECTRICITY AND MAGNETISM, James Clerk Maxwell. Important foundation work of modern physics. Brings to final form Maxwell's theory of electromagnetism and rigorously derives his general equations of field theory. 1,084pp. 5⅜ × 8½. 60636-8, 60637-6 Pa., Two-vol. set $19.00

AN INTRODUCTION TO THE CALCULUS OF VARIATIONS, Charles Fox. Graduate-level text covers variations of an integral, isoperimetrical problems, least action, special relativity, approximations, more. References. 279pp. 5⅜ × 8½. 65499-0 Pa. $6.95

HYDRODYNAMIC AND HYDROMAGNETIC STABILITY, S. Chandrasekhar. Lucid examination of the Rayleigh-Benard problem; clear coverage of the theory of instabilities causing convection. 704pp. 5⅜ × 8¼. 64071-X Pa. $12.95

CALCULUS OF VARIATIONS, Robert Weinstock. Basic introduction covering isoperimetric problems, theory of elasticity, quantum mechanics, electrostatics, etc. Exercises throughout. 326pp. 5⅜ × 8½. 63069-2 Pa. $7.95

DYNAMICS OF FLUIDS IN POROUS MEDIA, Jacob Bear. For advanced students of ground water hydrology, soil mechanics and physics, drainage and irrigation engineering and more. 335 illustrations. Exercises, with answers. 784pp. 6⅛ × 9¼. 65675-6 Pa. $19.95

NUMERICAL METHODS FOR SCIENTISTS AND ENGINEERS, Richard Hamming. Classic text stresses frequency approach in coverage of algorithms, polynomial approximation, Fourier approximation, exponential approximation, other topics. Revised and enlarged 2nd edition. 721pp. 5⅜ × 8½.
65241-6 Pa. $14.95

THEORETICAL SOLID STATE PHYSICS, Vol. I: Perfect Lattices in Equilibrium; Vol. II: Non-Equilibrium and Disorder, William Jones and Norman H. March. Monumental reference work covers fundamental theory of equilibrium properties of perfect crystalline solids, non-equilibrium properties, defects and disordered systems. Appendices. Problems. Preface. Diagrams. Index. Bibliography. Total of 1,301pp. 5⅜ × 8½. Two volumes.
Vol. I 65015-4 Pa. $12.95
Vol. II 65016-2 Pa. $12.95

OPTIMIZATION THEORY WITH APPLICATIONS, Donald A. Pierre. Broadspectrum approach to important topic. Classical theory of minima and maxima, calculus of variations, simplex technique and linear programming, more. Many problems, examples. 640pp. 5⅜ × 8½.
65205-X Pa. $12.95

THE MODERN THEORY OF SOLIDS, Frederick Seitz. First inexpensive edition of classic work on theory of ionic crystals, free-electron theory of metals and semiconductors, molecular binding, much more. 736pp. 5⅜ × 8½.
65482-6 Pa. $14.95

ESSAYS ON THE THEORY OF NUMBERS, Richard Dedekind. Two classic essays by great German mathematician: on the theory of irrational numbers; and on transfinite numbers and properties of natural numbers. 115pp. 5⅜ × 8½.
21010-3 Pa. $4.95

THE FUNCTIONS OF MATHEMATICAL PHYSICS, Harry Hochstadt. Comprehensive treatment of orthogonal polynomials, hypergeometric functions, Hill's equation, much more. Bibliography. Index. 322pp. 5⅜ × 8½.
65214-9 Pa. $8.95

NUMBER THEORY AND ITS HISTORY, Oystein Ore. Unusually clear, accessible introduction covers counting, properties of numbers, prime numbers, much more. Bibliography. 380pp. 5⅜ × 8½.
65620-9 Pa. $8.95

THE VARIATIONAL PRINCIPLES OF MECHANICS, Cornelius Lanczos. Graduate level coverage of calculus of variations, equations of motion, relativistic mechanics, more. First inexpensive paperbound edition of classic treatise. Index. Bibliography. 418pp. 5⅜ × 8½.
65067-7 Pa. $10.95

MATHEMATICAL TABLES AND FORMULAS, Robert D. Carmichael and Edwin R. Smith. Logarithms, sines, tangents, trig functions, powers, roots, reciprocals, exponential and hyperbolic functions, formulas and theorems. 269pp. 5⅜ × 8½.
60111-0 Pa. $5.95

THEORETICAL PHYSICS, Georg Joos, with Ira M. Freeman. Classic overview covers essential math, mechanics, electromagnetic theory, thermodynamics, quantum mechanics, nuclear physics, other topics. First paperback edition. xxiii + 885pp. 5⅜ × 8½.
65227-0 Pa. $17.95

HANDBOOK OF MATHEMATICAL FUNCTIONS WITH FORMULAS, GRAPHS, AND MATHEMATICAL TABLES, edited by Milton Abramowitz and Irene A. Stegun. Vast compendium: 29 sets of tables, some to as high as 20 places. 1,046pp. 8 × 10½. 61272-4 Pa. $21.95

MATHEMATICAL METHODS IN PHYSICS AND ENGINEERING, John W. Dettman. Algebraically based approach to vectors, mapping, diffraction, other topics in applied math. Also generalized functions, analytic function theory, more. Exercises. 448pp. 5⅜ × 8¼. 65649-7 Pa. $8.95

A SURVEY OF NUMERICAL MATHEMATICS, David M. Young and Robert Todd Gregory. Broad self-contained coverage of computer-oriented numerical algorithms for solving various types of mathematical problems in linear algebra, ordinary and partial, differential equations, much more. Exercises. Total of 1,248pp. 5⅜ × 8½. Two volumes. Vol. I 65691-8 Pa. $13.95
Vol. II 65692-6 Pa. $13.95

TENSOR ANALYSIS FOR PHYSICISTS, J.A. Schouten. Concise exposition of the mathematical basis of tensor analysis, integrated with well-chosen physical examples of the theory. Exercises. Index. Bibliography. 289pp. 5⅜ × 8½.
65582-2 Pa. $7.95

INTRODUCTION TO NUMERICAL ANALYSIS (2nd Edition), F.B. Hildebrand. Classic, fundamental treatment covers computation, approximation, interpolation, numerical differentiation and integration, other topics. 150 new problems. 669pp. 5⅜ × 8½. 65363-3 Pa. $13.95

INVESTIGATIONS ON THE THEORY OF THE BROWNIAN MOVEMENT, Albert Einstein. Five papers (1905–8) investigating dynamics of Brownian motion and evolving elementary theory. Notes by R. Fürth. 122pp. 5⅜ × 8½.
60304-0 Pa. $3.95

NUMERICAL METHODS FOR SCIENTISTS AND ENGINEERS, Richard Hamming. Classic text stresses frequency approach in coverage of algorithms, polynomial approximation, Fourier approximation, exponential approximation, other topics. Revised and enlarged 2nd edition. 721pp. 5⅜ × 8½. 65241-6 Pa. $14.95

AN INTRODUCTION TO STATISTICAL THERMODYNAMICS, Terrell L. Hill. Excellent basic text offers wide-ranging coverage of quantum statistical mechanics, systems of interacting molecules, quantum statistics, more. 523pp. 5⅜ × 8½. 65242-4 Pa. $10.95

ELEMENTARY DIFFERENTIAL EQUATIONS, William Ted Martin and Eric Reissner. Exceptionally, clear comprehensive introduction at undergraduate level. Nature and origin of differential equations, differential equations of first, second and higher orders. Picard's Theorem, much more. Problems with solutions. 331pp. 5⅜ × 8½. 65024-3 Pa. $8.95

STATISTICAL PHYSICS, Gregory H. Wannier. Classic text combines thermodynamics, statistical mechanics and kinetic theory in one unified presentation of thermal physics. Problems with solutions. Bibliography. 532pp. 5⅜ × 8½.
65401-X Pa. $10.95

ORDINARY DIFFERENTIAL EQUATIONS, Morris Tenenbaum and Harry Pollard. Exhaustive survey of ordinary differential equations for undergraduates in mathematics, engineering, science. Thorough analysis of theorems. Diagrams. Bibliography. Index. 818pp. 5⅜ × 8½. 64940-7 Pa. $15.95

STATISTICAL MECHANICS: Principles and Applications, Terrell L. Hill. Standard text covers fundamentals of statistical mechanics, applications to fluctuation theory, imperfect gases, distribution functions, more. 448pp. 5⅜ × 8½. 65390-0 Pa. $9.95

ORDINARY DIFFERENTIAL EQUATIONS AND STABILITY THEORY: An Introduction, David A. Sánchez. Brief, modern treatment. Linear equation, stability theory for autonomous and nonautonomous systems, etc. 164pp. 5⅜ × 8¼. 63828-6 Pa. $4.95

THIRTY YEARS THAT SHOOK PHYSICS: The Story of Quantum Theory, George Gamow. Lucid, accessible introduction to influential theory of energy and matter. Careful explanations of Dirac's anti-particles, Bohr's model of the atom, much more. 12 plates. Numerous drawings. 240pp. 5⅜ × 8½. 24895-X Pa. $5.95

ORDINARY DIFFERENTIAL EQUATIONS, I.G. Petrovski. Covers basic concepts, some differential equations and such aspects of the general theory as Euler lines, Arzel's theorem, Peano's existence theorem, Osgood's uniqueness theorem, more. 45 figures. Problems. Bibliography. Index. xi + 232pp. 5⅜ × 8½. 64683-1 Pa. $6.00

GREAT EXPERIMENTS IN PHYSICS: Firsthand Accounts from Galileo to Einstein, edited by Morris H. Shamos. 25 crucial discoveries: Newton's laws of motion, Chadwick's study of the neutron, Hertz on electromagnetic waves, more. Original accounts clearly annotated. 370pp. 5⅜ × 8½. 25346-5 Pa. $8.95

INTRODUCTION TO PARTIAL DIFFERENTIAL EQUATIONS WITH APPLICATIONS, E.C. Zachmanoglou and Dale W. Thoe. Essentials of partial differential equations applied to common problems in engineering and the physical sciences. Problems and answers. 416pp. 5⅜ × 8½. 65251-3 Pa. $9.95

BURNHAM'S CELESTIAL HANDBOOK, Robert Burnham, Jr. Thorough guide to the stars beyond our solar system. Exhaustive treatment. Alphabetical by constellation: Andromeda to Cetus in Vol. 1; Chamaeleon to Orion in Vol. 2; and Pavo to Vulpecula in Vol. 3. Hundreds of illustrations. Index in Vol. 3. 2,000pp. 6⅛ × 9¼. 23567-X, 23568-8, 23673-0 Pa., Three-vol. set $38.85

ASYMPTOTIC EXPANSIONS FOR ORDINARY DIFFERENTIAL EQUATIONS, Wolfgang Wasow. Outstanding text covers asymptotic power series, Jordan's canonical form, turning point problems, singular perturbations, much more. Problems. 384pp. 5⅜ × 8½. 65456-7 Pa. $8.95

AMATEUR ASTRONOMER'S HANDBOOK, J.B. Sidgwick. Timeless, comprehensive coverage of telescopes, mirrors, lenses, mountings, telescope drives, micrometers, spectroscopes, more. 189 illustrations. 576pp. 5⅜ × 8¼. 24034-7 Pa. $8.95

SPECIAL FUNCTIONS, N.N. Lebedev. Translated by Richard Silverman. Famous Russian work treating more important special functions, with applications to specific problems of physics and engineering. 38 figures. 308pp. 5⅜ × 8½.
60624-4 Pa. $6.95

OBSERVATIONAL ASTRONOMY FOR AMATEURS, J.B. Sidgwick. Mine of useful data for observation of sun, moon, planets, asteroids, aurorae, meteors, comets, variables, binaries, etc. 39 illustrations 384pp. 5⅜ × 8¼. (Available in U.S. only)
24033-9 Pa. $5.95

INTEGRAL EQUATIONS, F.G. Tricomi. Authoritative, well-written treatment of extremely useful mathematical tool with wide applications. Volterra Equations, Fredholm Equations, much more. Advanced undergraduate to graduate level. Exercises. Bibliography. 238pp. 5⅜ × 8½.
64828-1 Pa. $6.95

CELESTIAL OBJECTS FOR COMMON TELESCOPES, T.W. Webb. Inestimable aid for locating and identifying nearly 4,000 celestial objects. 77 illustrations. 645pp. 5⅜ × 8½.
20917-2, 20918-0 Pa., Two-vol. set $12.00

MODERN NONLINEAR EQUATIONS, Thomas L. Saaty. Emphasizes practical solution of problems; covers seven types of equations. ". . . a welcome contribution to the existing literature. . . ."—*Math Reviews.* 490pp. 5⅜ × 8½. 64232-1 Pa. $9.95

FUNDAMENTALS OF ASTRODYNAMICS, Roger Bate et al. Modern approach developed by U.S. Air Force Academy. Designed as a first course. Problems, exercises. Numerous illustrations. 455pp. 5⅜ × 8½.
60061-0 Pa. $8.95

INTRODUCTION TO LINEAR ALGEBRA AND DIFFERENTIAL EQUATIONS, John W. Dettman. Excellent text covers complex numbers, determinants, orthonormal bases, Laplace transforms, much more. Exercises with solutions. Undergraduate level. 416pp. 5⅜ × 8½.
65191-6 Pa. $8.95

INCOMPRESSIBLE AERODYNAMICS, edited by Bryan Thwaites. Covers theoretical and experimental treatment of the uniform flow of air and viscous fluids past two-dimensional aerofoils and three-dimensional wings; many other topics. 654pp. 5⅜ × 8½.
65465-6 Pa. $14.95

INTRODUCTION TO DIFFERENCE EQUATIONS, Samuel Goldberg. Exceptionally clear exposition of important discipline with applications to sociology, psychology, economics. Many illustrative examples; over 250 problems. 260pp. 5⅜ × 8½.
65084-7 Pa. $6.95

LAMINAR BOUNDARY LAYERS, edited by L. Rosenhead. Engineering classic covers steady boundary layers in two- and three-dimensional flow, unsteady boundary layers, stability, observational techniques, much more. 708pp. 5⅜ × 8½.
65646-2 Pa. $15.95

LECTURES ON CLASSICAL DIFFERENTIAL GEOMETRY, Second Edition, Dirk J. Struik. Excellent brief introduction covers curves, theory of surfaces, fundamental equations, geometry on a surface, conformal mapping, other topics. Problems. 240pp. 5⅜ × 8½.
65609-8 Pa. $6.95

ROTARY-WING AERODYNAMICS, W.Z. Stepniewski. Clear, concise text covers aerodynamic phenomena of the rotor and offers guidelines for helicopter performance evaluation. Originally prepared for NASA. 537 figures. 640pp. 6⅛ × 9¼.
64647-5 Pa. $14.95

DIFFERENTIAL GEOMETRY, Heinrich W. Guggenheimer. Local differential geometry as an application of advanced calculus and linear algebra. Curvature, transformation groups, surfaces, more. Exercises. 62 figures. 378pp. 5⅜ × 8½.
63433-7 Pa. $7.95

INTRODUCTION TO SPACE DYNAMICS, William Tyrrell Thomson. Comprehensive, classic introduction to space-flight engineering for advanced undergraduate and graduate students. Includes vector algebra, kinematics, transformation of coordinates. Bibliography. Index. 352pp. 5⅜ × 8½.   65113-4 Pa. $8.00

A SURVEY OF MINIMAL SURFACES, Robert Osserman. Up-to-date, in-depth discussion of the field for advanced students. Corrected and enlarged edition covers new developments. Includes numerous problems. 192pp. 5⅜ × 8½.
64998-9 Pa. $8.00

ANALYTICAL MECHANICS OF GEARS, Earle Buckingham. Indispensable reference for modern gear manufacture covers conjugate gear-tooth action, gear-tooth profiles of various gears, many other topics. 263 figures. 102 tables. 546pp. 5⅜ × 8½.   65712-4 Pa. $11.95

SET THEORY AND LOGIC, Robert R. Stoll. Lucid introduction to unified theory of mathematical concepts. Set theory and logic seen as tools for conceptual understanding of real number system. 496pp. 5⅜ × 8¼.   63829-4 Pa. $8.95

A HISTORY OF MECHANICS, René Dugas. Monumental study of mechanical principles from antiquity to quantum mechanics. Contributions of ancient Greeks, Galileo, Leonardo, Kepler, Lagrange, many others. 671pp. 5⅜ × 8½.
65632-2 Pa. $14.95

FAMOUS PROBLEMS OF GEOMETRY AND HOW TO SOLVE THEM, Benjamin Bold. Squaring the circle, trisecting the angle, duplicating the cube: learn their history, why they are impossible to solve, then solve them yourself. 128pp. 5⅜ × 8½.   24297-8 Pa. $3.95

MECHANICAL VIBRATIONS, J.P. Den Hartog. Classic textbook offers lucid explanations and illustrative models, applying theories of vibrations to a variety of practical industrial engineering problems. Numerous figures. 233 problems, solutions. Appendix. Index. Preface. 436pp. 5⅜ × 8½.   64785-4 Pa. $8.95

CURVATURE AND HOMOLOGY, Samuel I. Goldberg. Thorough treatment of specialized branch of differential geometry. Covers Riemannian manifolds, topology of differentiable manifolds, compact Lie groups, other topics. Exercises. 315pp. 5⅜ × 8½.   64314-X Pa. $6.95

HISTORY OF STRENGTH OF MATERIALS, Stephen P. Timoshenko. Excellent historical survey of the strength of materials with many references to the theories of elasticity and structure. 245 figures. 452pp. 5⅜ × 8½.   61187-6 Pa. $9.95

GEOMETRY OF COMPLEX NUMBERS, Hans Schwerdtfeger. Illuminating, widely praised book on analytic geometry of circles, the Moebius transformation, and two-dimensional non-Euclidean geometries. 200pp. 5⅜ × 8¼.
63830-8 Pa. $6.95

MECHANICS, J.P. Den Hartog. A classic introductory text or refresher. Hundreds of applications and design problems illuminate fundamentals of trusses, loaded beams and cables, etc. 334 answered problems. 462pp. 5⅜ × 8½. 60754-2 Pa. $8.95

TOPOLOGY, John G. Hocking and Gail S. Young. Superb one-year course in classical topology. Topological spaces and functions, point-set topology, much more. Examples and problems. Bibliography. Index. 384pp. 5⅜ × 8¼.
65676-4 Pa. $7.95

STRENGTH OF MATERIALS, J.P. Den Hartog. Full, clear treatment of basic material (tension, torsion, bending, etc.) plus advanced material on engineering methods, applications. 350 answered problems. 323pp. 5⅜ × 8½. 60755-0 Pa. $7.50

ELEMENTARY CONCEPTS OF TOPOLOGY, Paul Alexandroff. Elegant, intuitive approach to topology from set-theoretic topology to Betti groups; how concepts of topology are useful in math and physics. 25 figures. 57pp. 5⅜ × 8½.
60747-X Pa. $2.95

ADVANCED STRENGTH OF MATERIALS, J.P. Den Hartog. Superbly written advanced text covers torsion, rotating disks, membrane stresses in shells, much more. Many problems and answers. 388pp. 5⅜ × 8½. 65407-9 Pa. $8.95

COMPUTABILITY AND UNSOLVABILITY, Martin Davis. Classic graduate-level introduction to theory of computability, usually referred to as theory of recurrent functions. New preface and appendix. 288pp. 5⅜ × 8½. 61471-9 Pa. $6.95

GENERAL CHEMISTRY, Linus Pauling. Revised 3rd edition of classic first-year text by Nobel laureate. Atomic and molecular structure, quantum mechanics, statistical mechanics, thermodynamics correlated with descriptive chemistry. Problems. 992pp. 5⅜ × 8½. 65622-5 Pa. $18.95

AN INTRODUCTION TO MATRICES, SETS AND GROUPS FOR SCIENCE STUDENTS, G. Stephenson. Concise, readable text introduces sets, groups, and most importantly, matrices to undergraduate students of physics, chemistry, and engineering. Problems. 164pp. 5⅜ × 8½. 65077-4 Pa. $5.95

THE HISTORICAL BACKGROUND OF CHEMISTRY, Henry M. Leicester. Evolution of ideas, not individual biography. Concentrates on formulation of a coherent set of chemical laws. 260pp. 5⅜ × 8½. 61053-5 Pa. $6.00

THE PHILOSOPHY OF MATHEMATICS: An Introductory Essay, Stephan Körner. Surveys the views of Plato, Aristotle, Leibniz & Kant concerning propositions and theories of applied and pure mathematics. Introduction. Two appendices. Index. 198pp. 5⅜ × 8½. 25048-2 Pa. $5.95

THE DEVELOPMENT OF MODERN CHEMISTRY, Aaron J. Ihde. Authoritative history of chemistry from ancient Greek theory to 20th-century innovation. Covers major chemists and their discoveries. 209 illustrations. 14 tables. Bibliographies. Indices. Appendices. 851pp. 5⅜ × 8½. 64235-6 Pa. $15.95

THE FOUR-COLOR PROBLEM: Assaults and Conquest, Thomas L. Saaty and Paul G. Kainen. Engrossing, comprehensive account of the century-old combinatorial topological problem, its history and solution. Bibliographies. Index. 110 figures. 228pp. 5⅜ × 8½. 65092-8 Pa. $6.00

CATALYSIS IN CHEMISTRY AND ENZYMOLOGY, William P. Jencks. Exceptionally clear coverage of mechanisms for catalysis, forces in aqueous solution, carbonyl- and acyl-group reactions, practical kinetics, more. 864pp. 5⅜ × 8½. 65460-5 Pa. $18.95

PROBABILITY: An Introduction, Samuel Goldberg. Excellent basic text covers set theory, probability theory for finite sample spaces, binomial theorem, much more. 360 problems. Bibliographies. 322pp. 5⅜ × 8½. 65252-1 Pa. $7.95

LIGHTNING, Martin A. Uman. Revised, updated edition of classic work on the physics of lightning. Phenomena, terminology, measurement, photography, spectroscopy, thunder, more. Reviews recent research. Bibliography. Indices. 320pp. 5⅜ × 8¼. 64575-4 Pa. $7.95

PROBABILITY THEORY: A Concise Course, Y.A. Rozanov. Highly readable, self-contained introduction covers combination of events, dependent events, Bernoulli trials, etc. Translation by Richard Silverman. 148pp. 5⅜ × 8¼. 63544-9 Pa. $4.50

THE CEASELESS WIND: An Introduction to the Theory of Atmospheric Motion, John A. Dutton. Acclaimed text integrates disciplines of mathematics and physics for full understanding of dynamics of atmospheric motion. Over 400 problems. Index. 97 illustrations. 640pp. 6 × 9. 65096-0 Pa. $16.95

STATISTICS MANUAL, Edwin L. Crow, et al. Comprehensive, practical collection of classical and modern methods prepared by U.S. Naval Ordnance Test Station. Stress on use. Basics of statistics assumed. 288pp. 5⅜ × 8½. 60599-X Pa. $6.00

WIND WAVES: Their Generation and Propagation on the Ocean Surface, Blair Kinsman. Classic of oceanography offers detailed discussion of stochastic processes and power spectral analysis that revolutionized ocean wave theory. Rigorous, lucid. 676pp. 5⅜ × 8½. 64652-1 Pa. $14.95

STATISTICAL METHOD FROM THE VIEWPOINT OF QUALITY CONTROL, Walter A. Shewhart. Important text explains regulation of variables, uses of statistical control to achieve quality control in industry, agriculture, other areas. 192pp. 5⅜ × 8½. 65232-7 Pa. $6.00

THE INTERPRETATION OF GEOLOGICAL PHASE DIAGRAMS, Ernest G. Ehlers. Clear, concise text emphasizes diagrams of systems under fluid or containing pressure; also coverage of complex binary systems, hydrothermal melting, more. 288pp. 6½ × 9¼. 65389-7 Pa. $8.95

STATISTICAL ADJUSTMENT OF DATA, W. Edwards Deming. Introduction to basic concepts of statistics, curve fitting, least squares solution, conditions without parameter, conditions containing parameters. 26 exercises worked out. 271pp. 5⅜ × 8½. 64685-8 Pa. $7.95

DE RE METALLICA, Georgius Agricola. The famous Hoover translation of greatest treatise on technological chemistry, engineering, geology, mining of early modern times (1556). All 289 original woodcuts. 638pp. 6¾ × 11.
60006-8 Clothbd. $15.95

SOME THEORY OF SAMPLING, William Edwards Deming. Analysis of the problems, theory and design of sampling techniques for social scientists, industrial managers and others who find statistics increasingly important in their work. 61 tables. 90 figures. xvii + 602pp. 5⅜ × 8½. 64684-X Pa. $14.95

THE VARIOUS AND INGENIOUS MACHINES OF AGOSTINO RAMELLI: A Classic Sixteenth-Century Illustrated Treatise on Technology, Agostino Ramelli. One of the most widely known and copied works on machinery in the 16th century. 194 detailed plates of water pumps, grain mills, cranes, more. 608pp. 9 × 12.
25497-6 Clothbd. $34.95

LINEAR PROGRAMMING AND ECONOMIC ANALYSIS, Robert Dorfman, Paul A. Samuelson and Robert M. Solow. First comprehensive treatment of linear programming in standard economic analysis. Game theory, modern welfare economics, Leontief input-output, more. 525pp. 5⅜ × 8½. 65491-5 Pa. $12.95

ELEMENTARY DECISION THEORY, Herman Chernoff and Lincoln E. Moses. Clear introduction to statistics and statistical theory covers data processing, probability and random variables, testing hypotheses, much more. Exercises. 364pp. 5⅜ × 8½. 65218-1 Pa. $8.95

THE COMPLEAT STRATEGYST: Being a Primer on the Theory of Games of Strategy, J.D. Williams. Highly entertaining classic describes, with many illustrated examples, how to select best strategies in conflict situations. Prefaces. Appendices. 268pp. 5⅜ × 8½. 25101-2 Pa. $5.95

MATHEMATICAL METHODS OF OPERATIONS RESEARCH, Thomas L. Saaty. Classic graduate-level text covers historical background, classical methods of forming models, optimization, game theory, probability, queueing theory, much more. Exercises. Bibliography. 448pp. 5⅜ × 8¼. 65703-5 Pa. $12.95

CONSTRUCTIONS AND COMBINATORIAL PROBLEMS IN DESIGN OF EXPERIMENTS, Damaraju Raghavarao. In-depth reference work examines orthogonal Latin squares, incomplete block designs, tactical configuration, partial geometry, much more. Abundant explanations, examples. 416pp. 5⅜ × 8¼.
65685-3 Pa. $10.95

THE ABSOLUTE DIFFERENTIAL CALCULUS (CALCULUS OF TENSORS), Tullio Levi-Civita. Great 20th-century mathematician's classic work on material necessary for mathematical grasp of theory of relativity. 452pp. 5⅜ × 8½.
63401-9 Pa. $9.95

VECTOR AND TENSOR ANALYSIS WITH APPLICATIONS, A.I. Borisenko and I.E. Tarapov. Concise introduction. Worked-out problems, solutions, exercises. 257pp. 5⅜ × 8¼. 63833-2 Pa. $6.95

TENSOR CALCULUS, J.L. Synge and A. Schild. Widely used introductory text covers spaces and tensors, basic operations in Riemannian space, non-Riemannian spaces, etc. 324pp. 5⅜ × 8¼. 63612-7 Pa. $7.00

A CONCISE HISTORY OF MATHEMATICS, Dirk J. Struik. The best brief history of mathematics. Stresses origins and covers every major figure from ancient Near East to 19th century. 41 illustrations. 195pp. 5⅜ × 8½. 60255-9 Pa. $7.95

A SHORT ACCOUNT OF THE HISTORY OF MATHEMATICS, W.W. Rouse Ball. One of clearest, most authoritative surveys from the Egyptians and Phoenicians through 19th-century figures such as Grassman, Galois, Riemann. Fourth edition. 522pp. 5⅜ × 8½. 20630-0 Pa. $9.95

HISTORY OF MATHEMATICS, David E. Smith. Non-technical survey from ancient Greece and Orient to late 19th century; evolution of arithmetic, geometry, trigonometry, calculating devices, algebra, the calculus. 362 illustrations. 1,355pp. 5⅜ × 8½. 20429-4, 20430-8 Pa., Two-vol. set $21.90

THE GEOMETRY OF RENÉ DESCARTES, René Descartes. The great work founded analytical geometry. Original French text, Descartes' own diagrams, together with definitive Smith-Latham translation. 244pp. 5⅜ × 8½.
60068-8 Pa. $6.00

THE ORIGINS OF THE INFINITESIMAL CALCULUS, Margaret E. Baron. Only fully detailed and documented account of crucial discipline: origins; development by Galileo, Kepler, Cavalieri; contributions of Newton, Leibniz, more. 304pp. 5⅜ × 8½. (Available in U.S. and Canada only) 65371-4 Pa. $7.95

THE HISTORY OF THE CALCULUS AND ITS CONCEPTUAL DEVELOPMENT, Carl B. Boyer. Origins in antiquity, medieval contributions, work of Newton, Leibniz, rigorous formulation. Treatment is verbal. 346pp. 5⅜ × 8½.
60509-4 Pa. $6.95

THE THIRTEEN BOOKS OF EUCLID'S ELEMENTS, translated with introduction and commentary by Sir Thomas L. Heath. Definitive edition. Textual and linguistic notes, mathematical analysis. 2500 years of critical commentary. Not abridged. 1,414pp. 5⅜ × 8½. 60088-2, 60089-0, 60090-4 Pa., Three-vol. set $26.85

A HISTORY OF VECTOR ANALYSIS: The Evolution of the Idea of a Vectorial System, Michael J. Crowe. The first large-scale study of the history of vector analysis, now the standard on the subject. Unabridged republication of the edition published by University of Notre Dame Press, 1967, with second preface by Michael C. Crowe. Index. 278pp. 5⅜ × 8½. 64955-5 Pa. $7.00

THE HISTORICAL ROOTS OF ELEMENTARY MATHEMATICS, Lucas N.H. Bunt, Phillip S. Jones, and Jack D. Bedient. Fundamental underpinnings of modern arithmetic, algebra, geometry and number systems derived from ancient civilizations. 320pp. 5⅜ × 8½. 25563-8 Pa. $7.95

CALCULUS REFRESHER FOR TECHNICAL PEOPLE, A. Albert Klaf. Covers important aspects of integral and differential calculus via 756 questions. 566 problems, most answered. 431pp. 5⅜ × 8½. 20370-0 Pa. $7.95

CATALOG OF DOVER BOOKS

CHALLENGING MATHEMATICAL PROBLEMS WITH ELEMENTARY SOLUTIONS, A.M. Yaglom and I.M. Yaglom. Over 170 challenging problems on probability theory, combinatorial analysis, points and lines, topology, convex polygons, many other topics. Solutions. Total of 445pp. 5⅜ × 8½. Two-vol. set.
Vol. I 65536-9 Pa. $5.95
Vol. II 65537-7 Pa. $5.95

FIFTY CHALLENGING PROBLEMS IN PROBABILITY WITH SOLUTIONS, Frederick Mosteller. Remarkable puzzlers, graded in difficulty, illustrate elementary and advanced aspects of probability. Detailed solutions. 88pp. 5⅜ × 8½.
65355-2 Pa. $3.95

EXPERIMENTS IN TOPOLOGY, Stephen Barr. Classic, lively explanation of one of the byways of mathematics. Klein bottles, Moebius strips, projective planes, map coloring, problem of the Koenigsberg bridges, much more, described with clarity and wit. 43 figures. 210pp. 5⅜ × 8½. 25933-1 Pa. $4.95

RELATIVITY IN ILLUSTRATIONS, Jacob T. Schwartz. Clear non-technical treatment makes relativity more accessible than ever before. Over 60 drawings illustrate concepts more clearly than text alone. Only high school geometry needed. Bibliography. 128pp. 6⅛ × 9¼. 25965-X Pa. $5.95

AN INTRODUCTION TO ORDINARY DIFFERENTIAL EQUATIONS, Earl A. Coddington. A thorough and systematic first course in elementary differential equations for undergraduates in mathematics and science, with many exercises and problems (with answers). Index. 304pp. 5⅜ × 8¼. 65942-9 Pa. $7.95

FOURIER SERIES AND ORTHOGONAL FUNCTIONS, Harry F. Davis. An incisive text combining theory and practical example to introduce Fourier series, orthogonal functions and applications of the Fourier method to boundary-value problems. 570 exercises. Answers and notes. 416pp. 5⅜ × 8½. 65973-9 Pa. $8.95

THE THOERY OF BRANCHING PROCESSES, Theodore E. Harris. First systematic, comprehensive treatment of branching (i.e. multiplicative) processes and their applications. Galton-Watson model, Markov branching processes, electron-photon cascade, many other topics. Rigorous proofs. Bibliography. 240pp. 5⅜ × 8½. 65952-6 Pa. $6.95

AN INTRODUCTION TO ALGEBRAIC STRUCTURES, Joseph Landin. Superb self-contained text covers "abstract algebra": sets and numbers, theory of groups, theory of rings, much more. Numerous well-chosen examples, exercises. 247pp. 5⅜ × 8½. 65940-2 Pa. $6.95

GAMES AND DECISIONS: Introduction and Critical Survey, R. Duncan Luce and Howard Raiffa. Superb non-technical introduction to game theory, primarily applied to social sciences. Utility theory, zero-sum games, n-person games, decision-making, much more. Bibliography. 509pp. 5⅜ × 8½. 65943-7 Pa. $10.95

Prices subject to change without notice.

Available at your book dealer or write for free Mathematics and Science Catalog to Dept. GI, Dover Publications, Inc., 31 East 2nd St., Mineola, N.Y. 11501. Dover publishes more than 175 books each year on science, elementary and advanced mathematics, biology, music, art, literary history, social sciences and other areas.